Student's Solutions Manual

Michael Butros
Victor Valley College

Calculus for the Life Sciences

Marvin L. Bittinger
Indiana University Purdue University Indianapolis

Neal Brand
University of North Texas

John Quintanilla
University of North Texas

Boston San Francisco New York
London Toronto Sydney Tokyo Singapore Madrid
Mexico City Munich Paris Cape Town Hong Kong Montreal

Reproduced by Pearson Addison-Wesley from electronic files supplied by the author.

Publishing as Pearson Addison-Wesley, 75 Arlington Street, Boston, MA 02116.

ISBN 0-321-28605-7

1 2 3 HI 12 11 10

Contents

Chapter 1

Functions and Graphs

Exercise Set 1.1

1. Graph $y = -4$.

Note that y is constant and therefore any value of x we choose will yield the same value for y, which is -4. Thus, we will have a horizontal line at $y = -4$.

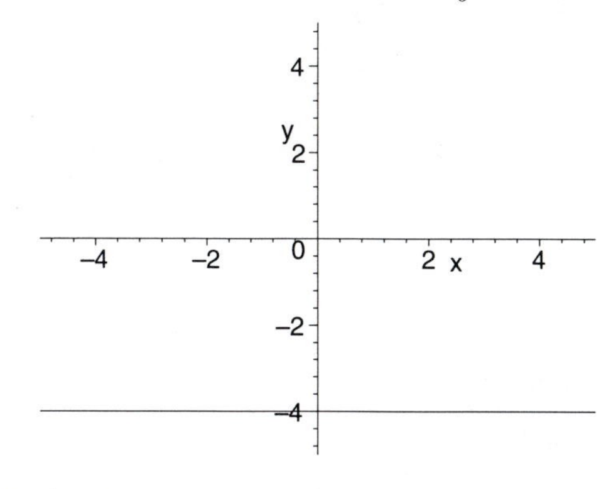

3. Graph $x = -4.5$.

Note that x is constant and therefore any value of y we choose will yield the same value for x, which is 4.5. Thus, we will have a vertical line at $x = -4.5$.

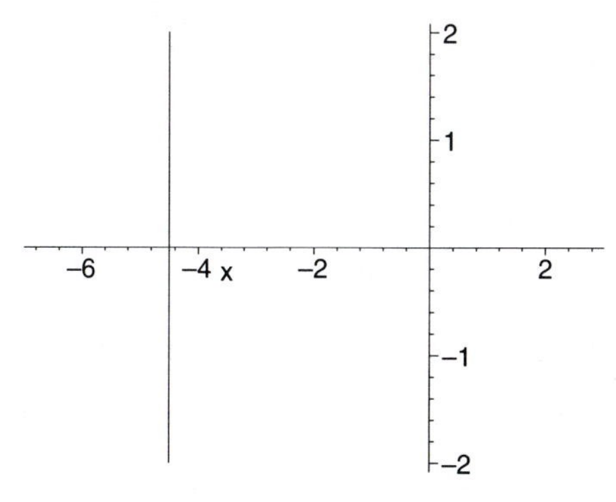

5. Graph. Find the slope and the y-intercept of $y = -3x$.

First, we find some points that satisfy the equation, then we plot the ordered pairs and connect the plotted points to get the graph.

When $x = 0$, $y = -3(0) = 0$, ordered pair $(0, 0)$

When $x = 1$, $y = -3(1) = -3$, ordered pair $(1, -3)$

When $x = -1$, $y = -3(-1) = 3$, ordered pair $(-1, 3)$

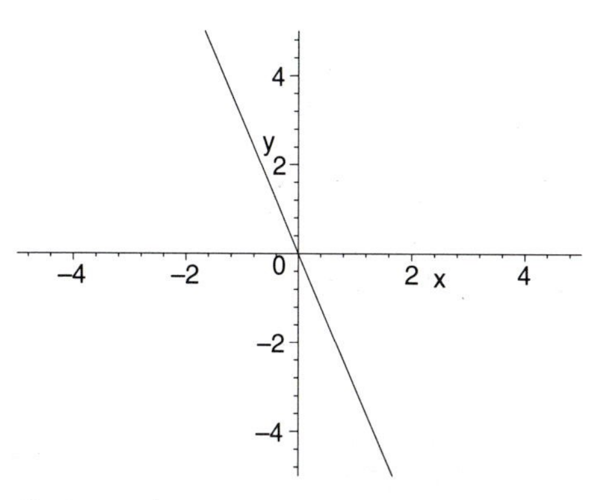

Compare the equation $y = -3x$ to the general linear equation form of $y = mx + b$ to conclude the equation has a slope of $m = -3$ and a y-intercept of $(0, 0)$.

7. Graph. Find the slope and the y-intercept of $y = 0.5x$.

First, we find some points that satisfy the equation, then we plot the ordered pairs and connect the plotted points to get the graph.

When $x = 0$, $y = 0.5(0) = 0$, ordered pair $(0, 0)$

When $x = 6$, $y = 0.5(6) = 3$, ordered pair $(6, 3)$

When $x = -2$, $y = 0.5(-2) = -1$, ordered pair $(-2, -1)$

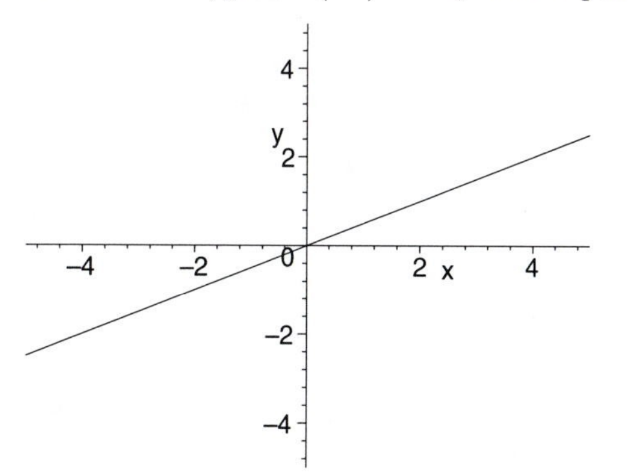

Compare the equation $y = 0.5x$ to the general linear equation form of $y = mx + b$ to conclude the equation has a slope of $m = 0.5$ and a y-intercept of $(0, 0)$.

9. Graph. Find the slope and the y-intercept of $y = -2x + 3$.

First, we find some points that satisfy the equation, then we plot the ordered pairs and connect the plotted points to get the graph.

When $x = 0$, $y = -2(0) + 3 = 3$, ordered pair $(0, 3)$

When $x = 2$, $y = -2(2) + 3 = -1$, ordered pair $(2, -1)$

When $x = -2$, $y = -2(-2) + 3 = 7$, ordered pair $(-2, 7)$

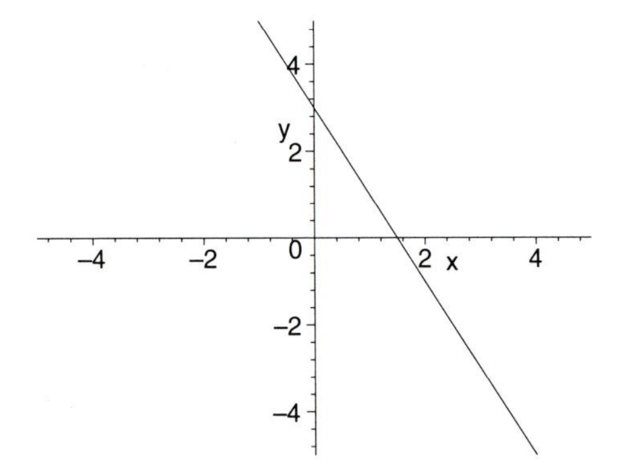

Compare the equation $y = -2x + 3$ to the general linear equation form of $y = mx + b$ to conclude the equation has a slope of $m = -2$ and a y-intercept of $(0, 3)$.

11. Graph. Find the slope and the y-intercept of $y = -x - 2$.

First, we find some points that satisfy the equation, then we plot the ordered pairs and connect the plotted points to get the graph.

When $x = 0$, $y = -(0) - 2 = -2$, ordered pair $(0, -2)$

When $x = 3$, $y = -(3) - 2 = -5$, ordered pair $(3, -5)$

When $x = -2$, $y = -(-2) - 2 = 0$, ordered pair $(-2, 0)$

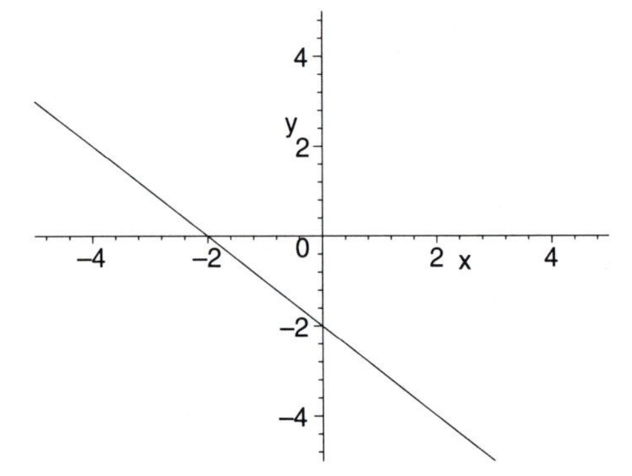

Compare the equation $y = -x - 2$ to the general linear equation form of $y = mx + b$ to conclude the equation has a slope of $m = -1$ and a y-intercept of $(0, -2)$.

13. Find the slope and y-intercept of $2x + y - 2 = 0$.

Solve the equation for y.

$$\begin{aligned} 2x + y - 2 &= 0 \\ y &= -2x + 2 \end{aligned}$$

Compare to $y = mx + b$ to conclude the equation has a slope of $m = -2$ and a y-intercept of $(0, 2)$.

15. Find the slope and y-intercept of $2x + 2y + 5 = 0$.

Solve the equation for y.

$$\begin{aligned} 2x + 2y + 5 &= 0 \\ 2y &= -2x - 5 \\ y &= -x - \frac{5}{2} \end{aligned}$$

Compare to $y = mx + b$ to conclude the equation has a slope of $m = -1$ and a y-intercept of $(0, -\frac{5}{2})$.

17. Find the slope and y-intercept of $x = 2y + 8$.

Solve the equation for y.

$$\begin{aligned} x &= 2y + 8 \\ x - 8 &= 2y \\ \frac{1}{2}x - 4 &= y \end{aligned}$$

Compare to $y = mx + b$ to conclude the equation has a slope of $m = \frac{1}{2}$ and a y-intercept of $(0, -4)$.

19. Find the equation of the line: with $m = -5$, containing $(1, -5)$

Plug the given information into equation $y - y_1 = m(x - x_1)$ and solve for y

$$\begin{aligned} y - y_1 &= m(x - x_1) \\ y - (-5) &= -5(x - 1) \\ y + 5 &= -5x + 5 \\ y &= -5x + 5 - 5 \\ y &= -5x \end{aligned}$$

21. Find the equation of line: with $m = -2$, containing $(2, 3)$

Plug the given information into the equation $y - y_1 = m(x - x_1)$ and solve for y

$$\begin{aligned} y - 3 &= -2(x - 2) \\ y - 3 &= -2x + 4 \\ y &= -2x + 4 + 3 \\ y &= -2x + 7 \end{aligned}$$

23. Find the equation of line: with $m = 2$, containing $(3, 0)$

Plug the given information into the equation $y - y_1 = m(x - x_1)$ and solve for y

$$\begin{aligned} y - 0 &= 2(x - 3) \\ y &= 2x - 6 \end{aligned}$$

25. Find the equation of line: with y-intercept $(0, -6)$ and $m = \frac{1}{2}$

Plug the given information into the equation $y = mx + b$

$$\begin{aligned} y &= mx + b \\ y &= \frac{1}{2}x + (-6) \\ y &= \frac{1}{2}x - 6 \end{aligned}$$

27. Find the equation of line: with $m = 0$, containing $(2, 3)$

Plug the given information into the equation $y - y_1 = m(x - x_1)$ and solve for y

$$\begin{aligned} y - 3 &= 0(x - 2) \\ y - 3 &= 0 \\ y &= 3 \end{aligned}$$

29. Find the slope given $(-4,-2)$ and $(-2,1)$

Use the slope equation $m = \frac{y_2-y_1}{x_2-x_1}$. **NOTE:** It does not matter which point is chosen as (x_1, y_1) and which is chosen as (x_2, y_2) as long as the order the point coordinates are subtracted in the same order as illustrated below

$$\begin{aligned} m &= \frac{1-(-2)}{-2-(-4)} \\ &= \frac{1+2}{-2+4} \\ &= \frac{3}{2} \end{aligned}$$

$$\begin{aligned} m &= \frac{-2-1}{-4-(-2)} \\ &= \frac{-3}{-2} \\ &= \frac{3}{2} \end{aligned}$$

31. Find the slope given $(\frac{2}{5}, \frac{1}{2})$ and $(-3, \frac{4}{5})$

$$\begin{aligned} m &= \frac{\frac{4}{5}-\frac{1}{2}}{-3-\frac{2}{5}} \\ &= \frac{\frac{8}{10}-\frac{5}{10}}{\frac{-15}{5}-\frac{10}{5}} \\ &= \frac{\frac{3}{10}}{\frac{17}{5}} \\ &= \frac{3}{10}\cdot\frac{5}{17} \\ &= \frac{15}{170} \\ &= \frac{3}{34} \end{aligned}$$

33. Find the slope given $(3,-7)$ and $(3,-9)$

$$\begin{aligned} m &= \frac{-9-(-7)}{3-3} \\ &= \frac{-2}{0} \quad \text{undefined quantity} \end{aligned}$$

This line has no slope

35. Find the slope given $(2,3)$ and $(-1,3)$

$$\begin{aligned} m &= \frac{3-3}{-1-2} \\ &= \frac{0}{-3} \\ &= 0 \end{aligned}$$

37. Find the slope given $(x, 3x)$ and $(x+h, 3(x+h))$

$$\begin{aligned} m &= \frac{3(x+h)-3x}{x+h-x} \\ &= \frac{3x+3h-3x}{h} \\ &= \frac{3h}{h} \\ &= 3 \end{aligned}$$

39. Find the slope given $(x, 2x+3)$ and $(x+h, 2(x+h)+3)$

$$\begin{aligned} m &= \frac{[2(x+h)+3]-(2x+3)}{x+h-x} \\ &= \frac{2x+2h+3-2x-3}{h} \\ &= \frac{2h}{h} \\ &= 2 \end{aligned}$$

41. Find equation of line containing $(-4,-2)$ and $(-2,1)$

From Exercise 29, we know that the slope of the line is $\frac{3}{2}$. Using the point$(-2,1)$ and the value of the slope in the point-slope formula $y - y_1 = m(x - x_1)$ and solving for y we get:

$$\begin{aligned} y-1 &= \frac{3}{2}(x-(-2)) \\ y-1 &= \frac{3}{2}(x+2) \\ y-1 &= \frac{3}{2}x+3 \\ y &= \frac{3}{2}x+3+1 \\ y &= \frac{3}{2}x+4 \end{aligned}$$

NOTE: You could use either of the given points and you would reach the final equation.

43. Find equation of line containing $(\frac{2}{5}, \frac{1}{2})$ and $(-3, \frac{4}{5})$

From Exercise 31, we know that the slope of the line is $-\frac{3}{34}$ and using the point $(-3, \frac{4}{5})$

$$\begin{aligned} y-\frac{4}{5} &= -\frac{3}{34}(x-(-3)) \\ y-\frac{4}{5} &= -\frac{3}{34}(x+3) \\ y-\frac{4}{5} &= -\frac{3}{34}x-\frac{9}{34} \\ y &= -\frac{3}{34}x-\frac{9}{34}+\frac{4}{5} \\ y &= -\frac{3}{34}x-\frac{45}{170}+\frac{136}{170} \\ y &= -\frac{3}{34}x+\frac{91}{170} \end{aligned}$$

45. Find equation of line containing $(3,-7)$ and $(3,-9)$

From Exercise 33, we found that the line containing $(3,-7)$ and $(3,-9)$ has no slope. We notice that the x-coordinate does not change regardless of the y-value. Therefore, the line in vertical and has the equation $x = 3$.

47. Find equation of line containing $(2,3)$ and $(-1,3)$

From Exercise 35, we found that the line containing $(2,3)$ and $(-1,3)$ has a slope of $m = 0$. We notice that the y-coordinate does not change regardless of the x-value. Therefore, the line in horizontal and has the equation $y = 3$.

49. Find equation of line containing $(x, 3x)$ and $(x+h, 3(x+h))$

From Exercise 37, we found that the line containing $(x, 3x)$ and $(x+h, 3(x+h))$ had a slope of $m = 3$. Using the point $(x, 3x)$ and the value of the slope in the point-slope formula

$$\begin{aligned} y - 3x &= 3(x - x) \\ y - 3x &= 3(0) \\ y - 3x &= = 0 \\ y &= 3x \end{aligned}$$

51. Find equation of line containing $(x, 2x+3)$ and $(x+h, 2(x+h)+3)$

From Exercise 37, we found that the line containing $(x, 2x+3)$ and $(x+h, 2(x+h)+3)$ had a slope of $m = 2$. Using the point $(x, 3x)$ and the value of the slope in the point-slope formula

$$\begin{aligned} y - (2x+3) &= 2(x - x) \\ y - (2x+3) &= 2(0) \\ y - (2x+3) &= 0 \\ y &= 2x + 3 \end{aligned}$$

53. Slope $= \frac{0.4}{5} = 0.08$. This means the treadmill has a grade of 8%.

55. The slope (or head) of the river is $\frac{43.33}{1238} = 0.035 = 3.5\%$

57. The average rate of change of life expectancy at birth is computed by finding the slope of the line containing the two points (1990, 73.7) and (2000, 76.9), which is given by

$$\begin{aligned} \text{Rate} &= \frac{\text{Change in Life expectancy}}{\text{Change in Time}} \\ &= \frac{76.9 - 73.7}{2000 - 1990} \\ &= \frac{3.2}{10} \\ &= 0.32 \text{ per year} \end{aligned}$$

59. a) Since R and T are directly proportional we can write that $R = kT$, where k is a constant of proportionality. Using $R = 12.51$ when $T = 3$ we can find k.

$$\begin{aligned} R &= kT \\ 12.51 &= k(3) \\ \frac{12.51}{3} &= k \\ 4.17 &= k \end{aligned}$$

Thus, we can write the equation of variation as $R = 4.17T$

b) This is the same as asking: find R when $T = 6$. So, we use the variation equation

$$\begin{aligned} R &= 4.17T \\ &= 4.17(6) \\ &= 25.02 \end{aligned}$$

61. a) Since B s directly proportional to W we can write $B = kW$.

b) When $W = 200$ $B = 5$ means that

$$\begin{aligned} B &= kW \\ 5 &= k(200) \\ \frac{5}{200} &= k \\ 0.025 &= k \\ 2.5\% &= k \end{aligned}$$

This means that the weight of the brain is 2.5% the weight of the person.

c) Find B when $W = 120$

$$\begin{aligned} B &= 0.025W \\ &= 0.025(120\ lbs) \\ &= 3\ lbs \end{aligned}$$

63. a)

$$\begin{aligned} D(0) &= 2(0) + 115 = 0 + 115\ ft \\ D(-20) &= 2(-20) + 115 = -40 + 115 = 75\ ft \\ D(10) &= 2(10) + 115 = 20 + 115 = 135\ ft \\ D(32) &= 2(32) + 115 = 64 + 115 = 179\ ft \end{aligned}$$

b) The stopping distance has to be a non-negative value. Therefore we need to solve the inequality

$$\begin{aligned} 0 &\leq 2F + 115 \\ -115 &\leq 2F \\ -57.5 &\leq F \end{aligned}$$

The 32° limit comes from the fact that for any temperature above that there would be no ice. Thus, the domain of the function is restricted in the interval $[-57.5, 32]$.

65. a)

$$\begin{aligned} M(x) &= 2.89x + 70.64 \\ M(26) &= 2.89(26) + 70.64 \\ &= 75.14 + 70.64 \\ &= 145.78 \end{aligned}$$

The male was 145.78 cm tall.

b)

$$\begin{aligned} F(x) &= 2.75x + 71.48 \\ F(26) &= 2.75(26) + 71.48 \\ &= 71.5 + 71.48 \\ &= 142.98 \end{aligned}$$

The female was 142.98 cm tall.

67. a)

$$\begin{aligned} A(0) &= 0.08(0) + 19.7 = 0 + 19.7 = 19.7 \\ A(1) &= 0.08(1) + 19.7 = 0.08 + 19.7 = 19.78 \\ A(10) &= 0.08(10) + 19.7 = 0.8 + 19.7 = 20.5 \\ A(30) &= 0.08(30) + 19.7 = 2.4 + 19.7 = 22.1 \\ A(50) &= 0.08(50) + 19.7 = 4 + 19.7 = 23.7 \end{aligned}$$

b) First we find the value of $t4$, which is $2003 - 1950 = 53$. So, we have to find $A(53)$.

$$A(53) = 0.08(53) + 19.7 = 4.24 + 19.8 = 23.94$$

The median age of women at first marriage in the year 2003 is 23.94 years.

c) $A(t) = 0.08t + 19.7$

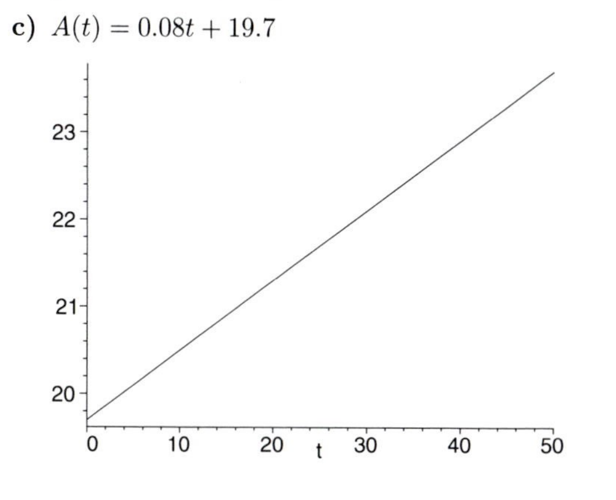

Exercise Set 1.2

1. $y = \frac{1}{2}x^2$ and $y = -\frac{1}{2}x^2$

3. $y = x^2$ and $y = (x-1)^2$

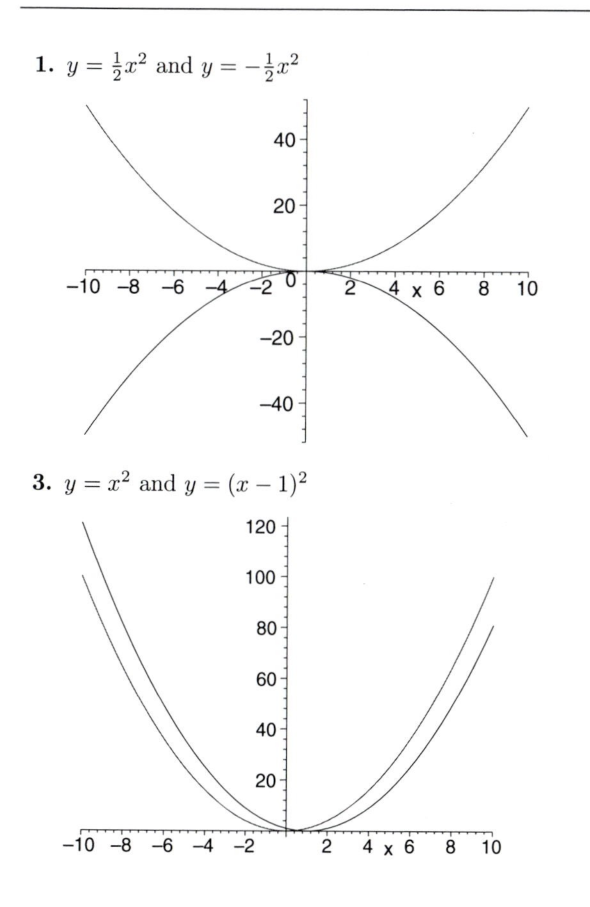

5. $y = x^2$ and $y = (x+1)^2$

7. $y = x^3$ and $y = x^3 + 1$

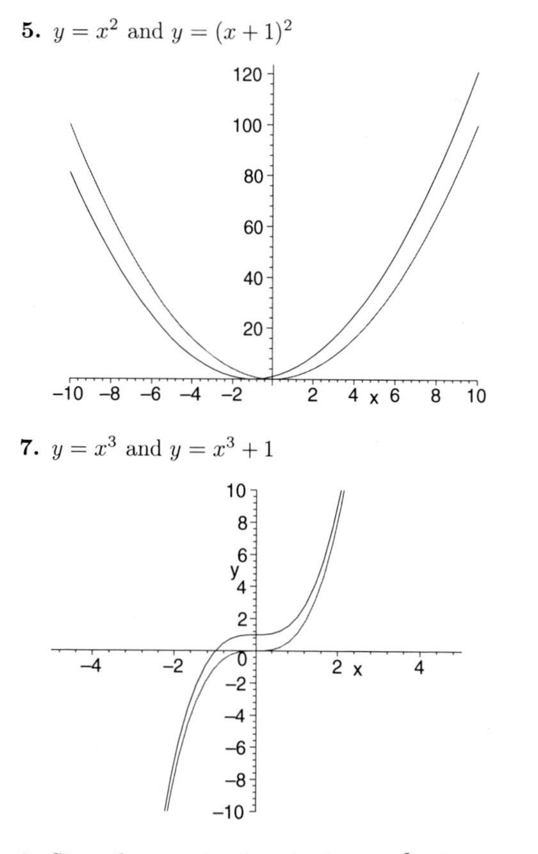

9. Since the equation has the form $ax^2 + bx + c$, with $a \neq 0$, the graph of the function is a parabola. The x-value of the vertex is given by

$$x = -\frac{b}{2a} = -\frac{4}{2(1)} = -2$$

The y-value of the vertex is given by

$$\begin{aligned} y &= (-2)^2 + 4(-2) - 7 \\ &= 4 - 8 - 7 \\ &= -11 \end{aligned}$$

Therefore, the vertex is $(-2, 11)$.

11. Since the equation is not in the form of $ax^2 + bx + c$, the graph of the function is not a parabola.

13. $y = x^2 - 4x + 3$

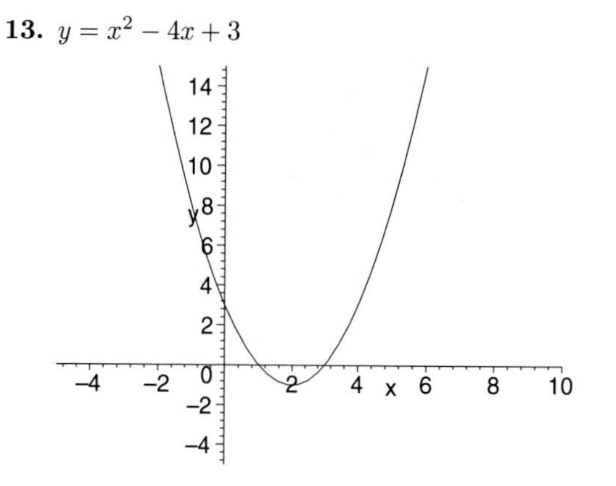

15. $y = -x^2 + 2x - 1$

17. $y = 2x^2 + 4x - 7$

19. $y = \frac{1}{2}x^2 + 3x - 5$

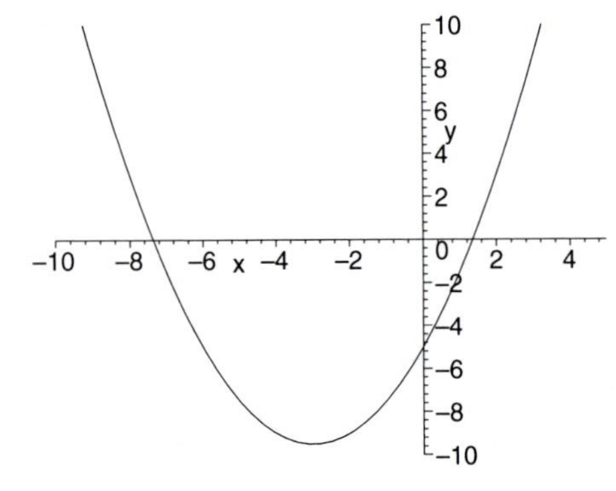

21. Solve $x^2 - 2x = 2$
Write the equation so that one side equals zero, that is $x^2 - 2x - 2 = 0$, then use the quadratic formula, with $a = 1$, $b = -2$, and $c = -2$, to solve for x.

$$\begin{aligned} x &= \frac{-b \pm \sqrt{b^2 - 4ac}}{2a} \\ x &= \frac{-(-2) \pm \sqrt{(-2)^2 - 4(1)(-2)}}{2(1)} \\ &= \frac{2 \pm \sqrt{4+8}}{2} \\ &= \frac{2 \pm \sqrt{12}}{2} \\ &= \frac{2 \pm 2\sqrt{3}}{2} \\ &= \frac{2(1 \pm \sqrt{3})}{2} \\ &= 1 \pm \sqrt{3} \end{aligned}$$

The solutions are $1 + \sqrt{3}$ and $1 - \sqrt{3}$

23. Solve $3y^2 + 8y + 2 = 0$
Use the quadratic formula, with $a = 3$, $b = 8$, and $c = 2$, to solve for y.

$$\begin{aligned} y &= \frac{-b \pm \sqrt{b^2 - 4ac}}{2a} \\ y &= \frac{-8 \pm \sqrt{(8)^2 - 4(3)(2)}}{2(3)} \\ &= \frac{-8 \pm \sqrt{64 - 24}}{6} \\ &= \frac{-8 \pm \sqrt{40}}{6} \\ &= \frac{-8 \pm 2\sqrt{10}}{6} \\ &= \frac{2(-4 \pm \sqrt{10})}{6} \\ &= \frac{-4 \pm \sqrt{10}}{3} \end{aligned}$$

The solutions are $\frac{-4+\sqrt{10}}{3}$ and $\frac{-4-\sqrt{10}}{3}$

25. Solve $x^2 - 2x + 10 = 0$
Using the quadratic formula with $a = 1$, $b = -2$, and $c = 10$

$$\begin{aligned} x &= \frac{-b \pm \sqrt{b^2 - 4ac}}{2a} \\ x &= \frac{-(-2) \pm \sqrt{(-2)^2 - 4(1)(10)}}{2(1)} \\ &= \frac{2 \pm \sqrt{4 - 40}}{2} \\ &= \frac{2 \pm \sqrt{-36}}{2} \\ &= \frac{2 \pm 6i}{2} \\ &= \frac{2(1 \pm 3i)}{2} \\ &= 1 \pm 3i \end{aligned}$$

The solutions are $1 + 3i$ and $1 - 3i$

27. Solve $x^2 + 6x = 1$
Write the equation so that one side equals zero, that is $x^2 + 6x - 1 = 0$, then use the quadratic formula, with $a = 1$, $b = 6$, and $c = -1$, to solve for x.

$$\begin{aligned} x &= \frac{-b \pm \sqrt{b^2 - 4ac}}{2a} \\ x &= \frac{-6 \pm \sqrt{(6)^2 - 4(1)(-1)}}{2(1)} \\ &= \frac{-6 \pm \sqrt{36+4}}{2} \\ &= \frac{-6 \pm \sqrt{40}}{2} \\ &= \frac{-6 \pm 2\sqrt{10}}{2} \end{aligned}$$

$$= \frac{2(-3 \pm \sqrt{10})}{2}$$
$$= -3 \pm \sqrt{10}$$

The solutions are $-3+\sqrt{10}$ and $-3-\sqrt{10}$

29. Solve $x^2 + 4x + 8 = 0$
Using the quadratic formula with $a = 1$, $b = 4$, and $c = 8$

$$\begin{aligned} x &= \frac{-b \pm \sqrt{b^2 - 4ac}}{2a} \\ x &= \frac{-4 \pm \sqrt{(4)^2 - 4(1)(8)}}{2(1)} \\ &= \frac{-4 \pm \sqrt{16 - 32}}{2} \\ &= \frac{-4 \pm \sqrt{-16}}{2} \\ &= \frac{-4 \pm 4i}{2} \\ &= \frac{4(1 \pm i)}{2} \\ &= 2(1 \pm i) = 2 \pm 2i \end{aligned}$$

The solutions are $2 + 2i$ and $2 - 2i$

31. Solve $4x^2 = 4x - 1$
Write the equation so that one side equals zero, that is $4x^2 - 4x - 1 = 0$, then use the quadratic formula, with $a = 4$, $b = -4$, and $c = -1$, to solve for x.

$$\begin{aligned} x &= \frac{-b \pm \sqrt{b^2 - 4ac}}{2a} \\ x &= \frac{-(-4) \pm \sqrt{(-4)^2 - 4(4)(-1)}}{2(4)} \\ &= \frac{4 \pm \sqrt{16 + 16}}{8} \\ &= \frac{4 \pm \sqrt{32}}{8} \\ &= \frac{4 \pm 4\sqrt{2}}{8} \\ &= \frac{4(1 \pm \sqrt{2})}{8} \\ &= \frac{1 \pm \sqrt{2}}{2} \end{aligned}$$

The solutions are $\frac{1+\sqrt{2}}{2}$ and $\frac{1-\sqrt{2}}{2}$

33. Find $f(7)$, $f(10)$, and $f(12)$

$$\begin{aligned} f(7) &= \frac{1}{6}(7)^3 + \frac{1}{2}(7)^2 + \frac{1}{2}(7) \\ &= \frac{343}{6} + \frac{49}{2} + \frac{7}{2} \\ &= \frac{343}{6} + \frac{147}{6} + \frac{21}{6} \\ &= \frac{511}{6} \approx 85.1\overline{6} \approx 85 \text{ oranges} \\ f(10) &= \frac{1}{6}(10)^3 + \frac{1}{2}(10)^2 + \frac{1}{2}(10) \\ &= \frac{1000}{6} + 50 + 5 \\ &= \frac{500}{3} + \frac{150}{3} + \frac{15}{3} \\ &= \frac{665}{3} \approx 221.\overline{6} \approx 222 \text{ oranges} \\ f(12) &= \frac{1}{6}(12)^3 + \frac{1}{2}(12)^2 + \frac{1}{2}(12) \\ &= 288 + 72 + 6 \\ &= 366 \text{ oranges} \end{aligned}$$

35. Solve $50 = 9.41 - 0.19x + 0.09x^2$. First, let us rewrite the equation as $0 = -40.59 - 0.19x + 0.09x^2$ then we can use the quadratic formula to solve for x

$$\begin{aligned} x &= \frac{-(-0.19) \pm \sqrt{(-0.19)^2 - 4(0.09)(-40.59)}}{2(0.09)} \\ &= \frac{0.19 \pm \sqrt{0.0361 + 14.6124}}{0.18} = \frac{0.19 \pm \sqrt{14.6485}}{0.18} \\ &= \frac{0.19 \pm 3.8273}{0.18} \\ &= \frac{0.19 + 3.8273}{0.18} = 22.3183 \end{aligned}$$

Therefore, the average price of a ticket will be \$50 will happen during the, $1990 + 22.3183 = 2012.3183$ 2012-13 season. **NOTE:** We could not choose the negative option of the quadratic formula since it would result in the result that is negative which corresponds to a year before 1990 and that does not make physical sense.

37. $f(x) = x^3 - x^2$

a) For large values of x, x^3 would be larger than x^2. $x^3 = x \cdot x \cdot x$ and $x^2 = x \cdot x$ so for very large values of x there is an extra factor of x in x^3 which causes x^3 to be larger than x^2.

b) As x gets very large the values of x^3 become much larger than those of x^2 and therefore we can "ignore" the effect of x^2 in the expression $x^3 - x^2$. Thus, we can approximate the function to look like x^3 for very large values of x.

c) Below is a graph of $x^3 - x^2$ and x^3 for $100 \le x \le 200$. It is hard to distinguish between the two graphs confirming the conclusion reached in part b).

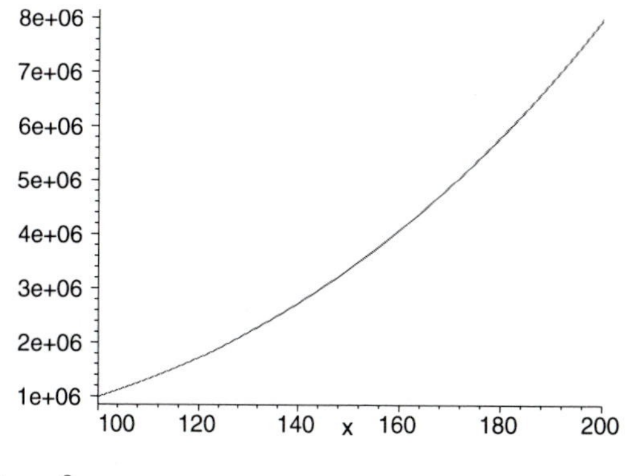

39. $f(x) = x^2 + x$

a) For values very close to 0, x is larger than x^2 since for values of x less than 1 $x^2 < x$.

b) For values of x very close to 0 $f(x)$ looks like x since the x^2 can be "ignored".

c) Below is a graph of x^2+x and x for $-0.01 \le x \le 0.01$. It is very hard to distinguish between the two graphs confirming our conclusion from part b).

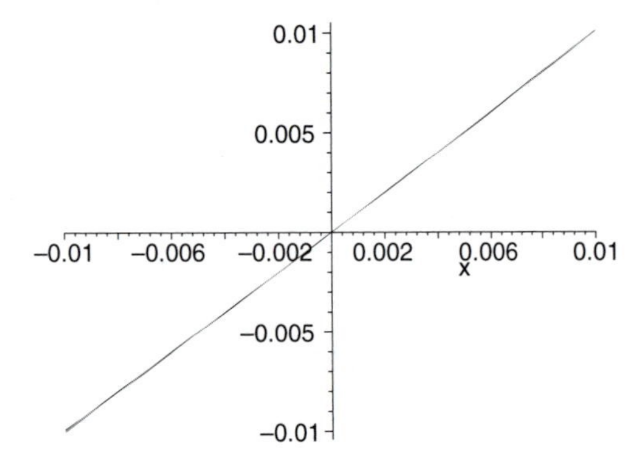

41. $f(x) = x^3 - x$

$$\begin{aligned} f(x) &= 0 \\ x^3 - x &= 0 \\ x(x^2-1) &= 0 \\ x(x-1)(x+1) &= 0 \\ x &= 0 \\ x &= 1 \\ x &= -1 \end{aligned}$$

43. $x = -1.831$, $x = -0.856$, and $x = 3.188$

45. $x = -10.153$, $x = -1.871$, $x = -0.821$, $x = -0.303$, $x = 0.098$, $x = 0.535$, $x = 1.219$, and $x = 3.297$

47. $y = -0.279x + 4.036$

49. $y = 0.942x^2 - 2.651x - 27.943$

51. $y = 0.237x^4 - 0.885x^3 - 29.224x^2 + 165.166x - 210.135$

Exercise Set 1.3

1. $y = |\, x\,|$ and $y = |\, x+3\,|$

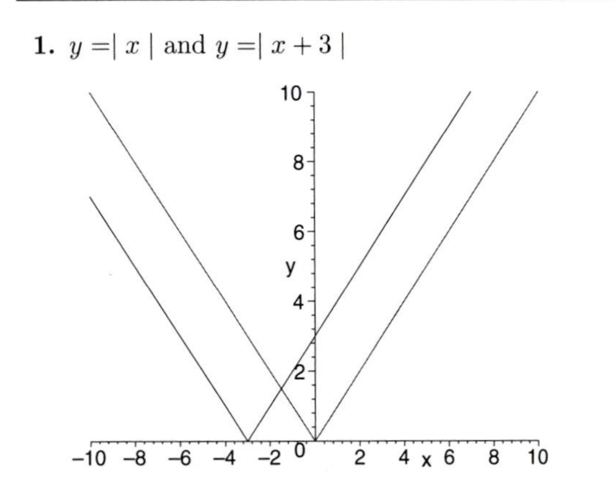

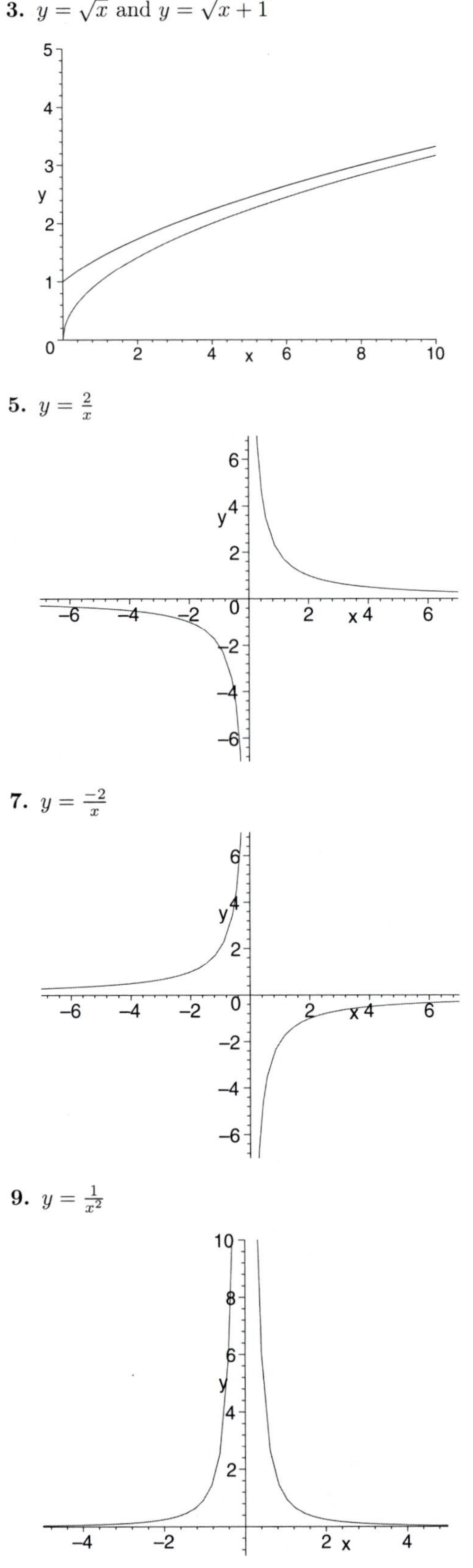

3. $y = \sqrt{x}$ and $y = \sqrt{x+1}$

5. $y = \frac{2}{x}$

7. $y = \frac{-2}{x}$

9. $y = \frac{1}{x^2}$

11. $y = \sqrt[3]{x}$

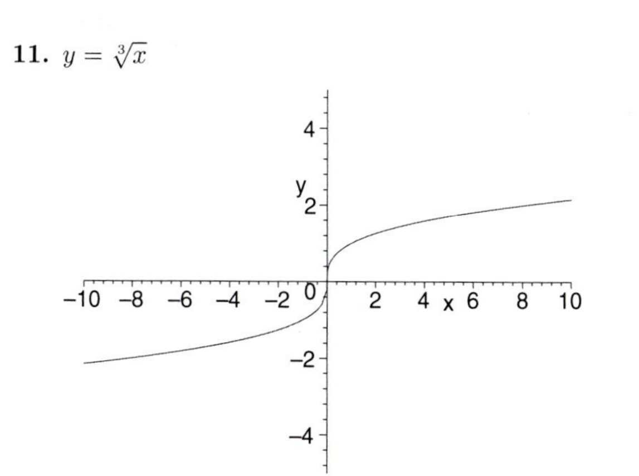

13. $y = \frac{x^2-9}{x+3}$. It is important to note here that $x = -3$ is not in the domain of the plotted function.

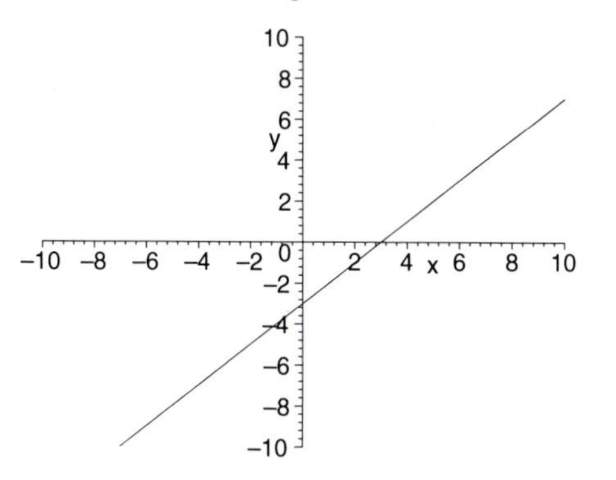

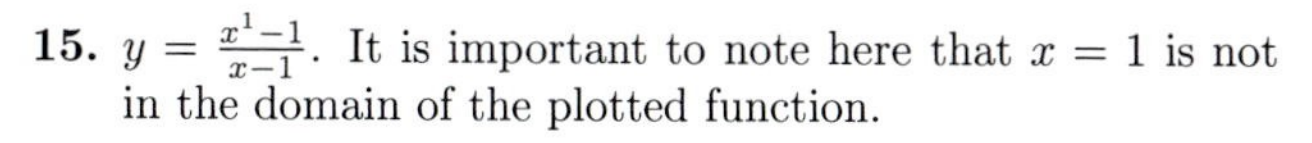

15. $y = \frac{x^1-1}{x-1}$. It is important to note here that $x = 1$ is not in the domain of the plotted function.

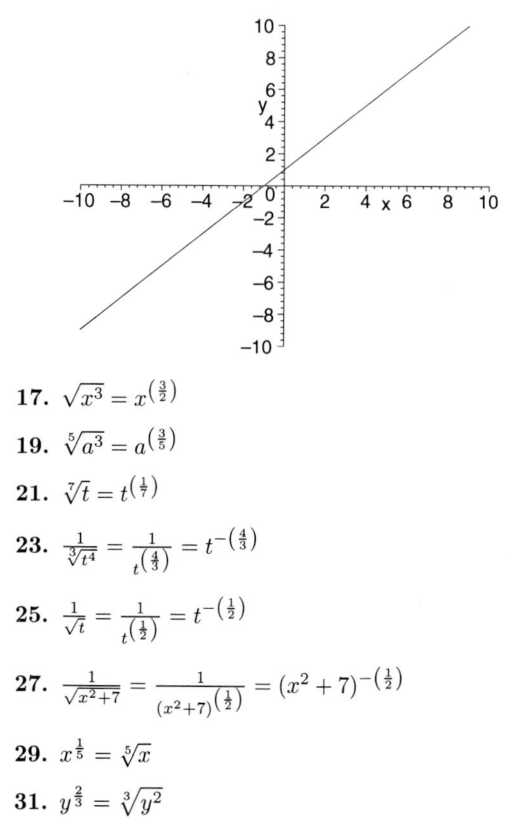

17. $\sqrt{x^3} = x^{\left(\frac{3}{2}\right)}$

19. $\sqrt[5]{a^3} = a^{\left(\frac{3}{5}\right)}$

21. $\sqrt[7]{t} = t^{\left(\frac{1}{7}\right)}$

23. $\frac{1}{\sqrt[3]{t^4}} = \frac{1}{t^{\left(\frac{4}{3}\right)}} = t^{-\left(\frac{4}{3}\right)}$

25. $\frac{1}{\sqrt{t}} = \frac{1}{t^{\left(\frac{1}{2}\right)}} = t^{-\left(\frac{1}{2}\right)}$

27. $\frac{1}{\sqrt{x^2+7}} = \frac{1}{(x^2+7)^{\left(\frac{1}{2}\right)}} = (x^2+7)^{-\left(\frac{1}{2}\right)}$

29. $x^{\frac{1}{5}} = \sqrt[5]{x}$

31. $y^{\frac{2}{3}} = \sqrt[3]{y^2}$

33. $t^{\frac{-2}{5}} = \frac{1}{t^{\frac{2}{5}}} = \frac{1}{\sqrt[5]{t^2}}$

35. $b^{\frac{-1}{3}} = \frac{1}{b^{\frac{1}{3}}} = \frac{1}{\sqrt[3]{b^2}}$

37. $e^{\frac{-17}{6}} = \frac{1}{e^{\frac{17}{6}}} = \frac{1}{\sqrt[6]{e^{17}}}$

39. $(x-3)^{\frac{-1}{2}} = \frac{1}{(x-3)^{\frac{1}{2}}} = \frac{1}{\sqrt{x-3}}$

41. $\frac{1}{t^23} = \frac{1}{\sqrt[3]{t^2}}$

43. $9^{3/2} = (\sqrt{9})^3 = (3)^3 = 27$

45. $64^{2/3} = (\sqrt[3]{64})^3 = (4)^2 = 16$

47. $16^{3/4} = (\sqrt[4]{16})^3 = (2)^3 = 8$

49. The domain consists of all x-values such that the denominator does not equal 0, that is $x - 5 \neq 0$, which leads to $x \neq 5$. Therefore, the domain is $\{x|x \neq 5\}$

51. Solving for the values of the x in the denominator that make it 0.

$$\begin{aligned} x^2 - 5x + 6 &= 0 \\ (x-3)(x-2) &= 0 \\ &\text{So} \\ x &= 3 \text{ and} \\ x &= 2 \end{aligned}$$

Which means that the domain is the set of all x -values such that $x \neq 3$ or $x \neq 2$

53. The domain of a square root function is restricted by the value where the radicant is positive. Thus, the domain of $f(x) = \sqrt{5x+4}$ can be found by finding the solution to the inequality $5x + 4 \geq 0$.

$$\begin{aligned} 5x + 4 &\geq 0 \\ 5x &\geq -4 \\ x &\geq \frac{-4}{5} \end{aligned}$$

55. To complete the table we will plug the given W values into the equation

$$\begin{aligned} T(20) &= (20)^{1.31} = 50.623 \approx 51 \\ T(30) &= (30)^{1.31} = 86.105 \approx 86 \\ T(40) &= (40)^{1.31} = 125.516 \approx 126 \\ T(50) &= (50)^{1.31} = 168.132 \approx 168 \\ T(100) &= (100)^{1.31} = 416.869 \approx 417 \\ T(150) &= (150)^{1.31} = 709.054 \approx 709 \end{aligned}$$

Therefore the table is given by

W	0	10	20	30	40	50	100	150
T	0	20	51	86	126	168	417	709

Now the graph

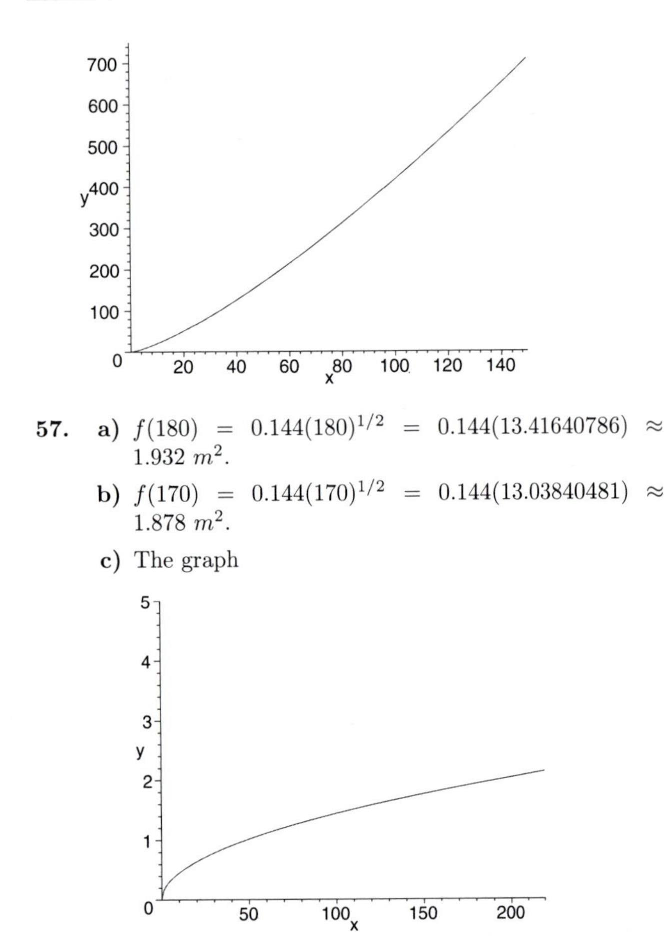

57. **a)** $f(180) = 0.144(180)^{1/2} = 0.144(13.41640786) \approx 1.932\ m^2$.

b) $f(170) = 0.144(170)^{1/2} = 0.144(13.03840481) \approx 1.878\ m^2$.

c) The graph

59. Let V be the velocity of the blood, and let A be the cross sectional area of the blood vessel. Then

$$V = \frac{k}{A}$$

Using $V = 30$ when $A = 3$ we can find k.

$$\begin{aligned} 30 &= \frac{k}{3} \\ (30)(3) &= k \\ 90 &= k \end{aligned}$$

Now we can write the proportial equation

$$V = \frac{90}{A}$$

we need to find A when $V = 0.026$

$$\begin{aligned} 0.026 &= \frac{90}{A} \\ 0.026A &= 90 \\ A &= \frac{90}{0.026} \\ &= 3461.538\ m^2 \end{aligned}$$

61.

$$\begin{aligned} x + 7 + \frac{9}{x} &= 0 \\ x(x + 7 + \frac{9}{x}) &= x(0) \\ x^2 + 7x + 9 &= 0 \\ x &= \frac{-7 \pm \sqrt{49 - 4(1)(9)}}{2} \\ &= \frac{-7 \pm \sqrt{13}}{2} \\ x &= \frac{-7 - \sqrt{13}}{2} \\ \text{and} & \\ x &= \frac{-7 + \sqrt{13}}{2} \end{aligned}$$

63. $P = 1000t^{5/4} + 14000$

a) $t = 37$, $P = 1000(37)^{5/4} + 14000 = 105254.0514$.
$t = 40$, $P = 1000(40)^{5/4} + 14000 = 114594.6744$
$t = 50$, $P = 1000(50)^{5/4} + 14000 = 146957.3974$

b) Below is the graph of P for $0 \le t \le 50$.

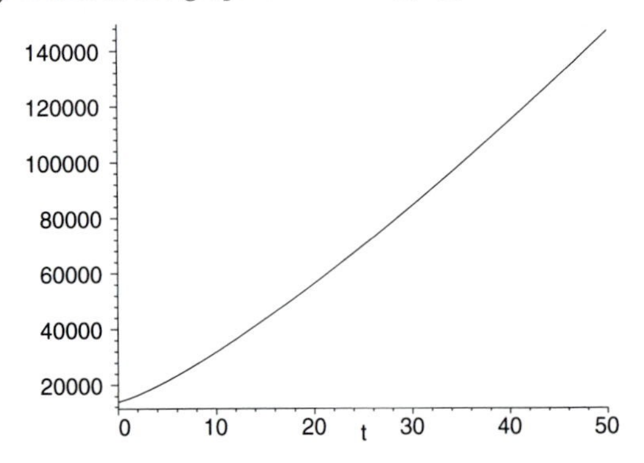

65. A rational function is a function given by the quotient of two polynomial functions while a polynomial function is a function that has the form $a_n x^n + a_{n-1}x^{n-1} + \cdots + a_1 x + a_0$. Since every polynomial function can be written as a quotient of two other polynomial function then every polynomial function is a rational function.

67. $x = 2.6458$ and $x = -2.6458$

69. The function has no zeros

Exercise Set 1.4

1. $(120^o)(\frac{\pi\ rad}{180^o}) = \frac{2\pi}{3}\ rad$

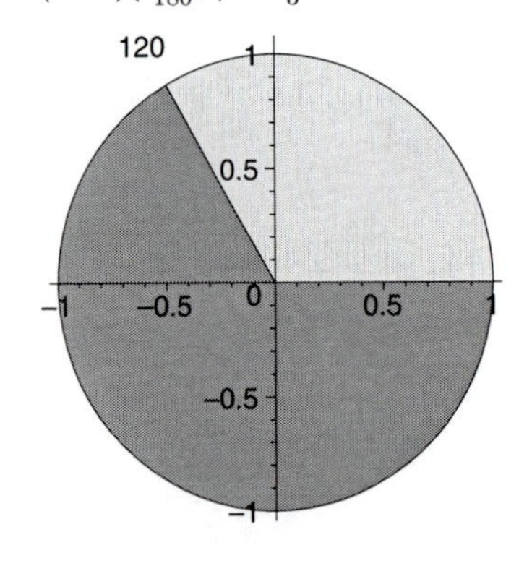

3. $(240^o)(\frac{\pi\ rad}{180^o}) = \frac{4\pi}{3}\ rad$

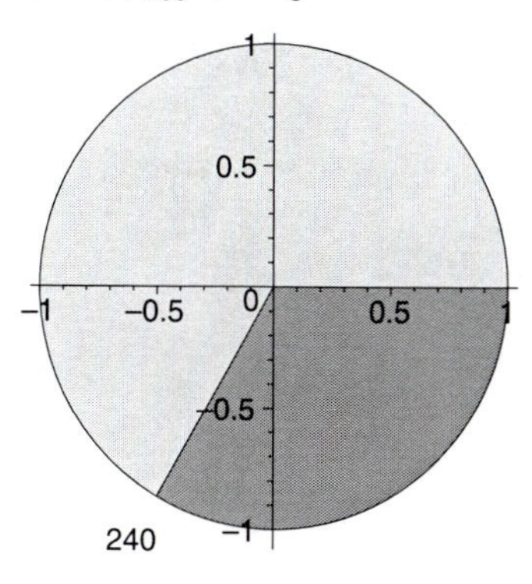

5. $(540^o)(\frac{\pi\ rad}{180^o}) = 3\pi\ rad$

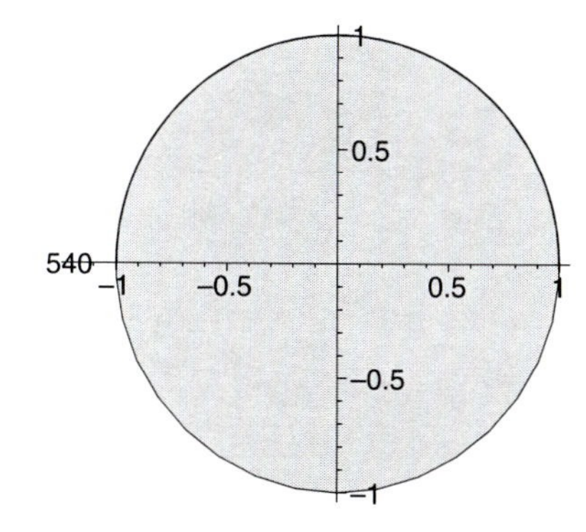

7. $(\frac{3\pi}{4})(\frac{180^o}{\pi\ rad}) = 135^o$

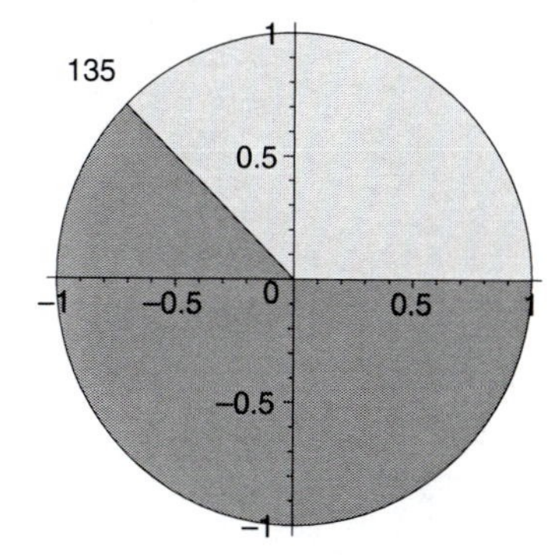

9. $(\frac{3\pi}{2})(\frac{180^o}{\pi\ rad}) = 270^o$

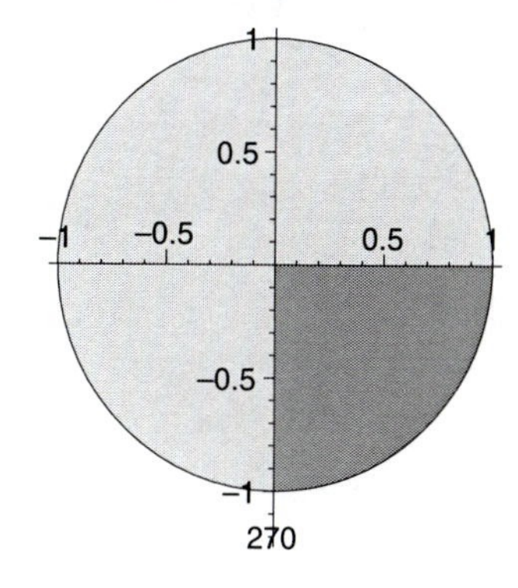

11. $(\frac{-\pi}{3})(\frac{180^o}{\pi\ rad}) = -60^o$

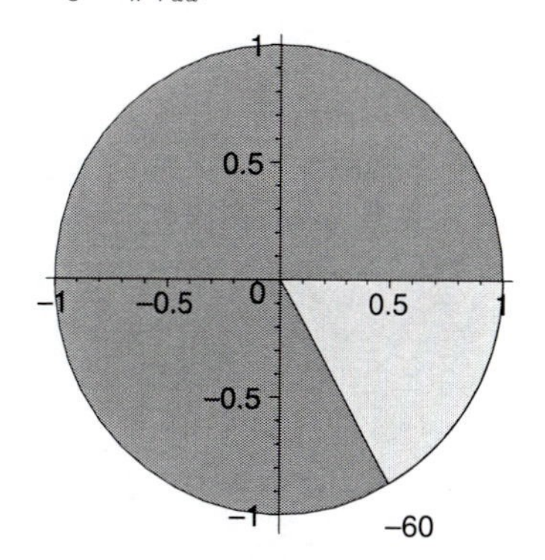

13. We need to solve $\theta_1 = \theta_2 + 360(k)$ for k. If the solution is an integer then the angles are coterminal otherwise they are not coterminal.

$$\begin{aligned} 395 &= 15 + 360(k) \\ 380 &= 360(k) \\ \frac{380}{360} &= k \\ 1.0\overline{5} &= k \end{aligned}$$

Since k is not an integer, we conclude that 15° and 395° are not coterminal.

15. We need to solve $\theta_1 = \theta_2 + 360(k)$ for k. If the solution is an integer then the angles are coterminal otherwise they are not coterminal.

$$\begin{aligned} 107 &= -107 + 360(k) \\ 214 &= 360(k) \\ \frac{214}{360} &= k \\ 0.59\overline{4} &= k \end{aligned}$$

Since k is not an integer, we conclude that 15° and 395° are not coterminal.

17. We need to solve $\theta_1 = \theta_2 + 2\pi(k)$ for k. If the solution is an integer then the angles are coterminal otherwise they are not coterminal.

$$\begin{aligned} \frac{\pi}{2} &= \frac{3\pi}{2} + 2\pi(k) \\ -\pi &= 2\pi(k) \\ \frac{-\pi}{2\pi} &= k \\ \frac{-1}{2} &= k \end{aligned}$$

Since k is not an integer, we conclude that $\frac{\pi}{2}$ and $\frac{3\pi}{2}$ are not coterminal.

19. We need to solve $\theta_1 = \theta_2 + 2\pi(k)$ for k. If the solution is an integer then the angles are coterminal otherwise they are not coterminal.

$$\begin{aligned} \frac{7\pi}{6} &= \frac{-5\pi}{6} + 2\pi(k) \\ 2\pi &= 2\pi(k) \\ \frac{2\pi}{2\pi} &= k \\ 1 &= k \end{aligned}$$

Since k is an integer, we conclude that $\frac{7\pi}{6}$ and $\frac{-5\pi}{6}$ are coterminal.

21. $sin\ 34^o = 0.5592$

23. $cos\ 12^o = 0.9781$

25. $tan\ 5^o = 0.0875$

27. $cot\ 34^o = \frac{1}{tan\ 34^o} = 1.4826$

29. $sec\ 23^o = \frac{1}{cos\ 23^o} = 1.0864$

31. $sin(\frac{\pi}{5}) = 0.5878$

33. $tan(\frac{\pi}{7}) = 0.4816$

35. $sec(\frac{3\pi}{8}) = \frac{1}{cos(\frac{3\pi}{8})} = 2.6131$

37. $sin(2.3) = 0.7457$

39. $t = sin^{-1}(0.45) = 26.7437^o$

41. $t = cos^{-1}(0.34) = 70.1231^o$

43. $t = tan^{-1}(2.34) = 66.8605^o$

45. $t = sin^{-1}(0.59) = 0.6311$

47. $t = cos^{-1}(0.60) = 0.9273$

49. $t = tan^{-1}(0.11) = 0.1096$

51.

$$\begin{aligned} sin\ 57^o &= \frac{x}{40} \\ x &= 40sin\ 57^o \\ x &= 33.5468 \end{aligned}$$

53.

$$\begin{aligned} cos\ 50^o &= \frac{15}{x} \\ x &= \frac{15}{cos\ 50^o} \\ x &= 23.3359 \end{aligned}$$

55.

$$\begin{aligned} cos\ t &= \frac{40}{60} \\ t &= cos^{-1}(\frac{40}{60}) \\ t &= 48.1897^o \end{aligned}$$

57.

$$\begin{aligned} tan\ t &= \frac{18}{9.3} \\ t &= tan^{-1}(\frac{18}{9.3}) \\ t &= 62.6761^o \end{aligned}$$

59. We can rewrite $75^o = 30^o + 45^o$ then use a sum identity

$$\begin{aligned} cos(A+B) &= cos\ Acos\ B - sin\ Asin\ B \\ cos\ 75^o &= cos(30^o + 45^o) \\ &= cos\ 30^o cos\ 45^o - sin\ 30^o sin\ 45^o \\ &= \frac{\sqrt{3}}{2} \cdot \frac{1}{\sqrt{2}} - \frac{1}{2} \cdot \frac{1}{\sqrt{2}} \\ &= \frac{\sqrt{3}}{2\sqrt{2}} - \frac{1}{2\sqrt{2}} \\ &= \frac{-1+\sqrt{3}}{2\sqrt{2}} \end{aligned}$$

61. Five miles is the same as $5 \cdot 5280\ ft = 26400\ ft$. The difference in elevation, y, is

$$\begin{aligned} sin\ 4^o &= \frac{y}{26400} \\ y &= 26400sin\ 4^o \\ &= 1841.57\ ft \end{aligned}$$

63. a)

$$\begin{aligned} cos\ 40^o &= \frac{x}{150} \\ x &= 150cos\ 40^o \\ &= 114.907 \end{aligned}$$

b)

$$\begin{aligned} sin\ 40^o &= \frac{y}{150} \\ y &= 150sin\ 40^o \\ &= 96.4181 \end{aligned}$$

c)

$$\begin{aligned} z^2 &= (x+180)^2 + y^2 \\ &= (114.907+180)^2 + (96.4181)^2 \\ z^2 &= 96266.58866 \\ z &= \sqrt{96266.58866} \\ &= 310.268 \end{aligned}$$

65.

$$\begin{aligned} v &= \frac{77000 \cdot 100 \cdot sec\ 65^o}{4000000} \\ &= \frac{7700000}{4000000cos\ 65^o} \\ &= 4.55494\ cm/sec \end{aligned}$$

67. a) When we consider the two triangles we have a new triangle that has three equal angles which is the definition of an equilateral triangle.

b) The short leg of each triangle is given by $2sin(30) = 2(\frac{1}{2}) = 1$

c) The long leg (L) is given by

$$\begin{aligned} 2^2 &= L^2 + 1^2 \\ 4 - 1 &= L^2 \\ \sqrt{3} &= L \end{aligned}$$

d) By considering all possible ratios between the long, short and hypotenuse of small triangles we obtain the trigonometric functions of $\frac{\pi}{6} = 30^o$ and $\frac{\pi}{3} = 60^o$

69. a) The tangent of an angle is equal to the ratio of the opposite side to the adjacent side (of a right triangle), and for the small triangle that ratio is $\frac{5}{7}$.

b) For the large right triangle, the opposite side is 10 and the adjacent side is $7+7=14$. Thus the tangent is $\frac{10}{14}$

c) Because the trigonometric functions depend on the ratios of the sides and not the size of triangle. Note that the answer in part b) is equivalent to that in part a) even though the triangle in part b) was larger that that used in part a)

71.

$$\begin{aligned} \frac{sin\ t}{cos\ t} &= \frac{y/r}{x/r} \\ &= \frac{y}{r} \div \frac{x}{r} \\ &= \frac{y}{r} \cdot \frac{r}{x} \\ &= \frac{y}{x} \\ &= tan\ t \end{aligned}$$

$$\begin{aligned} & Thus \\ \frac{sin\ t}{cos\ t} &= tan\ t \end{aligned}$$

and

$$\begin{aligned} \frac{cos\ t}{sin\ t} &= \frac{x/r}{y/r} \\ &= \frac{x}{r} \div \frac{y}{r} \\ &= \frac{x}{r} \cdot \frac{r}{y} \\ &= \frac{x}{y} \\ &= cot\ t \end{aligned}$$

$$\begin{aligned} & Thus \\ \frac{cos\ t}{sin\ t} &= cot\ t \end{aligned}$$

73. a) $sin(t) = \frac{u}{1} = u$, and $cos(t) = \frac{v}{1} = v/$

b) Consider the triangle made by the sides v, w, and y. The angle vw has a value of $90 - r$ (completes a straight angle). The sum of angles in any triangle is 180. Therefore

$$\begin{aligned} s + 90 + (90 - r) &= 180 \\ s + 180 - r &= 180 \\ s - r &= 0 \\ s &= r \end{aligned}$$

c) $cos(s) = \frac{y}{v}$, which means $y = cos(s)v$.
But from part a) $v = cos(t)$, therefore $y = cos(s)cos(t)$

d) $sin(r) = \frac{z}{u}$, which means $z = sin(r)u$.
Using results from part a) and part b) we get $sin(r) = sin(s)$ and $u = sin(t)$, therefore $z = sin(s)sin(t)$

e) $cos(s+t) = \frac{(y-z)}{1} = y - z$. Replacing ur results for y and z we get $cos(s+t) = cos(s)cos(t) - sin(s)sin(t)$

75. Use $cos^2t + sin^2t = 1$ as follows

$$\begin{aligned} cos^2t + sin^2t &= 1 \\ \frac{cos^2t}{cos^2t} + \frac{sin^2t}{cos^2t} &= \frac{1}{cos^2t} \\ 1 + tan^t &= sec^2t \end{aligned}$$

77. Let $2t = t + t$

$$\begin{aligned} sin(a+b) &= sin(a)cos(b) + cos(a)sin(b) \\ sin(2t) &= sin(t+t) \\ &= sin(t)cos(t) + cos(t)sin(t) \\ &= 2sin(t)cos(t) \end{aligned}$$

79. Using the result from Exercise 78 part (c)

$$\begin{aligned} cos(2t) &= 1 - 2sin^2(t) \\ cos(2t) - 1 &= -2sin^2(t) \\ \frac{cos(2t) - 1}{-2} &= sin^2(t) \\ \frac{1 - cos(2t)}{2} &= sin^2(t) \end{aligned}$$

81. a) $V(0) = sin^p(0)sin^q(0)sin^r(0)sin^s(0) = 0$
$V(1) = sin^p(\frac{\pi}{2})sin^q(\frac{\pi}{2})sin^r(\frac{\pi}{2})sin^s(\frac{\pi}{2}) = 1$

b) When $h = 0$ the volume of the tree is zero since there is no height and therefore the proportion of volume under that height is zero. While at the top of the tree, $h = 1$, the proportion of volume under the tree is 1 since the entire tree volume falls below its height.

83.

$$\begin{aligned} V(\frac{1}{2}) &= sin^{-5.621}(\frac{\pi}{4})sin^{74.831}(\frac{\pi}{2\sqrt{2}}) \\ & \times sin^{-195.644}(\frac{\pi}{2\sqrt[3]{2}})sin^{138.959}(\frac{\pi}{2\sqrt[4]{2}}) \\ &= 0.8219 \end{aligned}$$

Exercise Set 1.5

1. $5\pi/4$

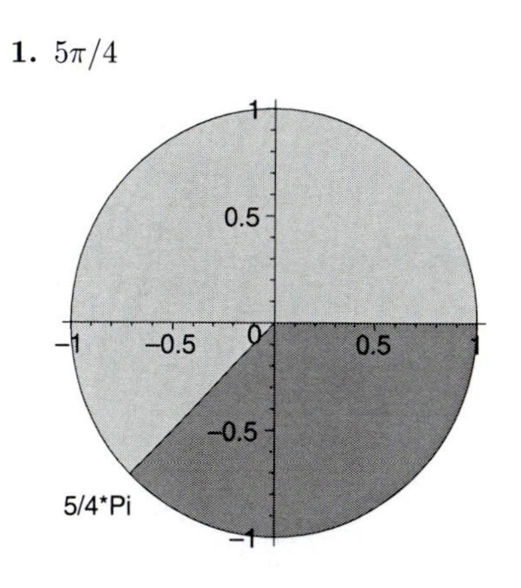

3. $-\pi$

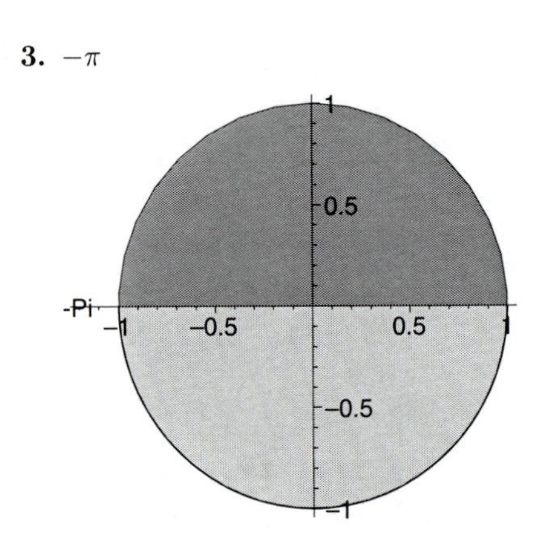

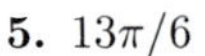

5. $13\pi/6$

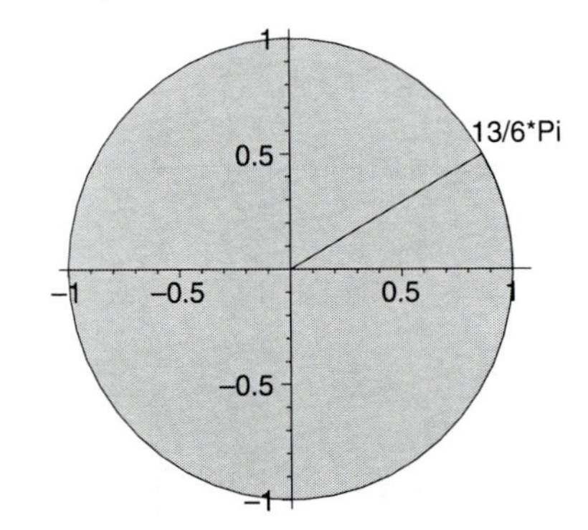

7. $cos(9\pi/2) = 0$

9. $sin(-5\pi/6) = \frac{-1}{2}$

11. $cos(5\pi) = -1$

13. $tan(-4\pi/3) = -\sqrt{3}$

15. $cos\ 125^o = -0.5736$

17. $tan(-220^o) = -0.8391$

19. $sec\ 286^o = \frac{1}{cos\ 286^o} = 3.62796$

21. $sin(1.2\pi) = -0.587785$

23. $cos(-1.91) = -0.332736$

25. $t = sin^{-1}(1/2) = \frac{\pi}{6} + 2n\pi$ and $\frac{5\pi}{6} + 2n\pi$

27. $2t = sin^{-1}(0) = n\pi$ so $t = \frac{n\pi}{2}$

29.

$$\begin{aligned} cos(3t + \frac{\pi}{4}) &= -\frac{1}{2} \\ 3t + \frac{\pi}{4} &= cos^{-1}(-\frac{1}{2}) \\ 3t &= -\frac{\pi}{4} + \frac{2\pi}{3} + 2n\pi \\ 3t &= \frac{5\pi}{12} 2n\pi \\ t &= \frac{5\pi}{36}\frac{2}{3}n\pi \\ \text{and} \\ 3t &= -\frac{\pi}{4} + \frac{4\pi}{3} + 2n\pi \\ 3t &= \frac{13\pi}{12} + 2n\pi \\ t &= \frac{13}{36}\pi + \frac{2}{3}n\pi \end{aligned}$$

31.

$$\begin{aligned} cos(3t) &= 1 \\ 3t &= cos^{-1}(1) \\ 3t &= 2n\pi \\ t &= \frac{2}{3}n\pi \end{aligned}$$

33.

$$\begin{aligned} 2sin^2 t - 5sin\ t - 3 &= 0 \\ (2sin\ t + 1)(sin\ t - 3) &= 0 \\ \text{The only solution comes from} \\ (2sin\ t + 1) &= 0 \\ sin\ t &= -\frac{1}{2} \\ t &= sin^{-1}(-\frac{1}{2}) \\ t &= \frac{7\pi}{6} + 2n\pi \\ \text{and} \\ t &= \frac{11\pi}{6} + 2n\pi \end{aligned}$$

35.

$$\begin{aligned} cos^2 x + 5cos\ x &= -6 \\ cos^2 x + 5cos\ x + 6 &= 0 \\ cos\ x &= \frac{-5 \pm \sqrt{25 - 4(1)(6)}}{2} \\ &= \frac{-5 \pm 1}{2} \\ &= \frac{-5 - 1}{2} = -3 \\ \text{and} \\ &= \frac{-5 + 1}{2} = -2 \end{aligned}$$

Since both values are larger than one, then the equation has no solutions.

37. $y = 2sin\ 2t + 4$
amplitude $= 2$, period $= \frac{2\pi}{2} = \pi$, mid-line $y = 4$
maximum $= 4 + 2 = 6$, minimum $= 4 - 2 = 2$

39. $y = 5cos(t/2) + 1$
amplitude $= 5$, period $= \frac{2\pi}{\frac{1}{2}} = 4\pi$, mid-line $y = 1$
maximum $= 1 + 5 = 6$, minimum $= 1 - 5 = -4$

41. $y = \frac{1}{2}sin(3t) - 3$
amplitude $= \frac{1}{2}$, period $= \frac{2\pi}{3}$, mid-line $y = -3$
maximum $-3 + \frac{1}{2} = \frac{-5}{2}$, minimum $= -3 - \frac{1}{2} = \frac{-7}{2}$

43. $y = 4sin(\pi t) + 2$
amplitude $= 4$, period $= \frac{2\pi}{\pi} = 2$, mid-line $y = 2$
maximum $= 2 + 4 = 6$, minimum $= 2 - 4 = -2$

45. The maximum is 10 and the minimum is -4 so the amplitude is $\frac{10-(-4)}{2} = 7$. The mid-line is $y = 10 - 7 = 3$, and the period is 2π (the distance from one peak to the next one) which means that $b = \frac{2\pi}{2\pi} = 1$. From the information above, and the graph, we conclude that the function is

$$y = 7sin\ t + 3$$

47. The maximum is 1 and the minimum is -3 so the amplitude is $\frac{1-(-3)}{2} = 2$. The mid-line is $y = 1 - 2 = -1$, and the period is 4π which means $b = \frac{\pi}{2}$. From the information above, and the graph, we conclude that the function is

$$y = 2cos(t/2) - 1$$

49.

$$\begin{aligned} R &= 0.339 + 0.808cos\ 40^o\ cos\ 30^o \\ &\quad -0.196sin\ 40^o\ sin\ 30^o - 0.482cos\ 0^o\ cos\ 30^o \\ &= 0.571045\ megajoules/m^2 \end{aligned}$$

51.

$$\begin{aligned} R &= 0.339 + 0.808cos\ 50^o\ cos\ 55^o \\ &\quad -0.196sin\ 50^o\ sin\ 55^o - 0.482cos\ 45^o\ cos\ 55^o \\ &= 0.234721\ megajoules/m^2 \end{aligned}$$

53. Period is 5 so $b = \frac{2\pi}{5}$, $k = 2500$, $a = 250$. Therefore, the function is

$$V(t) = 250cos\ \frac{2\pi t}{5} + 2500$$

55. Since our lungs increase and decrease as we breathe then there is a maximum and minimum volume for the air capacity in our lungs. We have a regular period of time at which we breathe (inhale and exhale). These facotrs are reasons why the cosine model is appropriate for describing lung capacity.

57. The frequency is the reciprocal of the period. Therefore, $f = \frac{b}{2\pi} = \frac{880\pi}{2\pi} = 440\ Hz$

59. The amplitude is given as 5.3. $b = f \cdot 2\pi$ where f is the frequency, $b = 0.172 \cdot 2\pi = 1.08071$, $k = 143$. Therefore, the function is

$$p(t) = 5.3cos(1.08071t) + 143$$

61. $x = cos(140^o)$, $y = sin(140^o)$, $(-0.76604, 0.64279)$

63. $x = cos(\frac{9\pi}{5})$, $y = sin(\frac{9\pi}{5})$, $(0.80902, -0.58779)$

65. Rewrite $105^o = 45^o + 60^o$ and use a sum identity.

$$\begin{aligned} sin\ 105^o &= sin(45^o + 60^o) \\ &= sin\ 45^o\ cos\ 60^o + cos\ 45^o\ sin\ 60^o \\ &= \frac{1}{\sqrt{2}} \cdot \frac{1}{2} + \frac{1}{\sqrt{2}} \cdot \frac{\sqrt{3}}{2} \\ &= \frac{1}{2\sqrt{2}} + \frac{\sqrt{3}}{2\sqrt{2}} \\ &= \frac{1+\sqrt{3}}{2\sqrt{2}} \end{aligned}$$

67. a) From the graph we can see that the point with angle t has an opposite x and y coordinate than the point with angle $t + \pi$. Since the x coordinate corresponds to the *cos* of the angle which the point makes and the y coordinate corresponds to the *sin* of the angle which the point makes it follows that $sin(t + \pi) = -sin(t)$ and $cos(t + \pi) = -cos(t)$.

b)

$$\begin{aligned} sin(t + \pi) &= sin\ tcos\ \pi + cos\ tsin\ \pi \\ &= sin\ t \cdot -1 + cos\ t \cdot 0 \\ &= -sin\ t \end{aligned}$$

and

$$\begin{aligned} cos(t + \pi) &= cos\ tcos\ \pi - sin\ tsin\ \pi \\ &= cos\ t \cdot -1 - sin\ t \cdot 0 \\ &= -cos\ t \end{aligned}$$

c)

$$\begin{aligned} tan(t + \pi) &= \frac{sin(t + \pi)}{cos(t + \pi)} \\ &= \frac{-sin\ t}{-cos\ t} \\ &= \frac{sin\ t}{cos\ t} \\ &= tan\ t \end{aligned}$$

69. a) Since the radius of a unit circle is 1, the circumference of the unit circle is 2π. Therefore any point $t + 2\pi$ will have exactly the same terminal side as the point t, that is to say that the points t and $t+2\pi$ are coterminal on the unit circle. Therefore, $sin\ t = sin(t + 2\pi)$ for all numbers t.

b)

$$\begin{aligned} g(t + 2\pi/b) &= asin[b(t + 2\pi/b)] + k \\ &= asin(bt + 2\pi) + k \end{aligned}$$

from part a)

$$= asin(bt) + k$$

by definition

$$g(t + 2\pi/b) = g(t)$$

c) Since the function evaluated at $t+2\pi/b$ has the same value as the function evaluated at t and $2\pi/b \neq 0$ then $t + 2\pi/b$ is evaluated after t. Since we have a periodic function in $g(t)$ it follows that the period of the function is implied to be $2\pi/b$.

71. Since at the base, L is small, T is large, and d is large, then the basilar membrane is affected mostly by high frequency sounds.

73. $f = \frac{880 \cdot 2^{-9/12}\pi}{2\pi} = 261.626$

75.

$$\begin{aligned} \frac{880(2^{n/12})\pi}{2\pi} &= 1760 \\ 2^{n/12-1} &= \frac{1760}{880} \\ 2^{n/12-1} &= 2 \end{aligned}$$

Comparing exponents we can conclude that

$$\begin{aligned} \frac{n}{12} - 1 = 1 & \\ \frac{n}{12} &= 2 \\ n &= 24 \end{aligned}$$

There are 24 notes above A above middle C.

77.

$$\begin{aligned} \frac{880(2^{n/12})\pi}{2\pi} &= 2200 \\ 2^{n/12-1} &= \frac{2200}{880} \\ 2^{n/12-1} &= 2.5 \\ (\frac{n}{12} - 1)ln(2) = ln(2.5) & \\ \frac{n}{12} - 1 &= \frac{ln(2.5)}{ln(2)} \\ \frac{n}{12} &= \frac{ln(2.5)}{ln(2)} + 1 \\ n &= 12\left(\frac{ln(2.5)}{ln(2)} + 1\right) \\ n &= 27.86314 \end{aligned}$$

There are 28 notes above A above middle C

79. **a)** Left to the student

b) $y = \frac{1}{2}cos(2t) + \frac{1}{2}$

c) We use the double angle identity obtained in Exercise 79 of Section 1.4 and solve for $cos^2(t)$ to obtain the model in part b).

81. **a)** Left to the student

b) Left to the student

c) The horizontal shift moves every point of the original graph $\frac{\pi}{4}$ units to the right.

83. Left to the student

85. Left to the student

Chapter 2

Differentiation

Exercise Set 2.1

1. The function is not continuous at $x = 1$ since the limit from the left of $x = 1$ is not equal to the limit from the right of $x = 1$ and therefore the limit of the function at $x = 1$ does not exist.

3. The function is continuous at every point in the given plot. Note that the graph can be traced without a jump from one point to another.

5. **a)** As we approach the x-value of 1 from the right we notice that the y-value is approaching a value of -1. Thus, $\lim_{x \to 1^+} f(x) = -1$. As we approach the x-value of 1 from the left we notice that the y-value is approaching a value of 2. Thus, $\lim_{x \to 1^-} f(x) = 2$. Since $\lim_{x \to 1^+} f(x) \neq \lim_{x \to 1^-} f(x)$ then $\lim_{x \to 1} f(x)$ does not exist.

b) Reading the value from the graph $f(1) = -1$.

c) Since the $\lim_{x \to 1^-} f(x)$ does not exist, then $f(x)$ is not continuous at $x = 1$.

d) As we approach the x-value of -2 from the right we notice that the y-value is approaching a value of 3. Thus, $\lim_{x \to -2^+} f(x) = 3$. As we approach the x-value of -2 from the left we notice that the y-value is approaching a value of 3. Thus, $\lim_{x \to -2^-} f(x) = 3$. Since $\lim_{x \to -2^+} f(x) = \lim_{x \to -2^-} f(x) = 3$ then $\lim_{x \to -2} f(x) = 3$.

e) Reading the value from the graph $f(-2) = 3$.

f) Since $\lim_{x \to -2} f(x) = 3$ and $f(-2) = 3$, then $f(x)$ is continuous at $x = -2$.

7. **a)** As we approach the x-value of 1 from the right we notice that the y-value is approaching a value of 2. Thus, $\lim_{x \to 1^+} h(x) = 2$. As we approach the x-value of 1 from the left we notice that the y-value is approaching a value of 2. Thus, $\lim_{x \to 1^-} h(x) = 2$. Since $\lim_{x \to 1^+} h(x) = \lim_{x \to 1^-} h(x) = 2$ then $\lim_{x \to 1} h(x) = 2$.

b) Reading the value from the graph $h(1) = 2$.

c) Since the $\lim_{x \to 1} h(x) = 2$ and $= h(1) = 2$ then $h(x)$ is continuous at $x = 1$.

d) As we approach the x-value of -2 from the right we notice that the y-value is approaching a value of 3. Thus, $\lim_{x \to -2^+} h(x) = 0$. As we approach the x-value of -2 from the left we notice that the y-value is approaching a value of 0. Thus, $\lim_{x \to -2^-} h(x) = 0$. Since $\lim_{x \to -2^+} h(x) = \lim_{x \to -2^-} h(x) = 0$ then $\lim_{x \to -2} h(x) = 0$.

e) Reading the value from the graph $h(-2) = 0$.

f) Since $\lim_{x \to -2} h(x) = 0$ and $h(-2) = 0$, then $h(x)$ is continuous at $x = -2$.

9. **a)** As we approach the x value of 1 from the right we find that the y value is approching 3. Thus $\lim_{x \to 1^+} f(x) = 3$

b) As we approach the x value of 1 from the left, we find that the y value is approching 3. Thus $\lim_{x \to 1^-} f(x) = 3$

c) Since $\lim_{x \to 1^+} f(x) = 3$ and $\lim_{x \to 1^-} f(x) = 3$ then $\lim_{x \to 1} f(x) = 3$

d) From the given conditions $f(1) = 2$

e) $f(x)$ is not continuous at $x = 1$ since $\lim_{x \to 3} f(x) \neq f(1)$

f) $f(x)$ is continuous at $x = 2$ since $\lim_{x \to 2^+} f(x) = \lim_{x \to 2^-} f(x) = 2 = f(2)$

11. **a)** True. The values of y as we approch $x = 0$ from the right is the same as the value of the function at $x = 0$, which is 0

b) True. The values of y as we approch $x = 0$ from the left is the same as the value of the function at $x = 0$, which is 0

c) True. Since $\lim_{x \to 0^+} f(x) = 0$ and $\lim_{x \to 0^-} f(x) = 0$

d) False. Since $\lim_{x \to 3^+} f(x) = 3$ and $\lim_{x \to 3^-} f(x) = 1$

e) True. Since $\lim_{x \to 0^+} f(x) = \lim_{x \to 0^-} f(x) = 0$

f) False. Since $\lim_{x \to 3^+} f(x) \neq \lim_{x \to 3^-} f(x)$

g) True. Since $\lim_{x \to 0} f(x) = 0 = f(0)$

h) False. Since $\lim_{x \to 3} f(x)$ does not exist

13. **a)** False. As we approach the x value of -2 from the right we find that the y value is approching 2.

b) True. As we approach the x value of -2 from the left, we find that the y value is approching 0.

c) False. Since $\lim_{x \to -2^+} f(x) = 1$ and $\lim_{x \to 1^-} f(x) = 0$

d) False. Since $\lim_{x \to -2^+} f(x) \neq \lim_{x \to -2^-} f(x)$

e) False. Since $\lim_{x \to -2} f(x)$ does not exist

f) True. Since $\lim_{x \to 0^+} f(x) = \lim_{x \to 0^-} f(x) = 0$

g) True. The graph indicate a point (solid dot) at $(0, 2)$

h) False. Since $\lim_{x \to -2^+} f(x) \neq \lim_{x \to -2^-} f(x)$

i) False. Since $\lim_{x \to 0} f(x) \neq f(0)$

j) True. Since $\lim_{x \to -1^+} f(x) = \lim_{x \to -1^-} f(x) = f(-1)$

15. **a)** True. As we approach the x value of 0 from the right we find that the y value is approching 0, which is the value of the function at $x = 0$.

b) False. As we approach the x value of 0 from the left, we find that the y value is approching 2 instead of 0.

c) False. Since $\lim_{x\to 0^+} f(x) = 0$ and $\lim_{x\to 0^-} f(x) = 2$

d) True. Since $\lim_{x\to 2^+} f(x) = 4 = \lim_{x\to 2^-} f(x)$

e) False. Since $\lim_{x\to 0^-} f(x) \neq \lim_{x\to 0^+} f(x)$

f) True. Since $\lim_{x\to 2^+} f(x) = \lim_{x\to 2^-} f(x) = 4$

g) False. Since $\lim_{x\to 0} f(x)$ does not exist

h) True. Since $\lim_{x\to 2} f(x) = 4 = f(2)$

17. The function p is not continuous at $x = 1$ since the $\lim_{x\to 1} p(x)$ does not exist. p is continuous at $x = 1.5$ since $\lim_{x\to 1.5} p(x) = 0.6 = p(1.5)$. p is not continous at $x = 1$ since the $\lim_{x\to 2} p(x)$ does not exist. p is continuous at $x = 2.01$ since $\lim_{x\to 2.01} p(x) = 0.8 = p(2.01)$.

19. $\lim_{x\to 1^-} p(x) = 0.4$, $\lim_{x\to 1^+} p(x) = 0.6$, therefore $\lim_{x\to 1} p(x)$ does not exist

21. $\lim_{x\to 2.6^-} p(x) = 0.8$, $\lim_{x\to 2.6^+} p(x) = 0.8$, therefore $\lim_{x\to 2.6} p(x) = 0.8$

23. $\lim_{x\to 3.4} p(x) = 1$ since $\lim_{x\to 3.4^-} p(x) = 1$, $\lim_{x\to 3.4^+} p(x) = 1$

25. If we continue the pattern used for the taxi fare function, we see that for $x = 2.3$, which falls in the range of 2.2 and 2.4 miles, the fare will be \$5.60, for $x = 5$, which falls between the range 2.4 and 2.6 miles, the fare is \$5.90. For $x = 2.6$ and $x = 3$ we need to be careful since they act as a boundary of two possible fares. Therefore C is continuous at $x = 2.3$ since $\lim_{x\to 2.3} C(x) = 5.60 = C(2.3)$. C is continuous at $x = 2.5$ since $\lim_{x\to 2.5} C(x) = 5.90 = C(2.5)$. C is not continuous at $x = 2.6$ since $\lim_{x\to 2.6^-} C(x) = 5.90$ and $\lim_{x\to 2.6^+} C(x) = 6.20$ thus, $\lim_{x\to 2.6} C(x)$ does not exist. C is not continuous at $x = 3$ since $\lim_{x\to 3^-} C(x) 6.50$ and $\lim_{x\to 3^+} C(x) = 6.80$ thus, $\lim_{x\to 3} C(x)$ does not exist.

27. $\lim_{x\to 0.2^-} C(x) = \$2.30$, $\lim_{x\to 0.2^+} C(x) = \$2.60$, therefore $\lim_{x\to 0.2} C(x)$ does not exist

29. $\lim_{x\to 0.5^-} C(x) = \$2.90$, $\lim_{x\to 0.5^+} C(x) = \$2.90$, therefore $\lim_{x\to 0.5} C(x) = \$2.90$

31. The population function, $p(t)$, is discontinuous at $t^* = 0.5$, $t^* = 0.75$, $t^* = 1.25$, $t^* = 1.5$, and at $t^* = 1.75$ since at these points the population function has a "jump" which means that the $\lim_{t\to t^*} p(t)$ does not exist

33. $\lim_{t\to 1.5^+} p(t) = 12$

35. The population function, $p(t)$, is discontinuous at $t^* = 0.1$, $t^* = 0.3$, $t^* = 0.4$, $t^* = 0.5$, $t^* = 0.6$, and at $t^* = 0.8$ since at these points the population function has a "jump" which means that the $\lim_{t\to t^*} p(t)$ does not exist

37. $\lim_{t\to 0.6^+} p(t) = 35$

39. From the graph, the "I've got it" experience seems to occur after spending 20 hours on the task.

41. $\lim_{t\to 20^+} N(t) = 100$, $\lim_{t\to 20^-} N(t) = 30$, therefore $\lim_{t\to} N(t)$ does not exist

43. $N(t)$ is discontinuous at $t = 20$ since $\lim_{t\to 20} N(t)$ does not exist. $N(t)$ is continuous at $t = 30$ since $\lim_{t\to 30} N(t) = 100 = N(30)$

45. A function may not be continuous if the function is not defined at one of the points in the domain, it also may not be continuous if the limit at a point does not exist, it also may not be continuous if the limit at a point is different than the value of the function at that point.
NOTE: See the graphs on page 77.

47. $f(x)$ is continuous by C1 and C2, $\lim_{x\to 1} f(x) = 0$

49. $g(x)$ is continuous by C4, $\lim_{x\to 1} g(x) = 1$

51. $cot\ x$ is continuous by C5, $\lim_{x\to \frac{\pi}{3}} cot\ x = \dfrac{1}{\sqrt{3}}$

53. $csc\ x$ is continuous by C5, $\lim_{x\to \frac{\pi}{4}} csc\ x = \sqrt{2}$

55. $f(x)$ is continuous by C5, $\lim_{x\to \frac{\pi}{3}} f(x) = \dfrac{\sqrt[4]{3}}{\sqrt{2}}$

57. $g(x)$ is continuous by C5, $dslim_{x\to \frac{\pi}{6}} g(x) = -\frac{1}{2}$

59. Limit approaches 0

61. Limit approaches 1

63. Limit does not exist

Exercise Set 2.2

1. $x^2 - 3$ is a continuous function (it is a polynomial). Therefore, we can use direct substitution

$$\begin{aligned} \lim_{x\to 1} (x^2 - 3) &= (1)^2 - 3 \\ &= 1 - 3 \\ &= -2 \end{aligned}$$

3. The function $f(x) = \frac{3}{x}$ is not continuous at $x - 0$ since the denominator equals zero. There are no algebraic simplifications that can be done to the function. To find the limit, we can either plug points that are approaching 0 from the right and the left and detrmine the limit from each side, or we can use the graph of the function to determine the limit (if it exists). Looking at the graph, we see that as x

approches 0 from the left the y values are becoming more and more negative, and as x approches 0 from the right, the y values are becoming more and more positive. Therefore, since $\lim_{x\to 0^+} \frac{3}{x} \neq \lim_{x\to 0^-} \frac{3}{x}$ then $\lim_{x\to 0} \frac{3}{x}$ does not exist.

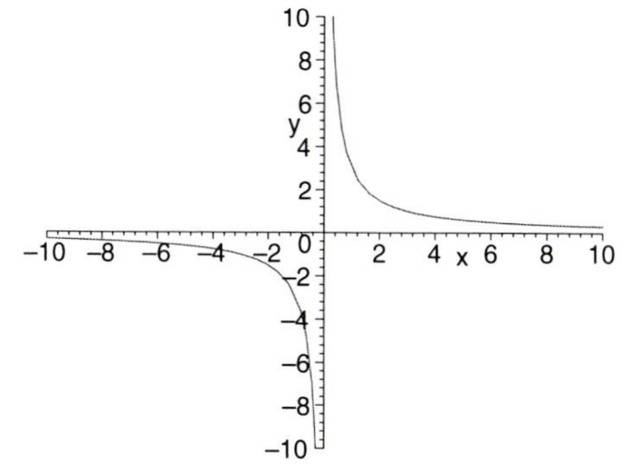

5. $2x+5$ is a continuous function (it is a polynomial). Therefore, we can use direct substitution

$$\begin{aligned}\lim_{x\to 3}(2x+5) &= 2(3)+5\\ &= 6+5\\ &= 11\end{aligned}$$

7. The function $\frac{x^2-25}{x+5}$ is discontinuous at $x=-5$, but it can be simplified algebraically.

$$\frac{x^2-25}{x+5} = \frac{(x-5)(x+5)}{x+5} = x-5$$

Therfore, $\lim_{x\to -5} \frac{x^2-25}{x+5} = \lim_{x\to -5}(x-5) = -5-5 = -10$

9. Since $\frac{5}{x}$ is continuous at $x=-2$ we can use direct substitution.
$\lim_{x\to -2} \frac{5}{x} = \frac{5}{-2} = -\frac{5}{2}$

11. The function $\frac{x^2+x-6}{x-2}$ is discontinuous at $x=2$, but it can be simplified algebraically. The limit is then computed as follows:

$$\begin{aligned}\lim_{x\to 2} \frac{x^2+x-6}{x-2} &= \lim_{x\to 2} \frac{(x-2)(x+3)}{x-2}\\ &= \lim_{x\to 2}(x+3)\\ &= 2+3\\ &= 5\end{aligned}$$

13. Since $\sqrt[3]{x^2-17}$ is continuous at $x=5$ we can use direct substitution

$$\begin{aligned}\lim_{x\to 5} \sqrt[3]{x^2-17} &= \sqrt[3]{5^2-17}\\ &= \sqrt[3]{25-17}\\ &= \sqrt[3]{8}\\ &= 2\end{aligned}$$

15. $\lim_{x\to \frac{\pi}{4}} (x + sin\ x) = \frac{\pi}{4} + sin\ \frac{\pi}{4} = \frac{\pi}{4} + \frac{1}{\sqrt{2}}$

17. $\lim_{x\to 0} \frac{1+sin\ x}{1-sin\ x} = \frac{1+0}{1-0} = 1$

19. Using the graph of $\frac{1}{x-2}$ we find that the limit as x approaches 2 does not exist since the limit from the left of $x=2$ does not equal the limit from the right of $x=2$

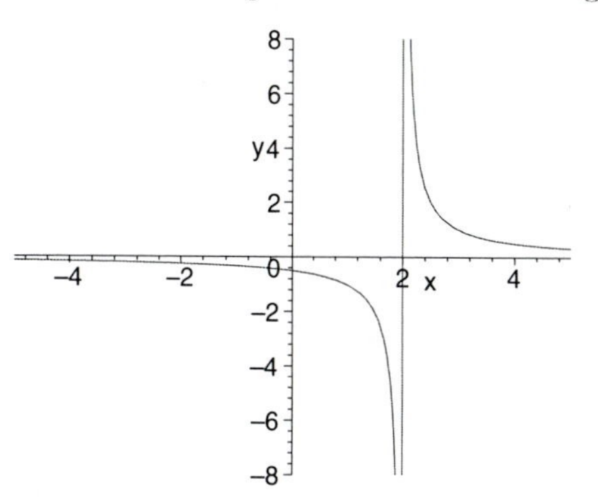

21. Since $\frac{3x^2-4x+2}{7x^2-5x+3}$ is continuous at $x=2$ we can use direct substitution

$$\begin{aligned}\lim_{x\to 2} \frac{3x^2-4x+2}{7x^2-5x+3} &= \frac{3(2)^2-4(2)+2}{7(2)^2-5(2)+3}\\ &= \frac{12-8+2}{28-10+3}\\ &= \frac{6}{21}\\ &= \frac{2}{7}\end{aligned}$$

23. The function $\frac{x^2+x-6}{x^2-4}$ is discontinuous at $x=2$. But we can simplify it algebraically first then find the limit as follows

$$\begin{aligned}\lim_{x\to 2} \frac{x^2+x-6}{x^2-4} &= \lim_{x\to 2} \frac{(x-2)(x+3)}{(x-2)(x+2)}\\ &= \lim_{x\to 2} \frac{(x+3)}{(x+2)}\\ &= \frac{2+3}{2+2}\\ &= \frac{5}{4}\end{aligned}$$

25. Since we have a limit in terms of h, we can treat x as a constant. To evaluate the limit we can use direct substitution (we have a polynomial in h, which is continuous for all values of h).
$\lim_{h\to 0}(6x^2+6xh+2h^2)$
$= 6x^2+6x(0)+2(0)^2$
$= 6x^2+0+0$
$= 6x^2$

27. Since we have a limit in terms of h, we can treat x as a constant. Since $\frac{-2x-h}{x^2(x+h)^2}$ is continuous at $h=0$ we can use direct substitution

$$\begin{aligned}\lim_{h\to 0} \frac{-2x-h}{x^2(x+h)^2} &= \frac{-2x-0}{x^2(x+0)^2}\\ &= \frac{-2x}{x^2(x)^2}\end{aligned}$$

$$= \frac{-2x}{x^4}$$
$$= \frac{-2}{x^3}$$

29. $\lim_{x\to 0} \frac{\tan x}{x} = \lim_{x\to 0} \frac{\sin x}{x} \cos x = 1 \cdot 1 = 1$ Recall that $\lim_{x\to 0} \frac{\sin x}{x} = 1.$

31. $\lim h \to 0 \frac{\sin x \sin h}{h} = \sin x \lim_{h\to 0} \frac{\sin h}{h}$
$= \sin x \cdot 1 = \sin x$

33.

$$\begin{aligned}\lim_{x\to 0} \frac{x^2+3x}{x-2x^4} &= \lim_{x\to 0} \frac{x(x+3)}{x(1-2x^3)} \\ &= \lim_{x\to 0} \frac{(x+3)}{(1-2x)} \\ &= \frac{(0+3)}{(1-0)} \\ &= \frac{3}{1} = 3\end{aligned}$$

35.

$$\begin{aligned}\lim_{x\to 0} \frac{x\sqrt{x}}{x+x^2} &= \lim_{x\to 0} \frac{x\sqrt{x}}{x(1+x)} \\ &= \lim_{x\to 0} \frac{\sqrt{x}}{(1+x)} \\ &= \frac{\sqrt{0}}{(1+0)} = 0\end{aligned}$$

37.

$$\begin{aligned}\lim_{x\to 2} \frac{x-2}{x^2-x-2} &= \lim_{x\to 2} \frac{x-2}{(x-2)(x+1)} \\ &= \lim_{x\to 2} \frac{1}{(x+1)} \\ &= \frac{1}{(2+1)} = \frac{1}{3}\end{aligned}$$

39.

$$\begin{aligned}\lim_{x\to 3} \frac{x^2-9}{2x-6} &= \lim_{x\to 3} \frac{(x-3)(x+3)}{2(x-3)} \\ &= \lim_{x\to 3} \frac{(x+3)}{2} \\ &= \frac{(3+3)}{2} \\ &= \frac{6}{2} = 3\end{aligned}$$

41.

$$\begin{aligned}\frac{a^2-4}{\sqrt{a^2+5}-3} &= \frac{a^2-4}{\sqrt{a^2+5}-3} \cdot \frac{\sqrt{a^2+5}+3}{\sqrt{a^2+5}+3} \\ &= \frac{(a^2-4)(\sqrt{a^2+5}+3)}{(a^2+5-9)} \\ &= \frac{(a^2-4)(\sqrt{a^2+5}+3)}{(a^2-4)} \\ &= \sqrt{a^2+5}+3\end{aligned}$$

Thus, $\lim_{a\to -2}(\sqrt{a^2+5}+3) = 6$

43.

$$\begin{aligned}\frac{\sqrt{3-x}-\sqrt{3}}{x} &= \frac{\sqrt{3-x}-\sqrt{3}}{x} \cdot \frac{\sqrt{3-x}+\sqrt{3}}{\sqrt{3-x}+\sqrt{3}} \\ &= \frac{3-x-3}{x(\sqrt{3-x}+\sqrt{3})} \\ &= \frac{-1}{\sqrt{3-x}+\sqrt{3}}\end{aligned}$$

Thus, $\lim_{x\to 0} \frac{-1}{\sqrt{3-x}+\sqrt{3}} = \frac{-1}{2\sqrt{3}}$

45. Limit approaches $\frac{3}{4}$

47.

$$\begin{aligned}\frac{2-\sqrt{x}}{4-x} &= \frac{2-\sqrt{x}}{4-x} \cdot \frac{2+\sqrt{x}}{2+\sqrt{x}} \\ &= \frac{4-x}{(4-x)(2+\sqrt{x})} \\ &= \frac{1}{2+\sqrt{x}}\end{aligned}$$

Thus, $\lim_{x\to 4} \frac{1}{2+\sqrt{x}} = \frac{1}{4}$

Exercise Set 2.3

1. a) First we obtain the expression for $f(x+h)$ with $f(x) = 7x^2$

$$\begin{aligned}f(x+h) &= 7(x+h)^2 \\ &= 7(x^2+2xh+h^2) \\ &= 7x^2+14xh+7h^2\end{aligned}$$

Then

$$\begin{aligned}\frac{f(x+h)-f(x)}{h} &= \frac{(7x^2+14xh+7h^2)-7x^2}{h} \\ &= \frac{14xh+7h^2}{h} \\ &= \frac{h(14x+7h)}{h} \\ &= 14x+7h\end{aligned}$$

b) For $x = 4$ and $h = 2$,

$$14x+7h = 14(4)+7(2) = 56+14 = 70$$

For $x = 4$ and $h = 1$,

$$14x+7h = 14(4)+7(1) = 56+7 = 63$$

For $x = 4$ and $h = 0.1$,

$$14x+7h = 14(4)+7(0.1) = 56+0.7 = 56.7$$

For $x = 4$ and $h = 0.01$,

$$14x+7h = 14(4)+7(0.01) = 56+0.07 = 56.07$$

3. **a)** First we obtain the expression for $f(x+h)$ with $f(x) = -7x^2$

$$\begin{aligned} f(x+h) &= -7(x+h)^2 \\ &= -7(x^2+2xh+h^2) \\ &= -7x^2-14xh-7h^2 \end{aligned}$$

Then

$$\begin{aligned} \frac{f(x+h)-f(x)}{h} &= \frac{(-7x^2-14xh-7h^2)-(-7x^2)}{h} \\ &= \frac{-14xh-7h^2}{h} \\ &= \frac{h(-14x-7h)}{h} \\ &= -14x-7h \end{aligned}$$

b) For $x = 4$ and $h = 2$,

$$-14x-7h = -14(4)-7(2) = -56-14 = -70$$

For $x = 4$ and $h = 1$,

$$-14x-7h = -14(4)-7(1) = -56-7 = -63$$

For $x = 4$ and $h = 0.1$,

$$-14x-7h = -14(4)-7(0.1) = -56-0.7 = -56.7$$

For $x = 4$ and $h = 0.01$,

$$-14x-7h = -14(4)-7(0.01) = -56-0.07 = -56.07$$

5. **a)** First we obtain the expression for $f(x+h)$ with $f(x) = 7x^3$

$$\begin{aligned} f(x+h) &= 7(x+h)^3 \\ &= 7(x^3+3x^2h+3xh^2+h^3) \\ &= 7x^3+21x^2h+21xh^2+7h^3 \end{aligned}$$

Then

$$\begin{aligned} \frac{f(x+h)-f(x)}{h} &= \frac{(7x^3+21x^2h+21xh^2+7h^3)-7x^3}{h} \\ &= \frac{21x^2h+21xh^2+7h^3}{h} \\ &= \frac{h(21x^2+21xh+7h^2}{h} \\ &= 21x^2+21xh+7h^2 \end{aligned}$$

b) For $x = 4$ and $h = 2$,

$$\begin{aligned} 21x^2+21xh+7h^2 &= 21(4)^2+21(4)(2)+7(2)^2 \\ &= 336+168+28 \\ &= 532 \end{aligned}$$

For $x = 4$ and $h = 1$,

$$\begin{aligned} 21x^2+21xh+7h^2 &= 21(4)^2+21(4)(1)+7(1)^2 \\ &= 336+84+7 \\ &= 427 \end{aligned}$$

For $x = 4$ and $h = 0.1$,

$$\begin{aligned} 21x^2+21xh+7h^2 &= 21(4)^2+21(4)(0.1)^2 \\ &= 336+8.4+0.07 \\ &= 344.47 \end{aligned}$$

For $x = 4$ and $h = 0.01$,

$$\begin{aligned} 21x^2+21xh+7h^2 &= 21(4)^2+21(4)(0.01)+7(0.01)^2 \\ &= 336+0.84+0.0007 \\ &= 336.8407 \end{aligned}$$

7. **a)** First we obtain the expression for $f(x+h)$ with $f(x) = \frac{5}{x}$

$$f(x+h) = \frac{5}{(x+h)}$$

Then

$$\begin{aligned} \frac{f(x+h)-f(x)}{h} &= \frac{\frac{5}{(x+h)}-\frac{5}{x}}{h} \\ &= \frac{\frac{5}{(x+h)}\cdot x(x+h)-\frac{5}{x}\cdot x(x+h)}{\frac{h}{1}\cdot x(x+h)} \\ &= \frac{5x-5(x+h)}{hx(x+h)} \\ &= \frac{-5h}{hx(x+h)} \\ &= \frac{-5}{x(x+h)} \end{aligned}$$

b) For $x = 4$ and $h = 2$,

$$\frac{-5}{x(x+h)} = \frac{-5}{4(4+2)} = \frac{-5}{26} \approx -0.208$$

For $x = 4$ and $h = 1$,

$$\frac{-5}{x(x+h)} = \frac{-5}{4(4+1)} = \frac{-5}{20} \approx -0.25$$

For $x = 4$ and $h = 0.1$,

$$\frac{-5}{x(x+h)} = \frac{-5}{4(4+0.1)} = \frac{-5}{16.4} \approx -0.305$$

For $x = 4$ and $h = 0.01$,

$$\frac{-5}{x(x+h)} = \frac{-5}{4(4+0.01)} = \frac{-5}{16.04} \approx -0.312$$

9. **a)** First we obtain the expression for $f(x+h)$ with $f(x) = -2x+5$

$$\begin{aligned} f(x+h) &= -2(x+h)+5 \\ &= -2x-2h+5 \end{aligned}$$

Then

$$\begin{aligned} \frac{f(x+h)-f(x)}{h} &= \frac{(-2x-2h+5)-(-2x+5)}{h} \\ &= \frac{-2h}{h} \\ &= -2 \end{aligned}$$

b) Since the difference quotient is a constant, then the value of the difference quotient will be -2 for all the values of x and h.

11. a) First we obtain the expression for $f(x+h)$ with $f(x) = x^2 - x$

$$\begin{aligned} f(x+h) &= (x+h)^2 - (x+h) \\ &= x^2 = 2xh + h^2 - x - h \end{aligned}$$

Then

$$\begin{aligned} \frac{f(x+h) - f(x)}{h} &= \frac{(x^2 + 2xh + h^2 - x - h) - (x^2 - x)}{h} \\ &= \frac{2xh + h^2 - h}{h} \\ &= \frac{h(2x + h - 1)}{h} \\ &= 2x + h - 1 \end{aligned}$$

b) For $x = 4$ and $h = 2$,

$$2x + h - 1 = 2(4) + 2 - 1 = 8 + 2 - 1 = 9$$

For $x = 4$ and $h = 1$,

$$2x + h - 1 = 2(4) + 2 - 1 = 8 + 1 - 1 = 8$$

For $x = 4$ and $h = 0.1$,

$$2x + h - 1 = 2(4) + 0.1 - 1 = 8 + 0.1 - 1 = 7.1$$

For $x = 4$ and $h = 0.01$,

$$2x + h - 1 = 2(4) + 2 - 1 = 8 + 0.01 - 1 = 7.01$$

13. a) For the average growth rate during the first year we use the points $(0, 7.9)$ and $(12, 22.4)$

$$\begin{aligned} \frac{y_2 - y_1}{x_2 - x_1} &= \frac{22.4 - 7.9}{12 - 0} \\ &= \frac{14.5}{12} \\ &\approx 1.20834 \text{ pounds per month} \end{aligned}$$

b) For the average growth rate during the second year we use the points $(12, 22.4)$ and $(24, 27.8)$

$$\begin{aligned} \frac{y_2 - y_1}{x_2 - x_1} &= \frac{27.8 - 22.4}{24 - 12} \\ &= \frac{5.4}{12} \\ &= 0.45 \text{ pounds per month} \end{aligned}$$

c) For the average growth rate during the third year we use the points $(24, 27.8)$ and $(36, 31.5)$

$$\begin{aligned} \frac{y_2 - y_1}{x_2 - x_1} &= \frac{31.5 - 27.8}{36 - 24} \\ &= \frac{3.7}{12} \\ &\approx 0.30834 \text{ pounds per month} \end{aligned}$$

d) For the average growth rate during his first three years we use the points $(0, 7.9)$ and $(36, 31.5)$

$$\begin{aligned} \frac{y_2 - y_1}{x_2 - x_1} &= \frac{31.5 - 7.9}{36 - 0} \\ &= \frac{23.6}{36} \\ &\approx 1.967 \text{ pounds per month} \end{aligned}$$

e) The graph indicates that the highest growth rate out of the first three years of a boy's life happens at birth (that is were the graph is the steepest).

15. a) For the average growth rate between ages 12 and 18 months

$$\begin{aligned} \frac{y_2 - y_1}{x_2 - x_1} &= \frac{25.9 - 22.4}{18 - 12} \\ &= \frac{3.5}{6} \\ &\approx 0.583 \text{ pounds per month} \end{aligned}$$

b) For the average growth rate between ages 12 and 14 (we use the point at 15 months)

$$\begin{aligned} \frac{y_2 - y_1}{x_2 - x_1} &= \frac{24.5 - 22.4}{15 - 12} \\ &= \frac{2.1}{3} \\ &= 0.7 \text{ pounds per month} \end{aligned}$$

c) For the average growth rate between ages 12 and 13 (we can approximate the value of y when $x = 13$ by reading it from the graph)

$$\begin{aligned} \frac{y_2 - y_1}{x_2 - x_1} &\approx \frac{23.23 - 22.4}{15 - 12} \\ &\approx \frac{0.83}{1} \\ &\approx 0.83 \text{ pounds per month} \end{aligned}$$

d) The average growth of a typical boy when he is 12 months old is about 0.9 pounds per month

17. a) Average rate of change from $t = 0$ to $t = 8$

$$\begin{aligned} \frac{N_2 - N_1}{t_2 - t_1} &= \frac{10 - 0}{8 - 0} \\ &= \frac{10}{8} = 1.25 \text{ words per minute} \end{aligned}$$

Average rate of change from $t = 8$ to $t = 16$

$$\begin{aligned} \frac{N_2 - N_1}{t_2 - t_1} &= \frac{20 - 10}{16 - 8} \\ &= \frac{10}{8} = 1.25 \text{ words per minute} \end{aligned}$$

Average rate of change from $t = 16$ to $t = 24$

$$\begin{aligned} \frac{N_2 - N_1}{t_2 - t_1} &= \frac{25 - 20}{24 - 16} \\ &= \frac{5}{8} = 0.625 \text{ words per minute} \end{aligned}$$

Average rate of change from $t = 24$ to $t = 32$

$$\frac{N_2 - N_1}{t_2 - t_1} = \frac{25 - 25}{32 - 24} = \frac{0}{8} = 0 \text{ words per minute}$$

Average rate of change from $t = 32$ to $t = 36$

$$\frac{N_2 - N_1}{t_2 - t_1} = \frac{25 - 25}{36 - 32} = \frac{0}{4} = 0 \text{ words per minute}$$

b) The rate of change becomes 0 after 24 minutes because the number of words memorized does not change and remains at 25 words, that means that there is no change in the number of words memorized after 24 minutes.

19. **a)** When $t = 3$, $s = 16(3)^2 = 16(9) = 144$ feet

b) When $t = 5$, $s = 16(5)^2 = 16(25) = 400$ feet

c) Average velocity $= \frac{400-144}{5-3} = \frac{256}{2} = 128$ feet per second

21. **a)** Population A: The average growth rate $= \frac{500-0}{4-0} = \frac{500}{4} = 125$ million per year
Population B: The average growth rate $= \frac{500-0}{4-0} =$ 125 million per year

b) We would not detect the fact that the population grow at different rates. The calculation shows the populations growing at the same average growth rate, since for either population we used the same points to calculate the average growth rate $(0, 0)$, and $(4, 500)$.

c) Population A:
Between $t = 0$ and $t = 1$, Average Growth Rate $= \frac{290-0}{1-0} = 290$ million people per year
Between $t = 1$ and $t = 2$, Average Growth Rate $= \frac{250-290}{2-1} = -40$ million people per year
Between $t = 2$ and $t = 3$, Average Growth Rate $= \frac{200-250}{3-2} = -50$ million people per year
Between $t = 3$ and $t = 4$, Average Growth Rate $= \frac{500-200}{4-3} = 300$ million people per year

Population B:
Between $t = 0$ and $t = 1$, Average Growth Rate $= \frac{125-0}{1-0} = 125$ million people per year
Between $t = 1$ and $t = 2$, Average Growth Rate $= \frac{250-125}{2-1} = 125$ million people per year
Between $t = 2$ and $t = 3$, Average Growth Rate $= \frac{375-250}{3-2} = 125$ million people per year
Between $t = 3$ and $t = 4$, Average Growth Rate $= \frac{400-375}{4-3} = 125$ million people per year

d) It is clear from part (c) that the first population has different growing rates depending on which interval of time we choose. Therefore, the statement "the population grew by 125 million each year" does not convey how population went through periods were the population increased and periods were the population decreased.

23. The rate of change in the period between 1800 and 1850 is similar to that of 1930 to 1950 is the sense that they both exhibit steady increase in the population. The drastic drop in the population shortly after 1850 is similar to the drop in population seen near 1975.

25.

$$\begin{aligned}
\frac{f(x+h) - f(x)}{h} &= \frac{a(x+h)^2 + b(x+h) + c}{h} - \frac{(ax^2 + bx + c)}{h} \\
&= \frac{ax^2 + 2axh + ah^2 + bx + bh + c}{h} \\
&\quad \frac{-ax^2 + bx + c}{h} \\
&= \frac{2axh + ah^2 + bh}{h} \\
&= 2ax + ah + b \\
&= a(2x + h) + b
\end{aligned}$$

27.

$$\begin{aligned}
\frac{f(x+h) - f(x)}{h} &= \frac{\sqrt{x+h} - \sqrt{x}}{h} \\
&= \frac{\sqrt{x+h} - \sqrt{x}}{h} \cdot \frac{\sqrt{x+h} + \sqrt{x}}{\sqrt{x+h} + \sqrt{x}} \\
&= \frac{x + h - x}{h\sqrt{x+h} + \sqrt{x}} \\
&= \frac{1}{\sqrt{x+h} + \sqrt{x}}
\end{aligned}$$

29.

$$\begin{aligned}
\frac{f(x+h) - f(x)}{h} &= \frac{\frac{1}{(x+h)^2} - \frac{1}{x^2}}{h} \\
&= \frac{x^2 - (x+h)^2}{hx^2(x+h)^2} \\
&= \frac{x^2 - x^2 - 2xh - h^2}{hx^2(x+h)^2} \\
&= \frac{-2x - h}{x^2(x+h)^2} \\
&= -\frac{2x + h}{x^2(x+h)^2}
\end{aligned}$$

31.

$$\begin{aligned}
\frac{f(x+h) - f(x)}{h} &= \frac{\frac{(x+h)}{(1+x+h)} - \frac{x}{1+x}}{h} \\
&= \frac{(x+h)(1+x) - x(1+x+h)}{h(1+x)(1+x+h)} \\
&= \frac{x^2 + x + xh + h - x - x^2 - xh}{h(1+x)(1+x+h)} \\
&= \frac{1}{(1+x)(1+x+h)}
\end{aligned}$$

33.

$$\begin{aligned}\frac{f(x+h)-f(x)}{h} &= \frac{\frac{1}{\sqrt{x+h}}-\frac{1}{\sqrt{x}}}{h}\\ &= \frac{\sqrt{x}-\sqrt{x+h}}{h\sqrt{x}\sqrt{x+h}}\\ &= \frac{\sqrt{x}-\sqrt{x+h}}{h\sqrt{x}\sqrt{x+h}}\frac{\sqrt{x}+\sqrt{x+h}}{\sqrt{x}+\sqrt{x+h}}\\ &= \frac{x-(x+h)}{h\sqrt{x}\sqrt{x+h}(\sqrt{x}+\sqrt{x+h})}\\ &= \frac{-1}{\sqrt{x}\sqrt{x+h}(\sqrt{x}+\sqrt{x+h})}\end{aligned}$$

Exercise Set 2.4

1. a-b) $f(x)=5x^2$

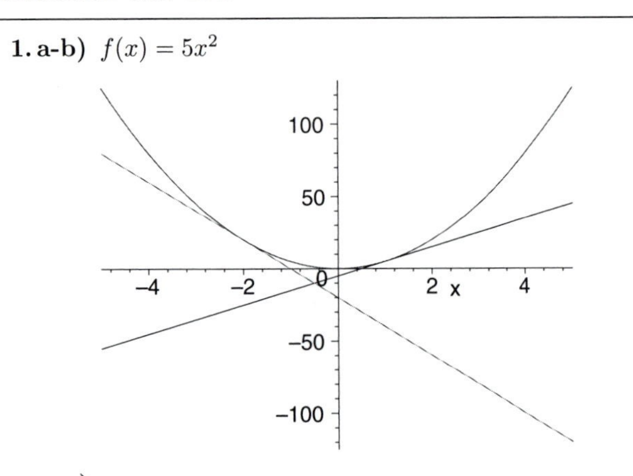

c)

$$\begin{aligned}f'(x) &= \lim_{h\to 0}\frac{f(x+h)-f(x)}{h}\\ &= \lim_{h\to 0}\frac{5(x+h)^2-5x^2}{h}\\ &= \lim_{h\to 0}\frac{5x^2+10xh+5h^2-5x^2}{h}\\ &= \lim_{h\to 0}\frac{h(10x+5h)}{h}\\ &= \lim_{h\to 0}10x+5h\\ &= 10x\end{aligned}$$

d) $f'(-2)=10(-2)=-20$
$f'(0)=10(0)=0$
$f'(1)=10(1)=10$. These slopes are in agreement with the slopes of the tangent lines drawn in part (b).

3. a-b) $f(x)=-5x^2$

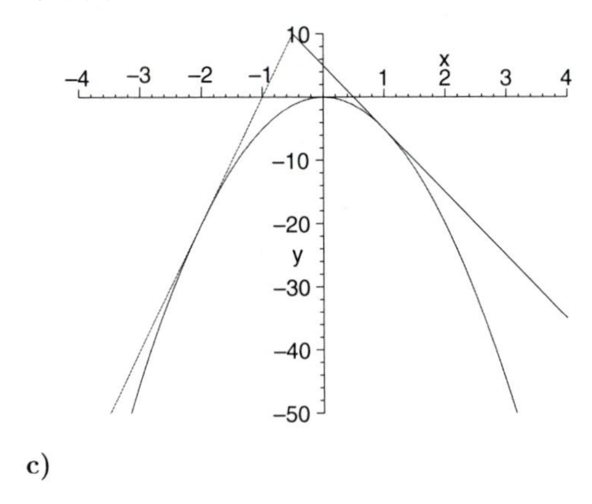

c)

$$\begin{aligned}f'(x) &= \lim_{h\to 0}\frac{f(x+h)-f(x)}{h}\\ &= \lim_{h\to 0}\frac{-5(x+h)^2-(-5x^2)}{h}\\ &= \lim_{h\to 0}\frac{-5x^2-10xh-5h^2-(-5x^2)}{h}\\ &= \lim_{h\to 0}\frac{h(-10x-5h)}{h}\\ &= \lim_{h\to 0}-10x-5h\\ &= -10x\end{aligned}$$

d) $f'(-2)=-10(-2)=20$
$f'(0)=-10(0)=0$
$f'(1)=-10(1)=-10$. These slopes are in agreement with the slopes of the tangent lines drawn in part (b).

5. a-b) $f(x)=x^3$

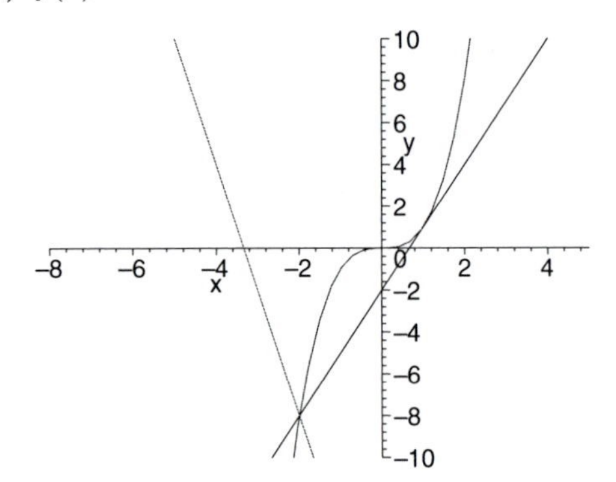

c)

$$\begin{aligned}f'(x) &= \lim_{h\to 0}\frac{f(x+h)-f(x)}{h}\\ &= \lim_{h\to 0}\frac{(x+h)^3-x^3}{h}\\ &= \lim_{h\to 0}\frac{x^3+3x^2h+3xh^2+h^3-x^3}{h}\\ &= \lim_{h\to 0}\frac{h(3x^2+3xh+h^2)}{h}\\ &= \lim_{h\to 0}3x^2+3xh+h^2\\ &= 3x^2\end{aligned}$$

d) $f'(-2) = 3(-2)^2 = 12$
$f'(0) = 3(0)^2 = 0$
$f'(1) = 3(1)^2 = 3$. These slopes are in agreement with the slopes of the tangent lines drawn in part (b).

7. a-b) $f(x) = 2x + 3$

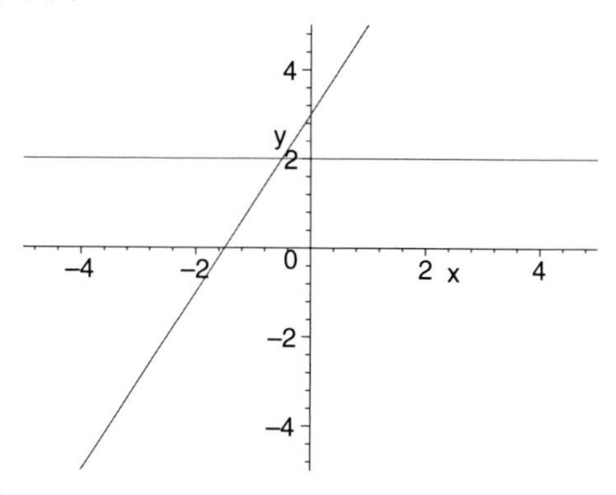

c)

$$\begin{aligned} f'(x) &= \lim_{h\to 0}\frac{f(x+h)-f(x)}{h} \\ &= \lim_{h\to 0}\frac{2(x+h)+3-(2x+3)}{h} \\ &= \lim_{h\to 0}\frac{2x+2h+3-2x-3}{h} \\ &= \lim_{h\to 0}\frac{2h}{h} \\ &= 2 \end{aligned}$$

d) $f'(-2) = 2$
$f'(0) = 2$
$f'(1) = 2$. These slopes are in agreement with the slopes of the tangent lines drawn in part (b).

9. a-b) $f(x) = -4x$

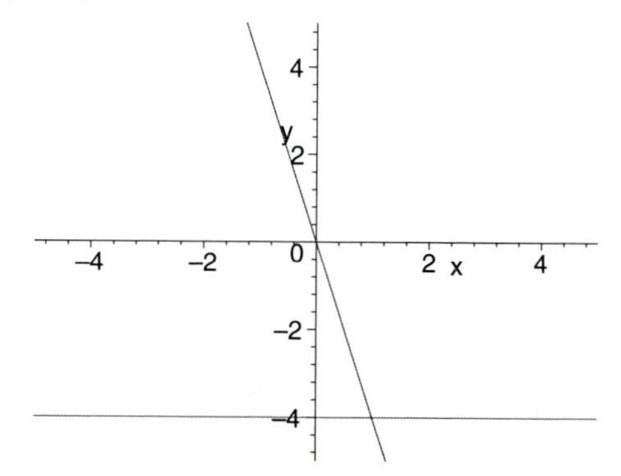

c)

$$\begin{aligned} f'(x) &= \lim_{h\to 0}\frac{f(x+h)-f(x)}{h} \\ &= \lim_{h\to 0}\frac{-4(x+h)-(-4x)}{h} \\ &= \lim_{h\to 0}\frac{-4x-4h+4x}{h} \\ &= \lim_{h\to 0}\frac{-4h}{h} \\ &= -4 \end{aligned}$$

d) $f'(-2) = -4$
$f'(0) = -4$
$f'(1) = -4$. These slopes are in agreement with the slopes of the tangent lines drawn in part (b).

11. a-b) $f(x) = x^2 + x$

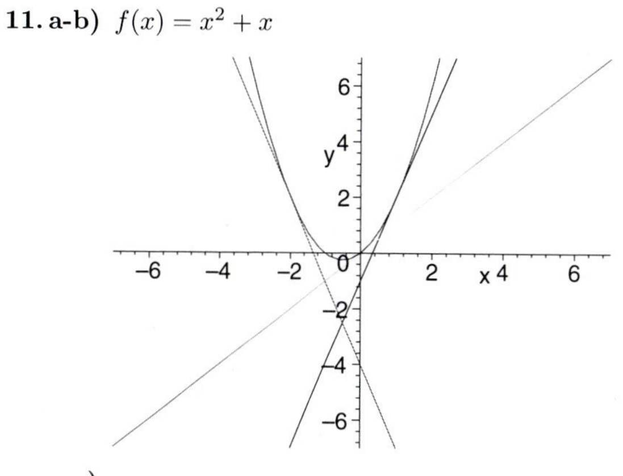

c)

$$\begin{aligned} f'(x) &= \lim_{h\to 0}\frac{f(x+h)-f(x)}{h} \\ &= \lim_{h\to 0}\frac{(x+h)^2+(x+h)-(x^2+x)}{h} \\ &= \lim_{h\to 0}\frac{x^2+2xh+h^2+x+h-x^2-x}{h} \\ &= \lim_{h\to 0}\frac{2xh+h^2+h}{h} \\ &= \lim_{h\to 0}\frac{h(2x+h+1)}{h} \\ &= 2x+1 \end{aligned}$$

d) $f'(-2) = 2(-2) + 1 = -3$
$f'(0) = 2(0) + 1 = 1$
$f'(1) = 2(1) + 1 = 3$. These slopes are in agreement with the slopes of the tangent lines drawn in part (b).

13. a-b) $f(x) = 2x^2 + 3x - 2$

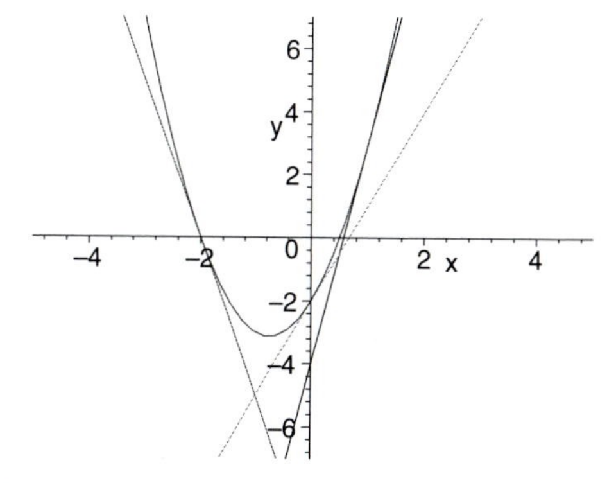

c)

$$\begin{aligned} f'(x) &= \lim_{h\to 0}\frac{f(x+h)-f(x)}{h} \\ &= \lim_{h\to 0}\frac{2(x+h)^2+3(x+h)-2-(2x^2+3x-2)}{h} \\ &= \lim_{h\to 0}\frac{4xh+2h^2+3h}{h} \end{aligned}$$

$$
\begin{aligned}
&= \lim_{h\to 0} \frac{h(4x+3)}{h} \\
&= 4x+3
\end{aligned}
$$

d) $f'(-2) = 4(-2)+3 = -5$
$f'(0) = 4(0)+3 = 3$
$f'(1) = 4(1)+3 = 7$. These slopes are in agreement with the slopes of the tangent lines drawn in part (b).

15. a-b) $f(x) = \frac{1}{x}$

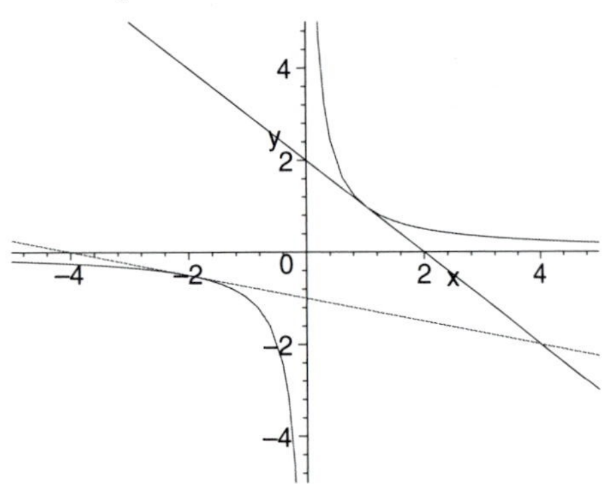

c)

$$
\begin{aligned}
f'(x) &= \lim_{h\to 0} \frac{f(x+h)-f(x)}{h} \\
&= \lim_{h\to 0} \frac{\frac{1}{(x+h)} - \frac{1}{x}}{h} \\
&= \lim_{h\to 0} \frac{\frac{1}{(x+h)}\cdot x(x+h) - \frac{1}{x}\cdot x(x+h)}{h\cdot x(x+h)} \\
&= \lim_{h\to 0} \frac{x-(x+h)}{hx(x+h)} \\
&= \lim_{h\to 0} \frac{-h)}{hx(x+h)} \\
&= \lim_{h\to 0} \frac{-1}{x(x+h)} \\
&= -\frac{1}{x^2}
\end{aligned}
$$

d) $f'(-2) = -\frac{1}{(-2)^2} = -\frac{1}{4}$
$f'(0) =$ does not exist
$f'(1) = -\frac{1}{(1)^2} = -1$. These slopes are in agreement with the slopes of the tangent lines drawn in part (b).

17. $f(x) = mx$

$$
\begin{aligned}
f'(x) &= \lim_{h\to 0} \frac{f(x+h)-f(x)}{h} \\
&= \lim_{h\to 0} \frac{m(x+h)-mx}{h} \\
&= \lim_{h\to 0} \frac{mx+mh-mx}{h} \\
&= \lim_{h\to 0} \frac{mh}{h} \\
&= m
\end{aligned}
$$

19. $f(x) = x^2$. From Example 3, $f'(x) = 2x$. For the point $(3,9)$ we have $f'(3) = 2(3) = 6 = m$. So the equation of the tangent line is

$$
\begin{aligned}
y - y_1 &= m(x-x_1) \\
y - 9 &= 6(x-3) \\
y &= 6x - 18 + 9 \\
y &= 6x - 9
\end{aligned}
$$

For the point $(-1,1)$ we have $f'(-1) = 2(-1) = -2$. So the equation of the tangent line is

$$
\begin{aligned}
y - y_1 &= m(x-x_1) \\
y - 1 &= -2(x-(-1)) \\
y - 1 &= -2x - 2 \\
y &= -2x - 2 + 1 \\
y &= -2x - 1
\end{aligned}
$$

For the point $(10,100)$ we have $f'(10) = 2(10) = 20$. So the equation of the tangent line is

$$
\begin{aligned}
y - y_1 &= m(x-x_1) \\
y - 100 &= 20(x-10) \\
y &= 20x - 200 + 100 \\
y &= 20x - 100
\end{aligned}
$$

21. From Exercise 14, $f'(x) = -\frac{5}{x^2}$. For the point $(1,5)$ we have $f'(1) = -\frac{5}{1^2} = -5 = m$. So the equation of the tangent line is

$$
\begin{aligned}
y - y_1 &= m(x-x_1) \\
y - 5 &= -5(x-1) \\
y &= -5x + 5 + 5 \\
y &= -5x + 10
\end{aligned}
$$

For the point $(-1,-5)$ we have $f'(-1) = -\frac{5}{(-1)^2} = -5$. So the equation of the tangent line is

$$
\begin{aligned}
y - y_1 &= m(x-x_1) \\
y - (-5) &= -5(x-(-1)) \\
y + 5 &= -5(x+1) \\
y &= -5x - 5 - 5 \\
y &= -5x - 10
\end{aligned}
$$

For the point $(100, 0.05)$, which can be rewritten as $(100, \frac{1}{20})$, we have $f'(10) = -\frac{5}{100^2} = -\frac{1}{2000}$. So the equation of the tangent line is

$$
\begin{aligned}
y - y_1 &= m(x-x_1) \\
y - \frac{1}{20} &= -\frac{1}{2000}(x-100) \\
y &= -\frac{x}{2000} + \frac{1}{20} + \frac{1}{20} \\
y &= -\frac{x}{2000} + \frac{2}{20} \\
y &= -\frac{x}{2000} + \frac{1}{10}
\end{aligned}
$$

23. First, let us find the expression for $f'(x)$.

$$\begin{aligned} f'(x) &= \lim_{h\to 0}\frac{f(x+h)-f(x)}{h} \\ &= \lim_{h\to 0}\frac{4-(x+h)^2-(4-x^2)}{h} \\ &= \lim_{h\to 0}\frac{4-x^2-2xh-h^2-4+x^2}{h} \\ &= \lim_{h\to 0}\frac{-2xh-h^2}{h} \\ &= \lim_{h\to 0}\frac{h(-2x-h)}{h} \\ &= \lim_{h\to 0}(-2x-h) \\ &= -2x \end{aligned}$$

For the point $(-1,3)$ we have $f'(-1)=-2(-1)=2=m$. So the equation of the tangent line is

$$\begin{aligned} y-y_1 &= m(x-x_1) \\ y-3 &= 2(x-(-1)) \\ y-3 &= -2x+2 \\ y &= 2x+2+3 \\ y &= 2x+5 \end{aligned}$$

For the point $(0,4)$ we have $f'(0)=-2(0)=0$. So the equation of the tangent line is

$$\begin{aligned} y-y_1 &= m(x-x_1) \\ y-4 &= 0(x-0) \\ y &= 0+4 \\ y &= 4 \end{aligned}$$

For the point $(5,-21)$ we have $f'(5)=-2(5)=-10$. So the equation of the tangent line is

$$\begin{aligned} y-y_1 &= m(x-x_1) \\ y-(-21) &= -10(x-5) \\ y+21 &= -10x+50 \\ y &= -10x+50-21 \\ y &= -10x+29 \end{aligned}$$

25. The function is not differentiable at x_0 since it is discontinuous, x_3 since it has a corner, x_4 since it has a corner, x_6 since it has a corner, and x_{12} since it has a vertical tangent.

27. The function is not differentiable at integer values of x since the function is not continuous at integer values of x.

29. The function is differentiable for all values in the domain.

31. The function is differentiable for all values in the domain.

33. As the points Q get closer to P the secant lines are getting closer to the tangent line at point P.

35.

$$\begin{aligned} f'(x) &= \lim_{h\to 0}\frac{f(x+h)-f(x)}{h} \\ &= \lim_{h\to 0}\frac{\frac{1}{x^2+2xh+h^2}-\frac{1}{x^2}}{h} \\ &= \lim_{h\to 0}\frac{x^2-x^2-2xh-h^2}{hx^2(x^2+2xh+h^2)} \\ &= \lim_{h\to 0}\frac{-h(2x-h)}{hx^2(x^2+2xh+h^2)} \\ &= \lim_{h\to 0}\frac{-2x+h}{x^2(x^2+2xh+h^2)} \\ &= \frac{-2x}{x^2(x^2)} \\ &= \frac{-2}{x^3} \end{aligned}$$

37.

$$\begin{aligned} f'(x) &= \lim_{h\to 0}\frac{f(x+h)-f(x)}{h} \\ &= \lim_{h\to 0}\frac{\frac{x+h}{1+x+h}-\frac{x}{1+x}}{h} \\ &= \lim_{h\to 0}\frac{(x+h)(1+x)-x(1+x+h)}{h(1+x)(1+x+h)} \\ &= \lim_{h\to 0}\frac{x+x^2+h+hx-x-x^2-xh}{h(1+x)(1+x+h)} \\ &= \lim_{h\to 0}\frac{h}{h(1+x)(1+x+h)} \\ &= \lim_{h\to 0}\frac{1}{(1+x)(1+x+h)} \\ &= \frac{1}{(1+x)^2} \end{aligned}$$

39.

$$\begin{aligned} f'(x) &= \lim_{h\to 0}\frac{f(x+h)-f(x)}{h} \\ &= \lim_{h\to 0}\frac{1}{\sqrt{x+h}}-\frac{1}{\sqrt{x}} \\ &= \lim_{h\to 0}\frac{\sqrt{x}-\sqrt{x+h}}{h\sqrt{x}\sqrt{x+h}} \\ &= \lim_{h\to 0}\frac{\sqrt{x}-\sqrt{x+h}}{h\sqrt{x}\sqrt{x+h}}\cdot\frac{\sqrt{x}+\sqrt{x+h}}{\sqrt{x}+\sqrt{x+h}} \\ &= \lim_{h\to 0}\frac{x-x-h}{h\sqrt{x}\sqrt{x+h}(\sqrt{x}+\sqrt{x+h})} \\ &= \lim_{h\to 0}\frac{-1}{\sqrt{x}\sqrt{x+h}(\sqrt{x}+\sqrt{x+h})} \\ &= \frac{-1}{2\sqrt{x^3}} \end{aligned}$$

41. The function $f(x)$ will is not differentiable at $x=-3$.

43. $f'(-2)=12$, $f'(0)=0$, $f'(4)=48$

45. $f'(-1)=-2$, $f'(2)=\frac{-1}{2}$, $f'(10)=\frac{-2}{100}$

47. $f'(-2)=-6$, $f'(1)=0$, $f'(4)=6$

Exercise Set 2.5

1. $\frac{dy}{dx} = 7x^{7-1} = 7x^6$

3. $\frac{dy}{dx} = 3 \cdot 2x^{2-1} = 6x$

5. $\frac{dy}{dx} = 4 \cdot 3x^{3-1} = 12x^2$

7. $\frac{dy}{dx} = 3 \cdot \frac{2}{3}x^{2/3-1} = 2x^{-1/3}$

9. Rewrite as $y = x^{3/4}$, $\frac{dy}{dx} = \frac{3}{4}x^{3/4-1} = \frac{3}{4}x^{-1/4}$

11. $\frac{dy}{dx} = 4cos\ x$

13. $\frac{dy}{dx} = cos\ x - 1x^{-2} = cos\ x - \frac{1}{x^2}$

15. $\frac{dy}{dx} = 2(2x+1)^{2-1}(2) = 4(2x+1)$

17. $f'(x) = 0.25(3.2x^{3.2-1}) = 0.8x^{2.2}$

19. $f'(x) = 10cos\ x + 12sin\ x$

21. $f'(x) = -\sqrt[3]{9}\ sin\ x$

23. Rewrite as $f(x) = 5x^{-1} + \frac{x}{5}$, $f'(x) = -5x^{-2} + \frac{1}{5}$

25. $f'(x) = -x^{-1/2} - x^{-3/4} - \frac{1}{2}x^{-5/4} + \frac{7}{2}x^{-3/2}$

27. $f(x) = x^{18/5} - x^{13/5}$
$f'(x) = \frac{18}{5}x^{13/5} - \frac{13}{5}x^{8/5}$

29. $f(x) = 3x^{-1} - 4x^{-2}$
$f'(x) = -3x^{-2} + 8x^{-3}$

31. $f(x) = 4x + 3 - 2x^{-1}$
$f'(x) = 4 + 2x^{-2}$

33. $p(x) = 6x^{-4} - 2x^{-3}$
$p'(x) = -24x^{-5} + 6x^{-4}$

35. $s(x) = 3\sqrt{2}cos\ x - 2\sqrt{2}sin\ x$
$s'(x) = -3\sqrt{2}sin\ x - 2\sqrt{2}cos\ x$

37. $q'(x) = 8\left(\frac{\sqrt{5}}{3}\right) cos\ x + 8\left(\frac{\sqrt[3]{5}}{7}\right) sin\ x$

39. $U(x) = sin\ x - 2x^{1/2} + 3x^{-1/2}$
$U'(x) = cos\ x - x^{-1/2} - \frac{3}{2}x^{-3/2}$

41. $f'(x) = -3sin\ x$
The slope at $(0, 4)$ is $f'(0) = 0$.
So the equation of the tangent line is

$$\begin{aligned} y - y_1 &= m(x - x_1) \\ y - 4 &= 0(x - 0) \\ y - 4 &= 0 \\ y &= 4 \end{aligned}$$

43. $f'(x) = \frac{4}{3}x^{1/3} + \frac{2}{3}x^{-2/3}$
The slope at $(8, 20)$ is
$f'(8) = \frac{4}{3}(8)^{1/3} + \frac{2}{3}(8)^{-2/3} = \frac{17}{6}$
So the equation of the line is

$$\begin{aligned} y - y_1 &= m(x - x_1) \\ y - 20 &= \frac{17}{6}(x - 8) \\ y &= \frac{17}{6}x - \frac{68}{3} + 20 \\ y &= \frac{17}{6}x - \frac{8}{3} \end{aligned}$$

45. Rewrite $y = \left(\frac{x^{2/3}}{x^{3/2}}\right)^{1/3} = x^{-5/18}$
$\frac{dy}{dx} = \frac{-5}{18}x^{-23/8}$

47. Use the trigonometric identity $1 + tan^2 x = sec^2 x$ to rewrite
$y = 3sec^2 x\ cos^3 x = 3cos\ x$
$\frac{dy}{dx} = -3sin\ x$

49. Use the sum identity for cosine to rewrite $y = \frac{\sqrt{3}}{2}cos\ x - \frac{1}{2}sin\ x$
$\frac{dy}{dx} = \frac{-\sqrt{3}}{2}sin\ x - \frac{1}{2}cos\ x$

51. $\frac{dy}{dx} = 2\left(\sqrt{x} - \frac{1}{\sqrt{x}}\right)\left(\frac{1}{2\sqrt{x}} + \frac{1}{2\sqrt{x^3}}\right)$
The slope at $(1, 0)$ is $\frac{dy}{dx}|_{x=1} = 0$
The equation of the line is

$$\begin{aligned} y - y_1 &= m(x - x_1) \\ y - 0 &= 0(x - 1) \\ y &= 0 \end{aligned}$$

53. Let $f(x) = cos\ x$ then

$$\begin{aligned} f'(x) &= lim_{h\to 0}\frac{f(x+h) - f(x)}{h} \\ &= \lim_{h\to 0}\frac{cos(x+h) - cos(x)}{h} \\ &= \lim_{h\to 0}\frac{cos\ xcos\ h - sin\ xsin\ h - cos\ x}{h} \\ &= \lim_{h\to 0} cos\ x\left(\frac{cos\ h - 1}{h}\right) - sin\ x\left(\frac{sin\ h}{h}\right) \\ &= cos\ x \cdot \lim_{h\to 0}\frac{cos\ h - 1}{h} - sin\ x \cdot \lim_{h\to 0}\frac{sin\ h}{h} \\ &= cos\ x \cdot 0 - sin\ x \cdot 1 \\ &= -sin\ x \end{aligned}$$

55. Left to the student

57. The tangent line is horizontal at $x = \pm\ 0.6922$

59. The tangent line is horizontal at $x = -0.3456$ and at $x = 1.9289$

Exercise Set 2.6

1. **a)** $v(t) = \frac{ds(t)}{dt} = 3t^2 + 1$

b) $a(t) = \frac{dv(t)}{dt} = 6t$

c) $v(4) = 3(4)^2 + 1 = 48 + 1 = 49$ feet per second
$a(4) = 6(4) = 24$ feet per squared seconds

3. **a)** $v(t) = \frac{ds(t)}{dt} = -20t + 2$

b) $a(t) = \frac{dv(t)}{dt} = -20$

c) $v(1) = -20(1) + 2 = -20 + 2 = 18$ feet per second
$a(1) = -20$ feet per squared seconds

5. **a)** $v(t) = \frac{ds(t)}{dt} = 5 + 2cos\ t$

b) $a(t) = \frac{dv(t)}{dt} = -2sin\ t$

c) Find $v(\frac{\pi}{4})$ and $a(\frac{\pi}{4})$

$$\begin{aligned} v(\frac{\pi}{4}) &= 5 + 2cos(\frac{\pi}{4}) \\ &= 5 + 2 \cdot \frac{\sqrt{2}}{2} \\ &= 5 + \sqrt{2} \\ &\approx 6.414\ m/sec \end{aligned}$$

$$\begin{aligned} a(\frac{\pi}{4}) &= -2sin(\frac{\pi}{4}) \\ &= 2 \cdot \frac{\sqrt{2}}{2} \\ &= \sqrt{2} \\ &\approx 1.414\ m/sec^2 \end{aligned}$$

d) Find t when $v(t) = 3$

$$\begin{aligned} 3 &= 5 + 2cos\ t \\ -2 &= 2cos\ t \\ -1 &= cos\ t \\ t &= cos^{-1}(-1) \\ &= \pi + 2n\pi \end{aligned}$$

7. **a)** $\frac{dN}{da} = -2a + 300$

b) Since a is counted in thousands we need to find

$$\begin{aligned} N(10) &= -(10)^2 + 300(10) + 6 \\ &= -100 + 3000 + 6 \\ &= 2996 \end{aligned}$$

There will be 2996 units sold after spending \$10000 on advertsing

c) At $a = 10$, $\frac{DN}{da} = -2(10) + 300 = -20 + 300 = 280$ units per thousand dollars spent on advertising

d) The rate of change of the number of units sold depends on the amount spent on advertising according to the equation $\frac{dN}{da} = -2a + 300$ which means that for every a thousand dollars spent on advertising, the change in the units solds is $-2a+300$ units. If \$10000 is spent on advertising, then there will be a 280 unit increase in the number of units sold.

9. **a)** $\frac{dw}{dt} = 1.61 - 0.0968t + 0.0018t^2$

b) $w(0) = 7.60+1.61(0)-0.0484(0)^2+0.0006(0)^3 = 7.60$ pounds

c) $\frac{dw}{dt}|_{t=0} = 1.61-0.0968(0)+0.0018(0)^2 = 1.61$ pounds per month

d) $w(12) = 7.60+1.61(12)-0.0484(12)^2+0.0006(12)^3 = 20.987$ pounds

e) $\frac{dw}{dt}|_{12} = 1.61 - 0.0968(12) + 0.0018(12)^2 = 0.708$ pounds per month

f) Average rate of change $= \frac{w(12)-w(0)}{12} = \frac{20.987-7.6}{12} = 1.1156$ pounds per month

g)

$$\begin{aligned} 1.61 - 0.0968t + 0.0018t^2 &= 1.1156 \\ 0.0018t^2 - 0.0968t + 0.4944 &= 0 \\ \frac{0.0968 + \sqrt{(0.0968)^2 - 4(0.0018)(0.4944)}}{2(0.0018)} &= t \\ 5.7147 &= t \end{aligned}$$

11. **a)** $\frac{dC}{dr} = 2\pi$

b) For every increase of one centimeter of the radius the healing wound circumference increases by 2π centimeters

13. **a)** $\frac{dT}{dt} = -0.2t + 1.2$

b) At $t = 1.5$

$$\begin{aligned} T(1.5) &= -0.1(1.5)^2 + 1.2(1.5) + 98.6 \\ &= 100.2 \text{ degrees} \end{aligned}$$

c) $\frac{dT}{dt}|_{t=1.5} = -0.2(1.5) + 1.2 = 0.9$ degrees per day

d) The sign of $T'(t)$ is significant because it indicates whether the rate of change in temperature is an increase (if positive) or a decrease (if negative). That is, whether the fever is increasing or decreasing during the illness

15. **a)** $\frac{dT}{dW} = 1.31W^{0.31}$

b) For every increase of W in body weight the territorial area of an animal increases by an amount equal to $1.31W^{0.31}$

17. First rewrite $R(Q)$ as follows $R(Q) = \frac{k}{2}Q^2 - \frac{1}{3}Q^3$

a) $\frac{dR}{dQ} = \frac{k}{2}(2Q) - \frac{1}{3}(3Q^2) = kQ - Q^2$

b) For every increase Q in the dosage there will be a change in the reaction of the body to that dosage change equal to the amount $kQ - Q^2$

19. **a)** $\frac{dA}{dt} = 0.08$

b) The rate of change for the median age of women at first marriage is constant at 0.08. That is, each year the median age of women at first marriage in increasing by 0.08 years

21. The average rate of change of a function is the value of the difference quotient evaluated over a period of time, while the instantaneous rate of change is the value of the slope of the tangent line at that particular instant.

23. **25.** **27.** a)

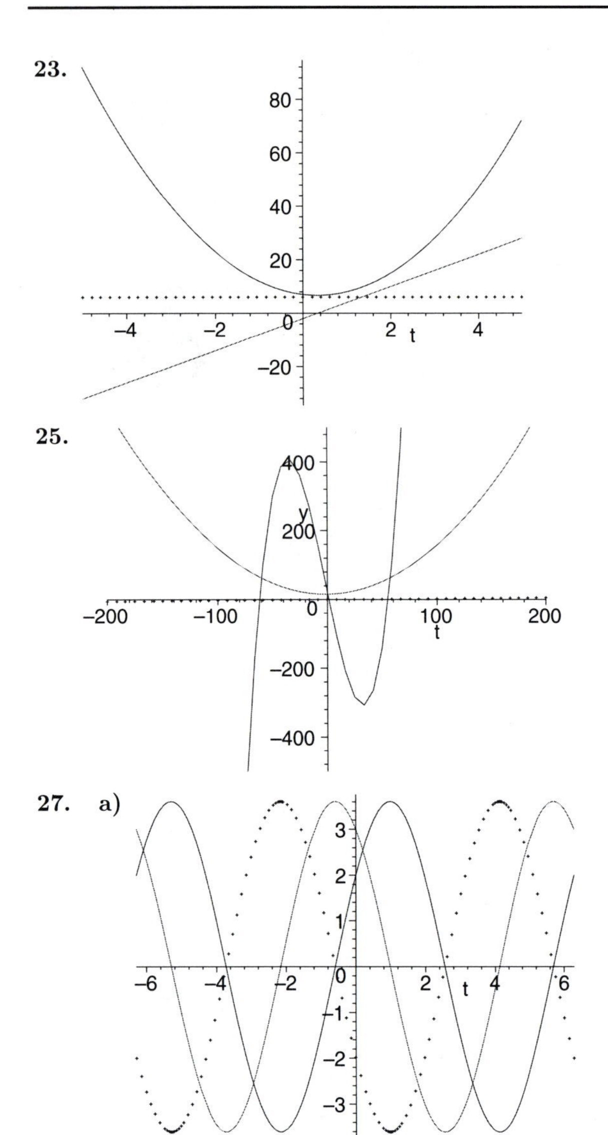

b) The acceleration function is the same as the product of the position function times -1. That is it is a reflection of the position function about the t axis.

Exercise Set 2.7

1. Method One: $x^3 \cdot x^8 = x^{3+8} = x^{11}$. $\frac{dy}{dx} = 11x^{10}$
Method Two (product rule):

$$\begin{aligned} \frac{dy}{dx} &= x^3 \cdot 8x^7 + 3x^2 \cdot x^8 \\ &= 8x^{10} + 3x^{10} \\ &= 11x^{10} \end{aligned}$$

3. Method One: $x\sqrt{x} = x^{3/2}$, $\frac{dy}{dx} = \frac{3}{2}x^{1/2}$
Method Two (product rule):

$$\begin{aligned} \frac{dy}{dx} &= x(\frac{1}{2}x^{-1/2}) + (1)x^{1/2} \\ &= \frac{1}{2}x^{1/2} + x^{1/2} \\ &= \frac{3}{2}x^{1/2} \end{aligned}$$

5. Method One: $\frac{x^8}{x^5} = x^3$, $\frac{dy}{dx} = 3x^2$
Method Two (quotient rule):

$$\begin{aligned} \frac{dy}{dx} &= \frac{x^5(8x^7) - 5x^4(x^8)}{(x^5)^2} \\ &= \frac{8x^{12} - 5x^{12}}{x^{10}} \\ &= \frac{3x^{12}}{x^{10}} \\ &= 3x^2 \end{aligned}$$

7. Method One: $y = x^2 - 25$, $\frac{dy}{dx} = 2x$
Method Two (product rule):

$$\begin{aligned} \frac{dy}{dx} &= (x+5)(1) + (1)(x-5) \\ &= x + 5 + x - 5 \\ &= 2x \end{aligned}$$

9.

$$\begin{aligned} y &= (8x^5 - 3x^2 + 20)(8x^4 - 3x^{1/2}) \\ \frac{dy}{dx} &= (8x^5 - 3x^2 + 20)(32x^3 - \frac{3}{2}x^{-1/2}) + \\ &\quad (40x^4 - 6x)(8x^4 - 3x^{1/2}) \\ &= (8x^5 - 3x^2 + 20)(32x^3 - \frac{3}{2\sqrt{x}}) + \\ &\quad (40x^4 - 6x)(8x^4 - 3\sqrt{x}) \end{aligned}$$

11. $f(x) = (x^{1/2} - x^{1/3})(2x + 3)$

$$\begin{aligned} f'(x) &= (\frac{1}{2}x^{-1/2} - \frac{1}{3}x^{-2/3})(2x+3) + (\sqrt{x} - \sqrt[3]{x})(2) \\ &= 2(\sqrt{x} - \sqrt[3]{x}) + (2x+3)\left(\frac{1}{2\sqrt{x}} - \frac{1}{3\sqrt[3]{x^2}}\right) \end{aligned}$$

13. $f(x) = x^{1/2} tan\ x$

$$\begin{aligned} f'(x) &= x^{1/2} sec^2 x + tan\ x\left(\frac{1}{2}x^{-1/2}\right) \\ &= \sqrt{x}\ sec^2 x + \left(\frac{1}{2\sqrt{x}}\right) tan\ x \end{aligned}$$

15. $(2t+3)^2 = (2t+3)(2t+3) = 4t^2 + 12t + 9$
$f'(t) = 8t + 12$

17. $g(x) = (0.02x^2 + 1.3x - 11.7)(4.1x + 11.3)$

$$\begin{aligned} g'(x) &= (0.02x^2 + 1.3x - 11.7)(4.1) + \\ &\quad (0.04x + 1.3)(4.1x + 11.3) \end{aligned}$$

19. $g(x) = sec\ x\ csc\ x$

$$\begin{aligned} g'(x) &= sec\ x(-csc\ x\ cot\ x) + csc\ x(sec\ x\ tan\ x) \\ &= \frac{-1}{sin^2 x} + \frac{1}{cos^2 x} \\ &= sec^2 x - csc^2 x \end{aligned}$$

21. $q(x) = \frac{sin\ x}{1+cos\ x}$

$$\begin{aligned} q'(x) &= \frac{(1+cos\ x)(cos\ x) - sin\ x(-sin\ x)}{(1+cos\ x)^2} \\ &= \frac{cos\ x + cos^2x + sin^2x}{(1+cos\ x)^2} \\ &= \frac{cos\ x + 1}{(1+cos\ x)^2} \\ &= \frac{1}{1+cos\ x} \end{aligned}$$

23. $s(t) = tan^2t$
$s'(t) = 2tan\ t\ sec^2t$

25. $f(x) = (x + 2x^{-1})(x^2 - 3)$

$$\begin{aligned} f'(x) &= (x+2x^{-1})(2x) + (1-2x^{-2})(x^2-3) \\ &= 2x^2 + 4 + x^2 - 3 - 2 + 6x^{-2} \\ &= 3x^2 - 1 + 6x^{-2} \\ &= 3x^2 - 1 + \frac{6}{x^2} \end{aligned}$$

27. You could use the quotient rule, but a better technique to use would be to rewrite the function as follows:$q(x) = \frac{3}{5}x^2 - \frac{6}{5}x + \frac{4}{5}$ then

$$q'(x) = \frac{6}{5}x - \frac{6}{5}$$

29. $y = \frac{x^2+3x-4}{2x-1}$

$$\begin{aligned} \frac{dy}{dx} &= \frac{(2x-1)(2x+3) - (x^2+3x-4)(2)}{(2x-1)^2} \\ &= \frac{2x^2+6x-2x-3-2x^2-6x+8}{(2x-1)^2} \\ &= \frac{-2x+5}{(2x-1)^2} \end{aligned}$$

31. $w = \frac{3t-1}{t^2-2t+6}$

$$\begin{aligned} \frac{dw}{dt} &= \frac{(t^2-2t+6)(3) - (3t-1)(2t-2)}{(t^2-2t+6)^2} \\ &= \frac{3t^2-6t+18-[6t^2-6t-2t+2]}{(t^2-2t+6)^2} \\ &= \frac{-3t^2+2t+16}{(t^2-2t+6)^2} \end{aligned}$$

33. $f(x) = \frac{x}{\frac{1}{x}+1} = \frac{x^2}{1+x}$

$$\begin{aligned} f'(x) &= \frac{(1+x)(2x) - (x^2)(1)}{(1+x)^2} \\ &= \frac{2x+2x^2-x^2}{(1+x)^2} \\ &= \frac{2x+x^2}{(1+x)^2} \end{aligned}$$

35. $y = \frac{tan\ t}{1+sec\ t}$ Which can be rewritten as $y = \frac{\frac{sin\ t}{cos\ t}}{1+\frac{1}{cos\ t}}$ multiplying every term by $cos\ t$ to clear the fractions gives $y = \frac{sin\ t}{cos\ t+1}$ which has a derivative of $\frac{dy}{dx} = \frac{1}{1+cos\ t}$ (see problem 21 for details)

37. $w = \frac{tan\ x + x\ sin\ x}{\sqrt{x}}$

$$\begin{aligned} \frac{dw}{dx} &= \frac{\sqrt{x}(sec^2x + xcos\ x + sin\ x) - \frac{(tan\ x + x\ sin\ x)}{2\sqrt{x}}}{(\sqrt{x})^2} \\ &= \frac{2x(sec^2x + xcos\ x + sin\ x) - tan\ x - x\ sin\ x}{2x^{3/2}} \\ &= \frac{2xsec^2x + 2x^2cos\ x + x\ sin\ x - tan\ x}{2x^{3/2}} \end{aligned}$$

39. $y = \frac{1+t^{1/2}}{1-t^{1/2}}$

$$\begin{aligned} \frac{dy}{dx} &= \frac{(1-t^{1/2})(\frac{1}{2t^{1/2}}) - (1+t^{1/2})(\frac{-1}{2t^{1/2}})}{(1-t^{1/2})^2} \\ &= \frac{1-t^{1/2}+1+t^{1/2}}{2t^{1/2}(1-\sqrt{t})^2} \\ &= \frac{2}{2\sqrt{t}(1-\sqrt{t})^2} \\ &= \frac{1}{\sqrt{t}(1-\sqrt{t})^2} \end{aligned}$$

41. $f(t) = t\ sin\ t\ tan\ t$

$$\begin{aligned} f'(t) &= (t\ sin\ t)(sec^2t) + (tan\ t)(t\ cos\ t + sin\ t) \\ &= t\ sin\ t\ sec^2t + t\ tan\ t\ cos\ t + sin\ t\ tan\ t \\ &= t\ sin\ t \frac{1}{cos^2t} + t\frac{sin\ t}{cos\ t}cos\ t + sin\ t\ tan\ t \\ &= t\ tan\ t\ sec\ t + t\ sin\ t + sin\ t\ tan\ t \end{aligned}$$

43. - 83. Left to the student

85. **a)** $f(x) = \frac{x}{x+1}$

$$\begin{aligned} f'(x) &= \frac{(x+1)(1) - x(1)}{(x+1)^2} \\ &= \frac{x+1-x}{(x+1)^2} \\ &= \frac{1}{(x+1)^2} \end{aligned}$$

b) $g(x) = \frac{-1}{x+1}$

$$\begin{aligned} g'(x) &= \frac{(x+1)(0) - (-1)(1)}{(x+1)^2} \\ &= \frac{1}{(x+1)^2} \\ &= \frac{1}{(x+1)^2} \end{aligned}$$

c) Since the graphs of both functions are similar then the average rate of change for the functions will be the same (that is why the answers in part (a) and part (b) are equal).

87. $f(x) = sin^2x + cos^2x$

a)

$$\begin{aligned} f'(x) &= 2sin\ x\ cos\ x + 2cos\ x(-sin\ x) \\ &= 2sin\ x\ cos\ x - 2sin\ x\ cos\ x \\ &= 0 \end{aligned}$$

b) Since $sin^2x + cos^2x = 1$ (fundamental trigonometric identity) we would expect the derivative to be zero since we are taking a derivative of a constant, which is always zero.

89. $y = \frac{8}{x^2+4}$

$$\begin{aligned} \frac{dy}{dx} &= \frac{(x^2+4)(0) - 8(2x)}{(x^2+4)^2} \\ &= \frac{0 - 16x}{(x^2+4)^2} \\ &= \frac{-16x}{(x^2+4)^2} \end{aligned}$$

For the point $(0, 2)$, $\frac{dy}{dx}|_{x=0} = m = \frac{-16(0)}{(0^2+4)^2} = 0$. The tangent line is

$$\begin{aligned} y - y_1 &= m(x - x_1) \\ y - 2 &= 0(x - 0) \\ y - 2 &= 0 \\ y &= 2 \end{aligned}$$

For the point $(-2, -1)$, $\frac{dy}{dx}|_{x=-2} = m = \frac{-16(-2)}{((-2)^2+4)^2} = \frac{32}{8} = 4$. The tangent line is

$$\begin{aligned} y - y_1 &= m(x - x_1) \\ y - 1) &= 4(x - (-2)) \\ y - 1 &= 4x + 8 \\ y &= 4x + +8 + 1 \\ y &= 4x + 9 \end{aligned}$$

91. $y = \frac{\sqrt{x}}{x+1} = \frac{x^{1/2}}{x+1}$

$$\begin{aligned} \frac{dy}{dx} &= \frac{(x+1)(\frac{1}{2}x^{-1/2}) - x^{1/2}(1)}{(x+1)^2} \\ &= \frac{\frac{1}{2}x^{1/2} + \frac{1}{2}x^{-1/2} - x^{1/2}}{(x+1)^2} \\ &= \frac{\frac{1}{2}x^{-1/2}}{(x+1)^2} \\ &= \frac{1}{2\sqrt{x}(x+1)^2} \end{aligned}$$

When $x = 1$, $y = \frac{\sqrt{1}}{1+1} = \frac{1}{2}$, and $\frac{dy}{dx}_{x=1} = m = \frac{1}{2\sqrt{1}(1+1)^2} = \frac{1}{8}$. The tangent line is

$$\begin{aligned} y - y_1 &= m(x - x_1) \\ y - \frac{1}{2} &= \frac{1}{8}(x - 1) \\ y - \frac{1}{2} &= \frac{1}{8}x - \frac{1}{8} \\ y &= \frac{1}{8}x - \frac{1}{8} + \frac{1}{2} \\ y &= \frac{1}{8}x - \frac{1}{8} + \frac{4}{8} \\ y &= \frac{1}{8}x + \frac{3}{8} \end{aligned}$$

When $x = \frac{1}{4}$, $y = \frac{\sqrt{\frac{1}{4}}}{1+\frac{1}{4}} = \frac{2}{5}$, and $\frac{dy}{dx}|_{x=\frac{1}{4}} = \frac{1}{2\sqrt{\frac{1}{4}}(1+\frac{1}{4})^2} = \frac{16}{25}$. The tangent line is

$$\begin{aligned} y - y_1 &= m(x - x_1) \\ y - \frac{2}{5} &= \frac{16}{25}(x - \frac{1}{4}) \\ y - \frac{2}{5} &= \frac{16}{25}x - \frac{4}{25} \\ y &= \frac{16}{25}x - \frac{4}{25} + \frac{2}{5} \\ y &= \frac{16}{25}x - \frac{4}{25} + \frac{10}{25} \\ y &= \frac{16}{25}x + \frac{6}{25} \end{aligned}$$

93. $y = x \ sin \ x$

$$\frac{dy}{dx} = x \ cos \ x + sin \ x$$

When $x = \frac{\pi}{4}$

$$\begin{aligned} \frac{dy}{dx}|_{x=\frac{\pi}{4}} &= m \\ &= \frac{\pi}{4}cos(\frac{\pi}{4}) + sin(\frac{\pi}{4}) \\ &= \frac{\pi}{4}\frac{\sqrt{2}}{2} + \frac{\sqrt{2}}{2} \\ &= \frac{\sqrt{2}(\pi+4)}{8} \end{aligned}$$

The tangent line is

$$\begin{aligned} y - y_1 &= m(x - x_1) \\ y - \frac{\sqrt{2}\pi}{8} &= \sqrt{2}\left(\frac{\pi+4}{8}\right)(x - \frac{\pi}{4}) \\ y &= \sqrt{2}\left(\frac{\pi+4}{8}\right)x - \frac{\sqrt{2}\pi}{4}\left(\frac{\pi+4}{8}\right) + \frac{\sqrt{2}\pi}{8} \\ y &= \sqrt{2}\left(\frac{\pi+4}{8}\right)x - \frac{\pi^2}{16\sqrt{2}} - \frac{\sqrt{2}\pi}{8} + \frac{\sqrt{2}\pi}{8} \\ y &= \sqrt{2}\left(\frac{\pi+4}{8}\right)x - \frac{\pi^2}{16\sqrt{2}} \end{aligned}$$

95. **a)** $T(t) = \frac{4t}{t^2+1} + 98.6$

$$\begin{aligned} \frac{dT}{dt} &= \frac{(t^2+1)(4) - 4t(2t)}{(t^2+1)^2} + 0 \\ &= \frac{4t^2 + 4 - 8t^2}{(t^2+1)^2} \\ &= \frac{-4t^2 + 4}{(t^2+1)^2} \end{aligned}$$

b) When $t = 2$ hours

$$\begin{aligned} T &= \frac{4(2)}{2^2+1} + 98.6 \\ &= \frac{8}{5} + 98.6 \\ &= 100.2 \text{ degrees} \end{aligned}$$

c) When $t = 2$ hours

$$\begin{aligned} \frac{dT}{dt} &= \frac{-4(2)^2+4}{(2^2+1)^2} \\ &= \frac{-12}{5} \\ &= -2.4 \text{ degrees per hour} \end{aligned}$$

97. **a)** Since $\frac{s(t)}{100} = tan\ t$ then $s(t) = 100\ tan\ t$

b) $\frac{d\ s(t)}{dt} = 100sec^2t$

c)

$$\begin{aligned} 100sec^2t &= 200 \\ sec^2t &= 2 \\ \frac{1}{cos^2t} &= 2 \\ cos^2t &= \frac{1}{2} \\ cos\ t &= \pm\frac{1}{\sqrt{2}} \\ t &= \pm\frac{\pi}{4}2n\pi \\ &= \frac{\pi}{4} + \frac{n\pi}{2} \end{aligned}$$

99. $g(x) = \frac{(x^2+1)tan\ x}{(x^2-1)}$

$$\begin{aligned} g'(x) &= \frac{(x^2-1)sec^2x(x^2+1)+2x\ tan\ x}{(x^2-1)^2} - \\ &\quad \frac{2x(x^2+1)tan\ x}{(x^2-1)^2} \\ &= \frac{(x^4-1)sec^2x}{(x^2-1)^2} + \frac{2x\ tan\ x}{(x^2-1)^2} - \\ &\quad \frac{2x\ tan\ x}{(x^2-1)^2} - \frac{2x^3tan\ x}{(x^2-1)^2} \\ &= \frac{(x^4-1)sec^2x}{(x^2-1)^2} - \frac{2x^3\ tan\ x}{(x^2-1)^2} \\ &= \frac{sec^2x(x^4-2x^3\ sin\ x\ cos\ x-1)}{(x^2-1)^2} \\ &= \frac{sec^2x(x^4-x^3\ sin(2x)-1)}{(x^2-1)^2} \end{aligned}$$

101. $s(t) = \frac{tan\ t}{t\ cos\ t}$

$$\begin{aligned} s'(t) &= \frac{t\ cos\ tsec^2t - tan\ t(-tsin\ t+cos\ t)}{(t\ cos\ t)^2} \\ &= \frac{t\ sec\ t + t\ sin\ t\ tan\ t - sin\ t}{t^2\ cos^2t} \\ &= \frac{sec^2t(t\ sec\ t + t\ sin\ t\ tan\ t - sin\ t}{t^2} \\ &= \frac{t\ sec^3t + t\ sin\ t\ tan\ tsec^2t - sin\ t\ sec^2t}{t^2} \\ &= \frac{sec\ t(t\ sec^2t + t\ sin\ t\ tan\ t\ sec\ t - sin\ t\ sec\ t)}{t^2} \\ &= \frac{sec\ t(t\ sec^2t + t\ tan^2\ t - tan\ t)}{t^2} \\ &= \frac{sec\ t(t\ sec^2t + tan\ t(t\ tan\ t-1))}{t^2} \end{aligned}$$

103.

$$\begin{aligned} g'(x) &= \frac{(x+cosx)[-xsinx+cosx(xcosx+sinx)-sec^2x]}{(x+cos\ x)^2} \\ &\quad -\frac{(1-sinx)[xsinxcosx-tanx]}{(x+cosx)^2} \\ &= \frac{-x^2sin^2x+x^2cos^2x+xcosxsinx-xsec^2x}{(x+cosx)^2} \\ &\quad -\frac{xcosxsin^2x+xcos^3x+cos^2xsinx-secx}{(x+cosx)^2} \\ &\quad -\frac{xsinxcosx+tanx+xsin^2xcosx-sinxtanx}{(x+cosx)^2} \\ &= \frac{x\ cos^2x+(x^2+sin\ x)cos^2x-x\ sec^2x}{(x+cos\ x)^2} \\ &\quad +\frac{-x^2sin^2x-sec\ x-sin\ x\ tan\ x+tan\ x}{(x+cos\ x)^2} \end{aligned}$$

105. Let $y = (x-1)(x-2)(x-3)$

a)

$$\begin{aligned} \frac{dy}{dx} &= (x-1)[(x-2)(1)+(x-3)(1)]+[(x-2)(x-3)](1 \\ &= (x-1)(x-2)+(x-1)(x-3)+(x-2)(x-3) \\ &= 3x^2-12x+12 \end{aligned}$$

b) $y = (2x+1)(3x-5)(-x+3)$

$$\begin{aligned} \frac{dy}{dx} &= (2x+1)(3x-5)(-1)+(2x+3)(-x+3)(3) \\ &\quad +(3x-5)(-x+3)(2) \\ &= -(2x+1)(3x-5)+3(2x+1)(-x+3) \\ &\quad +2(3x-5)(-x+3) \\ &= -18x^2+50x-16 \end{aligned}$$

c) The derivative of a product of three functions is the sum of all possible combinations consisting of the product of two functions and the derivative of the third function.

d) The derivative of more than three function is the sum of all possible combinations consisting of the product of three of the functions and the derivative of the fourth function.
Let $y = x(x+1)(2x+3)(-x+1) = -2x^4-3x^3+2x^2+3x$ which has a derivative $y' = -8x^3-9x^2+4x+3$. Let us use the rule to find the derivative of y:

$$\begin{aligned} y' &= -x(x+1)(2x+3)+x(2x+3)(-x+1)+ \\ &\quad 2x(x+1)(-x+1)+(x+1)(2x+3)(-x+1) \\ &= -2x^3-5x^2-3x-2x^3-x^2+3x \\ &\quad -2x^3+2x-2x^3-3x^2+2x+3 \\ &= -8x^3-9x^2+4x+3 \end{aligned}$$

107. No horizontal tangent lines

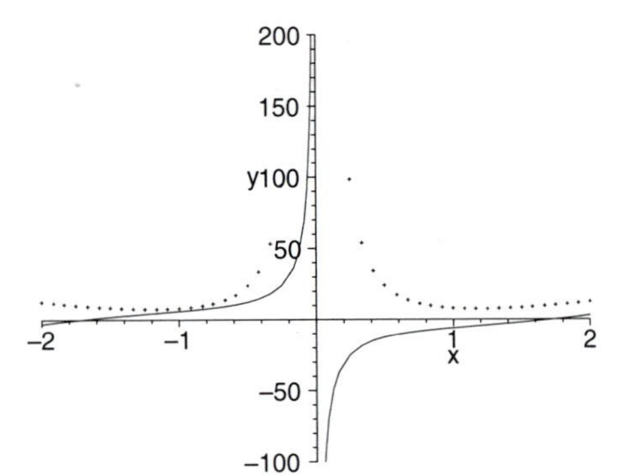

109. $(0.2, 0.75)$ and $(-0.2, -0.75)$

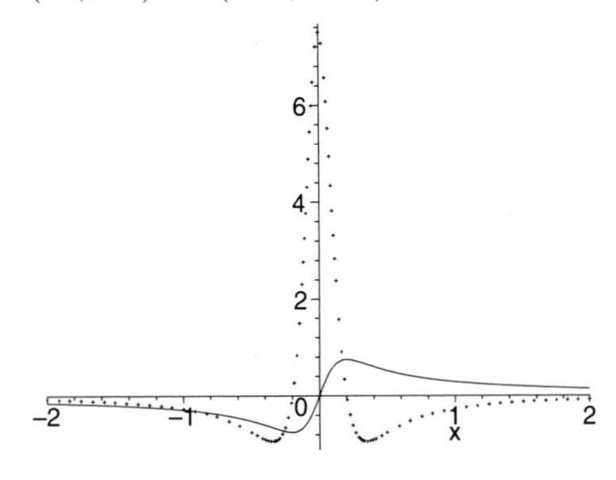

111. $(1, 2)$ and $(-1, -2)$

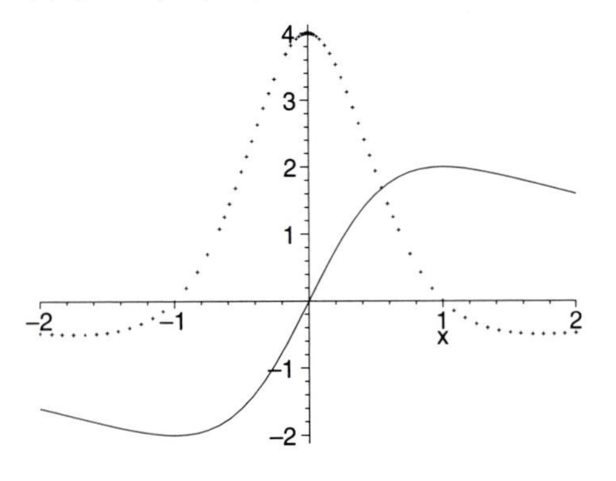

Exercise Set 2.8

1. $y = (2x+1)^2$

Method One (chain rule):

$$\begin{aligned} \frac{dy}{dx} &= 2(2x+1)(2) \\ &= 4(2x+1) \\ &= 8x+4 \end{aligned}$$

Method Two (product rule): $y = (2x+1)(2x+1)$

$$\begin{aligned} \frac{dy}{dx} &= (2x+1)(2) + (2x+1)(2) \\ &= 4x+2+4x+2 \\ &= 8x+4 \end{aligned}$$

Method Three (expand first):

$$\begin{aligned} y &= 4x^2+4x+1 \\ \frac{dy}{dx} &= 4(2x)+4 \\ &= 8x+4 \end{aligned}$$

3. $y = (1-x)^{55}$

$$\begin{aligned} \frac{dy}{dx} &= 55(1-x)^{54}(-1) \\ &= -55(1-x)^{54} \end{aligned}$$

5. $sec^2 x$

$$\begin{aligned} \frac{dy}{dx} &= 2\ sec\ x\ sec\ x\ tan\ x \\ &= 2\ tan\ x\ sec^2 x \end{aligned}$$

7. $y = \sqrt{1-3x} = (1-3x)^{1/2}$

$$\begin{aligned} \frac{dy}{dx} &= \frac{1}{2}(1-3x)^{-1/2}(-3) \\ &= \frac{-3}{2\sqrt{1-3x}} \end{aligned}$$

9. $y = \frac{2}{3x^2+1} = 2(3x^2+1)^{-1}$

$$\begin{aligned} \frac{dy}{dx} &= 2(-1)(3x^2+1)^{-2}(6x) \\ &= \frac{-12x}{(3x^2+1)^2} \end{aligned}$$

11. $s(t) = t(2t+3)^{1/2}$

$$\begin{aligned} s'(t) &= t\left(\frac{1}{2}(2t+3)^{-1/2}(2)\right) + (2t+3)^{1/2}(1) \\ &= \frac{t}{\sqrt{2t+3}} + \sqrt{2t+3} \end{aligned}$$

13. $s(t) = sin\left(\frac{\pi}{6}\ t + \frac{\pi}{3}\right)$

$$\begin{aligned} s'(t) &= cos\left(\frac{\pi}{6}\ t + \frac{\pi}{3}\right) \cdot \frac{\pi}{6} \\ &= \frac{\pi}{6}\ cos\left(\frac{\pi}{6}\ t + \frac{\pi}{3}\right) \end{aligned}$$

15. $g(x) = (1+x^3)^3 - (1+x^3)^4$

$$\begin{aligned} g'(x) &= 3(1+x^3)^2(3x^2) - 4(1+x^3)^3(3x^2) \\ &= 9x^2(1+x^3)^2 - 12x^2(1+x^3)^3 \end{aligned}$$

17. $y = \sqrt{1 - csc\ x} = (1 - csc\ x)^{1/2}$

$$\begin{aligned} \frac{dy}{dx} &= \frac{1}{2}(1 - csc\ x)^{-1/2}(-(-csc\ x\ cot\ x)) \\ &= \frac{csc\ x\ cot\ x}{2\sqrt{1 - csc\ x}} \end{aligned}$$

19. $g(x) = (2x-1)^{1/3} + (4-x)^2$

$$\begin{aligned} g'(x) &= \frac{1}{3}(2x-1)^{-2/3}(2) + 2(4-x)(-1) \\ &= \frac{2}{3\sqrt[3]{(2x-1)^2}} - 2(4-x) \end{aligned}$$

21. $y = x^3 sin\ x + 5x\ cos\ x + 4sec\ x$

$$\begin{aligned} \frac{dy}{dx} &= x^3 cos\ x + 3x^2 sin\ x + 5x(-sin\ x) \\ &\quad +5cos\ x + 4sec\ x\ tan\ x \\ &= x^3 cos\ x + 3x^2 sin\ x - 5x\ sin\ x \\ &\quad +5cos\ x + 4sec\ x\ tan\ x \end{aligned}$$

23. $y = (x^2+x^3)^{1/2}(2x^2+3x+5)$

$$\begin{aligned} \frac{dy}{dx} &= (x^2+x^3)^{1/2}(4x+3) \\ &\quad +\frac{1}{2}(x^2+x^3)^{-1/2}(2x+3x^2)(2x^2+3x+5) \\ &= (4x+3)\sqrt{x^2+x^3} + \frac{(2x+3x^2)(2x^2+3x+5)}{2\sqrt{x^2+x^3}} \end{aligned}$$

25. $f(t) = cos\sqrt{t}$

$$\begin{aligned} f'(t) &= -sin\sqrt{t}\left(\frac{1}{2}t^{-1/2}\right) \\ &= \frac{-sin\sqrt{t}}{2\sqrt{t}} \end{aligned}$$

27. $f(x) + (3x+2)(2x+5)^{1/2}$

$$\begin{aligned} f'(x) &= (3x+2)\left(\frac{1}{2}(2x+5)^{-1/2}(2)\right) + 3(2x+5)^{1/2} \\ &= \frac{3x+2}{\sqrt{2x+5}} + 3\sqrt{2x+5} \end{aligned}$$

29.

$$\begin{aligned} \frac{dy}{dx} &= cos(cos\ x)(-sin\ x) \\ &= -sin\ x\ cos(cos\ x) \end{aligned}$$

31. $y = (cos(4t))^{1/2}$

$$\begin{aligned} \frac{dy}{dx} &= \frac{1}{2}(cos(4t))^{-1/2}(-sin(4t)(4)) \\ &= \frac{-2sin(4t)}{\sqrt{cos(4t)}} \end{aligned}$$

33. $f(x) = \frac{(x^3+2x^2+3x-1)^3}{(2x^4+1)^2}$

$$\begin{aligned} f'(x) &= \frac{(2x^4+1)^2[3(x^3+2x+3x-1)^2(3x^2+4x+3)]}{(2x^4+1)^4} - \\ &\quad \frac{(x^3+2x^2+3x-1)^3[2(2x^4+1)(8x^3)]}{(2x^4+1)^4} \\ &= \frac{3(x^3+2x^2+3x-1)^2(2x^4+1)^2(3x^2+4x+3)}{(2x^4+1)^4} - \\ &\quad \frac{16x^3(x^3+2x^2+3x-1)^3(2x^4+1)}{(2x^4+1)^4} \end{aligned}$$

35. $y = \left(\frac{3x-4}{5x+3}\right)^{1/2}$

$$\begin{aligned} \frac{dy}{dx} &= \frac{1}{2}\left(\frac{3x-4}{5x+3}\right)^{-1/2}\left[\frac{(5x+3)(3)-(3x-4)(5)}{(5x+3)^2}\right] \\ &= \frac{15x+9-15x+20}{2\sqrt{3x-4}(5x+3)^{3/2}} \\ &= \frac{29}{2\sqrt{3x-4}(5x+3)^{3/2}} \end{aligned}$$

37. $r(x) = x(0.01x^2+2.391x-8.51)^5$

$$\begin{aligned} r'(x) &= x[5(0.01x^2+2.391x-8.51)^4(0.02x+2.391)] + \\ &\quad (0.01x^2+2.391x-8.51)^5(1) \\ &= 5x(0.02x+2.391)(0.01x^2+2.391x-8.51)^4 + \\ &\quad (0.01x^2+2.391x-8.51)^5 \\ &= (0.01x^2+2.391x-8.51)^4(5x(0.02x+2.391) + \\ &\quad (0.01x^2+2.391x-8.51)) \\ &= (0.01x^2+2.391x-8.51)^4(0.1x^2+11.955x + \\ &\quad 0.01x^2+2.391x-8.51) \\ &= (0.01x^2+2.391x-8.51)^4(0.11x^2 + \\ &\quad 14.346x-8.51) \end{aligned}$$

39. $y = (cot\ 5x - cos\ 5x)^{1/5}$

$$\begin{aligned} \frac{dy}{dx} &= \frac{1}{5}(cot\ 5x - cos\ 5x)^{-4/5}(-csc^2(5x)(5) + sin(5x)(5)) \\ &= \frac{sin\ 5x - csc\ 5x}{(cot\ 5x - cos\ 5x)^{4/5}} \end{aligned}$$

41. $y = sin(sec^4(x^2))$

$$\begin{aligned} \frac{dy}{dx} &= cos(sec^4(x^2)) \cdot 4sec^3(x^2)(sec\ x^2\ tan\ x^2) \cdot 2x \\ &= 8x\ sec^4(x^2)\ tan(x^2)\ cos(sec^4(x^2)) \end{aligned}$$

43. $y = ((x^2+2)^{1/4}+1)^{1/3}$

$$\begin{aligned} \frac{dy}{dx} &= \frac{1}{3}((x^2+2)^{1/4}+1)^{-2/3}\left(\frac{1}{4}(x^2+2)^{-3/4}(2x)\right) \\ &= \frac{x}{6\sqrt[4]{(x^2+2)^3}\sqrt[3]{(\sqrt[4]{x^2+2}+1)^2}} \end{aligned}$$

45. $y = \frac{sin^3 x}{x^2+5}$

$$\begin{aligned} \frac{dy}{dx} &= \frac{(x^2+5)3sin^2 x\ cos\ x - sin^3 x(2x)}{(x^2+5)^2} \\ &= \frac{3x^2\ sin^2 x\ cos\ x + 15sin^2 x\ cos\ x - 2x\ sin^3 x}{(x^2+5)^2} \\ &= \frac{sin^2 x(3x^2\ cos\ x - 2x\ sin\ x + 15\ cos\ x}{(x^2+5)^2} \end{aligned}$$

47. $f(x) = cot^3(x\ sin(2x+4))$

$$\begin{aligned} f'(x) &= 3cot^2(x\ sin(2x+4))(-csc^2(x\ sin(2x+4)) \times \\ &\quad [x\ cos(2x+4)(2) + sin(2x+4)] \\ &= -3cot^2(x\ sin(2x+4))csc^2(x\ sin(2x+4)) \times \\ &\quad [2x\ cos(2x+4) + sin(2x+4)] \end{aligned}$$

49. $y = \sqrt{sec^4 x + x}$

$$\frac{dy}{dx} = \frac{1}{2\sqrt{sec^4 x + x}} \cdot (4sec^3 x(sec\ x\ tan\ x) + 1) = \frac{2\ sec^4\ x\ tan\ x + 1}{\sqrt{sec^4 x + x}}$$

51. $y = \sqrt{u} = u^{1/2}, \quad u = x^2 - 1$

$$\frac{dy}{du} = \frac{1}{2}u^{1/2-1} = \frac{1}{2}u^{-1/2} = \frac{1}{2\sqrt{u}}$$

$$\frac{du}{dx} = 2x$$

$$\frac{dy}{dx} = \frac{dy}{du} \cdot \frac{du}{dx} = \frac{1}{2\sqrt{u}} \cdot 2x = \frac{2x}{2\sqrt{x^2 - 1}} = \frac{x}{\sqrt{x^2 - 1}}$$

53. $y = u^{50}$, $u = 4x^3 - 2x^2$

$$\frac{dy}{du} = 50u^{49}$$

$$\frac{du}{dx} = 12x^2 - 4x$$

$$\frac{dy}{dx} = \frac{dy}{du} \cdot \frac{du}{dx} = 50u^{49}(12x^2 - 4x) = 50(4x^3 - 2x^2)^{49}(12x^2 - 4x)$$

55. $y = u(u + 1)$, $u = x^3 - 2x$

$$\frac{dy}{du} = u \cdot 1 + 1 \cdot (u + 1) = u + u + 1 = 2u + 1$$

$$\frac{du}{dx} = 3x^2 - 2$$

$$\frac{dy}{dx} = \frac{dy}{du} \cdot \frac{du}{dx} = (2u + 1)(3x^2 - 2) = [2(x^3 - 2x) + 1](3x^2 - 2) = (2x^3 - 4x + 1)(3x^2 - 2)$$

57. $y = \sqrt{x^2 + 3x} = (x^2 + 3x)^{1/2}$

$$\frac{dy}{dx} = \frac{1}{2}(x^2 + 3x)^{-1/2}(2x + 3) = \frac{2x + 3}{2\sqrt{x^2 + 3x}}$$

When $x = 1$, $\frac{dy}{dx} = \frac{2 \cdot 1 + 3}{2\sqrt{1^2 + 3 \cdot 1}} = \frac{2 + 3}{2\sqrt{4}} = \frac{5}{2 \cdot 2} = \frac{5}{4}$

Thus, at $(1, 2)$, $m = \frac{5}{4}$. We use point-slope equation.

$$y - y_1 = m(x - x_1)$$
$$y - 2 = \frac{5}{4}(x - 1)$$
$$y - 2 = \frac{5}{4}x - \frac{5}{4}$$
$$y = \frac{5}{4}x + \frac{3}{4}$$

59. $y = x\sqrt{2x + 3} = x(2x + 3)^{1/2}$

$$\frac{dy}{dx} = x \cdot \frac{1}{2}(2x + 3)^{-1/2}(2) + 1 \cdot (2x + 3)^{1/2} = \frac{x}{\sqrt{2x + 3}} + \sqrt{2x + 3}$$

When $x = 3$, $\frac{dy}{dx} = \frac{3}{\sqrt{2 \cdot 3 + 3}} + \sqrt{2 \cdot 3 + 3} = \frac{3}{\sqrt{9}} + \sqrt{9} = \frac{3}{3} + 3 = 1 + 3 = 4$

Thus, at $(3, 9)$, $m = 4$. We use point-slope equation.

$$y - y_1 = m(x - x_1)$$
$$y - 9 = 4(x - 3)$$
$$y - 9 = 4x - 12$$
$$y = 4x - 3$$

61. $f(x) = sin^2 x$

$$\frac{dy}{dx} = 2\ sin\ x\ cos\ x$$

When $x = -\frac{\pi}{6}$, $\frac{dy}{dx} = 2sin(\frac{-\pi}{6})cos(\frac{-\pi}{6}) = \frac{-\sqrt{3}}{2}$

Use the point-slope equation:

$$y - \frac{1}{4} = \frac{-\sqrt{3}}{2}(x - (-\frac{\pi}{6}))$$
$$y - \frac{1}{4} = -\frac{\sqrt{3}}{2}x - \frac{\sqrt{3}}{12}$$
$$y = -\frac{\sqrt{3}}{2}x - \frac{\sqrt{3}}{12} + \frac{1}{4}$$
$$y = \frac{1}{12}(-6\sqrt{3}x - \sqrt{3}\pi + 3)$$

63. $f(x) = \frac{x^2}{(1+x)^5}$

a)

$$f'(x) = \frac{(1 + x)^5(2x) - x^2(5(1 + x)^4)}{(1 + x)^{10}} = \frac{(1 + x)^4[2x + 2x^2 - 5x^2]}{(1 + x)^{10}} = \frac{2x - 3x^2}{(1 + x)^6}$$

b)

$$f'(x) = x^2[-5(1 + x)^{-6}(1)] + 2x(1 + x)^{-5} = \frac{-5x^2}{(1 + x)^6} + \frac{2x}{(1 + x)^5} = \frac{-5x^2 + 2x(1 + x)}{(1 + x)^6} = \frac{2x - 3x^2}{(1 + x)^6}$$

c) The results in the previous parts are the same.

65. Using the Chain Rule:

Let $y = f(u)$. Then

$$\begin{aligned}\frac{dy}{dx} &= \frac{dy}{du} \cdot \frac{du}{dx} \\ &= 3u^2(8x^3) \\ &= 3(2x^4+1)^2(8x^3) \quad \text{Substituting } 2x^4+1 \text{ for } u\end{aligned}$$

When $x = -1$,
$$\begin{aligned}\frac{dy}{dx} &= 3[2(-1)^4+1]^2[8(-1)^3] \\ &= 3(2+1)^2(-8) \\ &= 3 \cdot 3^2(-8) \\ &= -216\end{aligned}$$

Finding $f(g(x))$:

$f \circ g(x) = f(g(x)) = f(2x^4+1) = (2x^4+1)^3$

Then $(f \circ g)'(x) = 3(2x^4+1)^2(8x^3)$ and

$(f \circ g)'(-1) = -216$ as above.

67. Using the Chain Rule:

Let $y = f(u) = \sqrt[3]{u} = u^{1/3}$. Then

$$\begin{aligned}\frac{dy}{dx} &= \frac{dy}{du} \cdot \frac{du}{dx} \\ &= \frac{1}{3}u^{-2/3} \cdot (-6x) \\ &= -2x \cdot u^{-2/3} \\ &= -2x(1-3x^2)^{-2/3} \quad \text{Substituting } 1-3x^2 \text{ for } u\end{aligned}$$

When $x = 2$,
$$\begin{aligned}\frac{dy}{dx} &= 2 \cdot 2(1-3 \cdot 2^2)^{-2/3} \\ &= -4(-11)^{-2/3} \approx -0.8087\end{aligned}$$

Finding $f(g(x))$:

$f \circ g(x) = f(g(x)) = f(1-3x^2) = \sqrt[3]{1-3x^2}$, or $(1-3x^2)^{1/3}$

Then $(f \circ g)'(x) = \frac{1}{3}(1-3x^2)^{-2/3}(-6x) = -2x(1-3x^2)^{-2/3}$ and

$(f \circ g)'(2) = -4(-11)^{-2/3} \approx -0.8087$ as above.

69. $A = 1000(1+i)^3$

a)

$$\begin{aligned}\frac{dA}{di} &= 1000(3(1+i)^2) \\ &= 3000(1+i)^2\end{aligned}$$

b) $\frac{dA}{di}$ represents the rate at which the amount of investment is changing with respect to an annual interest rate i.

71. $D = 0.85A(c+25)$, $c = (140-y)\frac{w}{72x}$

a) To find D as a function of c, we substitute 5 for A in the formula for D.

$$\begin{aligned}D &= 0.85A(c+25) \\ &= 0.85(5)(c+25) \\ &= 4.25(c+25) \\ &= 4.25(c+25) \\ &= 4.25c+106.25\end{aligned}$$

To find c as a function of w, we substitute 45 for y and 0.6 for x in the formula for c.

$$\begin{aligned}c &= (140-45)\frac{w}{72(0.6)} \\ &= 95 \cdot \frac{w}{43.2} \\ &\approx 2.199w\end{aligned}$$

b) $\frac{dD}{dc} = 4.25$

c) $\frac{dc}{dw} = 2.199$

d) First we find $D \circ c(w)$.

$$\begin{aligned}D \circ c(w) &= D(c(w)) \\ &= 4.25(2.199w)+106.25 \\ &= 9.34575w+106.25\end{aligned}$$

Then we have

$$\frac{dD}{dw} = 9.34575 \approx 9.346.$$

e) $\frac{dD}{dw}$ represents the rate of change of the dosage with respect to the patient's weight. For each additional kilogram of weight, the dosage is increased by about 9.35 mg.

73. **a)** January 2009 corresponds to $t = 52$

$$\begin{aligned}C'(t) &= 0.74+0.02376t-1.0814\pi\ cos(2\pi t) \\ C'(52) &= 0.74+0.02376(52)-1.0814\pi cos(104\pi) \\ &= -1.4218\ ppmv/yr\end{aligned}$$

b) July 2009 corresponds to $t = 52.5$

$$\begin{aligned}C'(52.5) &= 0.74+0.02376(52.5)-1.0814\pi cos(105\pi) \\ &= 5.3847\ ppmv/yr\end{aligned}$$

75. $y = ((x^2+4)^8+3\sqrt{x})^4$

$$\begin{aligned}\frac{dy}{dx} &= 4((x^2+4)^8+3\sqrt{x})^3[8(x^2+4)^7(2x)+\frac{3}{2\sqrt{x}}] \\ &= 4((x^2+4)^8+3\sqrt{x})^3\left(16x(x^2+4)^7+\frac{3}{2\sqrt{x}}\right)\end{aligned}$$

77. Let $y = sin(sin(sin\ x))$ then

$$\begin{aligned}\frac{dy}{dx} &= cos(sin(sin\ x)) \cdot cos(sin\ x) \cdot cos\ x \\ &= cos\ x\ cos(sin\ x)\ cos(sin(sin\ x))\end{aligned}$$

79. Let $y = tan(cot(sec\ 3x))$ then

$$\begin{aligned} \frac{dy}{dx} &= sec^2(cot(sec\ 3x)) \cdot -csc^2(sec\ 3x) \cdot 3\ sec\ 3x\ tan\ 3x \\ &= -3\ sec\ 3x\ tan\ 3x csc^2(sec\ 3x)\ sec^2(cot(sex\ 3x)) \end{aligned}$$

81. $y = \left(sin\left(\frac{3\pi}{2} + 3\right)\right)^{1/5}$ is a constant, which means $\frac{dy}{dx} = 0$.

83.

$$\begin{aligned} sin(a+x) &= sin\ a\ cos\ x + cos\ a\ sin\ x \\ \frac{d}{dx}(sin(a+x)) &= \frac{d}{dx}(sin\ a\ cos\ x + cos\ a\ sin\ x) \\ &= -sin\ a\ sin\ x + cos\ a\ cos\ x \\ &= cos\ a\ cos\ x - sin\ a\ sin\ x \\ &= cos(a+x) \end{aligned}$$

85. Let $Q(x) = \frac{N(x)}{D(x)}$. Then we can write

$$Q(x) = N(x) \cdot [D(x)]^{-1}$$

using the property of negative exponents. Now we use the product differentiation rule

$$\begin{aligned} Q'(x) &= N(x) \cdot -1[D(x)]^{-2} \cdot D'(x) + [D(x)]^{-1} \cdot N'(x) \\ &= \frac{-N(x) \cdot D'(x)}{[D(x)]^2} + \frac{N'(x)}{D(x)} \\ &= \frac{-N(x) \cdot D'(x)}{[D(x)]^2} + \frac{N'(x) \cdot D(x)}{[D(x)]^2} \\ &= \frac{N'(x) \cdot D(x) - N(x) \cdot D'(x)}{[D(x)]^2} \end{aligned}$$

87. $(-2.145, -7.728)$ and $(2.145, 7.728)$

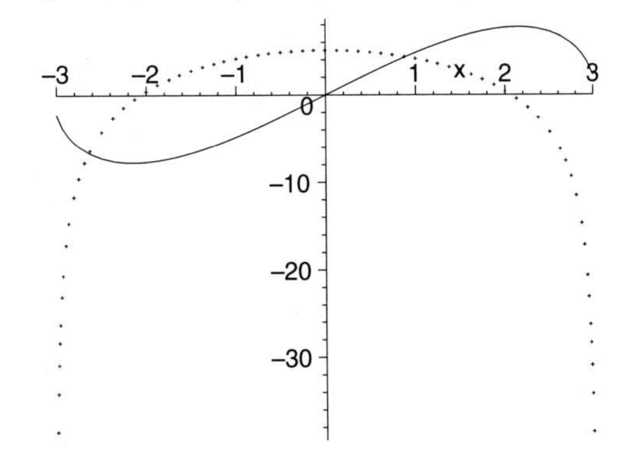

89.

$$\begin{aligned} f'(x) &= x \cdot \frac{-2x}{2\sqrt{4-x^2}} + \sqrt{4-x^2} \\ &= \frac{-x^2}{\sqrt{4-x^2}} + \frac{4-x^2}{\sqrt{4-x^2}} \\ &= \frac{4-2x^2}{\sqrt{4-x^2}} \end{aligned}$$

Exercise Set 2.9

1. $y = 3x + 5$

$$\begin{aligned} \frac{dy}{dx} &= 3 \\ \frac{d^2y}{dx^2} &= 0 \end{aligned}$$

3. $y = -3(2x+2)^{-1}$

$$\begin{aligned} \frac{dy}{dx} &= 3(2x+2)^{-2}(2) = 6(2x+2)^{-2} = \frac{6}{(2x+2)^2} \\ \frac{d^2y}{dx^2} &= -12(2x+2)^{-3}(2) = \frac{-24}{(2x+2)^3} \end{aligned}$$

5. $y = (2x+1)^{1/3}$

$$\begin{aligned} \frac{dy}{dx} &= \frac{1}{3}(2x+1)^{-2/3}(2) = \frac{2}{3(2x+1)^{2/3}} \\ \frac{d^2y}{dx^2} &= \frac{2}{3}(-\frac{2}{3}(2x+1)^{-5/3}(2)) = \frac{-8}{9(2x+1)^{5/2}} \end{aligned}$$

7. $f(x) = (4-3x)^{-4}$

$$\begin{aligned} f'(x) &= -4(4-3x)^{-5}(-3) = 12(4-3x)^{-5} \\ f''(x) &= -60(4-3x)^{-6}(-3) = \frac{180}{(4-3x)^6} \end{aligned}$$

9. $y = \sqrt{x+1} = (x+1)^{1/2}$

$$\begin{aligned} \frac{dy}{dx} &= \frac{1}{2}(x+1)^{-1/2} \cdot 1 \\ &= \frac{1}{2}(x+1)^{-1/2} \\ \frac{d^2y}{dx^2} &= -\frac{1}{4}(x+1)^{-3/2} \cdot 1 \\ &= -\frac{1}{4}(x+1)^{-3/2} \\ &= -\frac{1}{4(x+1)^{3/2}} \\ &= -\frac{1}{4\sqrt{(x+1)^3}} \end{aligned}$$

11. $f(x) = (2x+9)^{16}$

$$\begin{aligned} f'(x) &= 16(2x+9)^{15}(2) = 32(2x+9)^{15} \\ f''(x) &= 480(2x+9)^{14}(2) = 960(2x+9)^{14} \end{aligned}$$

13. $g(x) = sec(3x+1)$

$$\begin{aligned} g'(x) &= sec(3x+1)\ tan(3x+1)(3) \\ &= 3\ sec(3x+1)\ tan(3x+1) \\ g''(x) &= 3\ sec(3x+1)\ sec^{(}3x+1)(3) + \\ &\quad tan(3x+1)(3\ sec(3x+1)\ tan(3x+1)(3) \\ &= 9\ sec^3(3x+1) + 9\ sec(3x+1)\ tan^2(3x+1) \\ &= 9\ sec(3x+1)[sec^2(3x+1) + tan^2(3x+1)] \end{aligned}$$

15. $f(x) = sec(2x+3) + 4x^2 + 3x - 7$

$$\begin{aligned} f'(x) &= sec(2x+3)\ tan(2x+3)(2) + 8x + 3 \\ &= 2\ sec(2x+3)\ tan(2x+3) + 8x + 3 \\ f''(x) &= 2\ sec(2x+3)\ sec^2(2x+3)(2) + \\ &\quad tan(2x+3)\ sec(2x+3)\ tan(2x+3)(2) + 8 \\ &= 4\ sec^3(2x+3) + 2\ sec(2x+3)\ tan^2(2x+3) + 8 \end{aligned}$$

17. $y = ax^2 + bx + c$

$$\begin{aligned} \frac{dy}{dx} &= a \cdot 2x + b + 0 \\ &= 2ax + b \\ \frac{d^2y}{dx^2} &= 2a + 0 \\ &= 2a \end{aligned}$$

19. $y = \sqrt[4]{(x^2+1)^3} = (x^2+1)^{3/4}$

$$\begin{aligned} \frac{dy}{dx} &= \frac{3}{4}(x^2+1)^{-1/4}(2x) \\ &= \frac{3}{2}x(x^2+1)^{-1/4} \\ \frac{d^2y}{dx^2} &= \frac{3}{2}\left[x\left(-\frac{1}{4}\right)(x^2+1)^{-5/4}(2x) + 1 \cdot (x^2+1)^{-1/4}\right] \\ &= -\frac{3x^2}{4(x^2+1)^{5/4}} + \frac{3}{2(x^2+1)^{1/4}} \\ &= -\frac{3x^2}{4\sqrt[4]{(x^2+1)^5}} + \frac{3}{2\sqrt[4]{x^2+1}} \end{aligned}$$

21. $f(x) = (4x+3)\ cos\ x$

$$\begin{aligned} f'(x) &= (4x+3)\ (-sin\ x) + 4\ cos\ x \\ &= -(4x+3)\ sin\ x + 4\ cos\ x \end{aligned}$$

23. $s(t) = cos(at+b)$

$$\begin{aligned} s'(t) &= -sin(at+b)\ a = -a\ sin(at+b) \\ s''(t) &= -a(cos(at+b)\ a) \\ &= -a^2\ cos(at+b) \end{aligned}$$

25. $y = \frac{(t^2+3)^{1/2}}{7} + (3t^2+1)^{1/3}$

$$\begin{aligned} \frac{dy}{dx} &= \frac{1}{7} \cdot \frac{1}{2}(t^2+3)^{-1/2}(2t) + \frac{1}{3}(3t^2+1)^{-2/3}(6t) \\ &= \frac{t}{7}(t^2+3)^{-1/2} + 2t(3t^2+1)^{-2/3} \\ \frac{d^2y}{dx^2} &= \frac{t}{7}(\frac{-1}{2}(t^2+3)^{-3/2}(2t)) + (2)(t^2+3)^{-1/2} + \\ &\quad 2t(\frac{-2}{3}(3t^2+1)^{-5/3}(6t)) + (2)(3t^2+1)^{-2/3} \\ &= \frac{-t^2}{7(t^2+3)^{3/2}} + \frac{2}{(t^2+3)^{1/2}} \\ &\quad -\frac{8t^2}{(3t^2+1)^{5/3}} + \frac{2}{3t^2+1} \end{aligned}$$

27. $y = x^4$

$$\begin{aligned} \frac{dy}{dx} &= 4x^3 \\ \frac{d^2y}{dx^2} &= 4 \cdot 3x^2 \\ &= 12x^2 \\ \frac{d^3y}{dx^3} &= 12 \cdot 2x \\ &= 24x \\ \frac{d^4y}{dy^4} &= 24 \end{aligned}$$

29. $y = x^6 - x^3 + 2x$

$$\begin{aligned} \frac{dy}{dx} &= 6x^5 - 3x^2 + 2 \\ \frac{d^2y}{dx^2} &= 30x^4 - 6x \\ \frac{d^3y}{dx^3} &= 120x^3 - 6 \\ \frac{d^4y}{dx^4} &= 360x^2 \\ \frac{d^5y}{dx^5} &= 720x \end{aligned}$$

31. $y = (x^2-5)^{10}$

$$\begin{aligned} \frac{dy}{dx} &= 10(x^2-5)^9 \cdot 2x \\ &= 20x(x^2-5)^9 \\ \frac{d^2y}{dx^2} &= 20x \cdot 9(x^2-5)^8 \cdot 2x + 20(x^2-5)^9 \\ &= 360x^2(x^2-5)^8 + 20(x^2-5)^9 \\ &= 20(x^2-5)^8[18x^2 + (x^2-5)] \\ &= 20(x^2-5)^8(19x^2-5) \end{aligned}$$

33. $y = sec(2x+3)$

$$\begin{aligned} \frac{dy}{dx} &= sec(2x+3)\ tan(2x+3)\ (2) \\ &= 2\ sec(2x+3)\ tan(2x+3) \\ \frac{d^y}{dx^2} &= 2\ sec(2x+3)\ sec^2(2x+3)(2) + \\ &\quad [2\ sec(2x+3)\ tan(2x+3)(2)]\ tan(2x+3) \\ &= 4\ sec^3(2x+3) + 4\ sec(2x+3)\ tan^2(2x+3) \\ \frac{d^3y}{dx^3} &= 4(3sec^2(2x+3)\ sec(2x+3)\ tan(2x+3)(2)) + \\ &\quad 4\ sec(2x+3)[2\ tan(2x+3)\ sec^2(2x+3)(2)] + \\ &\quad sec(2x+3)\ tan(2x+3)\ (2)\ tan^2(2x+3) \\ &= 40\ sec^3(2x+3)\ tan(2x+3) + \\ &\quad 8\ sec(2x+3)\ tan^3(2x+3) \end{aligned}$$

35. $s(t)10\ cos(3t+2) - 4\ sin(3t+2)$

$$\begin{aligned} v(t) &= 10[-sin(3t+2)(3)] - 4[cos(3t+2)(3) \\ &= -30\ sin(3t+2) - 12\ cos(3t+2) \\ a(t) &= -30[cos(3t+2)(3)] - 12[-sin(3t+2)(3)] \\ &= -90\ cos(3t+2) - 36\ sin(3t+2) \\ &= 9[10\ cos(3t+2) - 4\ sin(3t+2)] \\ &= 9\ s(t) \end{aligned}$$

37. $s(t) = t^3 + t^2 + 2t$

$$\begin{aligned} v(t) &= s'(t) = 3t^2 + 2t + 2 \\ a(t) &= s''(t) = 6t + 2 \end{aligned}$$

39. $w(t) = 0.000758t^3 - 0.0596t^2 - 1.82t + 8.15$
The acceleration of a function that depends on time is the second derivative of the function with respect to time.

$$\begin{aligned} w'(t) &= 0.002274t^2 - 0.1192t + 1.82 \\ w''(t) &= 0.004548t - 0.1192 \end{aligned}$$

41. $P(t)100000(1 + 0.6t + t^2)$

$$\begin{aligned} P'(t) &= 100000(0.6 + 2t) \\ P''(t) &= 100000(2) \\ &= 200000 \end{aligned}$$

43. $y = \frac{x}{(x-1)^{1/2}}$

$$\begin{aligned} y' &= \frac{\sqrt{x-1}(1) - x \cdot \frac{1}{2\sqrt{x-1}}}{x-1} \\ &= \frac{2(x-1) - x}{2(x-1)\sqrt{x-1}} \\ &= \frac{x-2}{2(x-1)^{3/2}} \end{aligned}$$

$$\begin{aligned} y'' &= \frac{2(x-1)^{3/2}(1) - (x-2)\left[2 \cdot \frac{3}{2}(x-1)^{1/2}\right]}{4(x-1)^3} \\ &= \frac{2(x-1)^{3/2} - 3(x-2)(x-1)^{1/2}}{4(x-1)^3} \\ &= \frac{(x-1)^{1/2}[2(x-1) - 3(x-2)]}{(x-1)^3} \\ &= \frac{4-x}{(x-1)^{5/2}} \end{aligned}$$

$$\begin{aligned} y''' &= \frac{4(x-1)^{5/2}(-1) - (4-x)\left[4 \cdot \frac{5}{2}(x-1)^{3/2}\right]}{16(x-1)^5} \\ &= \frac{(x-1)^{3/2}[4(x-1) - 10(4-x)]}{16(x-1)^5} \\ &= \frac{4x - 4 - 40 + 10x}{16(x-1)^{7/2}} \\ &= \frac{3x-18}{16(x-1)^{7/2}} \end{aligned}$$

45. $f(x) = \frac{x}{x-1}$

$$\begin{aligned} f'(x) &= \frac{(x-1)(1) - x(1)}{(x-1)^2} \\ &= \frac{-1}{(x-1)^2} \\ &= -(x-1)^{-2} \end{aligned}$$

$$\begin{aligned} f''(x) &= -(-2(x-1)^{-3}) \\ &= \frac{2}{(x-1)^3} \end{aligned}$$

47. $y = sin\ x$

a) $\frac{dy}{dx} = cos\ x$

b) $\frac{d^2y}{dx^2} = -sin\ x$

c) $\frac{d^3y}{dx^3} = -cos\ x$

d) $\frac{d^4y}{dx^4} = sin\ x$

e) $\frac{d^8y}{dx^8} = sin\ x$

f) $\frac{d^{10}y}{dx^{10}} = -sin\ x$

g) $\frac{d^{837}y}{dx^{837}} = cos\ x$

49. Functions that have the form $f(x) = Asin\ x + Bcos\ x$ where A and B are constants, will satisfy the condition of their second derivative being the negative of the original function.

51. $f(x) = \frac{x+3}{x-2}$

$$\begin{aligned} f'(x) &= \frac{(x-2)(1) - (x+3)(1)}{(x-2)^2} \\ &= \frac{-5}{(x-2)^2} = -5(x-2)^{-2} \\ f''(x) &= 10(x-2)^{-3} = \frac{10}{(x-2)^3} \\ f'''(x) &= -30(x-2)^{-4} = \frac{-30}{(x-2)^4} \\ f^{(4)}(x) &= 120(x-2)^{-5} = \frac{120}{(x-2)^5} \\ f^{(5)}(x) &= -600(x-2)^{-6} = \frac{-600}{(x-2)^6} \end{aligned}$$

53.

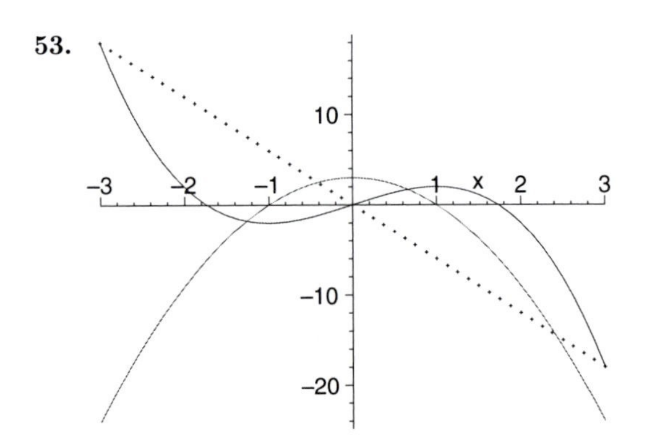

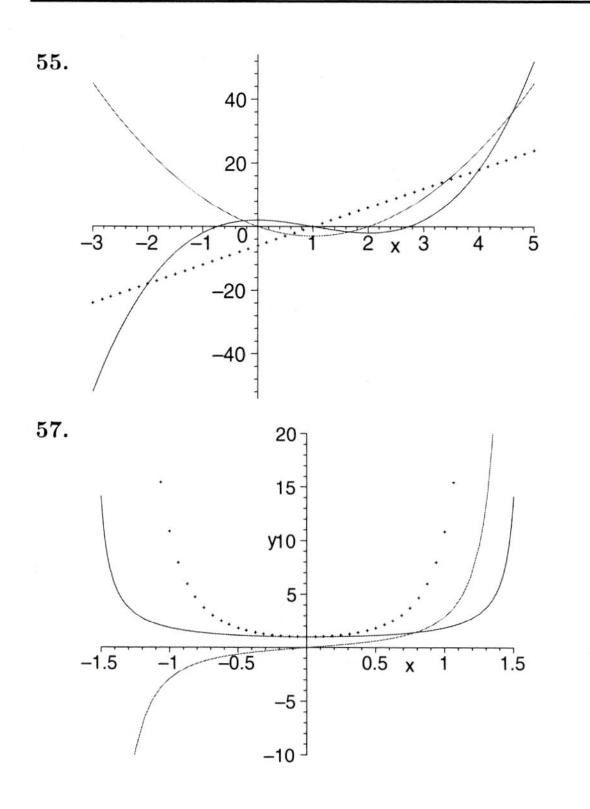
55.
40
20
–3
–2
–1
0
1
2
x
3
4
5
–20
–40
57.
20
15
y10
5
–1.5
–1
–0.5
0.5
x
1
1.5
–5
–10

Chapter 3

Application of Differentiation

Exercise Set 3.1

1. $f(x) = x^2 - 4x + 5$. First, find the critical points (values of x at which the derivative is zero or undefined).

$f'(x) = 2x - 4$

$f'(x)$ exists for all real numbers. We solve $f'(x) = 0$:

$$\begin{aligned} 2x - 4 &= 0 \\ 2x &= 4 \\ x &= 2 \end{aligned}$$

The only critical point is at $x = 2$. We use 2 to divide the real number line into two intervals, A: $(-\infty, 2)$ and B: $(2, \infty)$:

A B
2

We use a test value in each interval to determine the sign of the derivative in each interval.

A: Test 0, $f'(0) = 2 \cdot 0 - 4 = -4 < 0$

B: Test 3, $f'(3) = 2 \cdot 3 - 4 = 2 > 0$

We see that $f(x)$ is decreasing on $(-\infty, 2)$ and increasing on $(2, \infty)$, and the change from decreasing to increasing indicates that a relative minimum occurs at $x = 2$. We substitute into the original equation to find $f(2)$:

$f(2) = 2^2 - 4 \cdot 2 + 5 = 1$

Thus, there is a relative minimum at $(2, 1)$. We use the information obtained to sketch the graph.

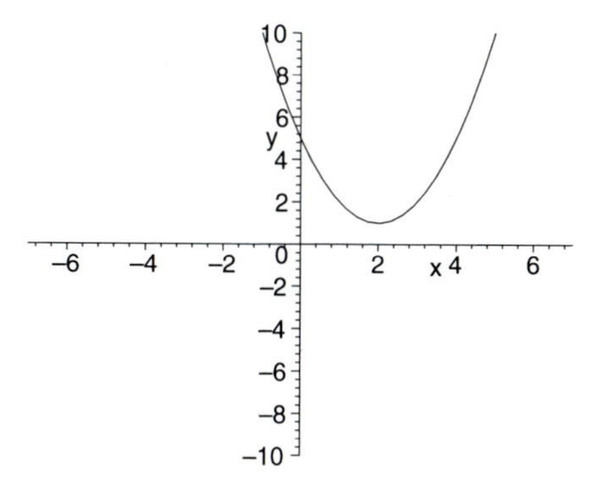

3. $f(x) = 5 + x - x^2$

First, find the critical points,

$f'(x) = 1 - 2x$

$f'(x)$ exists for all real numbers. We solve $f'(x) = 0$:

$$\begin{aligned} 1 - 2x &= 0 \\ 1 &= 2x \\ \frac{1}{2} &= x \end{aligned}$$

The only critical point is at $x = \frac{1}{2}$. We use $\frac{1}{2}$ to divide the real number line into two intervals, A: $\left(-\infty, \frac{1}{2}\right)$ and B: $\left(\frac{1}{2}, \infty\right)$:

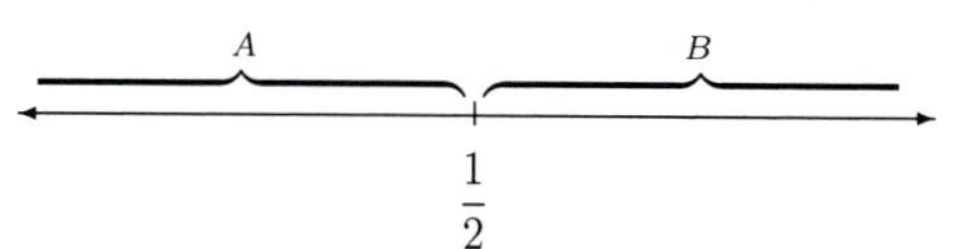

We use a test value in each interval to determine the sign of the derivative in each interval.

A: Test 0, $f'(0) = 1 - 2 \cdot 0 = 1 > 0$

B: Test 1, $f'(1) = 1 - 2 \cdot 1 = -1 < 0$

We see that $f(x)$ is increasing on $\left(-\infty, \frac{1}{2}\right)$ and decreasing on $\left(\frac{1}{2}, \infty\right)$, so there is a relative maximum at $x = \frac{1}{2}$. We find $f\left(\frac{1}{2}\right)$:

$$f\left(\frac{1}{2}\right) = 5 + \frac{1}{2} - \left(\frac{1}{2}\right)^2 = \frac{21}{4}$$

Thus, there is a relative maximum at $\left(\frac{1}{2}, \frac{21}{4}\right)$. We use the information obtained to sketch the graph.

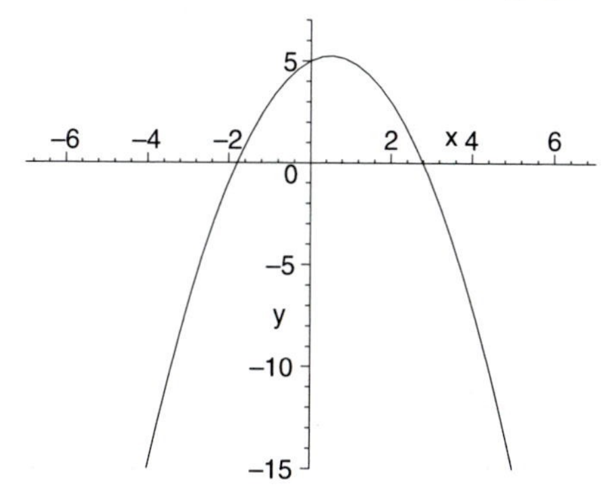

5. $f(x) = 1 + 6x + 3x^2$

First, find the critical points.

$f'(x) = 6 + 6x$

$f'(x)$ exists for all real numbers. We solve $f'(x) = 0$:

$$6 + 6x = 0$$

$$\begin{aligned} 6x &= -6 \\ x &= -1 \end{aligned}$$

The only critical point is at $x = -1$. We use -1 to divide the real number line into two intervals, A: $(-\infty, -1)$ and B: $(-1, \infty)$:

A B

-1

We use a test value in each interval to determine the sign of the derivative in each interval.

A: Test -2, $f'(-2) = 6 + 6(-2) = -6 < 0$

B: Test 0, $f'(0) = 6 + 6 \cdot 0 = 6 > 0$

We see that $f(x)$ is decreasing on $(-\infty, -1)$ and increasing on $(-1, \infty)$, so there is a relative minimum at $x = -1$. We find $f(-1)$:

$$f(-1) = 1 + 6(-1) + 3(-1)^2 = -2$$

Thus, there is a relative minimum at $(-1, -2)$. We use the information obtained to sketch the graph.

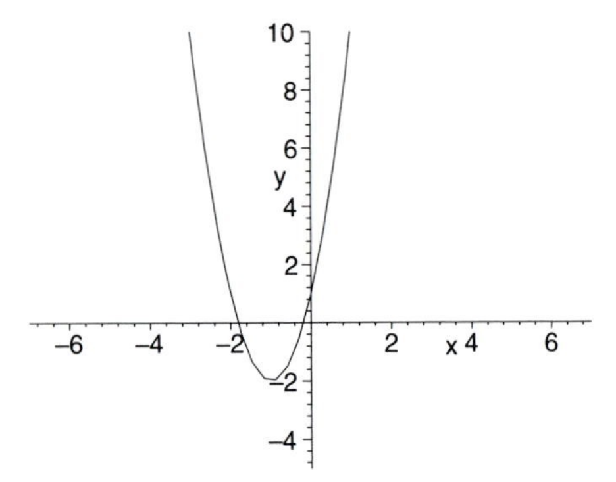

7. $f(x) = x^3 - x^2 - x + 2$

First, find the critical points.

$f'(x) = 3x^2 - 2x - 1$

$f'(x)$ exists for all real numbers. We solve $f'(x) = 0$:

$$\begin{aligned} 3x^2 - 2x - 1 &= 0 \\ (3x + 1)(x - 1) &= 0 \\ 3x + 1 &= 0 \\ 3x &= -1 \\ x &= -\frac{1}{3} \\ \text{Or} & \\ x - 1 &= 0 \\ x &= 1 \end{aligned}$$

The critical points are at $x = -\frac{1}{3}$ and $x = 1$. We use them to divide the real number line into three intervals, A: $\left(-\infty, -\frac{1}{3}\right)$, B: $\left(-\frac{1}{3}, 1\right)$, and C: $(1, \infty)$.

A B C

$-\frac{1}{3}$ 1

We use a test value in each interval to determine the sign of the derivative in each interval.

A: Test -1, $f'(-1) = 3(-1)^2 - 2(-1) - 1 = 3 + 2 - 1 = 4 > 0$

B: Test 0, $f'(0) = 3(0)^2 - 2(0) - 1 = -1 < 0$

C: Test 2, $f'(2) = 3(2)^2 - 2(2) - 1 = 12 - 4 - 1 = 7 > 0$

We see that $f(x)$ is increasing on $\left(-\infty, -\frac{1}{3}\right)$, decreasing on $\left(-\frac{1}{3}, 1\right)$, and increasing again on $(1, \infty)$, so there is a relative maximum at $x = -\frac{1}{3}$ and a relative minimum at $x = 1$. We find $f\left(-\frac{1}{3}\right)$:

$$\begin{aligned} f\left(-\frac{1}{3}\right) &= \left(-\frac{1}{3}\right)^3 - \left(-\frac{1}{3}\right)^2 - \left(-\frac{1}{3}\right) + 2 \\ &= -\frac{1}{27} - \frac{1}{9} + \frac{1}{3} + 2 \\ &= \frac{59}{27} \end{aligned}$$

Then we find $f(1)$:

$$\begin{aligned} f(1) &= 1^3 - 1^2 - 1 + 2 \\ &= 1 - 1 - 1 + 2 \\ &= 1 \end{aligned}$$

There is a relative maximum at $\left(-\frac{1}{3}, \frac{59}{27}\right)$, and there is a relative minimum at $(1, 1)$. We use the information obtained to sketch the graph.

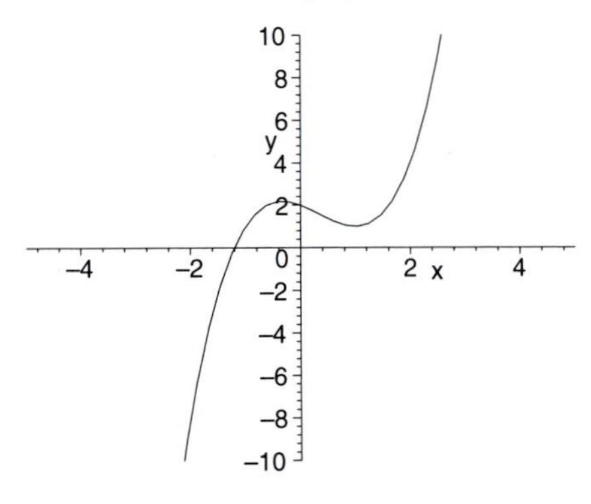

9. $f(x) = x^3 - 3x + 6$

First, find the critical points.

$f'(x) = 3x^2 - 3$

$f'(x)$ exists for all real numbers. We solve $f'(x) = 0$:

$$\begin{aligned} 3x^2 - 3 &= 0 \\ x^2 - 1 &= 0 \\ (x + 1)(x - 1) &= 0 \\ x + 1 &= 0 \\ x &= -1 \end{aligned}$$

Or

$$\begin{aligned} x-1 &= =0 \\ x &= 1 \end{aligned}$$

The critical points are at $x = -1$ and $x = 1$. We use them to divide the real number line into three intervals, A: $(-\infty, -1)$, B: $(-1, 1)$, and C: $(1, \infty)$.

A B C

-1 1

We use a test value in each interval to determine the sign of the derivative in each interval.

A: Test -2, $f'(-2) = 3(-2)^2 - 3 = 12 - 3 = 9 > 0$

B: Test 0, $f'(0) = 3 \cdot 0^2 - 3 = 0 - 3 = -3 < 0$

C: Test 2, $f'(2) = 3 \cdot 2^2 - 3 = 12 - 3 = 9 > 0$

We see that $f(x)$ is increasing on $(-\infty, -1)$, decreasing on $(-1, 1)$, and increasing again on $(1, \infty)$, so there is a relative maximum at $x = -1$ and a relative minimum at $x = 1$. We find $f(-1)$:

$$f(-1) = (-1)^3 - 3(-1) + 6 = -1 + 3 + 6 = 8$$

Then we find $f(1)$:

$$f(1) = 1^3 - 3 \cdot 1 + 6 = 1 - 3 + 6 = 4$$

There is a relative maximum at $(-1, 8)$, and there is a relative minimum at $(1, 4)$. We use the information obtained to sketch the graph.

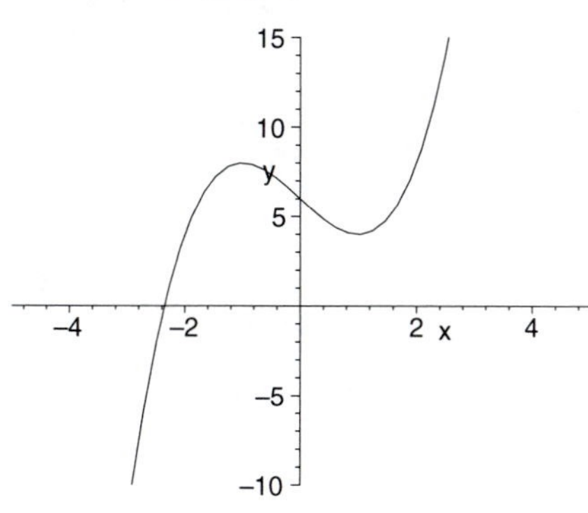

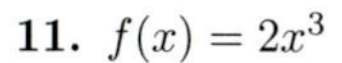

11. $f(x) = 2x^3$

First, find the critical points.

$f'(x) = 6x^2$

$f'(x)$ exists for all real numbers. We solve $f'(x) = 0$:

$$\begin{aligned} 6x^2 &= 0 \\ x^2 &= 0 \\ x &= 0 \end{aligned}$$

The only critical point is at $x = 0$. We use 0 to divide the real number line into two intervals, A: $(-\infty, 0)$ and B: $(0, \infty)$:

A B

0

We use a test value in each interval to determine the sign of the derivative in each interval.

A: Test -1, $f'(-1) = 6(-1)^2 = 6 > 0$

B: Test 1, $f'(1) = 6(1)^2 = 6 > 0$

We see that $f(x)$ is increasing on $(-\infty, 0)$ and increasing on $(0, \infty)$, therefore there is no change from decreasing to increasing or from increasing to decreasing. Therefore the function does not has a relative extrema.

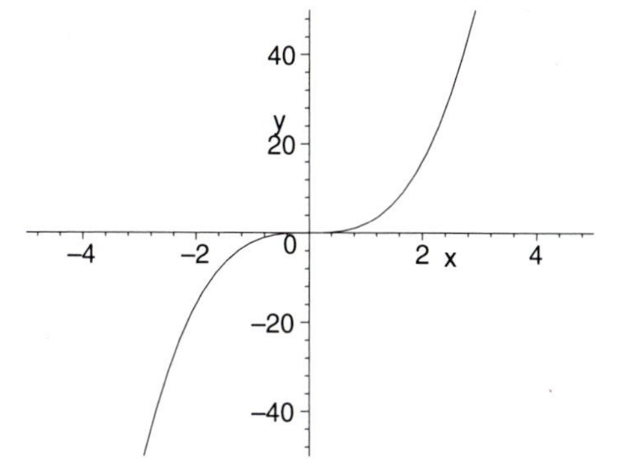

13. $f(x) = 0.02x^2 + 1.3x + 2.31$ First, find the critical points.

$f'(x) = 0.04x + 1.3$

$f'(x)$ exists for all real numbers. We solve $f'(x) = 0$:

$$\begin{aligned} 0.04x + 1.3 &= 0 \\ 0.04x &= -1.3 \\ x &= \frac{-1.3}{0.04} \\ x &= -32.5 \end{aligned}$$

The only critical point is at $x = -32.5$. We use -32.5 to divide the real number line into two intervals, A: $(-\infty, -32.5)$ and B: $(-32.5, \infty)$:

A B

-32.5

We use a test value in each interval to determine the sign of the derivative in each interval.

A: Test 0, $f'(0) = 0.04 \cdot 0 + 1.3 = 1.3 > 0$

B: Test -100, $f'(-100) = 0.04 \cdot (-100) + 1.3 = -2.7 < 0$

We see that $f(x)$ is decreasing on $(-\infty, -32.5)$ and increasing on $(-32.5, \infty)$, and the change from decreasing to increasing indicates that a relative minimum occurs at $x = -32.5$. We substitute into the original equation to find $f(-32.5)$:

$$\begin{aligned} f(-32.5) &= 0.02(-32.5)^2 + 1.3(-32.5) + 2.31 \\ &= 21.125 - 42.25 + 2.31 \\ &= -18.815 \end{aligned}$$

Thus, there is a relative minimum at $(-32.5, -18.815)$. We use the information obtained to sketch the graph.

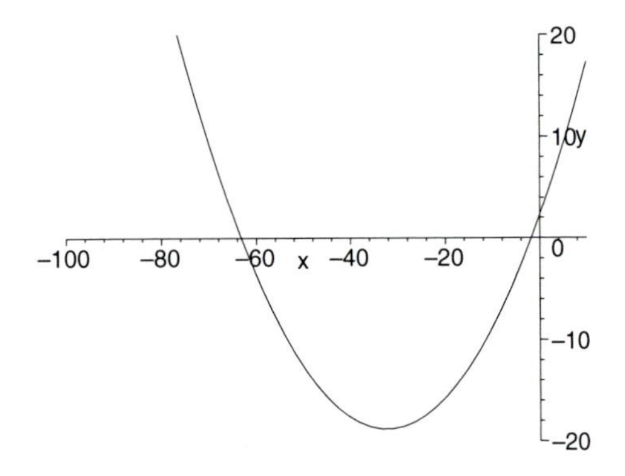

15. $f(x) = x^4 - 2x^3$

$f'(x) = 4x^3 - 6x^2$

$f'(x)$ exists for all real numbers. Solve $f'(x) = 0$.

$$\begin{aligned} 4x^3 - 6x^2 &= 0 \\ 2x^3 - 3x^2 &= 0 \\ x^2(2x-3) &= 0 \\ x &= 0 \\ \text{Or} & \\ x &= \frac{3}{2} \end{aligned}$$

The critical points are at $x = 0$ and $x = \frac{3}{2}$. Use them to divide the real number line into three intervals, A: $(-\infty, 0)$, B: $\left(0, \frac{3}{2}\right)$, and C: $\left(\frac{3}{2}, \infty\right)$.

A: Test -1, $f'(-1) = 4(-1)^3 - 6(-1)^2 = -10 < 0$

B: Test 1, $f'(1) = 4 \cdot 1^3 - 6 \cdot 1^2 = -2 < 0$

C: Test 2, $f'(2) = 4 \cdot 2^3 - 6 \cdot 2^2 = 8 > 0$

Since $f(x)$ is decreasing on both $(-\infty, 0)$ and $\left(0, \frac{3}{2}\right)$ and is increasing on $\left(\frac{3}{2}, \infty\right)$, there is no relative extremum at $x = 0$ but there is a relative minimum at $x = \frac{3}{2}$.

$$f\left(\frac{3}{2}\right) = \left(\frac{3}{2}\right)^4 - 2\left(\frac{3}{2}\right)^3 = \frac{81}{16} - \frac{27}{4} = -\frac{27}{16}$$

There is a realtive minimum at $\left(\frac{3}{2}, -\frac{27}{16}\right)$. We sketch the graph.

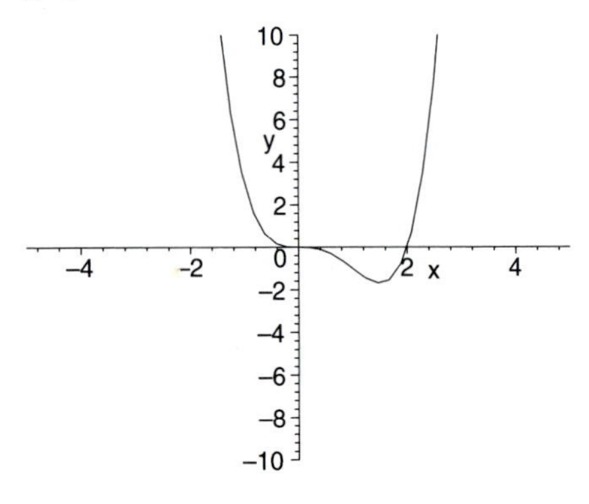

17. $f(x) = x\sqrt{8 - x^2}$

The domain of this function is between $[-\sqrt{8}, \sqrt{8}]$

First, find the critical points.

$$\begin{aligned} f'(x) &= x\left(\frac{1}{2}(8-x^2)^{-1/2}(-2x)\right) + (8-x^2)^{1/2}(1) \\ &= \frac{-x^2}{(8-x^2)^{1/2}} + (8-x^2)^{1/2} \end{aligned}$$

$f'(x)$ does not exist for $x = \pm\sqrt{8}$. We solve $f'(x) = 0$:

$$\begin{aligned} \frac{-x^2}{(8-x^2)^{1/2}} + (8-x^2)^{1/2} &= 0 \\ 8 - x^2 &= -x^2 \\ 8 &= 2x^2 \\ 4 &= x^2 \\ \pm 2 &= x \end{aligned}$$

The critical points are at $x = \pm\sqrt{8}$, $x = -2$ and $x = 2$. We use them to divide the real number line into three intervals, A: $(-\sqrt{8}, -2)$, B: $(-2, 2)$, C: $(2, \sqrt{8})$.

A B C

−2 2

We use a test value in each interval to determine the sign of the derivative in each interval.

A: Test -2.5, $f'(-2.5) = \frac{-(-2.5)^2}{(8-(-2.5)^2)^{1/2}} + (8-(-2.5)^2)^{1/2} < 0$

B: Test 0, $f'(0) = \frac{-(0)^2}{(8-0^2)^{1/2}} + (8-0^2)^{1/2} > 0$

C: Test 2.5, $f'(2.5) = \frac{-(2.5)^2}{(8-(2.5)^2)^{1/2}} + (8-(2.5)^2)^{1/2} < 0$

We see that $f(x)$ is decreasing on $(-\sqrt{8}, -2)$, and on $(2, \sqrt{8})$, and increasing on $(-2, 2)$,so there is a relative minimum at $x = -2$ and a relative maximum at $x = 2$.

We find $f(-2)$:

$f(-2) = (-2)\sqrt{8 - (-2)^2} = -4$

Then we find $f(2)$:

$f(2) = (2)\sqrt{8 - (2)^2} = 4$ There is a relative minimum at $(-2, -4)$, and there is a relative maximum at $(2, 4)$. We use the information obtained to sketch the graph.

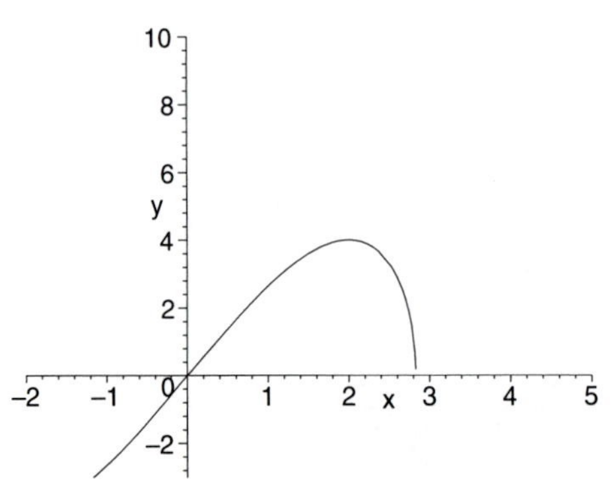

19. $f(x) = 1 - x^{2/3}$

First, find the critical points.

$$f'(x) = -\frac{2}{3}x^{-1/3} = -\frac{2}{3\sqrt[3]{x}}$$

$f'(x)$ does not exist for $x = 0$. The equation $f'(x) = 0$ has no solution, so the only critical point is at $x = 0$. We use it to divide the real number line into two intervals: A: $(-\infty, 0)$ and B: $(0, \infty)$.

We use a test value in each interval to determine the sign of the derivative in each interval.

A: Test -1, $f'(-1) = -\frac{2}{3\sqrt[3]{-1}} = -\frac{2}{3(-1)} = \frac{2}{3} > 0$

B: Test 1, $f'(1) = -\frac{2}{3\sqrt[3]{1}} = -\frac{2}{3 \cdot 1} = -\frac{2}{3} < 0$

We see that $f(x)$ is increasing on $(-\infty, 0)$ and decreasing on $(0, \infty)$, so there is a relative maximum at $x = 0$.

We find $f(0)$:

$$f(0) = 1 - 0^{2/3} = 1 - 0 = 1$$

There is a relative maximum at $(0, 1)$. We use the information obtained to sketch the graph.

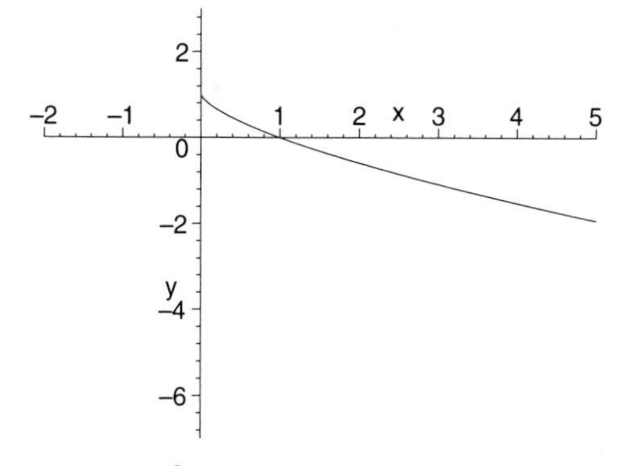

21. $f(x) = \frac{-8}{x^2+1} = -8(x^2+1)^{-1}$

First, find the critical points.

$$\begin{aligned} f'(x) &= -8(-1)(x^2+1)^{-2}(2x) \\ &= 16x(x^2+1)^{-2} \\ &= \frac{16x}{(x^2+1)^2} \end{aligned}$$

$f'(x)$ exists for all real numbers. We solve $f'(x) = 0$:

$$\begin{aligned} \frac{16x}{(x^2+1)^2} &= 0 \\ 16x &= 0 \\ x &= 0 \end{aligned}$$

The only critical point is at $x = 0$. We use it to divide the real number line into two intervals, A: $(-\infty, 0)$ and B: $(0, \infty)$.

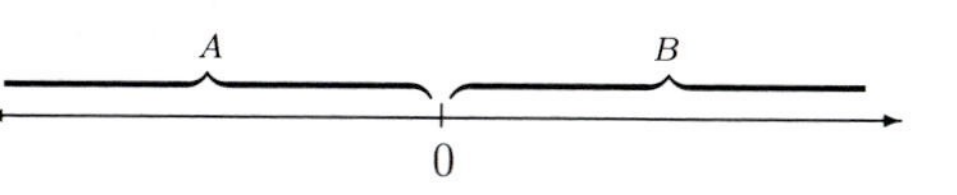

We use a test value in each interval to determine the sign of the derivative in each interval.

A: Test -1, $f'(-1) = \frac{16(-1)}{[(-1)^2+1]^2} = \frac{-16}{4} = -4 < 0$

B: Test 1, $f'(1) = \frac{16 \cdot 1}{(1^2+1)^2} = \frac{16}{4} = 4 > 0$

We see that $f(x)$ is decreasing on $(-\infty, 0)$ and increasing on $(0, \infty)$, so there is a relative minimum at $x = 0$.

We find $f(0)$:

$$f(0) = \frac{-8}{0^2+1} = \frac{-8}{1} = -8$$

There is a relative minimum at $(0, -8)$. We use the information obtained to sketch the graph.

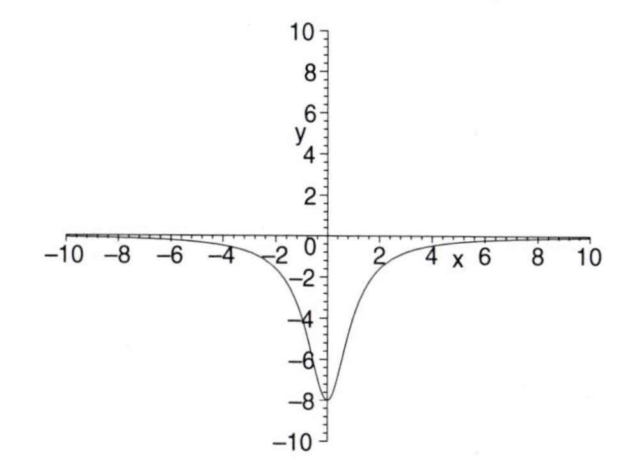

23. $f(x) = \frac{4x}{x^2+1}$

First, find the critical points.

$$\begin{aligned} f'(x) &= \frac{(x^2+1)(4) - 2x(4x)}{(x^2+1)^2} \\ &= \frac{4x^2 + 4 - 8x^2}{(x^2+1)^2} \\ &= \frac{4 - 4x^2}{(x^2+1)^2} \end{aligned}$$

$f'(x)$ exists for all real numbers. We solve $f'(x) = 0$:

$$\begin{aligned} \frac{4 - 4x^2}{(x^2+1)^2} &= 0 \\ 4 - 4x^2 &= 0 \\ 1 - x^2 &= 0 \\ (1-x)(1+x) &= 0 \\ 1 - x &= 0 \\ 1 &= x \end{aligned}$$

Or

$$\begin{aligned} 1 + x &= 0 \\ x &= -1 \end{aligned}$$

The critical points are at $x = -1$ and $x = 1$. We use them to divide the real number line into three intervals, A: $(-\infty, -1)$, B: $(-1, 1)$, and C: $(1, \infty)$.

A B C

$-1 \qquad 1$

We use a test value in each interval to determine the sign of the derivative in each interval.

A: Test -2, $f'(-2) = \dfrac{4 - 4(-2)^2}{[(-2)^2 + 1]^2} = \dfrac{-12}{25} < 0$

B: Test 0, $f'(0) = \dfrac{4 - 4 \cdot 0^2}{(0^2 + 1)^2} = \dfrac{4}{1} = 4 > 0$

C: Test 2, $f'(2) = \dfrac{4 - 4 \cdot 2^2}{(2^2 + 1)^2} = \dfrac{-12}{25} < 0$

We see that $f(x)$ is decreasing on $(-\infty, -1)$, increasing on $(-1, 1)$, and decreasing again on $(1, \infty)$, so there is a relative minimum at $x = -1$ and a relative maximum at $x = 1$.

We find $f(-1)$:

$$f(-1) = \frac{4(-1)}{(-1)^2 + 1} = \frac{-4}{2} = -2$$

Then we find $f(1)$:

$$f(1) = \frac{4 \cdot 1}{1^2 + 1} = \frac{4}{2} = 2$$

There is a relative minimum at $(-1, -2)$, and there is a relative maximum at $(1, 2)$. We use the information obtained to sketch the graph.

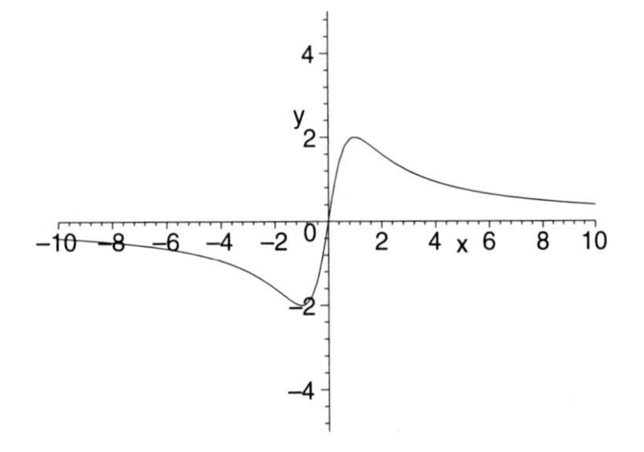

25. $f(x) = \sqrt[3]{x} = x^{1/3}$

First, find the critical points.

$$f'(x) = \frac{1}{3}x^{-2/3} = \frac{1}{3\sqrt[3]{x^2}}$$

$f'(x)$ does not exist for $x = 0$. The equation $f'(x) = 0$ has no solution, so the only critical point is at $x = 0$. We use it to divide the real number line into two intervals, A: $(-\infty, 0)$, and B: $(0, \infty)$.

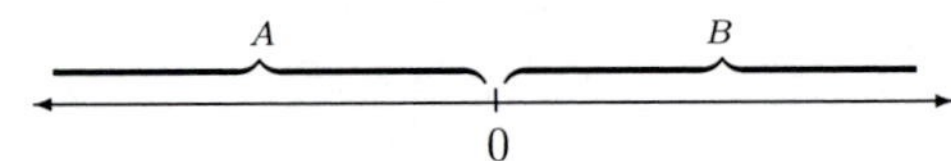

We use a test value in each interval to determine the sign of the derivative in each interval.

A: Test -1, $f'(-1) = \dfrac{1}{3\sqrt[3]{(-1)^2}} = \dfrac{1}{3 \cdot 1} = \dfrac{1}{3} > 0$

B: Test 1, $f'(1) = \dfrac{1}{3\sqrt[3]{1^2}} = \dfrac{1}{3 \cdot 1} = \dfrac{1}{3} > 0$

We see that $f(x)$ is increasing on both intervals, so the function has no relative extrema. We use the information obtained to sketch the graph.

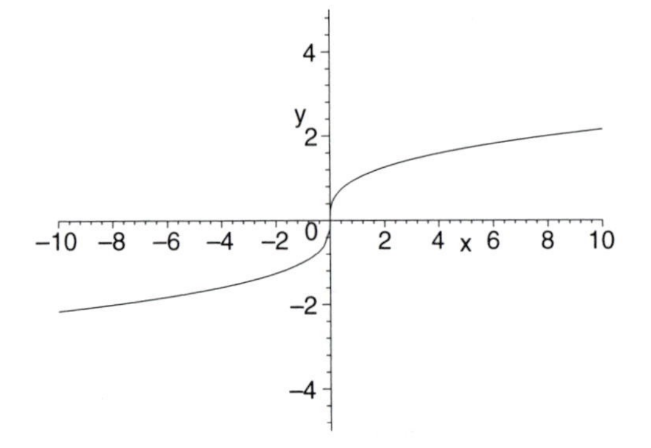

27. $f(x) = \sqrt{x^2 + 2x + 5} = (x^2 + 2x + 5)^{1/2}$

First, find the critical points.

$$f'(x) = \frac{1}{2}(x^2 + 2x + 5)^{-1/2}(2x + 2) = \frac{x + 1}{\sqrt{x^2 + 2x + 5}}$$

$f'(x)$ exists for all x values. We solve $f'(x) = 0$,

$$\begin{aligned} f'(x) &= 0 \\ \frac{x + 1}{\sqrt{x^2 + 2x + 5}} &= 0 \\ x + 1 &= 0 \\ x &= -1 \end{aligned}$$

So the only critical point is at $x = -1$. We use it to divide the real number line into two intervals, A: $(-\infty, -1)$, and B: $(-1, \infty)$.

A B

-1

We use a test value in each interval to determine the sign of the derivative in each interval.

A: Test -2, $f'(-2) = \dfrac{(-2) + 1}{\sqrt{(-2)^2 + 2(-2) + 5}} = \dfrac{-1}{\sqrt{5}} < 0$

B: Test 0, $f'(0) = \dfrac{0 + 1}{\sqrt{(0)^2 + 2(0) + 5}} = \dfrac{1}{\sqrt{5}} > 0$

We see that $f(x)$ is decreasing on $(-\infty, -1)$ and increasing on $(-1, \infty)$, so the function has a relative minimum at $x = -1$.

We find $f(-1)$:

$$\begin{aligned} f(-1) &= \sqrt{(-1)^2 + 2(-1) + 5} \\ &= \sqrt{1 - 2 + 5} \\ &= \sqrt{4} = 2 \end{aligned}$$

There is a relative minimum at $(-1, 2)$. We use the information obtained to sketch the graph.

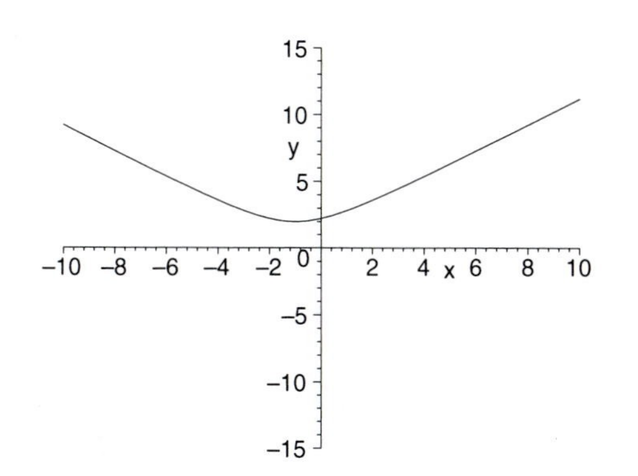

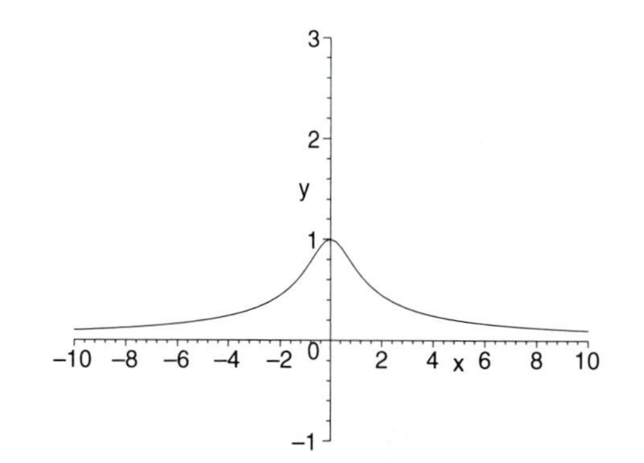

29. $f(x) = \frac{1}{\sqrt{x^2+1}} = (x^2+1)^{-1/2}$

First, find the critical points.

$f'(x) = \frac{-1}{2}(x^2+1)^{-3/2}(2x) = \frac{2x}{\sqrt{(x^2+1)^3}}$

$f'(x)$ exists for all x values. We solve $f'(x) = 0$,

$$\begin{aligned} f'(x) &= 0 \\ \frac{2x}{\sqrt{(x^2+1)^3}} &= 0 \\ 2x &= 0 \\ x &= 0 \end{aligned}$$

So the only critical point is at $x = 0$. We use it to divide the real number line into two intervals, A: $(-\infty, 0)$, and B: $(0, \infty)$.

A B

0

We use a test value in each interval to determine the sign of the derivative in each interval.

A: Test -1, $f'(-1) = \frac{2(-1)}{\sqrt{((-1)^2+1)^3}} = \frac{-2}{\sqrt{8}} < 0$

B: Test 1, $f'(1) = \frac{2(1)}{\sqrt{((1)^2+1)^3}} = \frac{2}{\sqrt{8}} > 0$

We see that $f(x)$ is decreasing on $(-\infty, 0)$ and increasing on $(0, \infty)$, so the function has a relative minimum at $x = 0$.

We find $f(0)$:

$$\begin{aligned} f(0) &= \sqrt{(0)^2+1} \\ &= \sqrt{0+1} \\ &= \sqrt{1} = 1 \end{aligned}$$

There is a relative minimum at $(0, 1)$. We use the information obtained to sketch the graph.

31. $f(x) = sin\ x$

First, find the critical points.

$f'(x) = cos\ x$

$f'(x)$ exists for all x values. We solve $f'(x) = 0$,

$$\begin{aligned} f'(x) &= 0 \\ cos\ x &= 0 \\ x &= \frac{\pi}{2} \\ \text{and} & \\ x &= \frac{3\pi}{2} \end{aligned}$$

So the only critical points are at $x = \frac{\pi}{2}$ and $x = \frac{3\pi}{2}$ and there might be extrema points at the end points $x = 0$ and $x = 2\pi$. We use them to divide the real number line into three intervals, A: $[0, \frac{\pi}{2})$, B: $(\frac{\pi}{2}, \frac{3\pi}{2})$, and C: $(\frac{3\pi}{2}, 2\pi]$.

A B C

$0 \quad \frac{\pi}{2} \quad \frac{3\pi}{2} \quad 2\pi$

We use a test value in each interval to determine the sign of the derivative in each interval.

A: Test 0, $f'(0) = cos(0) = 1 > 0$

B: Test π, $f'(\pi) = cos(\pi) = -1 < 0$

C: Test 2π, $f'(2\pi) = cos(2\pi) = 1 > 0$

We see that $f(x)$ is decreasing on $(-\infty, \frac{\pi}{2})$ and increasing on $(\frac{\pi}{2}, \frac{3\pi}{2})$, we also see that $f(x)$ is increasing on $(\frac{3\pi}{2}, \infty)$ so the function has a relative maximum at $x = \frac{\pi}{2}$ and a relative minimum at $x = \frac{3\pi}{2}$.

We find $f(\frac{\pi}{2})$:

$$\begin{aligned} f(\frac{\pi}{2}) &= sin(\frac{\pi}{2}) \\ &= 1 \end{aligned}$$

We find $f(\frac{3\pi}{2})$:

$$\begin{aligned} f(\frac{3\pi}{2}) &= sin(\frac{3\pi}{2}) \\ &= -1 \end{aligned}$$

There is a relative maximum at $(\frac{\pi}{2}, 1)$ and there is a relative minimum at $(\frac{3\pi}{2}, -1)$. We use the information obtained to sketch the graph.

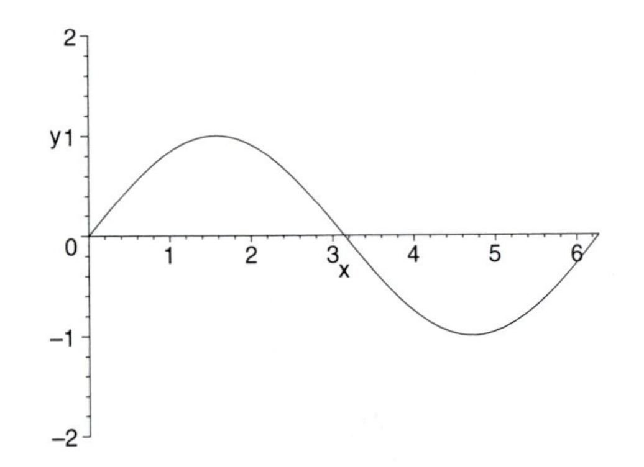

33. $f(x) = sin\ x - cos\ x$ First, find the critical points.

$f'(x) = cos\ x + sin\ x$

$f'(x)$ exists for all x values. We solve $f'(x) = 0$,

$$\begin{aligned} f'(x) &= 0 \\ cos\ x + sin\ x &= 0 \\ cos\ x &= -sin\ x \\ x &= \frac{3\pi}{4} \\ \text{and} & \\ x &= \frac{7\pi}{4} \end{aligned}$$

So the only critical points are at $x = \frac{3\pi}{4}$ and $x = \frac{7\pi}{4}$. We use them to divide the real number line into three intervals, A: $[0, \frac{3\pi}{4})$, B: $(\frac{3\pi}{4}, \frac{7\pi}{4})$, C: $(\frac{7\pi}{4}, 2\pi]$.

A B C

0 $\frac{3\pi}{4}$ $\frac{7\pi}{4}$ 2π

We use a test value in each interval to determine the sign of the derivative in each interval.

A: Test 0, $f'(0) = cos(0) + sin(0) = 1 > 0$

B: Test $\frac{5\pi}{4}$, $f'(\frac{5\pi}{2}) = cos(\frac{5\pi}{2}) + sin(\frac{5\pi}{2}) = -1.414 < 0$

C: Test 2π, $f'(2\pi) = cos(2\pi) + sin(2\pi) = 1 > 0$

We see that $f(x)$ is increasing on $[0, \frac{3\pi}{4})$ and on $(\frac{7\pi}{4}, 2\pi]$ and decreasing on $(\frac{3\pi}{4}, \frac{7\pi}{4})$ so the function has a relative maximum at $x = \frac{3\pi}{4}$ and a relative minimum at $x = \frac{7\pi}{4}$.

We find$f(\frac{3\pi}{4})$:

$$\begin{aligned} f(\frac{3\pi}{4}) &= sin(\frac{3\pi}{4}) - cos(\frac{3\pi}{4}) \\ &= \frac{1}{\sqrt{2}} - (-\frac{1}{\sqrt{2}}) \\ &= \frac{2}{\sqrt{2}} \approx 1.414 \end{aligned}$$

We find $f(\frac{7\pi}{4})$:

$$\begin{aligned} f(\frac{7\pi}{4}) &= sin(\frac{7\pi}{4}) - cos(\frac{7\pi}{4}) \\ &= -\frac{1}{\sqrt{2}} - (\frac{1}{\sqrt{2}}) \\ &= -\frac{2}{\sqrt{2}} \approx -1.414 \end{aligned}$$

There is a relative maximum at $(\frac{3\pi}{4}, \frac{2}{\sqrt{2}})$ and there is a relative minimum at $\frac{7\pi}{4}, \frac{2}{\sqrt{2}}$. We use the information obtained to sketch the graph.

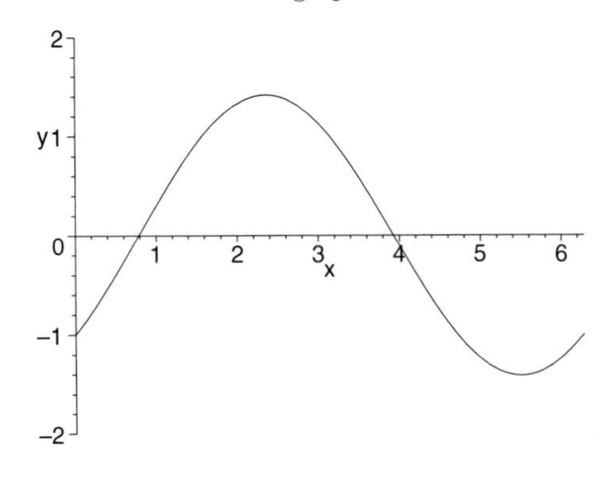

35. $f(x) = cos\ 2x$

First, find the critical points.

$f'(x) = -2sin\ 2x$

$f'(x)$ exists for all x values. We solve $f'(x) = 0$,

$$\begin{aligned} f'(x) &= 0 \\ -2sin\ 2x &= 0 \\ 2x &= 0 \\ x &= 0 \\ \text{and} & \\ x &= \frac{\pi}{2} \\ \text{and} & \\ x &= \pi \\ \text{and} & \\ x &= \frac{3\pi}{2} \\ \text{and} & \\ x &= 2\pi \end{aligned}$$

So the only critical points are at $x = 0$, $x = \frac{\pi}{2}$, $x = \pi$, $x = \frac{3\pi}{2}$ and $x = 2\pi$. We use them to divide the real number line into four intervals, A: $(0, \frac{\pi}{2})$, B: $(\frac{\pi}{2}, \pi)$, C: $(\pi, \frac{3\pi}{2})$ and D: $(\frac{3\pi}{2}, 2\pi)$.

A B C D

0 $\frac{\pi}{2}$ π $\frac{3\pi}{2}$ 2π

We use a test value in each interval to determine the sign of the derivative in each interval.

A: Test $\pi over4$, $f'(\pi over4) = -2sin(\pi over2) = -2 < 0$

B: Test $\frac{3\pi}{4}$, $f'(\frac{3\pi}{4}) = -2sin(\frac{3\pi}{2}) = 2 > 0$

C: Test $\frac{5\pi}{4}$, $f'(\frac{5\pi}{4}) = -2sin(\frac{3\pi}{2}) = -2 < 0$

D: Test $\frac{7\pi}{4}$, $f'(\frac{7\pi}{4}) = -2sin(\frac{5\pi}{2}) = 2 > 0$

We see that $f(x)$ is decreasing on $(0, \frac{\pi}{2})$ and $(\pi, \frac{3\pi}{2})$ and the function is increasing on $(\frac{\pi}{2}, \pi)$ and

$(\frac{3\pi}{2}, 2\pi)$ so the function has a relative maximum at $x = 0$ (end point), $x = \pi$ and $x = 2\pi$ (end point) and a relative minimum at $x = \frac{\pi}{2}$ and $x = \frac{3\pi}{2}$.

We find$f(0)$:

$$\begin{aligned} f(0) &= cos(2 \cdot 0) \\ &= cos(0) \\ &= 1 \end{aligned}$$

We find $f(\pi)$:

$$\begin{aligned} f(\pi) &= cos(2 \cdot \pi) \\ &= 1 \end{aligned}$$

We find $f(2\pi)$:

$$\begin{aligned} f(2\pi) &= cos(2 \cdot 2\pi) \\ &= 1 \end{aligned}$$

We find $f(\frac{\pi}{2})$:

$$\begin{aligned} f(2\pi) &= cos(2 \cdot \frac{\pi}{2}) \\ &= cos(\pi) \\ &= -1 \end{aligned}$$

We find $f(\frac{3\pi}{2})$:

$$\begin{aligned} f(2\pi) &= cos(2 \cdot \frac{3\pi}{2}) \\ &= cos(3\pi) \\ &= -1 \end{aligned}$$

There is a relative maximum at $(0, 1)$, $(\pi, 1)$ and $(2\pi, 1)$ and there is a relative minimum at $(\frac{\pi}{2}, -1)$ and $(\frac{3\pi}{2}, -1)$. We use the information obtained to sketch the graph.

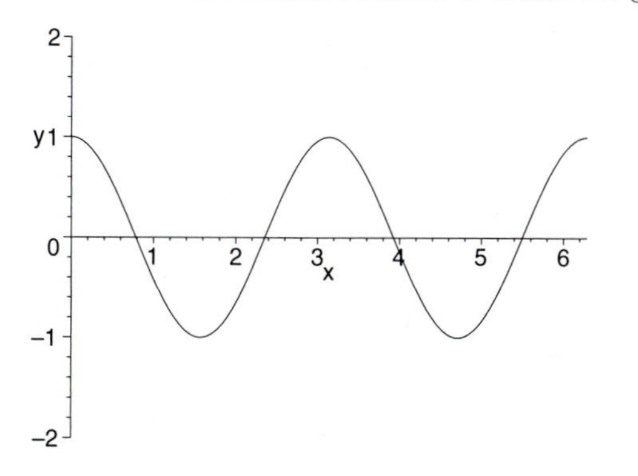

37. $f(x) = x + cos\ 2x$ First, find the critical points.

$f'(x) = 1 - 2sin\ 2x$

$f'(x)$ exists for all x values. We solve $f'(x) = 0$,

$$\begin{aligned} f'(x) &= 0 \\ 1 - 2sin\ 2x &= 0 \\ sin\ 2x &= \frac{1}{2} \\ x &= \frac{\pi}{12} \end{aligned}$$

and

$$x = \frac{5\pi}{12}$$

and

$$x = \frac{13\pi}{12}$$

and

$$x = \frac{17\pi}{12}$$

So the only critical points are at $x = \frac{\pi}{12}$, $x = \frac{5\pi}{12}$, $x = \frac{13\pi}{12}$, and $x = \frac{17\pi}{12}$ and there might be extrema points at the end points $x = 0$ and $x = 2\pi$. We use them to divide the real number line into five intervals, A: $[0, \frac{\pi}{12})$, B: $(\frac{\pi}{12}, \frac{5\pi}{12})$, C: $(\frac{5\pi}{12}, \frac{13\pi}{12})$, D: $(\frac{13\pi}{12}, \frac{17\pi}{12})$, and E: $(\frac{17\pi}{12}, 2\pi]$.

A B C D E

0 $\frac{\pi}{12}$ $\frac{5\pi}{12}$ $\frac{13\pi}{12}$ $\frac{17\pi}{12}$ 2π

We use a test value in each interval to determine the sign of the derivative in each interval.

A: Test 0, $f'(0) = 1 - 2sin(2 \cdot 0) = 1 > 0$

B: Test $\frac{\pi}{4}$, $f'(\frac{\pi}{4}) = 1 - 2sin(2 \cdot \frac{\pi}{4}) = -1 < 0$

C: Test π, $f'(\pi) = 1 - 2sin(2 \cdot \pi) = 1 > 0$

D: Test $\frac{15\pi}{12}$, $f'(\frac{15\pi}{12}) = 1 - 2sin(2 \cdot \frac{15\pi}{12}) = -1 < 0$

E: Test $2\pi f'(2\pi) = 1 - 2sin(2 \cdot 2\pi) = 1 > 0$

We see that $f(x)$ is increasing on $[0, \frac{\pi}{12})$, $(\frac{5\pi}{12}, \frac{13\pi}{12})$, and $(\frac{17\pi}{12}, 2\pi]$ and decreasing on $(\frac{\pi}{12}, \frac{5\pi}{12})$, $(\frac{13\pi}{12} and \frac{17\pi}{12})$ so the function has a relative maximum at $x = \frac{\pi}{12}$ and $x = \frac{13\pi}{12}$ and a relative minimum at $x = \frac{5\pi}{12}$, $x = \frac{17\pi}{12}$, and $x = 2\pi$.

We find $f(\frac{\pi}{12})$:

$$\begin{aligned} f(\frac{\pi}{12}) &= \frac{\pi}{12} + cos(2 \cdot \frac{\pi}{12}) \\ &\approx 1.128 \end{aligned}$$

We find $f(\frac{5\pi}{12})$:

$$\begin{aligned} f(\frac{5\pi}{12}) &= \frac{5\pi}{12} + cos(2 \cdot \frac{5\pi}{12}) \\ &\approx 0.443 \end{aligned}$$

We find $f(\frac{13\pi}{12})$:

$$\begin{aligned} f(\frac{13\pi}{12}) &= \frac{13\pi}{12} + cos(2 \cdot \frac{13\pi}{12}) \\ &\approx 4.269 \end{aligned}$$

We find $f(\frac{17\pi}{12})$:

$$\begin{aligned} f(\frac{17\pi}{12}) &= \frac{17\pi}{12} + cos(2 \cdot \frac{17\pi}{12}) \\ &\approx 3.585 \end{aligned}$$

We find $f(2\pi)$:

$$\begin{aligned} f(2pi) &= 2\pi + cos(2 \cdot 2\pi) \\ &\approx 7.2832 \end{aligned}$$

There is a relative maximum at $(\frac{\pi}{12}, 1.128)$ and $(\frac{13\pi}{12}, 4.269)$ and a relative minimum at $(\frac{5\pi}{12}, 0.443)$, $(\frac{17\pi}{12}, 3.585)$, and

$(2\pi, 7.283)$. We use the information obtained to sketch the graph.

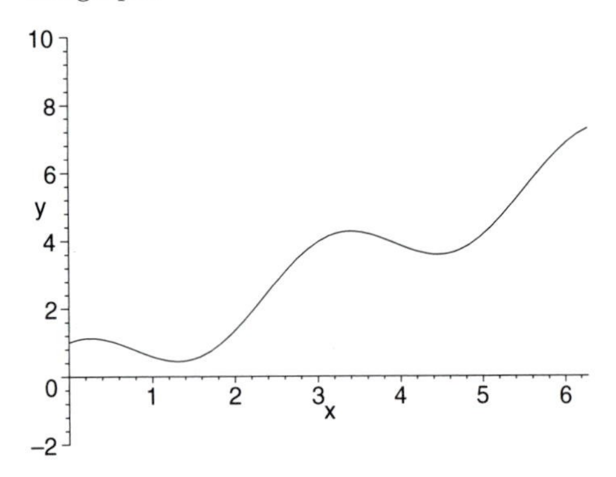

39. $f(x) = \frac{x}{3} + cos\frac{2x}{3}$. Find the critical values
$f'(x) = \frac{1}{3} - \frac{2}{3}\ sin\ \frac{2x}{3}$ We solve $f'(x) = 0$

$$\begin{aligned}
\frac{1}{3} - \frac{2}{3}\ sin\ \frac{2x}{3} &= 0 \\
\frac{2}{3} sin\ \frac{2x}{3} &= \frac{1}{3} \\
sin\ \frac{2x}{3} &= \frac{1}{2} \\
\frac{2x}{3} &= sin^{-1}(\frac{1}{2}) \\
\frac{2x}{3} &= \frac{pi}{6} \\
x &= \frac{\pi}{4}
\end{aligned}$$

and

$$\begin{aligned}
\frac{2x}{3} &= \frac{5\pi}{6} \\
x &= \frac{5\pi}{4}
\end{aligned}$$

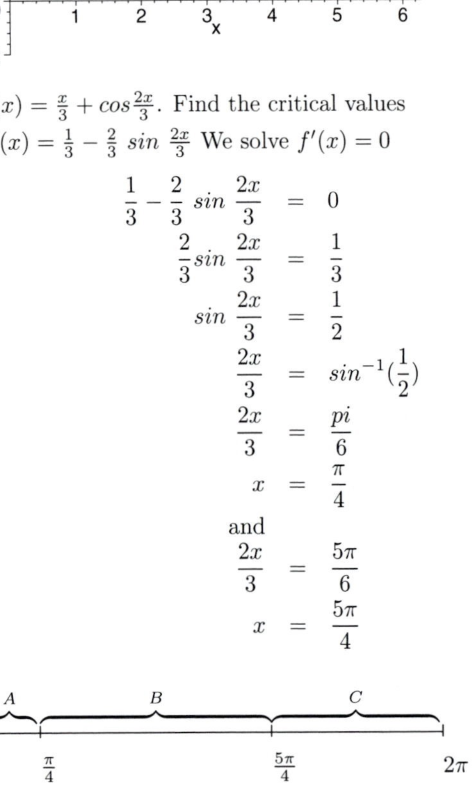

A B C

$0 \quad \frac{\pi}{4} \quad \frac{5\pi}{4} \quad 2\pi$

We use a test value in each interval to determine the sign of the derivative in each interval.

A: Test $\frac{\pi}{6}$, $f'(\frac{\pi}{6}) = \frac{1}{3} - \frac{2}{3} sin\ \frac{\pi}{9} = 0.105 > 0$

B: Test π, $f'(\pi) = \frac{1}{3} - \frac{2}{3} sin \frac{2\pi}{3} = -0.323 < 0$

C: Test $\frac{3\pi}{2}$, $f'(\frac{5\pi}{3}) = \frac{1}{3} - \frac{2}{3} sin\ \pi = 0.333 > 0$

We see that $f(x)$ is increasing on $[0, \frac{\pi}{4})$ and $(\frac{5\pi}{4}, 2\pi])$ and decreasing on $(\frac{\pi}{4}, \frac{5\pi}{4})$ so the function has a relative maximum at $x = \frac{\pi}{4}$ and a relative minimum at $x = \frac{5pi}{4}$. We find $f(\frac{\pi}{4})$

$$\begin{aligned}
f(\frac{\pi}{4}) &= \frac{\pi}{12} + cos(\frac{\pi}{6}) \\
&= 1.128
\end{aligned}$$

We find $f(\frac{5\pi}{4})$

$$\begin{aligned}
f(\frac{5\pi}{4}) &= \frac{5\pi}{12} + cos(\frac{5\pi}{12}) \\
&= 0.443
\end{aligned}$$

There is a relative maximum at $(\frac{\pi}{4}, 1.128)$ and a relative minimum at $(\frac{5\pi}{4}, 0.443)$. We sktch the graph

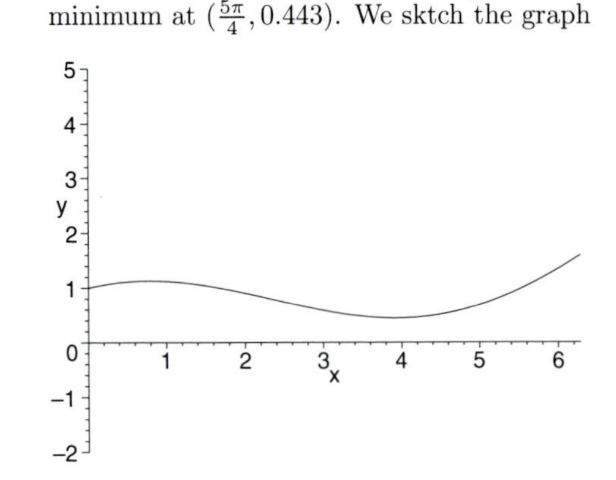

41. $f(x) = \frac{cos\ x}{2 - sin\ x}$

Find the critical values

$$\begin{aligned}
f'(x) &= \frac{(2 - sin\ x)(-sin\ x) - cos\ x(-cos\ x)}{(2 - sin\ x)^2} \\
&= \frac{-2\ sin\ x + sin^2 x + cos^2 x}{(2 - sin\ x)^2} \\
&= \frac{1 - 2\ sin\ x}{(2 - sin\ x)^2}
\end{aligned}$$

We solve $f'(x) = 0$

$$\begin{aligned}
\frac{1 - 2\ sin\ x}{(2 - sin\ x)^2} &= 0 \\
1 - 2\ sin\ x &= 0 \\
sin\ x &= \frac{1}{2} \\
x &= \frac{\pi}{6}
\end{aligned}$$

and

$$x = \frac{5\pi}{6}$$

A B C

$0 \quad \frac{\pi}{6} \quad \frac{5\pi}{6} \quad 2\pi$

We use a test value in each interval to determine the sign of the derivative in each interval.

A: Test $\frac{\pi}{18}$, $f'(\frac{\pi}{18}) = 0.196 > 0$

B: Test $\frac{\pi}{4}$, $f'(\frac{\pi}{4}) = -0.248 < 0$

C: Test $\frac{11\pi}{12}$, $f'(\frac{11\pi}{12}) = 0.159 > 0$

We see that $f(x)$ is increasing on $[0, \frac{\pi}{6})$ and $(\frac{5\pi}{6}, 2\pi])$ and decreasing on $(\frac{\pi}{6}, \frac{5\pi}{6})$ so the function has a relative maximum at $x = \frac{\pi}{3}$ and a relative minimum at $x = \frac{5\pi}{6}$.

We find $f(\frac{\pi}{6})$

$$f(\frac{\pi}{6}) = \frac{cos(\frac{\pi}{6})}{2 - sin(\frac{\pi}{6})}$$

$$= \frac{\frac{\sqrt{3}}{2}}{2-\frac{1}{2}}$$

$$= \frac{\sqrt{3}}{3}$$

We find $f(\frac{5\pi}{6})$

$$\begin{aligned} f(\frac{5\pi}{6}) &= \frac{cos(\frac{5\pi}{6})}{2 - sin(\frac{5\pi}{6})} \\ &= \frac{\frac{-\sqrt{3}}{2}}{2-\frac{1}{2}} \\ &= \frac{-\sqrt{3}}{3} \end{aligned}$$

There is a relative maximum at $(\frac{\pi}{6}, \frac{\sqrt{3}}{3})$ and a relative minimum at $(\frac{5\pi}{6}, -\frac{\sqrt{3}}{3})$. We sketch the graph

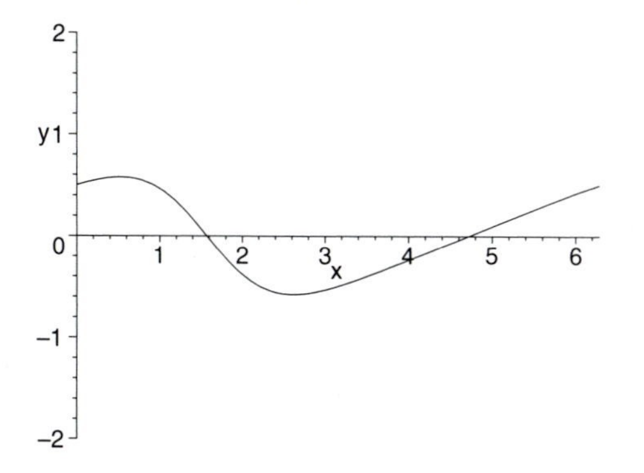

43. $f(x) = sin\ x - sin^2\ x$

Find the critical values

$$\begin{aligned} f'(x) &= cos\ x - 2\ sin\ x\ cos\ x \\ &= cos\ x(1 - 2\ sin\ x) \end{aligned}$$

We solve $f'(x) = 0$

$$\begin{aligned} cos\ x(1 - 2\ sin\ x) &= 0 \\ cos\ x &= 0 \\ x &= \frac{\pi}{2} \\ \text{and} & \\ x &= \frac{3\pi}{2} \\ 1 - 2\ sin\ x &= 0 \\ sin\ x &= \frac{1}{2} \\ x &= \frac{\pi}{6} \\ \text{and} & \\ x &= \frac{5\pi}{6} \end{aligned}$$

A B C D E

$0 \quad \frac{\pi}{6} \quad \frac{\pi}{2} \quad \frac{5\pi}{6} \quad \frac{3\pi}{2} \quad 2\pi$

We use a test value in each interval to determine the sign of the derivative in each interval.

A: Test $\frac{\pi}{18}$, $f'(\frac{\pi}{18}) = 0.643 > 0$

B: Test $\frac{\pi}{4}$, $f'(\frac{\pi}{4}) = -0.293 < 0$

C: Test $\frac{2\pi}{3}$, $f'(\frac{2\pi}{3}) = 0.366 > 0$

D: Test π, $f'(\pi) = -1 < 0$

E: Test $\frac{7\pi}{4}$, $f'(\frac{7\pi}{4}) = 1.707 < 0$

We see that $f(x)$ is increasing on $[0, \frac{\pi}{6})$, $(\frac{\pi}{2}, \frac{5\pi}{6})$, and $(\frac{3\pi}{2}, 2\pi]$ and decreasing on $(\frac{\pi}{6}, \frac{\pi}{2})$ and $(\frac{5\pi}{6}, \frac{3\pi}{2})$ so the function has a relative maximum at $x = \frac{\pi}{6}$ and $x = \frac{5\pi}{6}$ and a relative minimum at $x = \frac{\pi}{2}$ and $x = \frac{3\pi}{2}$.

We find $f(\frac{\pi}{6})$

$$\begin{aligned} f(\frac{\pi}{6}) &= sin(\frac{\pi}{6}) - sin^2(\frac{\pi}{6}) \\ &= \frac{1}{2} - \frac{1}{4} \\ &= \frac{1}{4} \end{aligned}$$

We find $f(\frac{5\pi}{6})$

$$\begin{aligned} f(\frac{5\pi}{6}) &= sin(\frac{5\pi}{6}) - sin^2(\frac{5\pi}{6}) \\ &= \frac{1}{2} - \frac{1}{4} \\ &= \frac{1}{4} \end{aligned}$$

We find $f(\frac{\pi}{2})$

$$\begin{aligned} f(\frac{\pi}{2}) &= sin(\frac{\pi}{2}) - sin^2(\frac{\pi}{2}) \\ &= 1 - 1 \\ &= 0 \end{aligned}$$

We find $f(\frac{3\pi}{2})$

$$\begin{aligned} f(\frac{3\pi}{2}) &= sin(\frac{3\pi}{2}) - sin^2(\frac{3\pi}{2}) \\ &= -1 - 1 \\ &= -2 \end{aligned}$$

There is a relative maximum at $(\frac{\pi}{6}, \frac{1}{4})$ and $(\frac{5\pi}{6}, \frac{1}{4})$ and a relative minimum at $(\frac{\pi}{2}, 0)$ and $(\frac{3\pi}{2}, -2)$. We sketch the graph

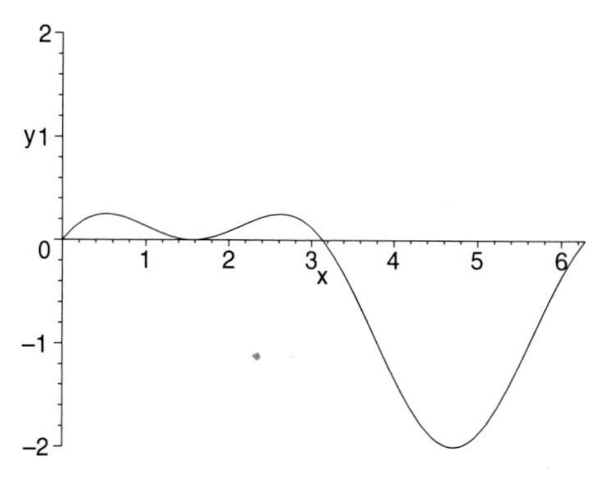

45. $f(x) = 9sin\ x - 4\ sin^3\ x$

Find the critical values

$$\begin{aligned} f'(x) &= 9\ cos\ x - 12\ sin^2x\ cos\ x \\ &= 3\ cos\ x(3 - 4\ sin^2x) \end{aligned}$$

We solve $f'(x) = 0$

$$\begin{aligned} 3\ cos\ x(3 - 4\ sin^2x) &= 0 \\ cos\ x &= 0 \\ x &= \frac{\pi}{2} \\ \text{and} & \\ x &= \frac{3\pi}{2} \\ 3 - 4\ sin^2x &= 0 \\ sin^2x &= \frac{3}{4} \\ sin\ x &= \pm\frac{\sqrt{3}}{2} \\ x &= \frac{\pi}{3} \\ \text{and} & \\ x &= \frac{2\pi}{3} \\ \text{and} & \\ x &= \frac{5\pi}{3} \end{aligned}$$

A B C D E F G

$0 \quad \frac{\pi}{3} \quad \frac{\pi}{2} \quad \frac{2\pi}{3} \quad \frac{4\pi}{3} \quad \frac{3\pi}{2} \quad \frac{5\pi}{3} \quad 2\pi$

We use a test value in each interval to determine the sign of the derivative in each interval.

A: Test $\frac{\pi}{4}$, $f'(\frac{\pi}{4}) = 2.121 > 0$

B: Test $\frac{5\pi}{12}$, $f'(\frac{5\pi}{12}) = -0.568 < 0$

C: Test $\frac{7\pi}{12}$, $f'(\frac{7\pi}{12}) = 0.568 > 0$

D: Test π, $f'(\pi) = -9 < 0$

E: Test $\frac{17\pi}{12}$, $f'(\frac{17\pi}{12}) = 0.568 > 0$

F: Test $\frac{19\pi}{12}$, $f'(\frac{19\pi}{12}) = -\ 0.568 < 0$

G: Test $\frac{11\pi}{6}$, $f'(\frac{11\pi}{6}) = 5.760 > 0$

We see that $f(x)$ is increasing on $[0, \frac{\pi}{3})$, $(\frac{\pi}{2}, \frac{2\pi}{3})$, $(\frac{4\pi}{3}, \frac{3\pi}{2})$ and $(\frac{5\pi}{3}, 2\pi]$ and decreasing on $(\frac{\pi}{3}, \frac{\pi}{2})$, $(\frac{2\pi}{3}, \frac{4\pi}{3})$ and $(\frac{3\pi}{2}, \frac{5\pi}{3})$ so the function has a relative maximum at $x = \frac{\pi}{3}$, $x = \frac{2\pi}{3}$ and $x = \frac{3\pi}{2}$ and a relative minimum at $x = \frac{\pi}{2}$, $x = \frac{4\pi}{3}$ and $x = \frac{5\pi}{3}$.

We find $f(\frac{\pi}{3})$

$$\begin{aligned} f(\frac{\pi}{3}) &= 9\ sin(\frac{\pi}{3}) - 4\ sin^3(\frac{\pi}{3}) \\ &= \frac{9\sqrt{3}}{2} - \frac{12\sqrt{3}}{8} \\ &= 3\sqrt{3} \end{aligned}$$

We find $f(\frac{2\pi}{3})$

$$\begin{aligned} f(\frac{2\pi}{3}) &= 9\ sin(\frac{2\pi}{3}) - 4\ sin^3(\frac{2\pi}{3}) \\ &= \frac{9\sqrt{3}}{2} - \frac{12\sqrt{3}}{8} \\ &= 3\sqrt{3} \end{aligned}$$

We find $f(\frac{3\pi}{2})$

$$\begin{aligned} f(\frac{3\pi}{2}) &= 9\ sin(\frac{3\pi}{2}) - 4\ sin^3(\frac{3\pi}{2}) \\ &= -9 - 4(-1) \\ &= -5 \end{aligned}$$

We find $f(\frac{\pi}{2})$

$$\begin{aligned} f(\frac{\pi}{2}) &= 9\ sin(\frac{\pi}{2}) - 4\ sin^3(\frac{\pi}{2}) \\ &= 9 - 4 \\ &= 5 \end{aligned}$$

We find $f(\frac{4\pi}{3})$

$$\begin{aligned} f(\frac{4\pi}{3}) &= 9\ sin(\frac{4\pi}{3}) - 4\ sin^3(\frac{4\pi}{3}) \\ &= -\frac{9\sqrt{3}}{2} + \frac{12\sqrt{3}}{8} \\ &= -3\sqrt{3} \end{aligned}$$

We find $f(\frac{5\pi}{3})$

$$\begin{aligned} f(\frac{5\pi}{3}) &= 9\ sin(\frac{5\pi}{3}) - 4\ sin^3(\frac{5\pi}{3}) \\ &= -\frac{9\sqrt{3}}{2} + \frac{12\sqrt{3}}{8} \\ &= -3\sqrt{3} \end{aligned}$$

There is a relative maximum at $(\frac{\pi}{3}, 3\sqrt{3})$, $(\frac{2\pi}{3}, 3\sqrt{3})$ and $(\frac{3\pi}{2}, -5)$ and a relative minimum at $(\frac{\pi}{2}, 5)$,$(\frac{4\pi}{3}, -3\sqrt{3})$, and $(\frac{5\pi}{3}, -3\sqrt{3})$. We sketch the graph

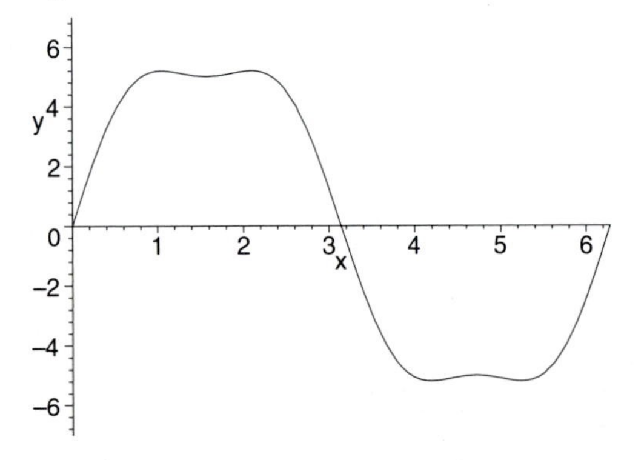

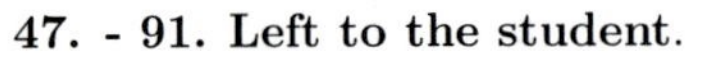

47. - 91. Left to the student.

93. $h(d) = -0.002d^2 + 0.8d + 6.6$

$h'(d) = -0.004d + 0.8$

Solve $h'(d) = 0$.

$$\begin{aligned} -0.004d + 0.8 &= 0 \\ -0.004d &= -0.8 \\ d &= 20 \end{aligned}$$

A: Test 100, $f'(50) = 0.4 > 0$

B: Test 300, $f'(50) = -0.4 < 0$

The function is increasing on to the left of $d = 200$ and decreasing to the right of $d = 200$ therefore there is a relative maximum at $d = 200$. We find $h(200)$

$$\begin{aligned} h(200) &= -0.002(200)^2 + 0.8(200) + 6.6 \\ &= -80 + 160 + 6.6 \\ &= 86.6 \end{aligned}$$

There is a relative maximum at $(200, 86.6)$.

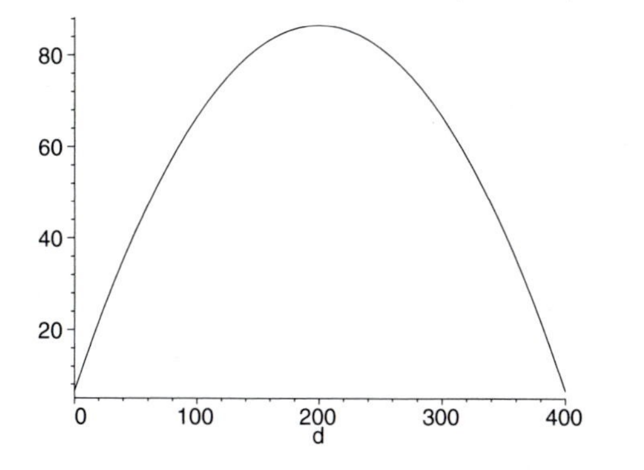

95. $T(t) = -0.1t^2 + 1.2t + 98.6,\ 0 \le t \le 12$

$T'(t) = -0.2t + 1.2$

$T'(t)$ exists for all real numbers. Solve $T'(t) = 0$.

$$\begin{aligned} -0.2t + 1.2 &= 0 \\ -0.2t &= -1.2 \\ t &= 6 \end{aligned}$$

The only critical point is at $t = 6$. We use it to divide the interval $[0, 12]$ (the domain of $T(t)$) into two intervals, A: $[0, 6)$ and B: $(6, 12]$.

A: Test 0, $T'(0) = -0.2(0) + 1.2 = 1.2 > 0$

B: Test 7, $T'(7) = -0.2(7) + 1.2 = -0.2 < 0$

Since $T(t)$ is increasing on $[0, 6)$ and decreasing on $(6, 12]$, there is a relative maximum at $x = 6$.

$$T(6) = -0.1(6)^2 + 1.2(6) + 98.6 = 102.2$$

There is a relative maximum at $(6, 102.2°)$. We sketch the graph.

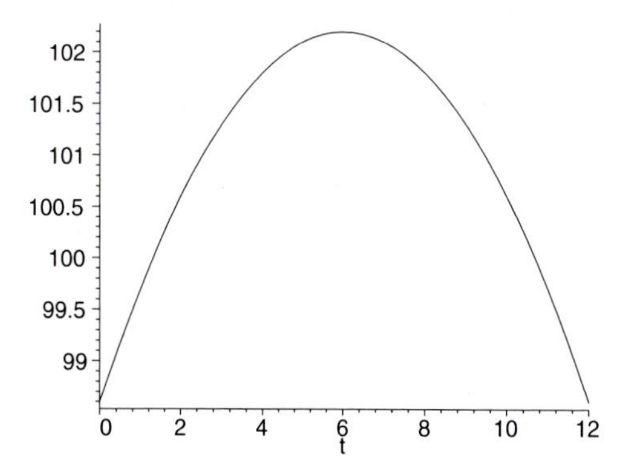

97. Let us consider the intervals $[a, b]$, $[c, d]$ and consider the sign of the slope of the tangent line. Next, consider the intervals $[b, c]$ and $[d, e]$ and again consider the sign of the slope of the tangent line. The sign of the slope of the tangent line at a point, which is the value of the derivative of the function at that point, informs us whether the function is increasing or decreasing.

99. The crtical values occur when the slope of the tangent line at the point is either zero or undefined (which usually indicates a maximum or a minimum).
$(2, 1)$ a relative minimum
$(4, 7)$ a relative maximum

101. Relative minimum at $(-5, 425)$ and $(4, -304)$ and a relative maximum at $(-2, 560)$.

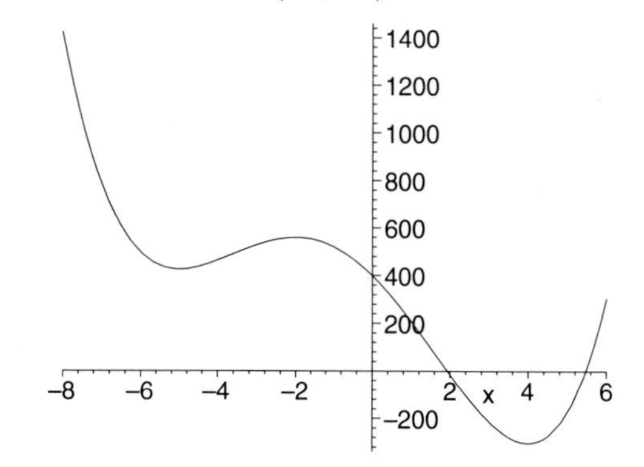

Exercise Set 3.2

1. $f(x) = 2 - x^2$

a) Find $f'(x)$ and $f''(x)$.

$f'(x) = -2x$

$f''(x) = -2$

b) Find the critical points.

Since $f'(x)$ exists for all values of x, the only critical points are where $-2x = 0$.

$-2x = 0$

$x = 0$

$f(0) = 2 - 0^2 = 2$

This gives the point $(0, 2)$ on the graph.

c) Use the Second-Derivative Test:

$f''(x) = -2$

$f''(0) = -2 < 0$

This tells us that $(0, 2)$ is a relative maximum. Then we can deduce that $f(x)$ is increasing on $(-\infty, 0)$ and decreasing on $(0, \infty)$.

The second derivative, $f''(x)$, exists and is -2 for all real numbers. Note that $f''(x)$ is never 0. Thus, there are no possible inflection points.

Since $f''(x)$ is always negative ($f''(x) = -2$), f is concave down on the interval $(-\infty, \infty)$.

d) Sketch the graph using the preceding information. By solving $2 - x^2 = 0$ we can easily find the x-intercepts. They are $(-\sqrt{2}, 0)$ and $(\sqrt{2}, 0)$.

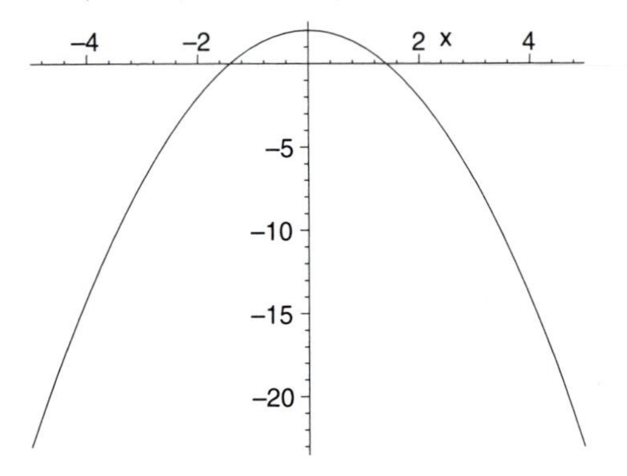

3. $f(x) = 2x^3 - 3x^2 - 36x + 28$

a) Find $f'(x)$ and $f''(x)$.

$f'(x) = 6x^2 - 6x - 36$

$f''(x) = 12x - 6$

b) Find the critical points of f.

Since $f'(x)$ exists for all values of x, the only critical points are where $6x^2 - 6x - 36 = 0$.

$6x^2 - 6x - 36 = 0$

$6(x + 2)(x - 3) = 0$

$x + 2 = 0 \quad or \quad x - 3 = 0$

$x = -2 \quad or \quad x = 3$ Critical points

Then $f(-2) = 2(-2)^3 - 3(-2)^2 - 36(-2) + 28$

$= -16 - 12 + 72 + 28$

$= 72,$

and $f(3) = 2 \cdot 3^3 - 3 \cdot 3^2 - 36 \cdot 3 + 28$

$= 54 - 27 - 108 + 28$

$= -53.$

These give the points $(-2, 72)$ and $(3, -53)$ on the graph.

c) Use the Second-Derivative Test:

$f''(-2) = 12(-2) - 6 = -30 < 0$, so $(-2, 72)$ is a relative maximum.

$f''(3) = 12 \cdot 3 - 6 = 30 > 0$, so $(3, -53)$ is a relative minimum.

Then if we use the points -2 and 3 to divide the real number line into three intervals, $(-\infty, -2)$, $(-2, 3)$, and $(3, \infty)$, we know that f is increasing on $(-\infty, -2)$, decreasing on $(-2, 3)$, and increasing again on $(3, \infty)$.

d) Find the possible inflection points.

$f''(x)$ exists for all valus of x, so we solve $f''(x) = 0$.

$12x - 6 = 0$

$12x = 6$

$x = \frac{1}{2}$ Possible inflection point

Then $f\left(\frac{1}{2}\right) = 2\left(\frac{1}{2}\right)^3 - 3\left(\frac{1}{2}\right)^2 - 36\left(\frac{1}{2}\right) + 28$

$= \frac{1}{4} - \frac{3}{4} - 18 + 28$

$= \frac{19}{2}.$

This gives the point $\left(\frac{1}{2}, \frac{19}{2}\right)$ on the graph.

e) To determine the concavity we use the possible inflection point, $\frac{1}{2}$, to divide the real number line into two invervals, A: $\left(-\infty, \frac{1}{2}\right)$ and B: $\left(\frac{1}{2}, \infty\right)$. Test a point in each interval.

A: Test 0, $f''(0) = 12 \cdot 0 - 6 = -6 < 0$

B: Test 1, $f''(1) = 12 \cdot 1 - 6 = 6 > 0$

Then f is concave down on $\left(-\infty, \frac{1}{2}\right)$ and concave up on $\left(\frac{1}{2}, \infty\right)$, so $\left(\frac{1}{2}, \frac{19}{2}\right)$ is an inflection point.

f) Sketch the graph using the preceding information.

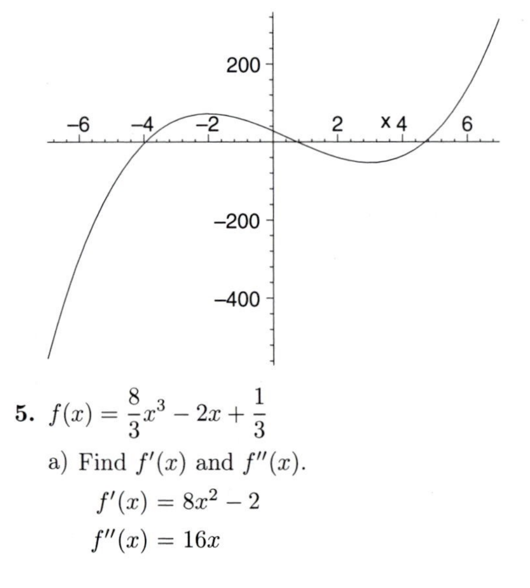

5. $f(x) = \frac{8}{3}x^3 - 2x + \frac{1}{3}$

a) Find $f'(x)$ and $f''(x)$.

$f'(x) = 8x^2 - 2$

$f''(x) = 16x$

b) Find the critical points of f.

Now $f'(x) = 8x^2 - 2$ exists for all values of x, so the only critical points of f are where $8x^2 - 2 = 0$.

$$8x^2 - 2 = 0$$
$$8x^2 = 2$$
$$x^2 = \frac{2}{8}$$
$$x^2 = \frac{1}{4}$$
$$x = \pm\frac{1}{2} \quad \text{Critical points}$$

Then $f\left(-\frac{1}{2}\right) = \frac{8}{3}\left(-\frac{1}{2}\right)^3 - 2\left(-\frac{1}{2}\right) + \frac{1}{3}$

$= -\frac{1}{3} + 1 + \frac{1}{3}$

$= 1,$

and $f\left(\frac{1}{2}\right) = \frac{8}{3}\left(\frac{1}{2}\right)^3 - 2\left(\frac{1}{2}\right) + \frac{1}{3}$

$= \frac{1}{3} - 1 + \frac{1}{3}$

$= -\frac{1}{3}$

These give the points $\left(-\frac{1}{2}, 1\right)$ and $\left(\frac{1}{2}, -\frac{1}{3}\right)$ on the graph.

c) Use the Second-Derivative Test:

$f''\left(-\frac{1}{2}\right) = 16\left(-\frac{1}{2}\right) - 8 < 0$, so $\left(-\frac{1}{2}, 1\right)$ is a relative maximum.

$f''\left(\frac{1}{2}\right) = 16 \cdot \frac{1}{2} = 8 > 0$, so $\left(\frac{1}{2}, -\frac{1}{3}\right)$ is a relative minimum.

Then if we use the points $-\frac{1}{2}$ and $\frac{1}{2}$ to divide the real number line into three intervals, A: $\left(-\infty, -\frac{1}{2}\right)$, B: $\left(-\frac{1}{2}, \frac{1}{2}\right)$, and C: $\left(\frac{1}{2}, \infty\right)$, we know that f is increasing on $\left(-\infty, -\frac{1}{2}\right)$, decreasing on $\left(-\frac{1}{2}, \frac{1}{2}\right)$, and increasing again on $\left(\frac{1}{2}, \infty\right)$.

d) Find the possible inflection points.

Now $f''(x) = 16x$ exists for all values of x, so the only critical points of f' are where $16x = 0$.

$$16x = 0$$
$$x = 0 \quad \text{Possible inflection point}$$

Then $f(0) = \frac{8}{3} \cdot 0^3 - 2 \cdot 0 + \frac{1}{3} = \frac{1}{3}$, so $\left(0, \frac{1}{3}\right)$ is another point on the graph.

e) To determine the concavity we use the possible inflection point, 0, to divide the real number line into two invervals, A: $(-\infty, 0)$ and B: $(0, \infty)$. Test a point in each interval.

A: Test -1, $f''(-1) = 16(-1) = -16 < 0$

B: Test 1, $f''(1) = 16 \cdot 1 = 16 > 0$

Then $f'(x)$ is concave down on $(-\infty, 0)$ and concave up on $(0, \infty)$, so $\left(0, \frac{1}{3}\right)$ is an inflection point.

f) Sketch the graph using the preceding information.

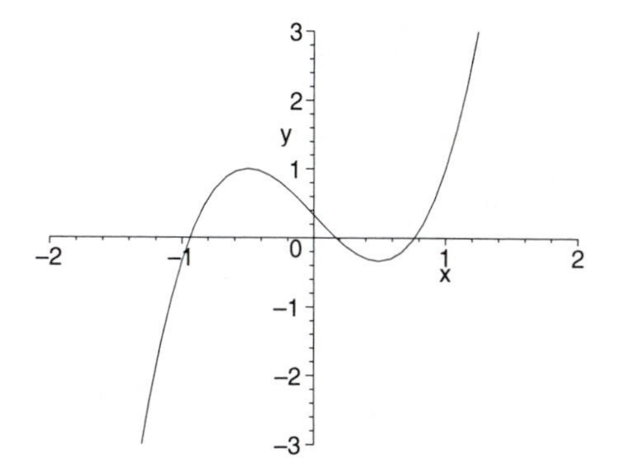

7. $f(x) = 3x^4 - 16x^3 + 18x^2$

a) $f'(x) = 12x^3 - 48x^2 + 36x$

$f''(x) = 36x^2 - 96x + 36$

b) Since $f'(x)$ exists for all values of x, the only critical points are where $f'(x) = 0$.

$$12x^3 - 48x^2 + 36x = 0$$
$$12x(x^2 - 4x + 3) = 0$$
$$12x(x - 1)(x - 3) = 0$$
$$12x = 0 \quad or \quad x - 1 = 0 \quad or \quad x - 3 = 0$$
$$x = 0 \quad or \quad x = 1 \quad or \quad x = 3$$

Then $f(0) = 3 \cdot 0^4 - 16 \cdot 0^3 + 18 \cdot 0^2 = 0$,

$f(1) = 3 \cdot 1^4 - 16 \cdot 1^3 + 18 \cdot 1^2 = 5$,

and $f(3) = 3 \cdot 3^4 - 16 \cdot 3^3 + 18 \cdot 3^2 = -27$.

These give the points $(0, 0)$, $(1, 5)$, and $(3, -27)$ on the graph.

c) Use the Second-Derivative Test:

$f''(0) = 36 \cdot 0^2 - 96 \cdot 0 + 36 = 36 > 0$, so $(0, 0)$ is a relative minimum.

$f''(1) = 36 \cdot 1^2 - 96 \cdot 1 + 36 = -24 < 0$, so $(1, 5)$ is a relative maximum.

$f''(3) = 36 \cdot 3^2 - 96 \cdot 3 + 36 = 72 > 0$, so $(3, -27)$ is a relative minimum.

Then if we use the points 0, 1, and 3 to divide the real number line into four intervals, $(-\infty, 0)$, $(0, 1)$, $(1, 3)$, and $(3, \infty)$, we know that f is decreasing on $(-\infty, 0)$ and on $(1, 3)$ and is increasing on $(0, 1)$ and $(3, \infty)$.

d) $f''(x)$ exists for all values of x, so the only possible inflection points are where $f''(x) = 0$.

$$36x^2 - 96x + 36 = 0$$
$$12(3x^2 - 8x + 3) = 0$$
$$3x^2 - 8x + 3 = 0$$

Using the quadratic formula, we find $x = \dfrac{4 \pm \sqrt{7}}{3}$, so $x \approx 0.45$ or $x \approx 2.22$ are possible inflection points.

Then $f(0.45) \approx 2.31$ and $f(2.22) \approx -13.48$, so $(0.45, 2.31)$ and $(2.22, -13.48)$ are two more points on the graph.

e) To determine the concavity we use the points 0.45 and 2.22 to divide the real number line into three invervals, A: $(-\infty, 0.45)$, B: $(0.45, 2.22)$, and C: $(2.22, \infty)$. Test a point in each interval.

A: Test 0, $f''(0) = 36 \cdot 0^2 - 96 \cdot 0 + 36 = 36 > 0$

B: Test 1, $f''(1) = 36 \cdot 1^2 - 96 \cdot 1 + 36 = -24 < 0$

C: Test 3, $f''(3) = 36 \cdot 3^2 - 96 \cdot 3 + 36 = 72 > 0$

Then f is concave up on $(-\infty, 0.45)$, concave down on $(0.45, 2.22)$, and concave up on $(2.22, \infty)$, so $(0.45, 2.31)$ and $(2.22, -13.48)$ are inflection points.

f) Sketch the graph.

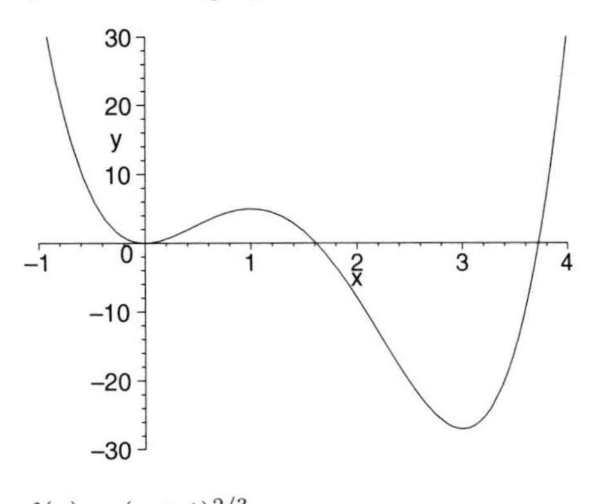

9. $f(x) = (x+1)^{2/3}$

a) $f'(x) = \dfrac{2}{3}(x+1)^{-1/3} = \dfrac{2}{3\sqrt[3]{x+1}}$

$f''(x) = -\dfrac{2}{9}(x+1)^{-4/3} = -\dfrac{2}{9\sqrt[3]{(x+1)^4}}$

b) Since $f'(-1)$ does not exist, -1 is a critical point. The equation $f'(x) = 0$ has no solution, so the only critical point is -1.

Now $f(-1) = (-1+1)^{2/3} = 0^{2/3} = 0$.

This gives the point $(-1, 0)$ on the graph.

c) We cannot use the Second-Derivative Test, because $f''(-1)$ is not defined. We will use the First-Derivative Test. Use -1 to divide the real number line into two intervals, A: $(-\infty, -1)$ and B: $(-1, \infty)$. Test a point in each interval.

A: Test -2, $f'(-2) = \dfrac{2}{3\sqrt[3]{-2+1}} = -\dfrac{2}{3} < 0$

B: Test 0, $f'(0) = \dfrac{2}{3\sqrt[3]{0+1}} = \dfrac{2}{3} > 0$

Since $f(x)$ is decreasing on $(-\infty, -1)$ and increasing on $(-1, \infty)$, there is a relative minimum at $(-1, 0)$.

d) Since $f''(-1)$ does not exist, -1 is a possible inflection point. The equation $f''(x) = 0$ has no solution, so the only possible inflection point is -1. We have already found $f(-1)$ in step (b).

e) To determine the concavity we use -1 to divide the real number line into two invervals as in step (c). Test a point in each interval.

A: Test -2, $f''(-2) = -\dfrac{2}{9\sqrt[3]{(-2+1)^4}} = -\dfrac{2}{9} < 0$

B: Test 0, $f''(0) = -\dfrac{2}{9\sqrt[3]{(0+1)^4}} = -\dfrac{2}{9} < 0$

Then f is concave down on both intervals, so there is no inflection point.

f) Sketch the graph using the preceding information.

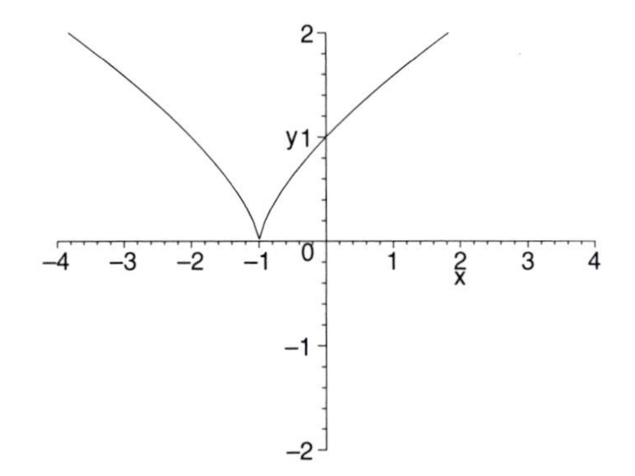

11. $f(x) = x^4 - 6x^2$

a) $f'(x) = 4x^3 - 12x$

$f''(x) = 12x^2 - 12$

b) Since $f'(x)$ exists for all values of x, the only critical points are where $4x^3 - 12x = 0$.

$$4x^3 - 12x = 0$$
$$4x(x^2 - 3) = 0$$
$$4x = 0 \quad or \quad x^2 - 3 = 0$$
$$x = 0 \quad or \quad x^2 = 3$$
$$x = 0 \quad or \quad x = \pm\sqrt{3}$$

The critical points are $-\sqrt{3}$, 0, and $\sqrt{3}$.

$$f(-\sqrt{3}) = (-\sqrt{3})^4 - 6(-\sqrt{3})^2 = 9 - 6 \cdot 3 = 9 - 18 = -9$$
$$f(0) = 0^4 - 6 \cdot 0^2 = 0 - 0 = 0$$
$$f(\sqrt{3}) = (\sqrt{3})^4 - 6(\sqrt{3})^2 = 9 - 6 \cdot 3 = 9 - 18 = -9$$

These give the points $(-\sqrt{3}, -9)$, $(0, 0)$, and $(\sqrt{3}, -9)$ on the graph.

c) Use the Second-Derivative Test:

$f''(-\sqrt{3}) = 12(-\sqrt{3})^2 - 12 = 12 \cdot 3 - 12 =$
$24 > 0$, so $(-\sqrt{3}, -9)$ is a relative minimum.

$f''(0) = 12 \cdot 0^2 - 12 = -12 < 0$, so $(0, 0)$ is a relative maximum.

$f''(\sqrt{3}) = 12(\sqrt{3})^2 - 12 = 12 \cdot 3 - 12 = 24 > 0$,
so $(\sqrt{3}, -9)$ is a relative minimum.

Then if we use the points $-\sqrt{3}$, 0, and $\sqrt{3}$ to divide the real number line into four intervals, $(-\infty, -\sqrt{3})$, $(-\sqrt{3}, 0)$, $(0, \sqrt{3})$, and $(\sqrt{3}, \infty)$, we know that f is decreasing on $(-\infty, -\sqrt{3})$ and on $(0, \sqrt{3})$ and is increasing on $(-\sqrt{3}, 0)$ and on $(\sqrt{3}, \infty)$.

d) Since $f''(x)$ exists for all values of x, the only possible inflection points are where $12x^2 - 12 = 0$.

$$12x^2 - 12 = 0$$
$$x^2 - 1 = 0$$
$$(x+1)(x-1) = 0$$
$$x + 1 = 0 \quad \text{or} \quad x - 1 = 0$$
$$x = -1 \quad \text{or} \quad x = 1$$

The possible inflection points are -1 and 1.

$$f(-1) = (-1)^4 - 6(-1)^2 = 1 - 6 \cdot 1 = -5$$
$$f(1) = 1^4 - 6 \cdot 1^2 = 1 - 6 \cdot 1 = -5$$

These give the points $(-1, -5)$ and $(1, -5)$ on the graph.

e) To determine the concavity we use the points -1 and 1 to divide the real number line into three invervals, A: $(-\infty, -1)$, B: $(-1, 1)$, and $(1, \infty)$. Test a point in each interval.

A: Test -2, $f''(-2) = 12(-2)^2 - 12 = 36 > 0$

B: Test 0, $f''(0) = 12 \cdot 0^2 - 12 = -12 < 0$

C: Test 2, $f''(2) = 12 \cdot 2^2 - 12 = 36 > 0$

We see that f is concave up on the intervals $(-\infty, -1)$ and $(1, \infty)$ and concave down on the interval $(-1, 1)$, so $(-1, -5)$ and $(1, -5)$ are inflection points.

f) Sketch the graph using the preceding information. By solving $x^4 - 6x^2 = 0$ we can find the x-intercepts. They are helpful in graphing.

$$x^4 - 6x^2 = 0$$
$$x^2(x^2 - 6) = 0$$
$$x^2 = 0 \quad \text{or} \quad x^2 - 6 = 0$$
$$x = 0 \quad \text{or} \quad x^2 = 6$$
$$x = 0 \quad \text{or} \quad x = \pm\sqrt{6}$$

The x-intercepts are $(0, 0)$, $(-\sqrt{6}, 0)$, and $(\sqrt{6}, 0)$.

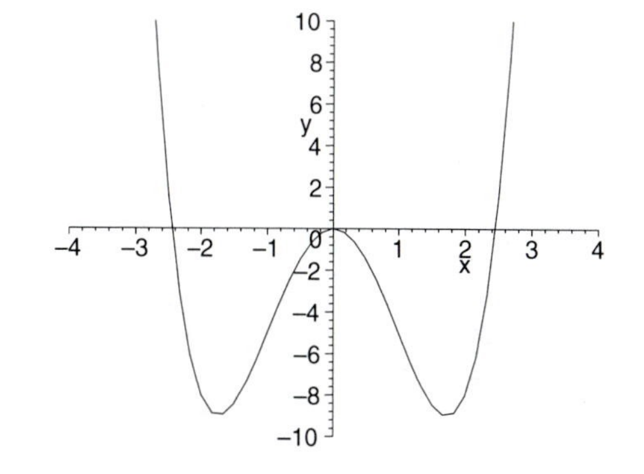

13. $f(x) = 3x^4 + 4x^3$

a) $f'(x) = 12x^3 + 12x^2$
$f''(x) = 36x^2 + 24x$

b) Since $f'(x)$ exists for all values of x, the only critical points of f are where $12x^3 + 12x^2 = 0$.

$$12x^3 + 12x^2 = 0$$
$$12x^2(x+1) = 0$$
$$12x^2 = 0 \quad \textit{or} \quad x + 1 = 0$$
$$x = 0 \quad \textit{or} \quad x = -1$$

The critical points are 0 and -1.

$$f(0) = 3 \cdot 0^4 + 4 \cdot 0^3 = 0 + 0 = 0$$
$$f(-1) = 3(-1)^4 + 4(-1)^3 = 3 \cdot 1 + 4(-1)$$
$$= 3 - 4$$
$$= -1$$

These give the points $(0, 0)$ and $(-1, -1)$ on the graph.

c) Use the Second-Derivative Test:

$f''(-1) = 36(-1)^2 + 24(-1) = 36 - 24 = 12 > 0$, so $(-1, -1)$ is a relative minimum.

$f''(0) = 36 \cdot 0^2 + 24 \cdot 0 = 0$, so this test fails. We will use the First-Derivative Test. Use 0 to divide the interval $(-1, \infty)$ into two intervals: A: $(-1, 0)$ and B: $(0, \infty)$. Test a point in each interval.

A: Test $-\frac{1}{2}$, $f'\left(-\frac{1}{2}\right) = 12\left(-\frac{1}{12}\right)^3 + 12\left(\frac{1}{2}\right)^2 =$
$\frac{3}{2} > 0$

B: Test 2, $f'(2) = 12 \cdot 2^3 + 12 \cdot 2^2 = 144 > 0$

Since f is increasing on both intervals, $(0, 0)$ is not a relative extremum. Since $(-1, -1)$ is a relative minimum, we know that f is decreasing on $(-\infty, -1)$.

d) Now $f''(x)$ exists for all values of x, so the only possible inflection points are where $36x^2 + 24x = 0$.

$$36x^2 + 24x = 0$$
$$12x(3x + 2) = 0$$
$$12x = 0 \quad \text{or} \quad 3x + 2 = 0$$
$$x = 0 \quad \text{or} \quad x = -\frac{2}{3}$$

The possible inflection points are 0 and $-\frac{2}{3}$.

$f(0) = 3 \cdot 0^4 + 4 \cdot 0^3 = 0$ Already found in step (b)

$$f\Big(-\frac{2}{3}\Big) = 3\Big(-\frac{2}{3}\Big)^4 + 4\Big(-\frac{2}{3}\Big)^3$$
$$= 3 \cdot \frac{16}{81} + 4 \cdot \Big(-\frac{8}{27}\Big)$$
$$= \frac{16}{27} - \frac{32}{27}$$
$$= -\frac{16}{27}$$

This gives one additional point $\Big(-\frac{2}{3}, -\frac{16}{27}\Big)$ on the graph.

e) To determine the concavity we use $-\frac{2}{3}$ and 0 to divide the real number line into three invervals, A: $\Big(-\infty, -\frac{2}{3}\Big)$, B: $\Big(-\frac{2}{3}, 0\Big)$, and C: $(0, \infty)$. Test a point in each interval.

A: Test -1, $f''(-1) = 36(-1)^2 + 24(-1) = 12 > 0$

B: Test $-\frac{1}{2}$, $f''\Big(-\frac{1}{2}\Big) = 36\Big(-\frac{1}{2}\Big)^2 + 24\Big(-\frac{1}{2}\Big) = -3 < 0$

C: Test 1, $f''(1) = 36 \cdot 1^2 + 24 \cdot 1 = 60 > 0$

We see that f is concave up on the intervals $\Big(-\infty, -\frac{2}{3}\Big)$ and $(0, \infty)$ and concave down on the interval $\Big(-\frac{2}{3}, 0\Big)$, so $\Big(-\frac{2}{3}, -\frac{16}{27}\Big)$ and $(0, 0)$ are both inflection points.

f) Sketch the graph using the preceding information. By solving $3x^4 + 4x^3 = 0$ we can find x-intercepts. They are helpful in graphing.

$3x^4 + 4x^3 = 0$

$x^3(3x + 4) = 0$

$x^3 = 0$ *or* $3x + 4 = 0$

$x = 0$ *or* $x = -\frac{4}{3}$

The intercepts are $(0, 0)$ and $\Big(-\frac{4}{3}, 0\Big)$.

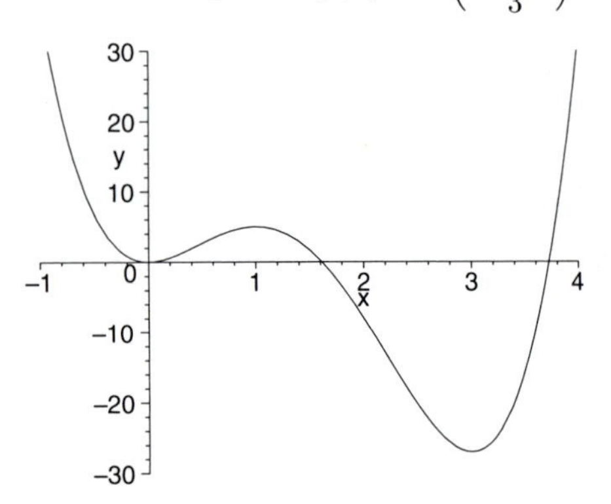

15. $f(x) = x^3 - 6x^2 - 135x$

a) $f'(x) = 3x^2 - 12x - 135$
$f''(x) = 6x - 12$

b) Since $f'(x)$ exists for all values of x, the only critical points of f are where $3x^2 - 12x - 135 = 0$.

$3x^2 - 12x - 135 = 0$

$x^2 - 4x - 45 = 0$

$(x - 9)(x + 5) = 0$

$x - 9 = 0$ *or* $x + 5 = 0$

$x = 9$ *or* $x = -5$

The critical points are 9 and -5.

$$f(9) = 9^3 - 6 \cdot 9^2 - 135 \cdot 9$$
$$= 729 - 486 - 1215$$
$$= -972$$

$$f(-5) = (-5)^3 - 6(-5)^2 - 135(-5)$$
$$= -125 - 150 + 675$$
$$= 400$$

These give the points $(9, -972)$ and $(-5, 400)$ on the graph.

c) Use the Second-Derivative Test:

$f''(-5) = 6(-5) - 12 = -30 - 12 = -42 < 0$, so $(-5, 400)$ is a relative maximum.

$f''(9) = 6 \cdot 9 - 12 = 54 - 12 = 42 > 0$, so $(9, -972)$ is a relative minimum.

Then if we use the points -5 and 9 to divide the real number line into three intervals, $(-\infty, -5)$, $(-5, 9)$, and $(9, \infty)$, we know that f is increasing on $(-\infty, -5)$ and on $(9, \infty)$ and is decreasing on $(-5, 9)$.

d) Now $f''(x)$ exists for all values of x, so the only possible inflection points are where $6x - 12 = 0$.

$6x - 12 = 0$

$6x = 12$

$x = 2$ Possible inflection point

$$f(2) = 2^3 - 6 \cdot 2^2 - 135 \cdot 2$$
$$= 8 - 24 - 270$$
$$= -286$$

This gives another point $(2, -286)$ on the graph.

e) To determine the concavity we use 2 to divide the real number line into two invervals, A: $(-\infty, 2)$ and B: $(2, \infty)$. Test a point in each interval.

A: Test 0, $f''(0) = 6 \cdot 0 - 12 = -12 < 0$

B: Test 3, $f''(3) = 6 \cdot 3 - 12 = 6 > 0$

We see that f is concave down on $(-\infty, 2)$ and concave up on $(2, \infty)$, so $(2, -286)$ is an inflection point.

f) Sketch the graph using the preceding information.

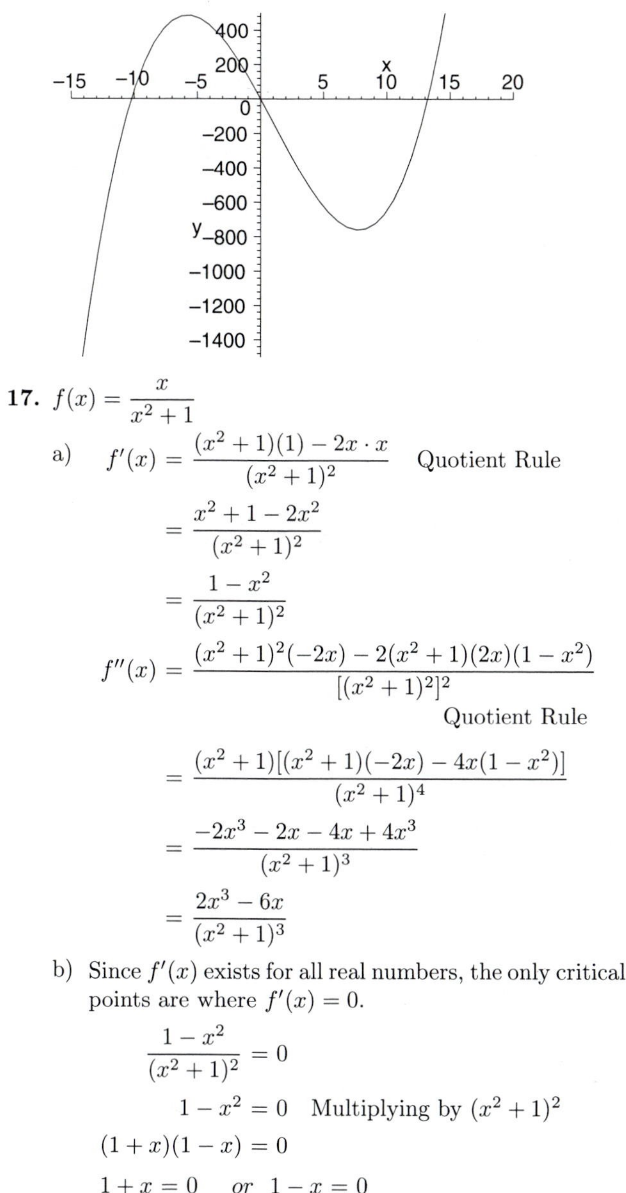

17. $f(x) = \dfrac{x}{x^2+1}$

a) $f'(x) = \dfrac{(x^2+1)(1) - 2x \cdot x}{(x^2+1)^2}$ Quotient Rule

$= \dfrac{x^2+1-2x^2}{(x^2+1)^2}$

$= \dfrac{1-x^2}{(x^2+1)^2}$

$f''(x) = \dfrac{(x^2+1)^2(-2x) - 2(x^2+1)(2x)(1-x^2)}{[(x^2+1)^2]^2}$

Quotient Rule

$= \dfrac{(x^2+1)[(x^2+1)(-2x) - 4x(1-x^2)]}{(x^2+1)^4}$

$= \dfrac{-2x^3-2x-4x+4x^3}{(x^2+1)^3}$

$= \dfrac{2x^3-6x}{(x^2+1)^3}$

b) Since $f'(x)$ exists for all real numbers, the only critical points are where $f'(x) = 0$.

$\dfrac{1-x^2}{(x^2+1)^2} = 0$

$1 - x^2 = 0$ Multiplying by $(x^2+1)^2$

$(1+x)(1-x) = 0$

$1 + x = 0$ *or* $1 - x = 0$

$x = -1$ *or* $1 = x$ Critical points

Then $f(-1) = \dfrac{-1}{(-1)^2+1} = -\dfrac{1}{2}$ and $f(1) = \dfrac{1}{1^2+1} = \dfrac{1}{2}$, so $\left(-1, -\dfrac{1}{2}\right)$ and $\left(1, \dfrac{1}{2}\right)$ are on the graph.

c) Use the Second-Derivative Test:

$f''(-1) = \dfrac{2(-1)^3 - 6(-1)}{[(-1)^2+1]^3}$

$= \dfrac{-2+6}{2^3}$

$= \dfrac{4}{8} = \dfrac{1}{2} > 0$, so $\left(-1, -\dfrac{1}{2}\right)$ is a relative minimum.

$f''(1) = \dfrac{2 \cdot 1^3 - 6 \cdot 1}{[(1)^2+1]^3}$

$= \dfrac{2-6}{2^3}$

$= \dfrac{-4}{8} = -\dfrac{1}{2} < 0$, so $\left(1, \dfrac{1}{2}\right)$ is a relative maximum.

Then if we use -1 and 1 to divide the real number line into three intervals, $(-\infty, -1)$, $(-1, 1)$, and $(1, \infty)$, we know that f is decreasing on $(-\infty, -1)$ and on $(1, \infty)$ and is increasing on $(-1, 1)$.

d) $f''(x)$ exists for all real numbers, so the only possible inflection points are where $f''(x) = 0$.

$\dfrac{2x^3-6x}{(x^2+1)^3} = 0$

$2x(x^2-3) = 0$

$2x = 0$ or $x^2 - 3 = 0$

$x = 0$ or $x^2 = 3$

$x = 0$ or $x = \pm\sqrt{3}$

Possible inflection points

$f(-\sqrt{3}) = \dfrac{-\sqrt{3}}{(-\sqrt{3})^2+1} = -\dfrac{\sqrt{3}}{4}$

$f(0) = \dfrac{0}{0^2+1} = 0$

$f(\sqrt{3}) = \dfrac{\sqrt{3}}{(\sqrt{3})^2+1} = \dfrac{\sqrt{3}}{4}$

These give the points $\left(-\sqrt{3}, -\dfrac{\sqrt{3}}{4}\right)$, $(0, 0)$, and $\left(\sqrt{3}, \dfrac{\sqrt{3}}{4}\right)$ on the graph.

e) To determine the concavity we use $-\sqrt{3}$, 0, and $\sqrt{3}$ to divide the real number line into four intervals, A: $(-\infty, -\sqrt{3})$, B: $(-\sqrt{3}, 0)$, C: $(0, \sqrt{3})$, and D: $(\sqrt{3}, \infty)$. Test a point in each interval.

A: Test -2, $f''(-2) = \dfrac{-4}{125} < 0$

B: Test -1, $f''(-1) = \dfrac{1}{2} > 0$

C: Test 1, $f''(1) = \dfrac{-1}{2} < 0$

D: Test 2, $f''(2) = \dfrac{4}{125} > 0$

Then f is concave down on $(-\infty, -\sqrt{3})$ and on $(0, \sqrt{3})$ and is concave up on $(-\sqrt{3}, 0)$ and on $(\sqrt{3}, \infty)$, so $\left(-\sqrt{3}, -\dfrac{\sqrt{3}}{4}\right)$, $(0, 0)$, and $\left(\sqrt{3}, \dfrac{\sqrt{3}}{4}\right)$ are all inflection points.

f) Sketch the graph using the preceding information.

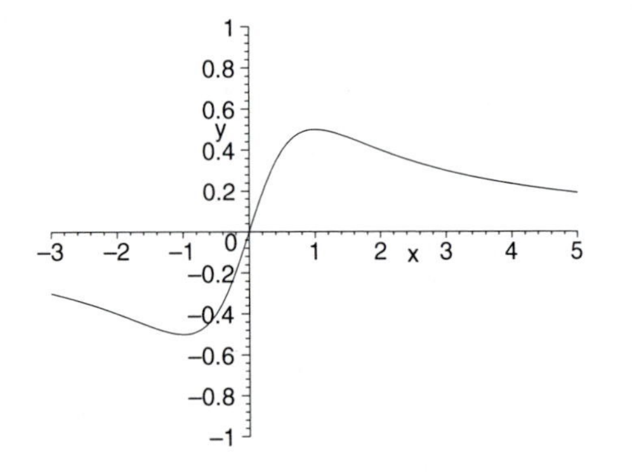

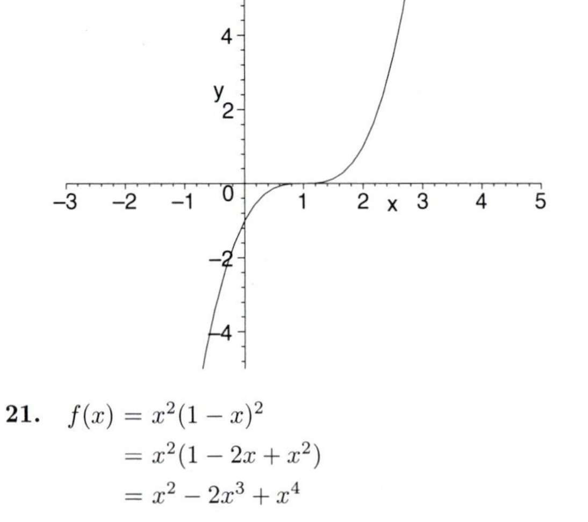

19. $f(x) = (x-1)^3$

a) $f'(x) = 3(x-1)^2(1) = 3(x-1)^2$

$f''(x) = 3 \cdot 2(x-1)(1) = 6(x-1)$

b) Since $f'(x)$ exists for all real numbers, the only critical points are where $f'(x) = 0$.

$$3(x-1)^2 = 0$$
$$(x-1)^2 = 0$$
$$x - 1 = 0$$
$$x = 1 \quad \text{Critical point}$$

Then $f(1) = (1-1)^3 = 0^3 = 0$, so $(1,0)$ is on the graph.

c) The Second-Derivative Test fails since $f''(1) = 0$, so we use the First-Derivative Test. Use 1 to divide the real number line into two intervals, A: $(-\infty, 1)$ and B: $(1, \infty)$. Test a point in each interval.

A: Test 0, $f'(0) = 3(0-1)^2 = 3 > 0$

B: Test 2, $f'(2) = 3(2-1)^2 = 3 > 0$

Since f is increasing on both intervals, $(1,0)$ is not a relative extremum.

d) $f''(x)$ exists for all real numbers, so the only possible inflection points are where $f''(x) = 0$.

$$6(x-1) = 0$$
$$x - 1 = 0$$
$$x = 1 \quad \text{Possible inflection point}$$

From step (b), we know that $(1,0)$ is on the graph.

e) To determine the concavity we use 1 to divide the real number line as in step (c).

A: Test 0, $f''(0) = 6(0-1) = -6 < 0$

B: Test 2, $f''(2) = 6(2-1) = 6 > 0$

Then f is concave down on $(-\infty, 1)$ and concave up on $(1, \infty)$, so $(1,0)$ is an inflection point.

f) Sketch the graph using the preceding information.

21. $f(x) = x^2(1-x)^2$

$= x^2(1 - 2x + x^2)$

$= x^2 - 2x^3 + x^4$

a) $f'(x) = 2x - 6x^2 + 4x^3$

$f''(x) = 2 - 12x + 12x^2$

b) Since $f'(x)$ exists for all real numbers, the only critical points are where $f'(x) = 0$.

$$2x - 6x^2 + 4x^3 = 0$$
$$2x(1 - 3x + 2x^2) = 0$$
$$2x(1-x)(1-2x) = 0$$
$$2x = 0 \;\; or \;\; 1 - x = 0 \;\; or \;\; 1 - 2x = 0$$
$$x = 0 \;\; or \;\; 1 = x \;\; or \;\; 1 = 2x$$
$$x = 0 \;\; or \;\; 1 = x \;\; or \;\; \frac{1}{2} = x$$

Critical points

$$f(0) = 0^2(1-0)^2 = 0$$
$$f(1) = 1^2(1-1)^2 = 0$$
$$f\left(\frac{1}{2}\right) = \left(\frac{1}{2}\right)^2\left(1 - \frac{1}{2}\right)^2 = \frac{1}{4} \cdot \frac{1}{4} = \frac{1}{16}$$

Thus, $(0,0)$, $(1,0)$, and $\left(\frac{1}{2}, \frac{1}{16}\right)$ are on the graph.

c) Use the Second-Derivative Test:

$f''(0) = 2 - 12 \cdot 0 + 12 \cdot 0^2 = 2 > 0$, so $(0,0)$ is a relative minimum.

$f''(\frac{1}{2}) = 2 - 12 \cdot \frac{1}{2} + 12\left(\frac{1}{2}\right)^2 = 2 - 6 + 3 =$

$-1 < 0$, so $\left(\frac{1}{2}, \frac{1}{16}\right)$ is a relative maximum.

$f''(1) = 2 - 12 \cdot 1 + 12 \cdot 1^2 = 2 > 0$, so $(1,0)$ is a relative minimum.

Then if we use the points 0, $\frac{1}{2}$, and 1 to divide the real number line into four intervals, $(-\infty, 0)$, $\left(0, \frac{1}{2}\right)$, $\left(\frac{1}{2}, 1\right)$ and $(1, \infty)$, we know that f is decreasing on $(-\infty, 0)$ and on $\left(\frac{1}{2}, 1\right)$ and is increasing on $\left(0, \frac{1}{2}\right)$ and on $(1, \infty)$.

d) $f''(x)$ exists for all real numbers, so the only possible inflection points are where $f''(x) = 0$.

$$2 - 12x + 12x^2 = 0$$
$$2(1 - 6x + 6x^2) = 0$$

Using the quadratic formula we find

$$x = \frac{3 \pm \sqrt{3}}{6}$$

$x \approx 0.21$ or $x \approx 0.79$ Possible inflection points

$f(0.21) \approx 0.03$ and $f(0.79) \approx 0.03$, so $(0.21, 0.03)$ and $(0.79, 0.03)$ are on the graph.

e) To determine the concavity we use 0.21 and 0.79 to divide the real number line into three intervals, A: $(-\infty, 0.21)$, B: $(0.21, 0.79)$, and C: $(0.79, \infty)$.

A: Test 0, $f''(0) = 2 - 12 \cdot 0 + 12 \cdot 0^2 = 2 > 0$

B: Test 0.5, $f''(0.5) = 2 - 12(0.5) + 12(0.5)^2 = -1 < 0$

C: Test 1, $f''(1) = 2 - 12 \cdot 1 + 12 \cdot 1^2 = 2 > 0$

Then f is concave up on $(-\infty, 0.21)$ and on $(0.79, \infty)$ and is concave down on $(0.21, 0.79)$, so $(0.21, 0.03)$ and $(0.79, 0.03)$ are both inflection points.

f) Sketch the graph using the preceding information.

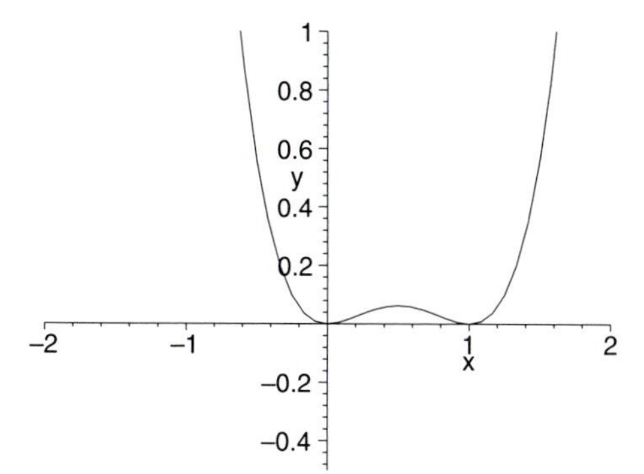

23. $f(x) = 20x^3 - 3x^5$

a) $f'(x) = 60x^2 - 15x^4$

$f''(x) = 120x - 60x^3$

b) Since $f'(x)$ exists for all real numbers, the only critical points are where $f'(x) = 0$.

$$60x^2 - 15x^4 = 0$$
$$15x^2(4 - x^2) = 0$$
$$15x^2(2 + x)(2 - x) = 0$$

$15x^2 = 0$ *or* $2 + x = 0$ *or* $2 - x = 0$

$x = 0$ *or* $x = -2$ *or* $2 = x$

Critical points

$f(0) = 20 \cdot 0^3 - 3 \cdot 0^5 = 0$

$f(-2) = 20(-2)^3 - 3(-2)^5 = -160 + 96 = -64$

$f(2) = 20 \cdot 2^3 - 3 \cdot 2^5 = 160 - 96 = 64$

Thus, $(0, 0)$, $(-2, -64)$, and $(2, 64)$ are on the graph.

c) Use the Second-Derivative Test:

$f''(-2) = 120(-2) - 60(-2)^3 = -240 + 480 = 240 > 0$, so $(-2, -64)$ is a relative minimum.

$f''(2) = 120 \cdot 2 - 60 \cdot 2^3 = 240 - 480 = -240 < 0$, so $(2, 64)$ is a relative maximum.

$f''(0) = 120 \cdot 0 - 60 \cdot 0^3 = 0$, so we will use the First-Derivative Test on $x = 0$. Use 0 to divide the interval $(-2, 2)$ into two intervals, A: $(-2, 0)$, and B: $(0, 2)$. Test a point in each interval.

A: Test -1, $f'(-1) = 60(-1)^2 - 15(-1)^4 =$

$60 - 15 = 45 > 0$

B: Test 1, $f'(1) = 60 \cdot 1^2 - 15 \cdot 1^4 = 60 - 15 =$

$45 > 0$

Then f is increasing on both intervals, so $(0, 0)$ is not a relative extremum.

If we use the points -2 and 2 to divide the real number line into three intervals, $(-\infty, -2)$, $(-2, 2)$, and $(2, \infty)$, we know that f is decreasing on $(-\infty, -2)$ and on $(2, \infty)$ and is increasing on $(-2, 2)$.

d) $f''(x)$ exists for all real numbers, so the only possible inflection points are where $f''(x) = 0$.

$$120x - 60x^3 = 0$$
$$60x(2 - x^2) = 0$$

$60x = 0$ or $2 - x^2 = 0$

$x = 0$ or $2 = x^2$

$x = 0$ or $\pm\sqrt{2} = x$ Possible inflection points

$f(-\sqrt{2}) = 20(-\sqrt{2})^3 - 3(-\sqrt{2})^5 =$

$-40\sqrt{2} + 12\sqrt{2} = -28\sqrt{2}$

$f(0) = 0$ from step (b)

$f(\sqrt{2}) = 20(\sqrt{2})^3 - 3(\sqrt{2})^5 =$

$40\sqrt{2} - 12\sqrt{2} = 28\sqrt{2}$

Thus, $(-\sqrt{2}, -28\sqrt{2})$ and $(\sqrt{2}, 28\sqrt{2})$ are also on the graph.

e) To determine the concavity we use $-\sqrt{2}$, 0, and $\sqrt{2}$ to divide the real number line into four intervals, A: $(-\infty, -\sqrt{2})$, B: $(-\sqrt{2}, 0)$, C: $(0, \sqrt{2})$, and D: $(\sqrt{2}, \infty)$.

A: Test -2, $f''(-2) = 120(-2) - 60(-2)^3 =$

$240 > 0$

B: Test -1, $f''(-1) = 120(-1) - 60(-1)^3 =$

$-60 < 0$

C: Test 1, $f''(1) = 120 \cdot 1 - 60 \cdot 1^3 = 60 > 0$

D: Test 2, $f''(2) = 120 \cdot 2 - 60 \cdot 2^3 = -240 < 0$

Then f is concave up on $(-\infty, -\sqrt{2})$ and on $(0, \sqrt{2})$ and is concave down on $(-\sqrt{2}, 0)$ and on $(\sqrt{2}, \infty)$, so $(-\sqrt{2}, -28\sqrt{2})$, $(0, 0)$, and $(\sqrt{2}, 28\sqrt{2})$ are all inflection points.

f) Sketch the graph using the preceding information.

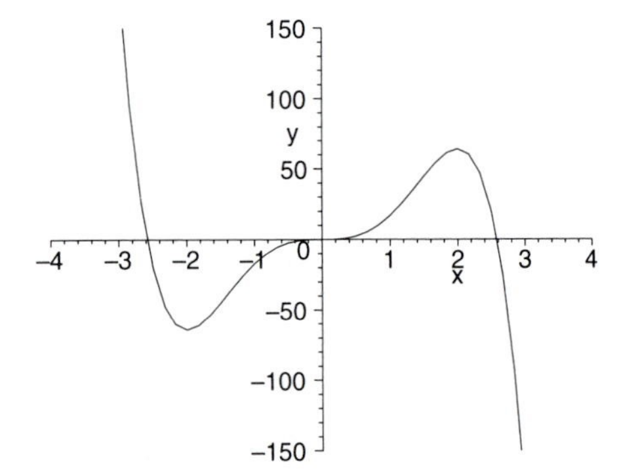

25. $f(x) = x\sqrt{4-x^2} = x(4-x^2)^{1/2}$

a) $f'(x) = x\cdot\frac{1}{2}(4-x^2)^{-1/2}(-2x) + 1\cdot(4-x^2)^{1/2}$

$$= \frac{-x^2}{\sqrt{4-x^2}} + \sqrt{4-x^2}$$

$$= \frac{-x^2+4-x^2}{\sqrt{4-x^2}}$$

$$= \frac{4-2x^2}{\sqrt{4-x^2}}, \text{ or } (4-2x^2)(4-x^2)^{-1/2}$$

$$f''(x) = (4-2x^2)\left(-\frac{1}{2}\right)(4-x^2)^{-3/2}(-2x) + (-4x)(4-x^2)^{-1/2}$$

$$= \frac{x(4-2x^2)}{(4-x^2)^{3/2}} - \frac{4x}{(4-x^2)^{1/2}}$$

$$= \frac{x(4-2x^2)-4x(4-x^2)}{(4-x^2)^{3/2}}$$

$$= \frac{4x-2x^3-16x+4x^3}{(4-x^2)^{3/2}}$$

$$= \frac{2x^3-12x}{(4-x^2)^{3/2}}$$

b) $f'(x)$ does not exist where $4-x^2=0$. Solve:

$$4-x^2=0$$
$$(2+x)(2-x)=0$$
$$2+x=0 \quad or \quad 2-x=0$$
$$x=-2 \quad or \quad 2=x$$

Note that $f(x)$ is not defined for $x < -2$ or $x > 2$. (For these values $4-x^2 < 0$.) Therefore, relative extrema canot occur at $x=-2$ or $x=2$, because there is no open interval containing -2 or 2 on which the function is defined. For this reason, we do not consider -2 and 2 further in our discussion of relative extrema.

Critical points occur where $f'(x)=0$. Solve:

$$\frac{4-2x^2}{\sqrt{4-x^2}} = 0$$
$$4-2x^2=0$$
$$4=2x^2$$
$$2=x^2$$
$$\pm\sqrt{2}=x \quad \text{Critical points}$$

$$f(-\sqrt{2}) = -\sqrt{2}\sqrt{4-(-\sqrt{2})^2} = -\sqrt{2}\cdot\sqrt{2} = -2$$

$$f(\sqrt{2}) = \sqrt{2}\sqrt{4-(\sqrt{2})^2} = \sqrt{2}\cdot\sqrt{2} = 2$$

Then $(-\sqrt{2},-2)$ and $(\sqrt{2},2)$ are on the graph.

c) Use the Second-Derivative Test:

$f''(-\sqrt{2}) = \frac{2(-\sqrt{2})^3-12(-\sqrt{2})}{[4-(-\sqrt{2})^2]^{3/2}} =$

$\frac{-4\sqrt{2}+12\sqrt{2}}{2^{3/2}} = \frac{8\sqrt{2}}{2\sqrt{2}} = 4 > 0$, so $(-\sqrt{2},-2)$ is a relative minimum.

$f''(\sqrt{2}) = \frac{2(\sqrt{2})^3-12(\sqrt{2})}{[4-(\sqrt{2})^2]^{3/2}} =$

$\frac{4\sqrt{2}-12\sqrt{2}}{2^{3/2}} = \frac{-8\sqrt{2}}{2\sqrt{2}} = -4 < 0$, so $(\sqrt{2},2)$ is a relative maximum.

If we use the points $-\sqrt{2}$ and $\sqrt{2}$ to divide the interval $(-2,2)$ into three intervals, $(-2,-\sqrt{2})$, $(-\sqrt{2},\sqrt{2})$, and $(\sqrt{2},2)$, we know that f is decreasing on $(-2,-\sqrt{2})$ and on $(\sqrt{2},2)$ and is increasing on $(-\sqrt{2},\sqrt{2})$.

d) $f''(x)$ does not exist where $4-x^2=0$. From step (b) we know that this occurs at $x=-2$ and at $x=2$. However, just as relative extrema cannot occur at $(-2,0)$ and $(2,0)$, they cannot be inflection points either. Inflection points could occur where $f''(x) = 0$.

$$\frac{2x^3-12x}{(4-x^2)^{3/2}} = 0$$
$$2x^3-12x=0$$
$$2x(x^2-6)=0$$
$$2x=0 \text{ or } x^2-6=0$$
$$x=0 \text{ or } x^2=6$$
$$x=0 \text{ or } x=\pm\sqrt{6}$$

Note that $f(x)$ is not defined for $x=\pm\sqrt{6}$. Therefore, the only possible inflection point is $x=0$.

$f(0) = 0\sqrt{4-0^2} = 0\cdot 2 = 0$

Then $(0,0)$ is on the graph.

e) To determine the concavity we use 0 to divide the interval $(-2,2)$ into two intervals, A: $(-2,0)$ and B: $(0,2)$.

A: Test -1, $f''(-1) = \dfrac{2(-1)^3 - 12(-1)}{[4-(-1)^2]^{3/2}} =$

$\dfrac{10}{3^{3/2}} > 0$

B: Test 1, $f''(1) = \dfrac{2\cdot 1^3 - 12\cdot 1}{(4-1^2)^{3/2}} = \dfrac{-10}{3^{3/2}} < 0$

Then f is concave up on $(-2,0)$ and concave down on $(0,2)$, so $(0,0)$ is an inflection point.

f) Sketch the graph using the preceding information.

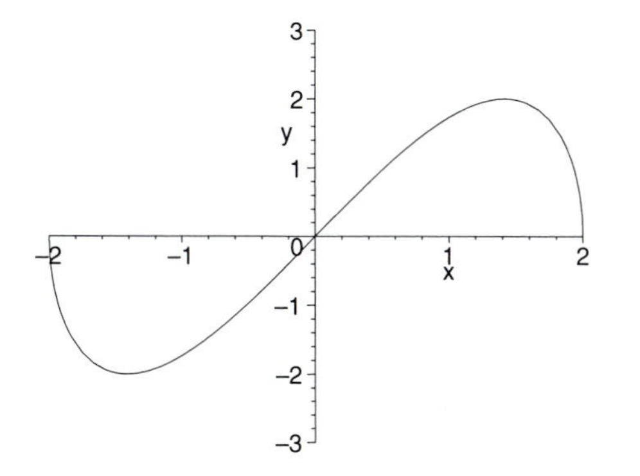

27. $f(x) = (x-1)^{1/3} - 1$

a) $f'(x) = \dfrac{1}{3}(x-1)^{-2/3}$, or $\dfrac{1}{3(x-1)^{2/3}}$

$f''(x) = \dfrac{1}{3}\Big(-\dfrac{2}{3}\Big)(x-1)^{-5/3}$

$= -\dfrac{2}{9}(x-1)^{-5/3}$, or $-\dfrac{2}{9(x-1)^{5/3}}$

b) $f'(x)$ does not exist for $x=1$. The equation $f'(x) = 0$ has no solution, so $x=1$ is the only critical point. $f(1) = (1-1)^{1/3} - 1 = 0 - 1 = -1$, so $(1,-1)$ is on the graph.

c) Use the First-Derivative Test: Use 1 to divide the real number line into two intervals, A: $(-\infty,1)$ and B: $(1,\infty)$. Test a point in each interval.

A: Test 0, $f'(0) = \dfrac{1}{3(0-1)^{2/3}} = \dfrac{1}{3\cdot 1} = \dfrac{1}{3} > 0$

B: Test 2, $f'(2) = \dfrac{1}{3(2-1)^{2/3}} = \dfrac{1}{3\cdot 1} = \dfrac{1}{3} > 0$

Then f is increasing on both intervals, so $(1,-1)$ is not a relative extremum.

d) $f''(x)$ does not exist for $x=1$. The equation $f''(x) = 0$ has no solution, so $x=1$ is the only possible inflection point. From step (b) we know $(1,-1)$ is on the graph.

e) To determine the concavity we use 1 to divide the real number line as in step (c). Test a point in each interval.

A: Test 0, $f''(0) = -\dfrac{2}{9(0-1)^{5/3}} = -\dfrac{2}{9(-1)} =$

$\dfrac{2}{9} > 0$

B: Test 2, $f''(2) = -\dfrac{2}{9(2-1)^{5/3}} = -\dfrac{2}{9\cdot 1} =$

$-\dfrac{2}{9} < 0$

Then f is concave up on $(-\infty,1)$ and concave down on $(1,\infty)$, so $(1,-1)$ is an inflection point.

f) Sketch the graph using the preceding information.

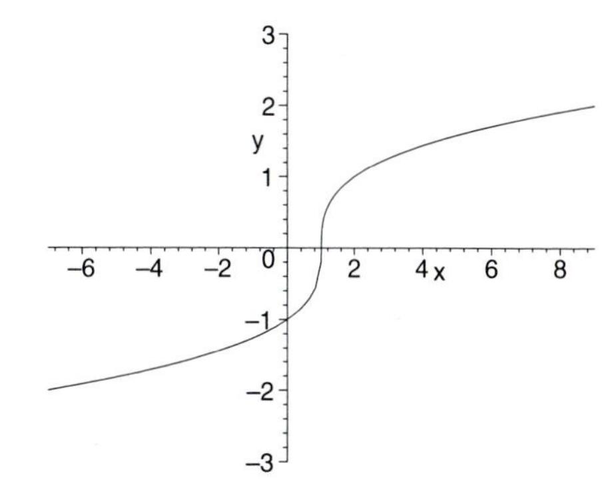

29. $f(x) = x + \cos 2x$

$f'(x) = 1 - 2\sin 2x$

We solve $f'(x) = 0$

$$\begin{aligned} 1 - 2\sin 2x &= 0 \\ \sin 2x &= \frac{1}{2} \\ 2x &= \frac{\pi}{6} \\ x &= \frac{\pi}{12} \\ &\text{and} \\ 2x &= \frac{5\pi}{6} \\ x &= \frac{5\pi}{12} \\ &\text{and} \\ 2x &= \frac{13\pi}{6} \\ x &= \frac{13\pi}{12} \\ &\text{and} \\ 2x &= \frac{17\pi}{6} \\ x &= \frac{17\pi}{12} \end{aligned}$$

The values above divide the number line into five intervals. We apply test points to check the sign of the first derivative in each of those intervals.

$f'(0) = 1 - 2\ sin(2(0)) = 1 > 0$

$f'(\frac{\pi}{3}) = 1 - 2\ sin(\frac{2\pi}{3}) = -0.7321 < 0$

$f'(\pi) = 1 - 2\ sin(2\pi) = 1 > 0$

$f'(\frac{5\pi}{4}) = 1 - 2\ sin(\frac{5\pi}{2}) = -1 < 0$

$f'(\frac{5\pi}{3}) = 1 - 2\ sin(\frac{10\pi}{3}) = 2.7321 > 0$

Thus, there is a relative maximum at $x = \frac{\pi}{12}$ and $x = \frac{13\pi}{12}$ and a relative minimum at $x = \frac{5\pi}{12}$ and $x = \frac{17\pi}{12}$.

$f(\frac{\pi}{12}) = \frac{\pi}{12} + cos\ \frac{\pi}{6} = 1.128$

$f(\frac{13\pi}{12}) = \frac{13\pi}{12} + cos(\frac{13\pi}{6}) = 4.269$

$f(\frac{5\pi}{12}) = \frac{5\pi}{12} + cos(\frac{5\pi}{6}) = 0.443$

$f(\frac{17\pi}{12}) = \frac{17\pi}{12} + cos(\frac{17\pi}{6}) = 3.585$

$$\begin{aligned} f''(x) &= -4\ cos\ 2x \\ f''(x) &= 0 \\ -4\ cos\ 2x &= 0 \\ cos\ 2x &= 0 \\ 2x &= \frac{\pi}{2} \\ x &= \frac{\pi}{4} \\ \text{and} & \\ 2x &= \frac{3\pi}{2} \\ x &= \frac{3\pi}{4} \\ \text{and} & \\ 2x &= \frac{5\pi}{2} \\ x &= \frac{5\pi}{4} \\ \text{and} & \\ 2x &= \frac{7\pi}{2} \\ x &= \frac{7\pi}{4} \end{aligned}$$

The second derivative zeros divide the number line into five intervals, we determine the sign of the second derivative in each of those intervals:

$f''(\frac{\pi}{6}) = -4\ cos(\frac{\pi}{3}) = -2 < 0$
$f''(\frac{\pi}{2}) = -4\ cos(\pi) = 4 > 0$ $f''(\pi) = -4\ cos(2\pi) = -4 < 0$
$f''(\frac{3\pi}{2}) = -4\ cos(3\pi) = 4 > 0$ $f''(\frac{15\pi}{8}) = -4\ cos(\frac{15\pi}{4}) = -2.828 < 0$

This means that we have inflection points at $x = \frac{\pi}{4}$, $x = \frac{\pi}{4}$, $x = \frac{5\pi}{4}$ and $x = \frac{7\pi}{4}$

$f(\frac{\pi}{4}) = \frac{\pi}{4} + cos(\frac{\pi}{2}) = 0.785$

$f(\frac{3\pi}{4}) = \frac{3\pi}{4} + cos(\frac{3\pi}{2}) = 2.356$

$f(\frac{\pi}{4}) = \frac{5\pi}{4} + cos(\frac{5\pi}{2}) = 3.927$

$f(\frac{\pi}{4}) = \frac{7\pi}{4} + cos(\frac{7\pi}{2}) = 5.498$

we sketch the graph

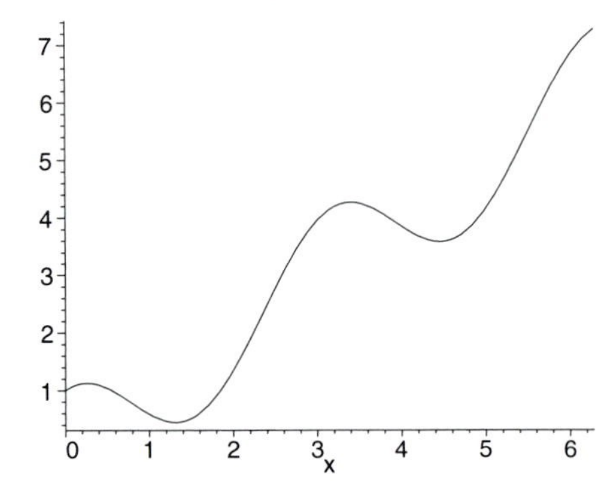

31. $f(x) = \frac{x}{3} - sin\ \frac{2x}{3}$

$f'(x) = \frac{1}{3} - \frac{2}{3}\ cos\ \frac{2x}{3}$
We solve $f'(x) = 0$

$$\begin{aligned} \frac{1}{3} - \frac{2}{3}\ cos\ \frac{2x}{3} &= 0 \\ cos\ \frac{2x}{3} &= \frac{1}{2} \\ \frac{2x}{3} &= \frac{\pi}{3} \\ x &= \frac{\pi}{2} \end{aligned}$$

The zero of the first derivative divides the number line into two intervals. We apply test points to check the sign of the first derivative in each of those intervals.

$f'(\frac{\pi}{4}) = \frac{1}{3} - \frac{2}{3}\ sin\ (\frac{\pi}{12}) = -0.244 < 0$

$f'(\pi) = \frac{1}{3} - \frac{2}{3}\ sin\ (\frac{2\pi}{3}) = 0.667 > 0$

Thus, there is a relative minimum at $x = \frac{\pi}{2}$.

$f(\frac{\pi}{2}) = \frac{\pi}{2} - sin\ (\frac{\pi}{3}) = -0.342$

$$\begin{aligned} f''(x) &= \frac{4}{9}\ sin\ \frac{2x}{3} \\ f''(x) &= 0 \\ \frac{4}{9}\ sin\ \frac{2x}{3} &= 0 \\ sin\ \frac{2x}{3} &= 0 \end{aligned}$$

$$\begin{aligned}\frac{2x}{3} &= \pi \\ x &= \frac{3\pi}{2}\end{aligned}$$

The second derivative zero divides the number line into two intervals, we determine the sign of the second derivative in each of those intervals:

$f''(\pi) = \frac{4}{9}\ sin\ (\frac{2\pi}{3}) = 0.385 > 0$

$f''(2\pi) = \frac{4}{9}\ sin\ (\frac{4\pi}{3}) = -0.385 < 0$

This means that we have an inflection points at $x = \frac{3\pi}{2}$. $f(\frac{3\pi}{2}) = \frac{\pi}{2} - sin\ (\pi) = \frac{\pi}{2} = 1.571$

We sketch the graph

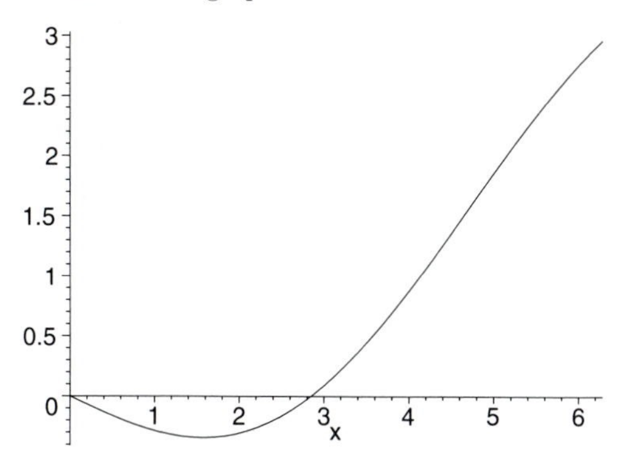

33. $f(x) = sin\ x + cos\ x$

$f'(x) = cos\ x - sin\ x$
We solve$f'(x) = 0$

$$\begin{aligned}cos\ x - sin\ x &= 0 \\ cos\ x &= sin\ x \\ x &= \frac{\pi}{4} \\ \text{and} & \\ x &= \frac{5\pi}{4}\end{aligned}$$

The zeros of the first derivative divide the number line into three intervals. We apply test points to check the sign of the first derivative in each of those intervals.

$f'(0) = cos(0) - sin(0) = 1 > 0$

$f'(\pi) = cos(\pi) - sin(\pi) = -1 < 0$

$f'(\frac{3\pi}{2}) = cos(\frac{3\pi}{2}) - sin(\frac{3\pi}{2}) = 1 > 0$
Thus, there is a relative maximum at $x = \frac{\pi}{4}$ and a relative minimum at $x = \frac{5\pi}{4}$.
$f(\frac{\pi}{4}) = sin(\frac{\pi}{4}) + cos(\frac{\pi}{4}) = 1.414$
$f(\frac{5\pi}{4}) = sin\frac{5\pi}{4}) + cos(\frac{5\pi}{4}) = -1.414$

$$\begin{aligned}f''(x) &= -sin\ x - cos\ x \\ f''(x) &= 0\end{aligned}$$

$$\begin{aligned}-sin\ x - cos\ x &= 0 \\ -sin\ x &= cos\ x \\ x &= \frac{3\pi}{4} \\ \text{and} & \\ x &= \frac{7\pi}{4}\end{aligned}$$

The second derivative zero divides the number line into three interval, we dtermine the sign of the second derivative in each of those intervals:

$f''(0) = -sin(0) - cos(0) = -1 < 0$
$f''(\pi) = -sin(\pi) - cos(\pi) = 1 > 0$
$f''(\frac{11\pi}{6}) = -sin(\frac{11\pi}{6}) - cos(\frac{11\pi}{6}) = -0.366 < 0$
This means that we have an inflection point at $x = \frac{3\pi}{4}$ and at $x = \frac{7\pi}{4}$.
$f(\frac{3\pi}{4}) = sin(\frac{3\pi}{4}) + cos(\frac{3\pi}{4}) = 0$

$f(\frac{3\pi}{4}) = sin(\frac{7\pi}{4}) + cos(\frac{7\pi}{4}) = 0$
We sketch the graph

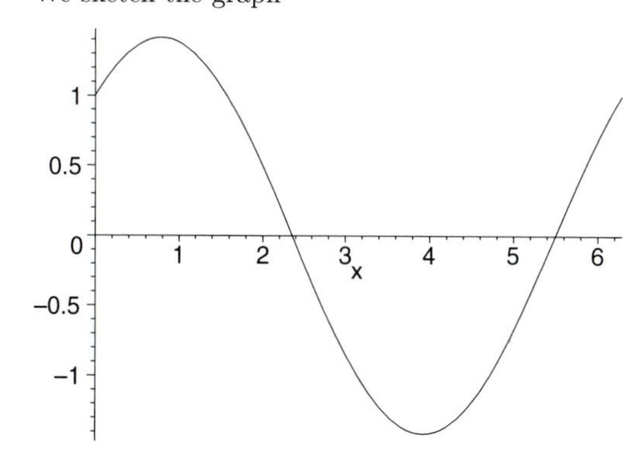

35. $f(x) = \sqrt{3}\ sin\ x + cos\ x$

$f'(x) = \sqrt{3}\ cos\ x - sin\ x$
We solve $f'(x) = 0$

$$\begin{aligned}\sqrt{3}\ cos\ x - sin\ x &= 0 \\ \sqrt{3}\ cos\ x &= sin\ x \\ tan\ x &= \sqrt{3} \\ x &= \frac{\pi}{3} \\ \text{and} & \\ x &= \frac{4\pi}{3}\end{aligned}$$

The zeros of the first derivative divide the number line into three intervals. We apply test points to check the sign of the first derivative in each of those intervals. $f'(0) = \sqrt{3}\ cos(0) + sin(0) = 1.732 > 0$
$f'(\pi) = \sqrt{3}\ cos(\pi) + sin(\pi) = -1.732 < 0$
$f'(\frac{5\pi}{3}) = \sqrt{3}\ cos(\frac{5\pi}{3}) + sin(\frac{5\pi}{3}) = 1.732$

Thus, we have a relative maximum at $x = \frac{\pi}{3}$ and a relative minimum at $x = \frac{4\pi}{3}$.
$f(\frac{\pi}{3}) = \sqrt{3}\ sin(\frac{\pi}{3}) + cos(\frac{\pi}{3}) = 2$

$f(\frac{4\pi}{3}) = \sqrt{3}\ sin(\frac{4\pi}{3}) + cos(\frac{4\pi}{3}) = -2$

$$\begin{aligned} f''(x) &= -\sqrt{3}\ sin\ x - cos\ x \\ f''(x) &= 0 \\ -\sqrt{3}\ sin\ x - cos\ x &= 0 \\ tan\ x &= \frac{-1}{\sqrt{3}} \\ x &= \frac{5\pi}{6} \\ \text{and} & \\ x &= \frac{11\pi}{6} \end{aligned}$$

The second derivative zeros divide the number line into three intervals, we determine the sign of the second derivative in each of those intervals:

$f''(0) = -\sqrt{3}\ sin(0) - cos(0) = -1 < 0$

$f''(\pi) = -\sqrt{3}\ sin(\pi) - cos(\pi) = 1 < 0$

$f''(\frac{35\pi}{18}) = -\sqrt{3}\ sin(\frac{35\pi}{18}) - cos(\frac{35\pi}{18}) = -0.845 < 0$

This means that we have an inflection point at $x = \frac{5\pi}{6}$ and at $x = \frac{11\pi}{6}$.

$f(\frac{5\pi}{6}) = \sqrt{3}\ sin(\frac{5\pi}{6}) + cos(\frac{5\pi}{6}) = 0$

$f(\frac{11\pi}{6}) = \sqrt{3}\ sin(\frac{11\pi}{6}) + cos(\frac{11\pi}{6}) = 0$

We sketch the graph

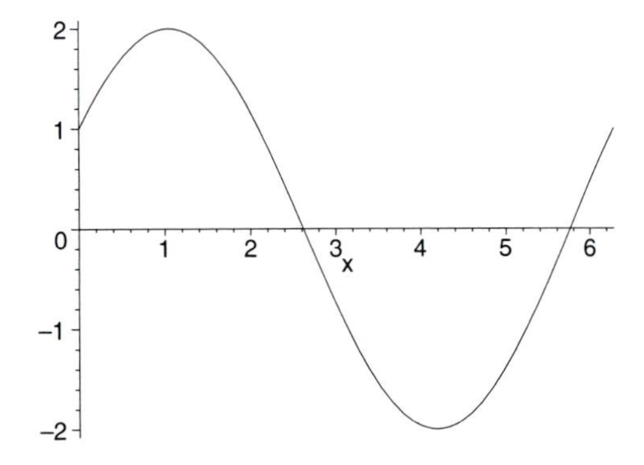

37. $f(x) = \frac{sin\ x}{2 - cos\ x}$

$f'(x) = \frac{2\ cos\ x - 1}{(cos\ x - 2)^2}$

We solve $f'(x) = 0$

$$\begin{aligned} \frac{2\ cos\ x - 1}{(cos\ x - 2)^2} &= 0 \\ cos\ x &= \frac{1}{2} \\ x &= \frac{\pi}{3} \\ \text{and} & \\ x &= \frac{5\pi}{3} \end{aligned}$$

The zeros of the first derivative divide the number line into three intervals. We apply test points to check the sign of the first derivative in each of those intervals. $f'(0) = \frac{2\ cos(0) - 1}{(cos(0) - 2)^2} = 1 > 0$

$f'(\pi) = \frac{2\ cos(\pi) - 1}{(cos(\pi) - 2)^2} = -\frac{1}{3} < 0$

$f'(2\pi) = \frac{2\ cos(2\pi) - 1}{(cos(2\pi) - 2)^2} = 1$

Thus, we have a relative maximum at $x = \frac{\pi}{3}$ and a relative minimum at $x = \frac{5\pi}{3}$.

$f(\frac{\pi}{3}) = \frac{sin(\frac{\pi}{3})}{2 - cos(\frac{\pi}{3})} = 0.577$

$f(\frac{5\pi}{3}) = \frac{sin(\frac{5\pi}{3})}{2 - cos(\frac{5\pi}{3})} = 0.174$

$$\begin{aligned} f''(x) &= \frac{2\ sin\ x(1 + cos\ x)}{(cos(x) - 2)^3} \\ f''(x) &= 0 \\ \frac{2\ sin\ x(1 + cos\ x)}{(cos(x) - 2)^3} &= 0 \\ \text{either} & \\ sin\ x &= 0 \\ x &= 0 \\ x &= \pi \\ x &= 2\pi \\ \text{or} & \\ 1 + cos\ x &= 0 \\ cos\ x &= -1 \\ x &= \pi \end{aligned}$$

The second derivative zeros divide the number line into two intervals (with the end points at 0 and 2π as potential inflection points), we determine the sign of the second derivative in each of those intervals:

$f''(\frac{\pi}{4}) = \frac{2\ sin\ x(1 + cos(\frac{\pi}{4}))}{(cos(\frac{\pi}{4}) - 2)^3} = -1.117 < 0$

$f''(\frac{3\pi}{2}) = \frac{2\ sin\ x(1 + cos(\frac{3\pi}{2}))}{(cos(\frac{3\pi}{2}) - 2)^3} = \frac{1}{4} > 0$

Note: checking a test point smaller than 0 will yield a positive value and testing a point larger than 2π will yield a negative value for the second derivative.

This means that we have an inflection point at $x = 0$, $x = \pi$ and at $x = 2\pi$.

$f(0) = \frac{sin(0)}{2 - cos(0)} = 0$

$f(\pi) = \frac{sin(\pi)}{2 - cos(\pi)} = 0$

$f(2\pi) = \frac{sin(2\pi)}{2 - cos(2\pi)} = 0$

We sketch the graph

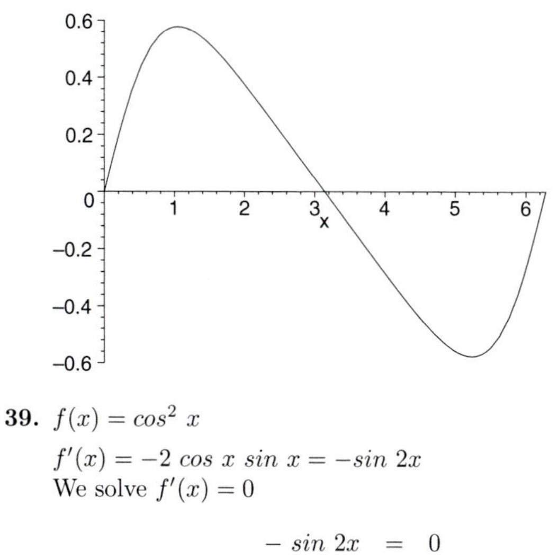

39. $f(x) = cos^2\ x$

$f'(x) = -2\ cos\ x\ sin\ x = -sin\ 2x$
We solve $f'(x) = 0$

$$\begin{aligned} -\ sin\ 2x &= 0 \\ x &= 0 \\ x &= \frac{\pi}{2} \\ x &= \pi \\ x &= \frac{3\pi}{2} \\ x &= 2\pi \end{aligned}$$

The zeros of the first derivative divide the number line into four intervals, plus the two endpoints. We apply test points to check the sign of the first derivative in each of those intervals. $f'(\frac{\pi}{4}) = -sin(\frac{\pi}{2}) = -1 < 0$

$f'(\frac{3\pi}{4}) = -sin(\frac{3\pi}{2}) = 1 > 0$

$f'(\frac{5\pi}{4}) = -sin(\frac{5\pi}{2}) = -1 < 0$

$f'(\frac{7\pi}{4}) = -sin(\frac{7\pi}{2}) = 1 > 0$

Thus, we have a relative maximum at $x = \pi$ and a relative minimum at $x = \frac{\pi}{2}$, and $x = \frac{3\pi}{2}$.
$f(\pi) = cos^2(\pi) = 1$

$f(\frac{\pi}{2}) = cos^2(\frac{\pi}{2}) = 0$

$f(\frac{3\pi}{2}) = cos^2(\frac{3\pi}{2}) = 0$

$f(0) = cos^2(0) = 1$ and $f(2\pi) = cos^2(2\pi) = 1$ which means that there is a relative maximum at $x = 0$ and $x = 2\pi$ as well.

$$\begin{aligned} f''(x) &= -2\ cos\ 2x \\ f''(x) &= 0 \\ -2\ cos\ 2x &= 0 \\ x &= \frac{\pi}{4} \\ x &= \frac{3\pi}{4} \\ x &= \frac{5\pi}{4} \\ x &= \frac{7\pi}{4} \end{aligned}$$

The second derivative zeros divide the number line into five intervals, we determine the sign of the second derivative in each of those intervals:

$f''(0) = -2\ cos(0) = -2 < 0$

$f''(\frac{\pi}{2}) = -2\ cos(\pi) = 2 > 0$

$f''(\pi) = -2\ cos(2\pi) = -2 < 0$

$f''(\frac{3\pi}{2}) = -2\ cos(3\pi) = 2 > 0$

$f''(2\pi) = -2\ cos(4\pi) = -2 < 0$

This means that we have an inflection point at $x = \frac{\pi}{4}$, $x = \frac{3\pi}{4}$, $x = \frac{5\pi}{4}$, and at $x = \frac{7\pi}{4}$.

$f(\frac{\pi}{4}) = -2\ cos^2(\frac{\pi}{4}) = 0.5$

$f(\frac{3\pi}{4}) = -2\ cos^2(\frac{3\pi}{4}) = 0.5$

$f(\frac{5\pi}{4}) = -2\ cos^2(\frac{5\pi}{4}) = 0.5$

$f(\frac{7\pi}{4}) = -2\ cos^2(\frac{7\pi}{4}) = 0.5$

We sketch the graph

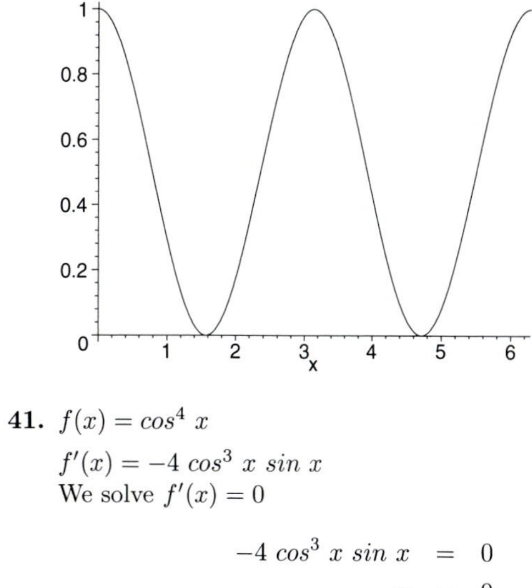

41. $f(x) = cos^4\ x$

$f'(x) = -4\ cos^3\ x\ sin\ x$
We solve $f'(x) = 0$

$$\begin{aligned} -4\ cos^3\ x\ sin\ x &= 0 \\ x &= 0 \\ x &= \frac{\pi}{2} \\ x &= \pi \\ x &= \frac{3\pi}{2} \\ x &= 2\pi \end{aligned}$$

The zeros of the first derivative divide the number line into four intervals, plus the two endpoints. We apply test points to check the sign of the first derivative in each of those intervals. $f'(\frac{\pi}{4}) = -4\ cos^3(\frac{\pi}{4})\ sin(\frac{\pi}{4}) = -1 < 0$

$f'(\frac{3\pi}{4}) = -4\ cos^3(\frac{3\pi}{4})\ sin(\frac{3\pi}{4}) = 1 > 0$

$f'(\frac{5\pi}{4}) = -4\ cos^3(\frac{5\pi}{4})\ sin(\frac{5\pi}{4}) = -1 < 0$

$f'(\frac{7\pi}{4}) = -4\ cos^3(\frac{7\pi}{4})\ cos(\frac{7\pi}{4}) = 1 > 0$

Thus, we have a relative minimum at $x = \frac{\pi}{2}$, and at $x = \frac{3\pi}{2}$ and a relative maximum at $x = \pi$

$f(\frac{\pi}{2}) = cos^4(\frac{\pi}{2}) = 0$

$f(\frac{3\pi}{2}) = cos^4(\frac{3\pi}{2}) = 0$

$f(\pi) = cos^4(\pi) = 1$

$f(0) = cos^4(0) = 1$ and $f(2\pi) = cos^4(2\pi) = 1$ which means that there is a relative maximum at $x = 0$ and $x = 2\pi$ as well.

$$\begin{aligned} f''(x) &= 12sin^2\ x\ cos^2\ x - 4\ cos^4\ x \\ &= 4\ cos^2\ x(3sin^2\ x - \ cos^2\ x) \\ f''(x) &= 0 \\ \text{either} & \\ 4\ cos^2\ x &= 0 \\ x &= \frac{\pi}{2} \\ x &= \frac{3\pi}{2} \\ \text{or} & \\ 3sin^2\ x - \ cos^2\ x &= 0 \\ tan^2\ x &= \frac{1}{3} \\ x &= \frac{\pi}{6} \\ x &= \frac{5\pi}{6} \\ x &= \frac{7\pi}{6} \\ x &= \frac{5\pi}{3} \\ x &= \frac{11\pi}{6} \end{aligned}$$

The second derivative zeros divide the number line into seven intervals, we determine the sign of the second derivative in each of those intervals:

$f''(0) = 4\ cos^2(0)(3\ sin^2(0) - cos^2(0)) = -4 < 0$

$f''(\frac{\pi}{4}) = 4\ cos^2(\frac{\pi}{4})(3\ sin^2(\frac{\pi}{4}) - cos^2(\frac{\pi}{4})) = 2 > 0$

$f''(\frac{3\pi}{4}) = 4\ cos^2(\frac{3\pi}{4})(3\ sin^2(\frac{3\pi}{4}) - cos^2(\frac{3\pi}{4})) = 2 > 0$

$f''(\pi) = 4\ cos^2(\pi)(3\ sin^2(\pi) - cos^2(\pi)) = -4 < 0$

$f''(\frac{5\pi}{4}) = 4\ cos^2(\frac{5\pi}{4})(3\ sin^2(\frac{5\pi}{4}) - cos^2(\frac{5\pi}{4})) = -4 < 0$

$f''(\frac{7\pi}{4}) = 4\ cos^2(\frac{7\pi}{4})(3\ sin^2(\frac{7\pi}{4}) - cos^2(\frac{7\pi}{4})) = -4 < 0$

$f''(2\pi) = 4\ cos^2(2\pi)(3\ sin^2(2\pi) - cos^2(2\pi)) = -4 < 0$

This means that we have an inflection point at $x = \frac{\pi}{6}$, $x = \frac{5\pi}{6}$, $x = \frac{7\pi}{6}$, and at $x = \frac{11\pi}{6}$.

$f(\frac{\pi}{6}) = cos^4(\frac{\pi}{6}) = 0.5625$

$f(\frac{5\pi}{6}) = cos^4(\frac{5\pi}{6}) = 0.5625$

$f(\frac{7\pi}{6}) = cos^4(\frac{7\pi}{3}) = 0.5625$

$f(\frac{11\pi}{3}) = cos^4(\frac{11\pi}{3}) = 0.5625$

We sketch the graph

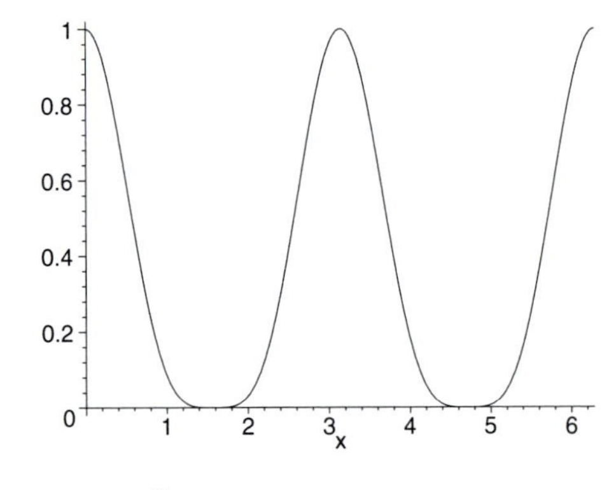

43. $f(x) = x^3 + 3x + 1$

$f'(x) = 3x^2 + 3$

$f''(x) = 6x$

$f''(x)$ exists for all values of x, so we solve $f''(x) = 0$.

$6x = 0$

$x = 0$ Possible inflection point

We use the possible inflection point, 0, to divide the real number line into two intervals, A: $(-\infty, 0)$ and B: $(0, \infty)$. Test a point in each interval.

A: Test -1, $f''(-1) = 6(-1) = -6 < 0$

B: Test 1, $f''(1) = 6 \cdot 1 = 6 > 0$

Then f is concave down on $(-\infty, 0)$ and concave up on $(0, \infty)$. We find that $f(0) = 0^3 + 3 \cdot 0 + 1 = 1$, so $(0, 1)$ is an inflection point.

45. $f(x) = \frac{4}{3}x^3 - 2x^2 + x$

$f'(x) = 4x^2 - 4x + 1$

$f''(x) = 8x - 4$

$f''(x)$ exists for all values of x, so we solve $f''(x) = 0$.

$$8x - 4 = 0$$
$$8x = 4$$
$$x = \frac{1}{2} \quad \text{Possible inflection point}$$

We use the possible inflection point, $\frac{1}{2}$, to divide the real number line into two intervals, A: $\left(-\infty, \frac{1}{2}\right)$ and B: $\left(\frac{1}{2}, \infty\right)$. Test a point in each interval.

A: Test 0, $f''(0) = 8 \cdot 0 - 4 = -4 < 0$

B: Test 1, $f''(1) = 8 \cdot 1 - 4 = 4 > 0$

Then f is concave down on $\left(-\infty, \frac{1}{2}\right)$ and concave up on $\left(\frac{1}{2}, \infty\right)$. We find that $f\left(\frac{1}{2}\right) = \frac{4}{3}\left(\frac{1}{2}\right)^3 - 2\left(\frac{1}{2}\right)^2 + \frac{1}{2} = \frac{1}{6}$, so $\left(\frac{1}{2}, \frac{1}{6}\right)$ is an inflection point.

47. $f(x) = x - sin\ x$

$f'(x) = 1 - cos\ x$

$f''(x) = sin\ x$

$f''(x)$ exists for all real numbers. Solve:

$sin\ x = 0$

$x = n\ \pi$ Possible inflection points

Use -2π, $-\pi$, 0, π, and 2π to divide the real number line,

A: Test $-\frac{3\pi}{2}$, $f''(\frac{-3\pi}{2}) = sin(\frac{-3\pi}{2}) = 1 > 0$

B: Test $-\frac{\pi}{2}$, $f''(-\frac{\pi}{2}) = sin(\frac{-\pi}{2}) = -1 < 0$

C: Test $\frac{\pi}{2}$, $f''(\frac{\pi}{2}) = sin(\frac{\pi}{2}) = 1 > 0$

D: Test $\frac{3\pi}{2}$, $f''(\frac{3\pi}{2}) = sin(\frac{3\pi}{2}) = -1 < 0$

Note that since sin/x is a periodic function this patern of changing signs will continue indefinitely. Therefore, the inflection points will occur at $n\pi$.

49. $f(x) = tan\ x$

$f'(x) = sec^2\ x$

$f''(x) = 2\ sec^2\ x\ tan\ x$

$f''(x)$ does not exist for $x = \frac{\pi}{2} + n\pi$.

$$\begin{aligned} f''(x) &= 0 \\ 2\ sec^2\ x\ tan\ x &= 0 \\ x &= n\pi \end{aligned}$$

Use -2π, $-\pi$, 0, π, and 2π to divide the real number line,

A: Test $-\frac{5\pi}{4}$, $f''(\frac{-5\pi}{4}) = 2\ sec^2(\frac{-5\pi}{4})\ tan(\frac{-5\pi}{4}) = 4 > 0$

B: Test $-\frac{\pi}{4}$, $f''(-\frac{\pi}{4}) = 2\ sec^2(\frac{-\pi}{4})\ tan(\frac{-\pi}{4}) = -4 < 0$

C: Test $\frac{\pi}{4}$, $f''(\frac{\pi}{4}) = 2\ sec^2(\frac{\pi}{4})\ tan(\frac{\pi}{4}) = 4 > 0$

D: Test $\frac{5\pi}{4}$, $f''(\frac{5\pi}{4}) = 2\ sec^2(\frac{5\pi}{4})\ tan(\frac{5\pi}{4}) = -4 < 0$

Note that since $2\ sec^2\ \ tan\ x$ is a periodic function this patern of changing signs will continue indefinitely. Therefore, the inflection points will occur at $n\pi$.

51. $f(x) = tan\ x + sec\ x$

$f'(x) = sec^2\ x + sec\ x\ tan\ x$

$$\begin{aligned} f''(x) &= 2\ sec^2\ x\ tan\ x + sec^3\ x + tan^2\ x\ sec\ x \\ &= \frac{-cos^2\ x + 2\ sin\ x + 2}{cos^3\ x} \\ &= \frac{sin^2\ x - 1 + 2\ sin\ x + 2}{cos^3\ x} \\ &= \frac{(sin\ x + 1)^2}{cos^3\ x} \end{aligned}$$

The values that make $f''(x) = 0$ are not in the domain of the second derivative, therefore there are no inflection points.

53. - 103 Left to the student.

105. $V(r) = k(20r^2 - r^3) = 20kr^2 - kr^3$, $0 \leq r \leq 20$

The maximum occurs at the critical value of the function, since the endpoints result in a value of zero.

$$\begin{aligned} V'(r) &= 40kr - 3kr^2 \\ V'(r) &= 0 \\ 40kr - 3kr^2 &= 0 \\ kr(40 - 3r) &= 0 \\ r &= 0 \text{ not acceptable} \\ 40 - 3r &= 0 \\ r &= \frac{40}{3} \end{aligned}$$

107. $T(x) = 0.0338x^4 - 0.996x^3 + 8.57x^2 - 18.4x + 43.5$

a)

$$\begin{aligned} T'(x) &= 0.1352x^3 - 2.988x^2 + 17.14x - 18.4 \\ T''(x) &= 0.4056x^2 - 5.976x + 17.14 \end{aligned}$$

Using the quadratic formula we solve $T''(x) = 0$ to get

$$\begin{aligned} x &= 10.833 \approx 11 \\ x &= 3.901 \approx 4 \end{aligned}$$

The zeros of the second derivative (middle od April and middle of November) divide the number line into three intervals. We next use test points to check the sign of the second derivative at each of those intervals.

$T''(0) = 0.4056(0)^2 - 5.976(0) + 17.14 = 17.14 > 0$

$T''(10) = 0.4056(10)^2 - 5.976(10) + 17.14 = -2.06 < 0$

$T''(12) = 0.4056(12)^2 - 5.976(12) + 17.14 = 3.83 > 0$

This means that both zeros of the second derivative are inflection points.

b) The inflection points represent the values at which the rate of change of the temperature is changing the fastest or slowest.

109. $N(x) = -0.00006x^3 + 0.006x^2 - 0.1x + 1.9$

a)

$$\begin{aligned} -0.00018x^2 + 0.012x - 0.1 &= N'(x) \\ N'(x) &= 0 \\ -0.00018x^2 + 0.012x - 0.1 &= 0 \end{aligned}$$

Using the quadratic formula

$$\begin{aligned} x &= 9.763 \\ x &= 56.904 \end{aligned}$$

The zeros of the first derivative (ages 9.763 and 56.904) divide the number line into three intervals. We next check the sign of the first derivative in each of those intervals.
$N'(0) = -0.00018(0)^2 + 0.012(0) - 0.1 = -0.1 < 0$
$N'(10) = -0.00018(10)^2 + 0.012(10) - 0.1 = 0.002 > 0$
$N'(80) = -0.00018(80)^2 + 0.012(80) - 0.1 = -0.292 < 0$
Thus, we have a relative maximum at $x = 56.904$ and a relative minim at $x = 9.763$.

$$\begin{aligned} N(9.763) &= -0.00006(9.763)^3 + 0.006(9.763)^2 - 0.1(9.763) + 1.9 \\ &= 1.440 \\ N(56.904) &= -0.00006(56.904)^3 + 0.006(56.904)^2 - 0.1(56.904) + 1.9 \\ &= 4.582 \end{aligned}$$

b)

$$\begin{aligned} N''(x) &= -0.00036x + 0.012 \\ N''(x) &= 0 \\ -0.00036x + 0.012 &= 0 \\ x &= \frac{100}{3} \end{aligned}$$

The zero of the second derivative divides the number line into two intervals. We check the sign of the second derivative in each of those of intervals.
$N''(10) = -0.00036(10) + 0.012 = 0.0084 > 0$
$N''(50) = -0.00036(50) + 0.012 = -0.006 < 0$
Thus, there is an inflection point at $x = \frac{100}{3}$ that has a value of

$$\begin{aligned} N\left(\frac{100}{3}\right) &= -0.00006\left(\frac{100}{3}\right)^3 + 0.006\left(\frac{100}{3}\right)^2 - 0.1\left(\frac{100}{3}\right) + 1.9 \\ &= 3.011 \end{aligned}$$

c)

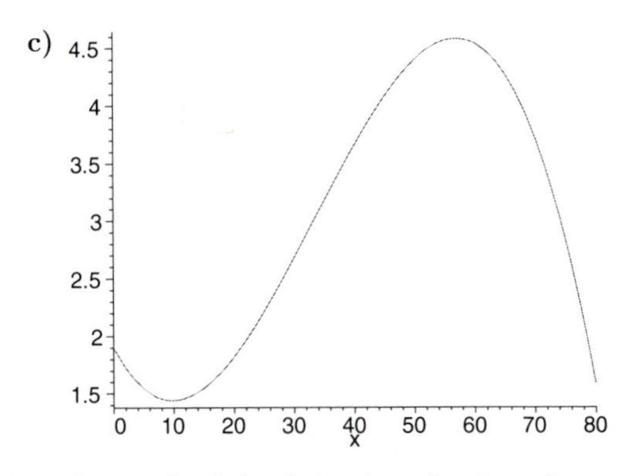

111. h is the graph of the derivative of g since the maximums and minumums occur where h is zero.

113.
a) Relative maximum at approximately $(2, 7)$
b) Relative mu=inimum at $(8, 0)$
c) Inflection points at $(4, 1)$
d) Increasing on $(0, 2)$ and $(8, 12)$
e) Decreasing on $(2, 8)$
f) Concave up on $(4, 12)$
g) Concave down on $(0, 4)$

115. Left to the student. (Answers vary)

117. Relative minimum at $(0, 0)$, and a relative maximum at $(1, 1)$

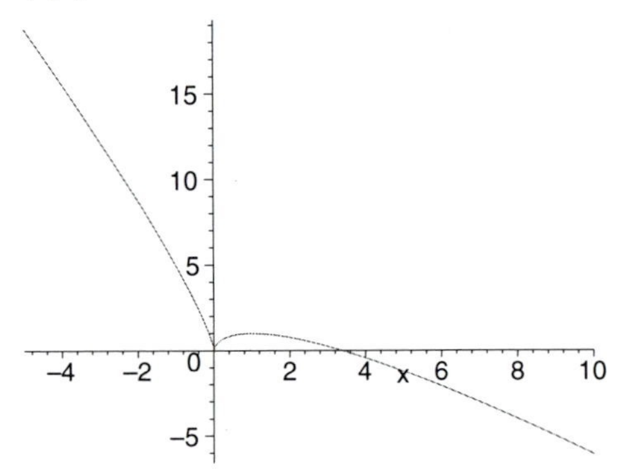

119. Relative maximum at $(0, 0)$ and a relative minimum at $(0.8, -1.1)$

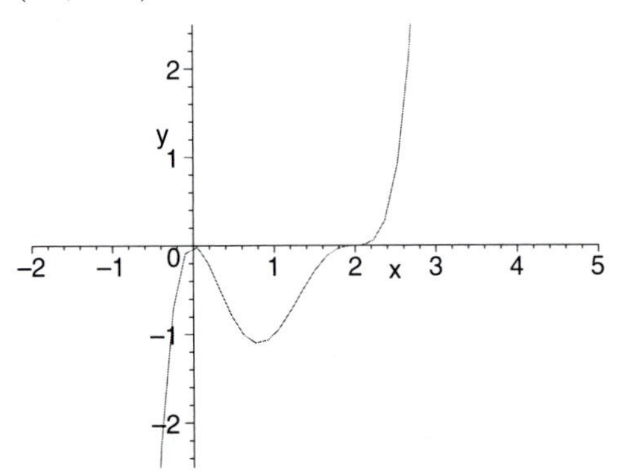

121. Relative minimum at $(0.25, -0.25)$

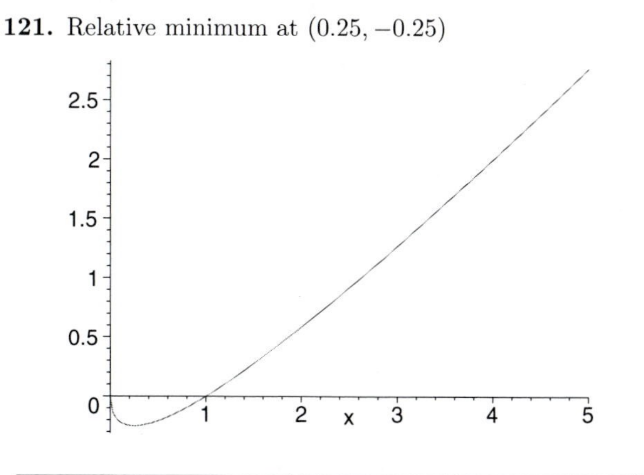

Exercise Set 3.3

1. Find $\lim_{x\to\infty} \frac{2x-4}{5x}$.

We will use some algebra and the fact that as $x \to \infty$, $\frac{b}{ax^n} \to 0$, for any positive integer n.

$$\lim_{x\to\infty} \frac{2x-4}{5x}$$

$$= \lim_{x\to\infty} \frac{2x-4}{5x} \cdot \frac{(1/x)}{(1/x)} \quad \text{Multiplying by a form of 1}$$

$$= \lim_{x\to\infty} \frac{2x \cdot \frac{1}{x} - 4 \cdot \frac{1}{x}}{5x \cdot \frac{1}{x}}$$

$$= \lim_{x\to\infty} \frac{2 - \frac{4}{x}}{5}$$

$$= \frac{2-0}{5} \quad \text{As } x \to \infty, \frac{4}{x} \to 0.$$

$$= \frac{2}{5}$$

3. Find $\lim_{x\to\infty} \left(5 - \frac{2}{x}\right)$.

We will use the fact that as $x \to \infty$, $\frac{b}{ax^n} \to 0$, for any positive integer n.

$$\lim_{x\to\infty} \left(5 - \frac{2}{x}\right)$$

$$= 5 - 0 \quad \text{As } x \to \infty, \frac{2}{x} \to 0.$$

$$= 5$$

5. Find $\lim_{x\to\infty} \frac{2x-5}{4x+3}$.

We will use some algebra and the fact that as $x \to \infty$, $\frac{b}{ax^n} \to 0$, for any positive integer n.

$$\lim_{x\to\infty} \frac{2x-5}{4x+3}$$

$$= \lim_{x\to\infty} \frac{2x-5}{4x+3} \cdot \frac{(1/x)}{(1/x)} \quad \text{Multiplying by a form of 1}$$

$$= \lim_{x\to\infty} \frac{2x \cdot \frac{1}{x} - 5 \cdot \frac{1}{x}}{4x \cdot \frac{1}{x} + 3 \cdot \frac{1}{x}}$$

$$= \lim_{x\to\infty} \frac{2 - \frac{5}{x}}{4 + \frac{3}{x}}$$

$$= \frac{2-0}{4+0} \quad \text{As } x \to \infty, \frac{5}{x} \to 0 \text{ and } \frac{3}{x} \to 0.$$

$$= \frac{2}{4} = \frac{1}{2}$$

7. Find $\lim_{x\to\infty} \frac{2x^2-5}{3x^2-x+7}$.

We will use some algebra and the fact that as $x \to \infty$, $\frac{b}{ax^n} \to 0$, for any positive integer n.

$$\lim_{x\to\infty} \frac{2x^2-5}{3x^2-x+7}$$

$$= \lim_{x\to\infty} \frac{2x^2-5}{3x^2-x+7} \cdot \frac{(1/x^2)}{(1/x^2)} \quad \text{Multiplying by a form of 1}$$

$$= \lim_{x\to\infty} \frac{2x^2 \cdot \frac{1}{x^2} - 5 \cdot \frac{1}{x^2}}{3x^2 \cdot \frac{1}{x^2} - x \cdot \frac{1}{x^2} + 7 \cdot \frac{1}{x^2}}$$

$$= \lim_{x\to\infty} \frac{2 - \frac{5}{x^2}}{3 - \frac{1}{x} + \frac{7}{x^2}}$$

$$= \frac{2-0}{3-0+0} \quad \text{As } x \to \infty, \frac{5}{x^2} \to 0, \frac{1}{x} \to 0, \text{ and } \frac{7}{x^2} \to 0$$

$$= \frac{2}{3}$$

9. Find $\lim_{x\to\infty} \frac{4-3x}{5-2x^2}$.

We divide the numerator and the denominator by x^2, the highest power of x in the denominator.

$$\lim_{x\to\infty} \frac{4-3x}{5-2x^2} = \lim_{x\to\infty} \frac{\frac{4}{x^2} - \frac{3}{x}}{\frac{5}{x^2} - 2}$$

$$= \frac{0-0}{0-2}$$

$$= 0$$

11. Find $\lim\limits_{x\to\infty} \dfrac{8x^4 - 3x^2}{5x^2 + 6x}$.

We divide the numerator and the denominator by x^2, the highest power of x in the denominator.

$$\lim_{x\to\infty} \frac{8x^4 - 3x^2}{5x^2 + 6x} = \lim_{x\to\infty} \frac{8x^2 - 3}{5 + \dfrac{6}{x}}$$

$$= \frac{\lim\limits_{x\to\infty} 8x^2 - 3}{5 + 0} = \infty$$

13. Find $\lim\limits_{x\to\infty} \dfrac{6x^4 - 5x^2 + 7}{8x^6 + 4x^3 - 8x}$.

We divide the numerator and the denominator by x^6, the highest power of x in the denominator.

$$\lim_{x\to\infty} \frac{6x^4 - 5x^4 + 7}{8x^6 + 4x^3 - 8x} = \lim_{x\to\infty} \frac{\dfrac{6}{x^2} - \dfrac{5}{x^4} + \dfrac{7}{x^6}}{8 + \dfrac{4}{x^3} - \dfrac{8}{x^5}}$$

$$= \frac{0 - 0 + 0}{8 + 0 - 0}$$

$$= 0$$

15. Find $\lim\limits_{x\to\infty} \dfrac{11x^5 + 4x^3 - 6x^2 + 2}{6x^3 + 5x^2 + 3x - 1}$.

We divide the numerator and the denominator by x^3, the highest power of x in the denominator.

$$\lim_{x\to\infty} \frac{11x^5+4x^3-6x+2}{6x^3+5x^2+3x-1} = \lim_{x\to\infty} \frac{11x^2+4-\dfrac{6}{x^2}+\dfrac{2}{x^3}}{6+\dfrac{5}{x}+\dfrac{3}{x^2}-\dfrac{1}{x^3}}$$

$$= \frac{\lim\limits_{x\to\infty} 11x^2+4-0+0}{6+0+0-0}$$

$$= \infty$$

17. $f(x) = \dfrac{4}{x}$, or $4x^{-1}$

a) *Intercepts.* Since the numerator is the constant 4, there are no x-intercepts. The number 0 is not in the domain of the function, so there are no y-intercepts.

b) *Asymptotes.*

Vertical. The denominator is 0 for $x = 0$, so the line $x = 0$ is a vertical asymptote.

Horizontal. The degree of the numerator is less than the degree of the denominator, so $y = 0$ is a horizontal asymptote.

Oblique. There is no oblique asymptote since the degree of the numerator is not one more than the degree of the denominator.

c) *Derivatives.*

$$f'(x) = -4x^{-2} = -\frac{4}{x^2}$$

$$f''(x) = 8x^{-3} = \frac{8}{x^3}$$

d) *Critical points.* The number 0 is not in the domain of f. Now $f'(x)$ exists for all values of x except 0. The equation $f'(x) = 0$ has no solution, so there are no critical points.

e) *Increasing, decreasing, relative extrema.* Use 0 to divide the real number line into two intervals, A: $(-\infty, 0)$ and B: $(0, \infty)$. Test a point in each interval.

A: Test -1, $f'(-1) = -\dfrac{4}{(-1)^2} = -4 < 0$

B: Test 1, $f'(1) = -\dfrac{4}{1^2} = -4 < 0$

Then f is decreasing on both intervals. Since there are no critical points, there are no relative extrema.

f) *Inflection points.* $f''(0)$ does not exist, but because $f(0)$ does not exist there cannot be an inflection point at 0. The equation $f''(x) = 0$ has no solution, so there are no inflection points.

g) *Concavity.* Use 0 to divide the real number line as in step (e). Note that for any $x < 0$, $x^3 < 0$, so

$$f''(x) = \frac{8}{x^3} < 0$$

and for any $x > 0$, $x^3 > 0$, so

$$f''(x) = \frac{8}{x^3} > 0.$$

Then f is concave down on $(-\infty, 0)$ and concave up on $(0, \infty)$.

h) *Sketch.* Use the preceding information to sketch the graph. Compute function values as needed.

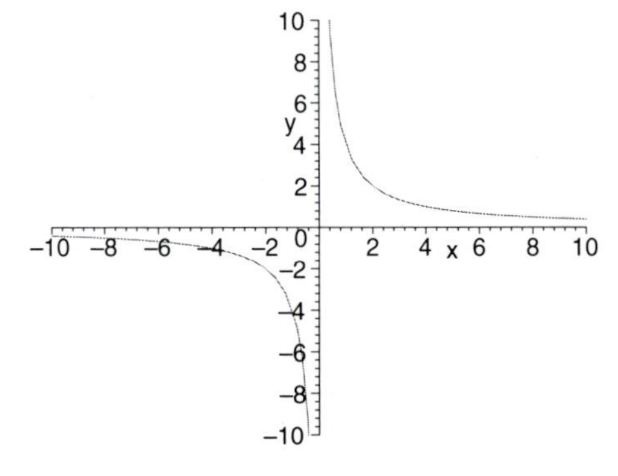

19. $f(x) = \dfrac{-2}{x - 5}$

a) *Intercepts.* Since the numerator is the constant -2, there are no x-intercepts. To find the y-intercepts we compute $f(0)$:

$$f(0) = \frac{-2}{0 - 5} = \frac{-2}{-5} = \frac{2}{5}$$

Then $\left(0, \dfrac{2}{5}\right)$ is the y-intercept.

b) *Asymptotes.*

Vertical. The denominator is 0 for $x = 5$, so the line $x = 5$ is a vertical asymptote.

Horizontal. The degree of the numerator is less than the degree of the denominator, so $y = 0$ is a horizontal asymptote.

Oblique. There is no oblique asymptote since the degree of the numerator is not one more than the degree of the denominator.

c) *Derivatives.*

$$f'(x) = 2(x-5)^{-2} = \frac{2}{(x-5)^2}$$

$$f''(x) = -4(x-5)^{-3} = -\frac{4}{(x-5)^3}$$

d) *Critical points.* $f'(5)$ does not exist, but because $f(5)$ does not exist, $x = 5$ is not a critical point. The equation $f'(x) = 0$ has no solution, so there are no critical points.

e) *Increasing, decreasing, relative extrema.* Use 5 to divide the real number line into two intervals, A: $(-\infty, 5)$ and B: $(5, \infty)$. Test a point in each interval.

A: Test 0, $f'(0) = \dfrac{2}{(0-5)^2} = \dfrac{2}{25} > 0$

B: Test 6, $f'(6) = \dfrac{2}{(6-5)^2} = 2 > 0$

Then f is increasing on both intervals. Since there are no critical points, there are no relative extrema.

f) *Inflection points.* $f''(5)$ does not exist, but because $f(5)$ does not exist there cannot be an inflection point at 5. The equation $f''(x) = 0$ has no solution, so there are no inflection points.

g) *Concavity.* Use 5 to divide the real number line as in step (e). Note that for any $x < 5$, $(x-5)^3 < 0$, so

$$f''(x) = -\frac{4}{(x-5)^3} > 0$$

and for any $x > 5$, $(x-5)^3 > 0$, so

$$f''(x) = -\frac{4}{(x-5)^3} < 0.$$

Then f is concave up on $(-\infty, 5)$ and concave down on $(5, \infty)$.

h) *Sketch.* Use the preceding information to sketch the graph. Compute function values as needed.

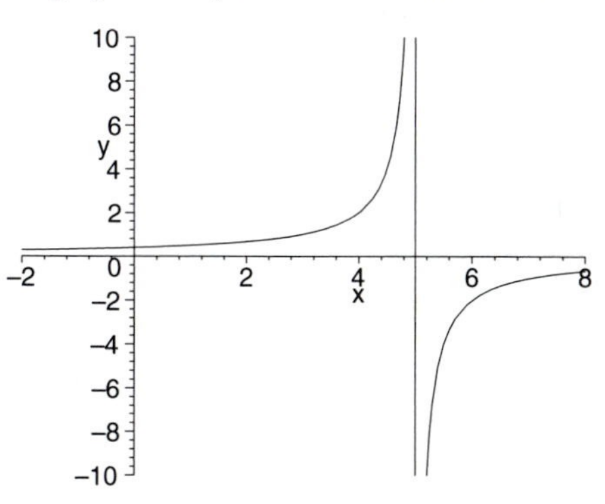

21. $f(x) = \dfrac{1}{x-3}$

a) *Intercepts.* Since the numerator is the constant 1, there are no x-intercepts.

$f(0) = \dfrac{1}{0-3} = -\dfrac{1}{3}$, so $\left(0, -\dfrac{1}{3}\right)$ is the y-intercept.

b) *Asymptotes.*

Vertical. The denominator is 0 for $x = 3$, so the line $x = 3$ is a vertical asymptote.

Horizontal. The degree of the numerator is less than the degree of the denominator, so $y = 0$ is a horizontal asymptote.

Oblique. There is no oblique asymptote since the degree of the numerator is not one more than the degree of the denominator.

c) *Derivatives.*

$$f'(x) = -(x-3)^{-2} = -\frac{1}{(x-3)^2}$$

$$f''(x) = 2(x-3)^{-3} = \frac{2}{(x-3)^3}$$

d) *Critical points.* $f'(3)$ does not exist, but because $f(3)$ does not exist, $x = 3$ is not a critical point. The equation $f'(x) = 0$ has no solution, so there are no critical points.

e) *Increasing, decreasing, relative extrema.* Use 3 to divide the real number line into two intervals, A: $(-\infty, 3)$ and B: $(3, \infty)$. Test a point in each interval.

A: Test 0, $f'(0) = -\dfrac{1}{(0-3)^2} = -\dfrac{1}{9} < 0$

B: Test 4, $f'(4) = -\dfrac{1}{(4-3)^2} = -1 < 0$

Then f is decreasing on both intervals. Since there are no critical points, there are no relative extrema.

f) *Inflection points.* $f''(3)$ does not exist, but because $f(3)$ does not exist there cannot be an inflection point at 3. The equation $f''(x) = 0$ has no solution, so there are no inflection points.

g) *Concavity.* Use 3 to divide the real number line as in step (e). Note that for any $x < 3$, $(x-3)^3 < 0$, so

$$f''(x) = \frac{2}{(x-3)^3} < 0$$

and for any $x > 3$, $(x-3)^3 > 0$, so

$$f''(x) = \frac{2}{(x-3)^3} > 0.$$

Then f is concave down on $(-\infty, 3)$ and concave up on $(3, \infty)$.

h) *Sketch.* Use the preceding information to sketch the graph. Compute function values as needed.

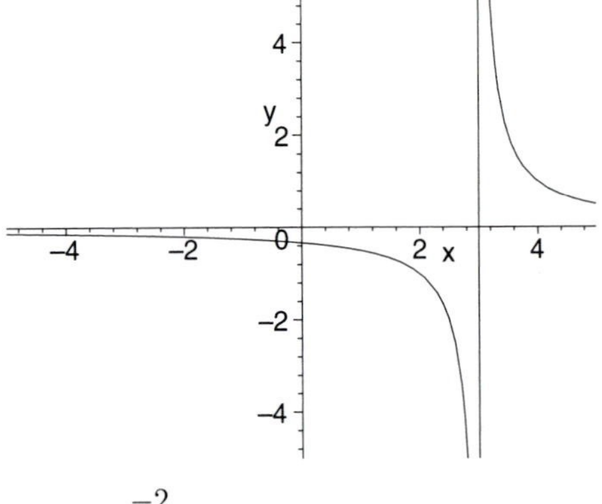

23. $f(x) = \dfrac{-2}{x+5}$

a) *Intercepts.* Since the numerator is the constant -2, there are no x-intercepts.

$f(0) = \dfrac{-2}{0+5} = -\dfrac{2}{5}$, so $\left(0, -\dfrac{2}{5}\right)$ is the y-intercept.

b) *Asymptotes.*

Vertical. The denominator is 0 for $x = -5$, so the line $x = -5$ is a vertical asymptote.

Horizontal. The degree of the numerator is less than the degree of the denominator, so $y = 0$ is a horizontal asymptote.

Oblique. There is no oblique asymptote since the degree of the numerator is not one more than the degree of the denominator.

c) *Derivatives.*

$$f'(x) = 2(x+5)^{-2} = \frac{2}{(x+5)^2}$$

$$f''(x) = -4(x+5)^{-3} = \frac{-4}{(x+5)^3}$$

d) *Critical points.* $f'(-5)$ does not exist, but because $f(-5)$ does not exist, $x = -5$ is not a critical point. The equation $f'(x) = 0$ has no solution, so there are no critical points.

e) *Increasing, decreasing, relative extrema.* Use -5 to divide the real number line into two intervals, A: $(-\infty, -5)$ and B: $(-5, \infty)$. Test a point in each interval.

A: Test -6, $f'(0) = \dfrac{2}{(-6+5)^2} = 2 > 0$

B: Test 0, $f'(0) = \dfrac{2}{(0+5)^2} = \dfrac{2}{25} > 0$

Then f is increasing on both intervals. Since there are no critical points, there are no relative extrema.

f) *Inflection points.* $f''(-5)$ does not exist, but because $f(-5)$ does not exist there cannot be an inflection point at -5. The equation $f''(x) = 0$ has no solution, so there are no inflection points.

g) *Concavity.* Use -5 to divide the real number line as in step (e). Test a point in each interval.

A: Test -6, $f''(-6) = \dfrac{-4}{(-6+5)^3} = 4 > 0$

B: Test 0, $f''(0) = \dfrac{-4}{(0+5)^3} = -\dfrac{4}{125} < 0$

Then f is concave up on $(-\infty, -5)$ and concave down on $(-5, \infty)$.

h) *Sketch.* Use the preceding information to sketch the graph. Compute function values as needed.

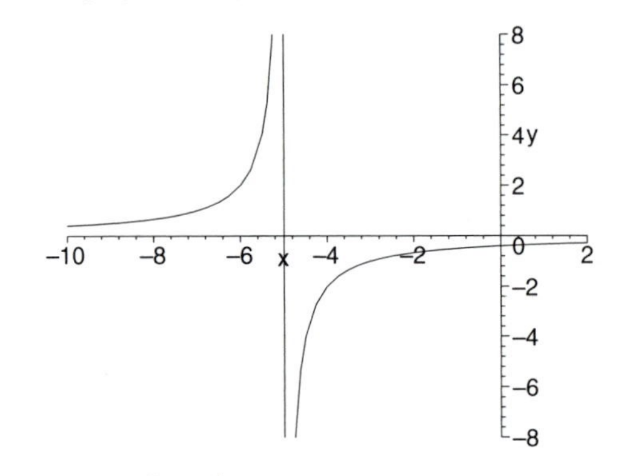

25. $f(x) = \dfrac{2x+1}{x}$

a) *Intercepts.* To find the x-intercepts, solve $f(x) = 0$.

$$\frac{2x+1}{x} = 0$$
$$2x+1 = 0$$
$$2x = -1$$
$$x = -\frac{1}{2}$$

Since $x = -\dfrac{1}{2}$ does not make the denominator 0, the x-intercept is $\left(-\dfrac{1}{2}, 0\right)$. The number 0 is not in the domain of f, so there are no y-intercepts.

b) *Asymptotes.*

Vertical. The denominator is 0 for $x = 0$, so the line $x = 0$ is a vertical asymptote.

Horizontal. The numerator and denominator have the same degree, so $y = \dfrac{2}{1}$, or $y = 2$, is a horizontal asymptote.

Oblique. There is no oblique asymptote since the degree of the numerator is not one more than the degree of the denominator.

c) *Derivatives.*

$$f'(x) = -\frac{1}{x^2}$$

$$f''(x) = 2x^{-3}, \text{ or } \frac{2}{x^3}$$

d) *Critical points.* $f'(0)$ does not exist, but because $f(0)$ does not exist $x = 0$ is not a critical point. The equation $f'(x) = 0$ has no solution, so there are no critical points.

e) *Increasing, decreasing, relative extrema.* Use 0 to divide the real number line into two intervals, A: $(-\infty, 0)$ and B: $(0, \infty)$. Test a point in each interval.

A: Test -1, $f'(-1) = -\dfrac{1}{(-1)^2} = -1 < 0$

B: Test 1, $f'(1) = -\dfrac{1}{1^2} = -1 < 0$

Then f is decreasing on both intervals. Since there are no critical points, there are no relative extrema.

f) *Inflection points.* $f''(0)$ does not exist, but because $f(0)$ does not exist there cannot be an inflection point at 0. The equation $f''(x) = 0$ has no solution, so there are no inflection points.

g) *Concavity.* Use 0 to divide the real number line as in step (e). Test a point in each interval.

A: Test -1, $f''(-1) = \dfrac{2}{(-1)^3} = -2 < 0$

B: Test 1, $f''(1) = \dfrac{2}{1^3} = 2 > 0$

Then f is concave down on $(-\infty, 0)$ and concave up on $(0, \infty)$.

h) *Sketch.* Use the preceding information to sketch the graph. Compute function values as needed.

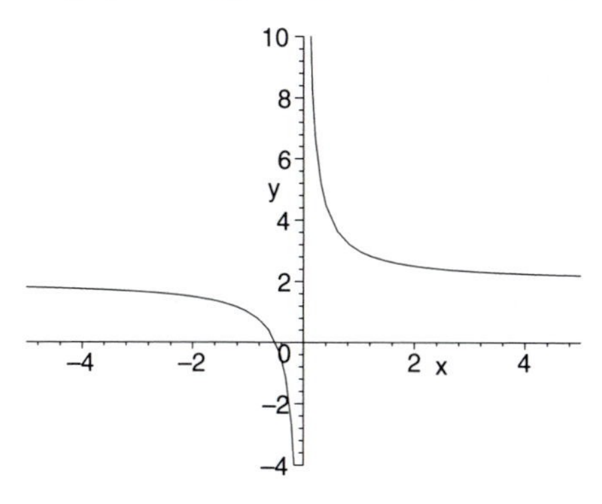

27. $f(x) = x + \dfrac{9}{x} = \dfrac{x^2+9}{x}$

a) *Intercepts.* The equation $f(x) = 0$ has no real number solution, so there are no x-intercepts. The number 0 is not in the domain of the function, so there are no y-intercepts.

b) *Asymptotes.*

Vertical. The denominator is 0 for $x = 0$, so the line $x = 0$ is a vertical asymptote.

Horizontal. The degree of the numerator is greater than the degree of the denominator, so there are no horizontal asymptotes.

Oblique. As $|x|$ gets very large, $f(x) = x + \dfrac{9}{x}$ approaches x, so $y = x$ is an oblique asymptote.

c) *Derivatives.*

$$f'(x) = 1 - 9x^{-2} = 1 - \frac{9}{x^2}$$

$$f''(x) = 18x^{-3} = \frac{18}{x^3}$$

d) *Critical points.* $f'(0)$ does not exist, but because $f(0)$ does not exist $x = 0$ is not a critical point. Solve $f'(x) = 0$.

$$1 - \frac{9}{x^2} = 1$$

$$1 = \frac{9}{x^2}$$

$$x^2 = 9$$

$$x = \pm 3$$

Thus, -3 and 3 are critical points. $f(-3) = -6$ and $f(3) = 6$, so $(-3, -6)$ and $(3, 6)$ are on the graph.

e) *Increasing, decreasing, relative extrema.* Use -3, 0, and 3 to divide the real number line into four intervals, A: $(-\infty, -3)$, B: $(-3, 0)$, C: $(0, 3)$, and D: $(3, \infty)$. Test a point in each interval.

A: Test -4, $f'(-4) = 1 - \dfrac{9}{(-4)^2} = \dfrac{7}{16} > 0$

B: Test -1, $f'(-1) = 1 - \dfrac{9}{(-1)^2} = -8 < 0$

C: Test 1, $f'(1) = 1 - \dfrac{9}{1^2} = -8 < 0$

D: Test 4, $f'(4) = 1 - \dfrac{9}{4^2} = \dfrac{7}{16} > 0$

Then f is increasing on $(-\infty, -3)$ and on $(3, \infty)$ and is decreasing on $(-3, 0)$ and on $(0, 3)$. Thus, there is a relative maximum at $(-3, -6)$ and a relative minimum at $(3, 6)$.

f) *Inflection points.* $f''(0)$ does not exist, but because $f(0)$ does not exist there cannot be an inflection point at 0. The equation $f''(x) = 0$ has no solution, so there are no inflection points.

g) *Concavity.* Use 0 to divide the real number line into two intervals, A: $(-\infty, 0)$ and B: $(0, \infty)$. Test a point in each interval.

A: Test -1, $f''(-1) = \dfrac{18}{(-1)^3} = -18 < 0$

B: Test 1, $f''(1) = \dfrac{18}{1^3} = 18 > 0$

Then f is concave down on $(-\infty, 0)$ and concave up on $(0, \infty)$.

h) *Sketch.* Use the preceding information to sketch the graph. Compute other function values as needed.

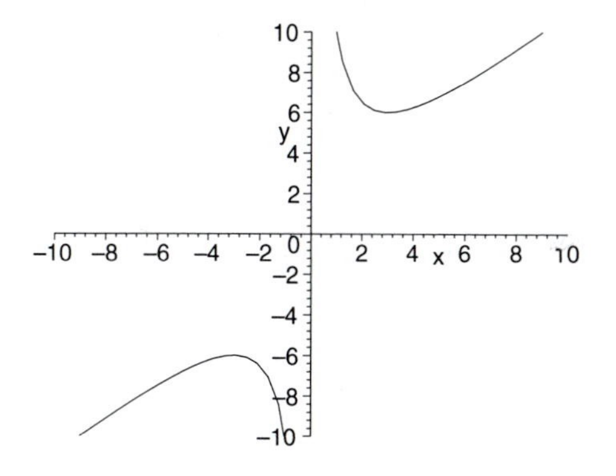

29. $f(x) = \frac{2}{x^2}$

a) *Intercepts.* Since the numerator is the constant 2, there are no x-intercepts. The number 0 is not in the domain of the function, so there are no y-intercepts.

b) *Asymptotes.*

Vertical. The denominator is 0 for $x = 0$, so the line $x = 0$ is a vertical asymptote.

Horizontal. The degree of the numerator is less than the degree of the denominator, so $y = 0$ is a horizontal asymptote.

Oblique. There is no oblique asymptote since the degree of the numerator is not one more than the degree of the denominator.

c) *Derivatives.*

$$f'(x) = -4x^{-3} = -\frac{4}{x^3}$$

$$f''(x) = 12x^{-4} = \frac{12}{x^4}$$

d) *Critical points.* $f'(0)$ does not exist, but because $f(0)$ does not exist $x = 0$ is not a critical point. The equation $f'(x) = 0$ has no solution, so there are no critical points.

e) *Increasing, decreasing, relative extrema.* Use 0 to divide the real number line into two intervals, A: $(-\infty, 0)$ and B: $(0, \infty)$.

A: Test -1, $f'(-1) = -\frac{4}{(-1)^3} = 4 > 0$

B: Test 1, $f'(1) = -\frac{4}{1^3} = -4 < 0$

Then f is increasing on $(-\infty, 0)$ and is decreasing on $(0, \infty)$. Since there are no critical points, there are no relative extrema.

f) *Inflection points.* $f''(0)$ does not exist, but because $f(0)$ does not exist there cannot be an inflection point at 0. The equation $f''(x) = 0$ has no solution, so there are no inflection points.

g) *Concavity.* Use 0 to divide the real number line as in step (e). Test a point in each interval.

A: Test -1, $f''(-1) = \frac{12}{(-1)^4} = 12 > 0$

B: Test 1, $f''(1) = \frac{12}{1^4} = 12 > 0$

Then f is concave up on both intervals.

h) *Sketch.* Use the preceding information to sketch the graph. Compute function values as needed.

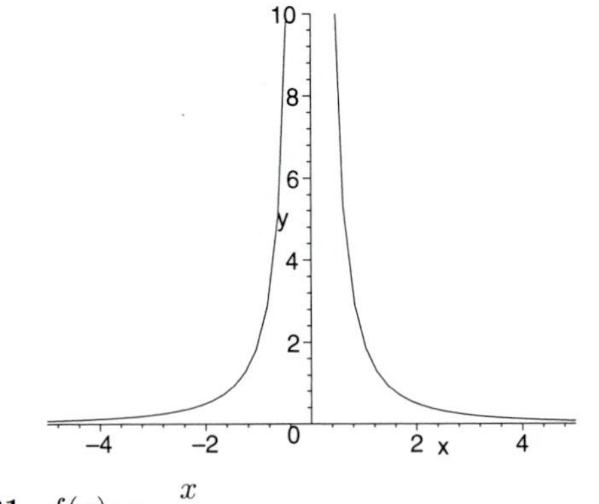

31. $f(x) = \frac{x}{x-3}$

a) *Intercepts.* The numerator is 0 for $x = 0$ and this value of x does not make the denominator 0, so $(0, 0)$ is the x-intercept. $f(0) = \frac{0}{0-3} = 0$, so the y-intercept is the x-intercept $(0, 0)$.

b) *Asymptotes.*

Vertical. The denominator is 0 for $x = 3$, so the line $x = 3$ is a vertical asymptote.

Horizontal. The numerator and the denominator have the same degree, so $y = \frac{1}{1}$, or $y = 1$, is a horizontal asymptote.

Oblique. There is no oblique asymptote since the degree of the numerator is not one more than the degree of the denominator.

c) *Derivatives.*

$$f'(x) = -\frac{3}{(x-3)^2}$$

$$f''(x) = 6(x-3)^{-3} = \frac{6}{(x-3)^3}$$

d) *Critical points.* $f'(3)$ does not exist, but because $f(3)$ does not exist $x = 3$ is not a critical point. The equation $f'(x) = 0$ has no solution, so there are no critical points.

e) *Increasing, decreasing, relative extrema.* Use 3 to divide the real number line into two intervals, A: $(-\infty, 3)$ and B: $(3, \infty)$. Test a point in each interval.

A: Test 0, $f'(0) = -\frac{1}{3} < 0$

B: Test 4, $f'(4) = -3 < 0$

Then f is decreasing on both intervals. Since there are no critical points, there are no relative extrema.

f) *Inflection points.* $f''(3)$ does not exist, but because $f(3)$ does not exist there cannot be an inflection point at 3. The equation $f''(x) = 0$ has no solution, so there are no inflection points.

g) *Concavity.* Use 3 to divide the real number line as in step (e).

A: Test 0, $f''(0) = -\frac{2}{9} < 0$

B: Test 4, $f''(4) = 6 > 0$

Then f is concave down on $(-\infty, 3)$ and concave up on $(3, \infty)$.

h) *Sketch.* Use the preceding information to sketch the graph. Compute function values as needed.

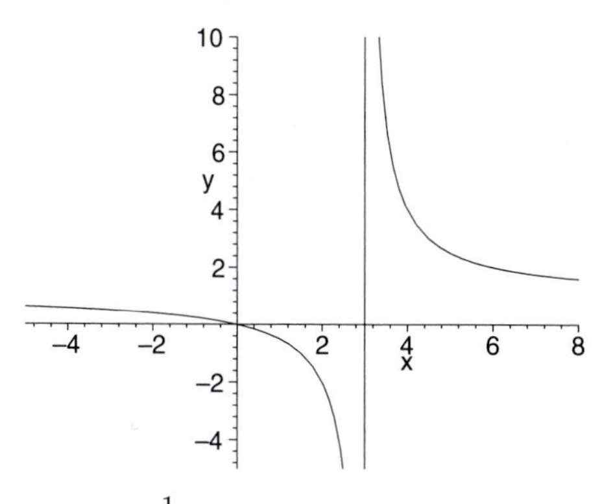

33. $f(x) = \dfrac{1}{x^2+3}$

a) *Intercepts.* Since the numerator is the constant 1, there are no x-intercepts.

$f(0) = \dfrac{1}{0^2+3} = \dfrac{1}{3}$, so $\left(0, \dfrac{1}{3}\right)$ is the y-intercept.

b) *Asymptotes.*

Vertical. $x^2+3=0$ has no real number solutions, so there are no vertical asymptotes.

Horizontal. The degree of the numerator is less than the degree of the denominator, so $y=0$ is a horizontal asymptote.

Oblique. There is no oblique asymptote since the degree of the numerator is not one more than the degree of the denominator.

c) *Derivatives.*

$$f'(x) = -\frac{2x}{(x^2+3)^2}$$

$$f''(x) = \frac{6x^2-6}{(x^2+3)^3}$$

d) *Critical points.* $f'(x)$ exists for all real numbers. Solve $f'(x)=0$.

$$-\frac{2x}{(x^2+3)^2} = 0$$
$$-2x = 0$$
$$x = 0 \quad \text{Critical point}$$

From step (a) we already know $\left(0, \dfrac{1}{3}\right)$ is on the graph.

e) *Increasing, decreasing, relative extrema.* Use 0 to divide the real number line into two intervals, A: $(-\infty, 0)$ and B: $(0, \infty)$. Test a point in each interval.

A: Test -1, $f'(-1) = \dfrac{1}{8} > 0$

B: Test 1, $f'(1) = -\dfrac{1}{8} < 0$

Then f is increasing on $(-\infty, 0)$ and decreasing on $(0, \infty)$. Thus, $\left(0, \dfrac{1}{3}\right)$ is a relative maximum.

f) *Inflection points.* $f''(x)$ exists for all real numbers. Solve $f''(x)=0$.

$$\frac{6x^2-6}{(x^2+3)^3} = 0$$
$$6x^2-6 = 0$$
$$6(x+1)(x-1) = 0$$

$x=-1$ or $x=1$ Possible inflection points

$f(-1) = \dfrac{1}{4}$ and $f(1) = \dfrac{1}{4}$, so $\left(-1, \dfrac{1}{4}\right)$ and $\left(1, \dfrac{1}{4}\right)$ are on the graph.

g) *Concavity.* Use -1 and 1 to divide the real number line into three intervals, A: $(-\infty, -1)$, B: $(-1, 1)$, and C: $(1, \infty)$. Test a point in each interval.

A: Test -2, $f''(-2) = \dfrac{18}{343} > 0$

B: Test 0, $f''(0) = -\dfrac{2}{9} < 0$

C: Test 2, $f''(2) = \dfrac{18}{343} > 0$

Then f is concave up on $(-\infty, -1)$ and on $(1, \infty)$ and is concave down on $(-1, 1)$. Thus, $\left(-1, \dfrac{1}{4}\right)$ and $\left(1, \dfrac{1}{4}\right)$ are both inflection points.

h) *Sketch.* Use the preceding information to sketch the graph. Compute other function values as needed.

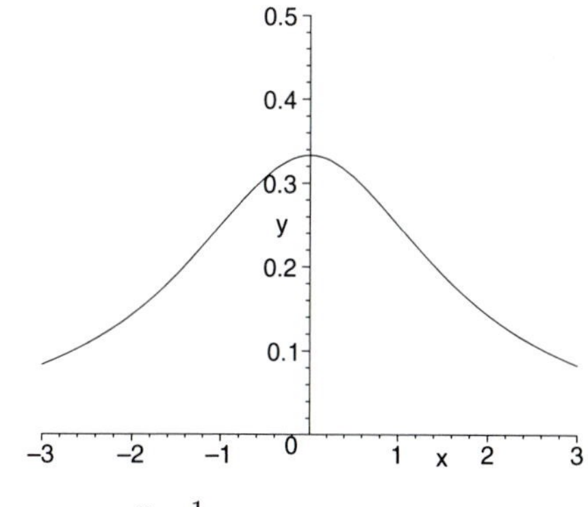

35. $f(x) = \dfrac{x-1}{x+2}$

a) *Intercepts.* The numerator is 0 for $x=1$ and this value of x does not make the denominator 0, so $(1, 0)$ is the x-intercept.

$f(0) = \dfrac{0-1}{0+2} = -\dfrac{1}{2}$, so $\left(0, -\dfrac{1}{2}\right)$ is the y-intercept.

b) *Asymptotes.*

Vertical. The denominator is 0 for $x=-2$, so the line $x=-2$ is a vertical asymptote.

Horizontal. The numerator and the denominator have the same degree, so $y = \dfrac{1}{1}$, or $y=1$, is a horizontal asymptote.

Oblique. There is no oblique asymptote since the degree of the numerator is not one more than the degree of the denominator.

c) *Derivatives.*

$$f'(x) = \frac{3}{(x+2)^2}$$

$$f''(x) = \frac{-6}{(x+2)^3}$$

d) *Critical points.* $f'(-2)$ does not exist, but because $f(-2)$ does not exist $x = -2$ is not a critical point. The equation $f'(x) = 0$ has no solution, so there are no critical points.

e) *Increasing, decreasing, relative extrema.* Use -2 to divide the real number line into two intervals, A: $(-\infty, -2)$ and B: $(-2, \infty)$. Test a point in each interval.

A: Test -3, $f'(-3) = 3 > 0$

B: Test -1, $f'(-1) = 3 > 0$

Then f is increasing on both intervals. Since there are no critical points, there are relative extrema.

f) *Inflection points.* $f''(-2)$ does not exist, but because $f(-2)$ does not exist there cannot be an inflection point at -2. The equation $f''(x) = 0$ has no solution, so there are no inflection points.

g) *Concavity.* Use -2 to divide the real number line as in step (e). Test a point in each interval.

A: Test -3, $f''(-3) = 6 > 0$

B: Test -1, $f''(-1) = -6 < 0$

Then f is concave up on $(-\infty, -2)$ and concave down on $(-2, \infty)$.

h) *Sketch.* Use the preceding information to sketch the graph. Compute function values as needed.

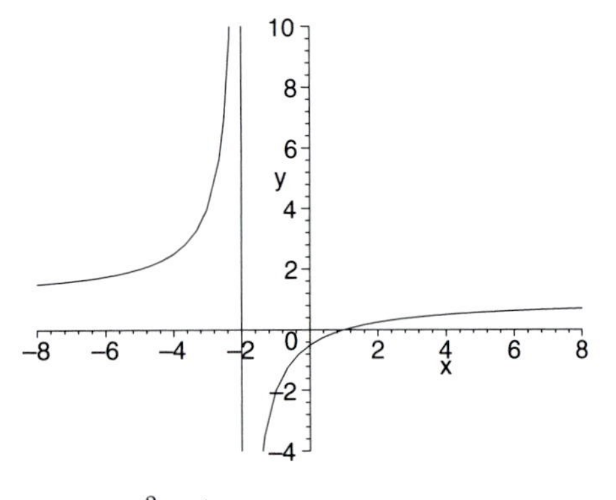

37. $f(x) = \dfrac{x^2 - 4}{x + 3}$

a) *Intercepts.* The numerator $x^2 - 4 = (x+2)(x-2)$ is 0 for $x = -2$ and or $x = 2$, and neither of these values makes the denominator 0. Thus, the x-intercepts are $(-2, 0)$ and $(2, 0)$.

$f(0) = \dfrac{0^2 - 4}{0 + 3} = -\dfrac{4}{3}$, so $\left(0, -\dfrac{4}{3}\right)$ is the y-intercept.

b) *Asymptotes.*

Vertical. The denominator is 0 for $x = -3$, so the line $x = -3$ is a vertical asymptote.

Horizontal. The degree of the numerator is greater than the degree of the denominator, so there are no horizontal asymptotes.

Oblique.

$$f(x) = x - 3 + \frac{5}{x+3}$$

$$\begin{array}{r|l} & x - 3 \\ \hline x+3 & x^2 \qquad\quad - 4 \\ & x^2 + 3x \\ \hline & \quad -3x - 4 \\ & \quad -3x - 9 \\ \hline & \qquad\quad 5 \end{array}$$

As $|x|$ gets very large, $f(x)$ approaches $x - 3$, so $y = x - 3$ is an oblique asymptote.

c) *Derivatives.*

$$f'(x) = \frac{x^2 + 6x + 4}{(x+3)^2}$$

$$f''(x) = \frac{10}{(x+3)^3}$$

d) *Critical points.* $f'(-3)$ does not exist, but because $f(-3)$ does not exist $x = -3$ is not a critical point. Solve $f'(x) = 0$.

$$\frac{x^2 + 6x + 4}{(x+3)^2} = 0$$

$$x^2 + 6x + 4 = 0$$

$x = -3 \pm \sqrt{5}$ Using the quadratic formula

$x \approx -5.24$ or $x \approx -0.76$ Critical points

$f(-5.24) \approx -10.47$ and $f(-0.76) \approx -1.53$, so $(-5.24, -10.47)$ and $(-0.76, -1.53)$ are on the graph.

e) *Increasing, decreasing, relative extrema.* Use -5.24, -3, and -0.76 to divide the real number line into four intervals, A: $(-\infty, -5.24)$, B: $(-5.24, -3)$, C: $(-3, -0.76)$, and D: $(-0.76, \infty)$. Test a point in each interval.

A: Test -6, $f'(-6) = \dfrac{4}{9} > 0$

B: Test -4, $f'(-4) = -4 < 0$

C: Test -2, $f'(-2) = -4 < 0$

D: Test 0, $f'(0) = \dfrac{4}{9} > 0$

Then f is increasing on $(-\infty, -5.24)$ and on $(-0.76, \infty)$ and is decreasing on $(-5.24, -3)$ and on $(-3, -0.76)$. Thus, $(-5.24, -10.47)$ is a relative maximum and $(-0.76, -1.53)$ is a relative minimum.

f) *Inflection points.* $f''(-3)$ does not exist, but because $f(-3)$ does not exist there cannot be an inflection point at -3. The equation $f''(x) = 0$ has no solution, so there are no inflection points.

g) *Concavity.* Use -3 to divide the real number line into two intervals, A: $(-\infty, -3)$ and B: $(-3, \infty)$. Test a point in each interval.

A: Test -4, $f''(-4) = -10 < 0$

B: Test -2, $f''(-2) = 10 > 0$

Then f is concave down on $(-\infty, -3)$ and concave up on $(-3, \infty)$.

h) *Sketch.* Use the preceding information to sketch the graph. Compute other function values as needed.

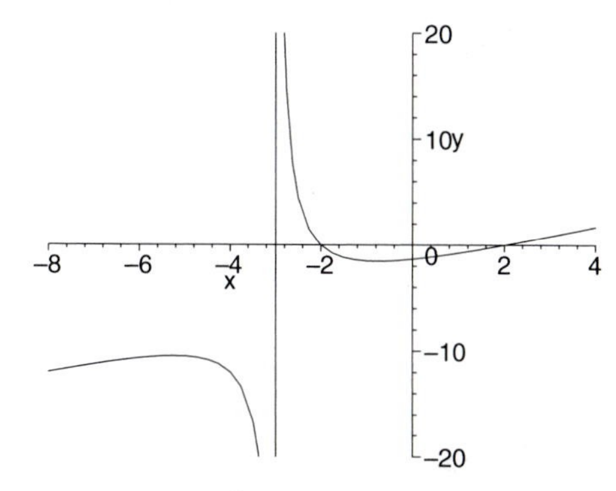

39. $f(x) = \dfrac{x-1}{x^2-2x-3}$

a) *Intercepts.* The numerator is 0 for $x = 1$, and this value of x does not make the denominator 0. Then $(1, 0)$ is the x-intercept.

$f(0) = \dfrac{0-1}{0^2-2\cdot 0-3} = \dfrac{1}{3}$, so $\left(0, \dfrac{1}{3}\right)$ is the y-intercept.

b) *Asymptotes.*

Vertical. The denominator $x^2 - 2x - 3 = (x+1)(x-3)$ is 0 for $x = -1$ or $x = 3$. Then the lines $x = -1$ and $x = 3$ are vertical asymptotes.

Horizontal. The degree of the numerator is less than the degree of the denominator, so $y = 0$ is a horizontal asymptote.

Oblique. There is no oblique asymptote since the degree of the numerator is not one more than the degree of the denominator.

c) *Derivatives.*

$$f'(x) = \frac{-x^2+2x-5}{(x^2-2x-3)^2}$$

$$f''(x) = \frac{2x^3-6x^2+30x-26}{(x^2-2x-3)^3}$$

d) *Critical points.* $f'(-1)$ and $f'(3)$ do not exist, but because $f(-1)$ and $f(3)$ do not exist $x = -1$ and $x = 3$ are not critical points. The equation $f'(x) = 0$ has no real number solution, so there are no critical points.

e) *Increasing, decreasing, relative extrema.* Use -1 and 3 to divide the real number line into three intervals, A: $(-\infty, -1)$, B: $(-1, 3)$, and C: $(3, \infty)$. Test a point in each interval.

A: Test -2, $f'(-2) = -\dfrac{13}{25} < 0$

B: Test 0, $f'(0) = -\dfrac{5}{9} < 0$

C: Test 4, $f'(4) = -\dfrac{13}{25} < 0$

Then f is decreasing on all three intervals. Since there are no critical points, there are no relative extrema.

f) *Inflection points.* $f''(-1)$ and $f''(3)$ do not exist, but because $f(-1)$ and $f(3)$ do not exist there cannot be an inflection point at -1 or at 3. Solve $f''(x) = 0$.

$$\frac{2x^3-6x^2+30x-26}{(x^2-2x-3)^3} = 0$$
$$2x^3-6x^2+30x-26 = 0$$
$$(x-1)(2x^2-4x+26) = 0$$
$$x-1 = 0 \quad\text{or}\quad 2x^2-4x+26 = 0$$
$$x = 1 \qquad \text{No real number solution}$$

$f(1) = 0$, so $(1, 0)$ is on the graph and is a possible inflection point.

g) *Concavity.* Use -1, 1, and 3 to divide the real number line into four intervals, A: $(-\infty, -1)$, B: $(-1, 1)$, C: $(1, 3)$, and D $(3, \infty)$. Test a point in each interval.

A: Test -2, $f''(-2) = -\dfrac{126}{125} < 0$

B: Test 0, $f''(0) = \dfrac{26}{27} > 0$

C: Test 2, $f''(2) = -\dfrac{26}{27} < 0$

D: Test 4, $f''(4) = \dfrac{126}{125} > 0$

Then f is concave down on $(-\infty, -1)$ and on $(1, 3)$ and is concave up on $(-1, 1)$ and on $(3, \infty)$. Thus, $(1, 0)$ is an inflection point.

h) *Sketch.* Use the preceding information to sketch the graph. Compute other function values as needed.

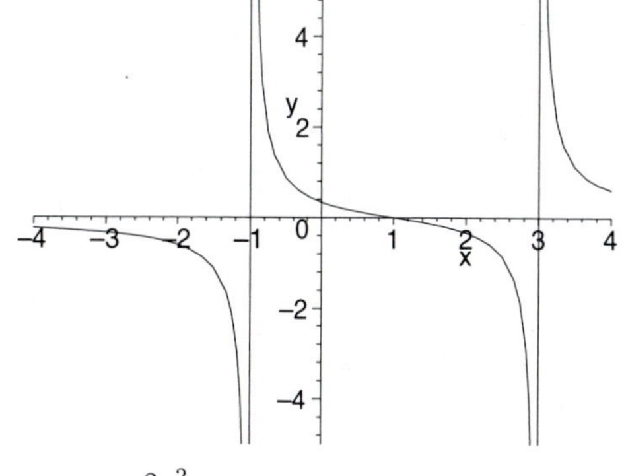

41. $f(x) = \dfrac{2x^2}{x^2-16}$

a) *Intercepts.* The numerator is 0 for $x = 0$, and this value of x does not make the denominator 0, so $(0, 0)$ is the x-intercept.

$f(0) = 0$, so the y-intercept is the x-intercept $(0, 0)$.

b) *Asymptotes.*

Vertical. The denominator $x^2 - 16 = (x+4)(x-4)$ is 0 for $x = -4$ or $x = 4$, so the lines $x = -4$ and $x = 4$ are vertical asymptotes.

Horizontal. The numerator and denominator have the same degree, so $y = \dfrac{2}{1}$, or $y = 2$, is a horizontal asymptote.

Oblique. There is no oblique asymptote since the degree of the numerator is not one more than the degree of the denominator.

c) *Derivatives.*

$$f'(x) = \frac{-64x}{(x^2-16)^2}$$

$$f''(x) = \frac{192x^2+1024}{(x^2-16)^3}$$

d) *Critical points.* $f'(-4)$ and $f'(4)$ do not exist, but because $f(-4)$ and $f(4)$ do not exist $x = -4$ and $x = 4$ are not critical points. Solve $f'(x) = 0$.

$$\frac{-64x}{(x^2-16)^2} = 0$$
$$-64x = 0$$
$$x = 0 \quad \text{Critical point}$$

From step (a) we already know that $(0,0)$ is on the graph.

e) *Increasing, decreasing, relative extrema.* Use -4, 0, and 4 to divide the real number line into four intervals, A: $(-\infty,-4)$, B: $(-4,0)$, C: $(0,4)$, and D: $(4,\infty)$.

A: Test -5, $f'(-5) = \frac{320}{81} > 0$

B: Test -1, $f'(-1) = \frac{64}{225} > 0$

C: Test 1, $f'(1) = -\frac{64}{225} < 0$

D: Test 5, $f'(5) = -\frac{320}{81} < 0$

Then f is increasing on $(-\infty,-4)$ and on $(-4,0)$ and is decreasing on $(0,4)$ and on $(4,\infty)$. Thus, there is a relative maximum at $(0,0)$.

f) *Inflection points.* $f''(-4)$ and $f''(4)$ do not exist, but because $f(-4)$ and $f(4)$ do not exist there cannot be an inflection point at -4 or 4. The equation $f''(x) = 0$ has no real-number solution, so there are no inflection points.

g) *Concavity.* Use -4 and 4 to divide the real number line into three intervals, A: $(-\infty,-4)$, B: $(-4,4)$, and C: $(4,\infty)$. Test a point in each interval.

A: Test -5, $f''(-5) = \frac{5824}{729} > 0$

B: Test 0, $f''(0) = -\frac{1}{4} < 0$

C: Test 5, $f''(5) = -\frac{5824}{729} > 0$

Then f is concave up on $(-\infty,-4)$ and on $(4,\infty)$ and is concave down on $(-4,4)$.

h) *Sketch.* Use the preceding information to sketch the graph. Compute other function values as needed.

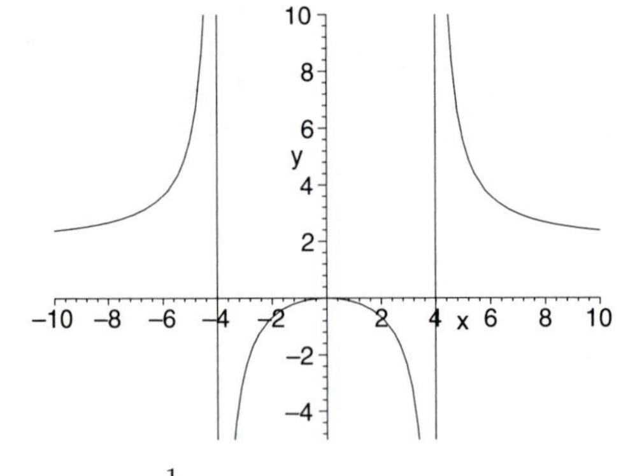

43. $f(x) = \frac{1}{x^2-1}$

a) *Intercepts.* Since the numerator is the constant 1, there are no x-intercepts.

$f(0) = \frac{1}{0^2-1} = -1$, so the y-intercept is $(0,-1)$.

b) *Asymptotes.*

Vertical. The denominator $x^2 - 1 = (x+1)(x-1)$ is 0 for $x = -1$ or $x = 1$, so the lines $x = -1$ and $x = 1$ are vertical asymptotes.

Horizontal. The degree of the numerator is less than the degree of the denominator, so $y = 0$ is a horizontal asymptote.

Oblique. There is no oblique asymptote since the degree of the numerator is not one more than the degree of the denominator.

c) *Derivatives.*

$$f'(x) = \frac{-2x}{(x^2-1)^2}$$

$$f''(x) = \frac{2(3x^2+1)}{(x^2-1)^3}$$

d) *Critical points.* $f'(-1)$ and $f'(1)$ do not exist, but because $f(-1)$ and $f(1)$ do not exist $x = -1$ and $x = 1$ are not critical points. Solve $f'(x) = 0$.

$$\frac{-2x}{(x^2-1)^2} = 0$$
$$-2x = 0$$
$$x = 0 \quad \text{Critical point}$$

From step (a) we already know that $(0,-1)$ is on the graph.

e) *Increasing, decreasing, relative extrema.* Use -1, 0, and 1 to divide the real number line into four intervals, A: $(-\infty,-1)$, B: $(-1,0)$, C: $(0,1)$, and D: $(1,\infty)$. Test a point in each interval.

A: Test -2, $f'(-2)=\frac{4}{9}>0$

B: Test $-\frac{1}{2}$, $f'\left(-\frac{1}{2}\right)=\frac{16}{9}>0$

C: Test $\frac{1}{2}$, $f'\left(\frac{1}{2}\right)=-\frac{16}{9}<0$

D: Test 2, $f'(2)=-\frac{4}{9}<0$

Then f is increasing on $(-\infty,-1)$ and on $(-1,0)$ and is decreasing on $(0,1)$ and on $(1,\infty)$. Thus, there is a relative maximum at $(0,-1)$.

f) *Inflection points.* $f''(-1)$ and $f''(1)$ do not exist, but because $f(-1)$ and $f(1)$ do not exist there cannot be an inflection point at -1 or at 1. The equation $f''(x)=0$ has no real-number solution, so there are no inflection points.

g) *Concavity.* Use -1 and 1 to divide the real number line into three intervals, A: $(-\infty,-1)$, B: $(-1,1)$, and C: $(1,\infty)$. Test a point in each interval.

A: Test -2, $f''(-2)=\frac{26}{9}>0$

B: Test 0, $f''(0)=-2<0$

C: Test 2, $f''(2)=\frac{26}{9}>0$

Then f is concave up on $(-\infty,-1)$ and on $(1,\infty)$ and is concave down on $(-1,1)$.

h) *Sketch.* Use the preceding information to sketch the graph. Compute other function values as needed.

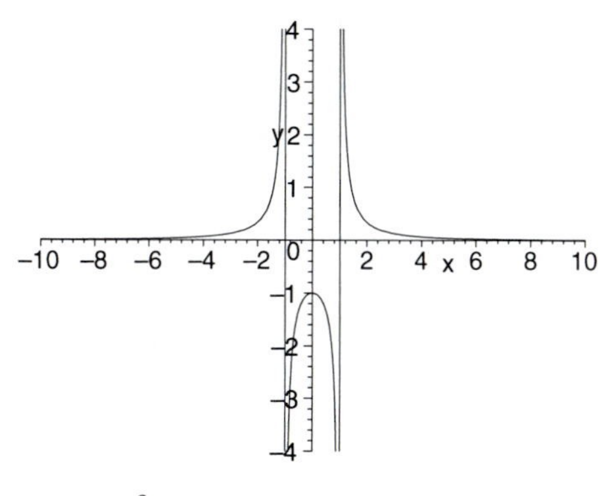

45. $f(x)=\dfrac{x^2+1}{x}$

a) *Intercepts.* Since the numerator has no real-number solutions, there are no x-intercepts.

$f(0)$ does not exist, so there is no y-intercept.

b) *Asymptotes.*

Vertical. The denominator x is 0 for $x=0$, so the line $x=0$ is a vertical asymptote.

Horizontal. The degree of the numerator is not less than or equal to the degree of the denominator, so there is no horizontal asymptote.

Oblique. The degree of the numerator is one more than the degree of the denominator, so there is an oblique asymptote. When we divide x^2+1 by x we have $f(x)=\frac{x^2+1}{x}=x+\frac{1}{x}$. As $|x|$ gets very large, $\frac{1}{x}$ approaches 0. Thus, $y=x$ is an oblique asymptote.

c) *Derivatives.*

$$f'(x)=\frac{x^2-1}{x^2}$$

$$f''(x)=\frac{2}{x^3}$$

d) *Critical points.* $f'(0)$ does not exist, but because $f(0)$ does not exist 0 is not a critical point. Solve $f'(x)=0$.

$$\frac{x^2-1}{x^2}=0$$

$$x^2-1=0$$

$$(x+1)(x-1)=0$$

$x=-1$ or $x=1$ Critical points

$f(-1)=-2$ and $f(1)=2$, so $(-1,-2)$ and $(1,2)$ are on the graph.

e) *Increasing, decreasing, relative extrema.* Use -1, 0, and 1 to divide the real number line into four intervals, A: $(-\infty,-1)$, B: $(-1,0)$, C: $(0,1)$, and D: $(1,\infty)$. Test a point in each interval.

A: Test -2, $f'(-2)=\frac{3}{4}>0$

B: Test $-\frac{1}{2}$, $f'\left(-\frac{1}{2}\right)=-3<0$

C: Test $\frac{1}{2}$, $f'\left(\frac{1}{2}\right)=-3<0$

D: Test 2, $f'(2)=\frac{3}{4}>0$

Then f is increasing on $(-\infty,-1)$ and on $(1,\infty)$ and is decreasing on $(-1,0)$ and $(0,1)$. Thus, there is a relative maximum at $(-1,-2)$ and a relative minimum at $(1,2)$.

f) *Inflection points.* $f''(0)$ does not exist, but because $f(0)$ does not exist there cannot be an inflection point at 0. The equation $f''(x)=0$ has no solution, so there are no inflection points.

g) *Concavity.* Use 0 to divide the real number line into two intervals, A: $(-\infty,0)$ and B: $(0,\infty)$. Test a point in each interval.

A: Test -1, $f''(-1)=-2<0$

B: Test 1, $f''(1)=2>0$

Then f is concave down on $(-\infty,0)$ and is concave up on $(0,\infty)$.

h) *Sketch.* Use the preceding information to sketch the graph. Compute other function values as needed.

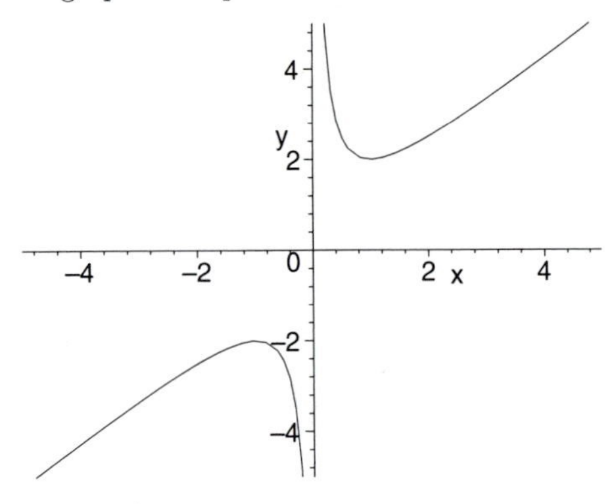

47. $C(p) = \dfrac{\$48,000}{100-p}$

We will only consider the interval $[0, 100)$ since it is not possible to remove less than 0% or more than 100% of the pollutants and $C(p)$ is not defined for $p = 100$.

a) $C(0) = \dfrac{\$48,000}{100-0} = \$480$

$C(20) = \dfrac{\$48,000}{100-20} = \$600$

$C(80) = \dfrac{\$48,000}{100-80} = \$2400$

$C(90) = \dfrac{\$48,000}{100-90} = \$4800$

b) $\lim\limits_{p\to 100^-} C(p) = \lim\limits_{p\to 100^-} \dfrac{\$48,000}{100-p} = \infty$

c) The cost of removing 100% of the pollutants is infinitely high.

d) Using the techniques of this section we find the following additional information.

Intercepts. No p-intercept; $(0, 480)$ is the C-intercept.

Asymptotes. *Vertical.* $p = 100$

Horizontal. $C = 0$

Oblique. None

Increasing, decreasing, relative extrema. $C(p)$ is increasing on $[0, 100)$. There are no relative extrema.

Inflection points, concavity. $C(p)$ is concave up on $[0, 100)$. There is no inflection point.

We use this information and compute other function values as needed to sketch the graph.

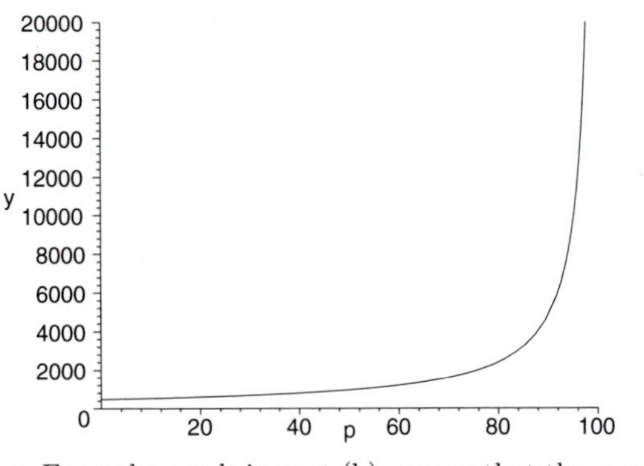

e) From the result in part (b), we see that the company cannot afford to remove 100% of the pollutants.

49. $T(t) = \dfrac{6t}{t^2+1} + 98.6$

a)

$$\begin{aligned} T(0) &= \frac{6(0)}{0+1} + 98.6 = 98.6 \\ T(1) &= \frac{6(1)}{1+1} + 98.6 = 101.6 \\ T(2) &= \frac{6(2)}{4+1} + 98.6 = 101 \\ T(5) &= \frac{6(5)}{25+1} + 98.6 = 99.8 \\ T(10) &= \frac{6(10)}{100+1} + 98.6 = 99.2 \end{aligned}$$

b)

$$\begin{aligned} \lim_{t\to\infty} T(t) &= \lim_{t\to\infty} \frac{6t}{t^2+1} + 98.6 \\ &= \lim_{t\to\infty} \frac{\frac{6}{t}}{1+\frac{1}{t^2}} + 98.6 \\ &= 0 + 98.6 \\ &= 98.6 \end{aligned}$$

c) The maximum occurs at the critical value of the function

$$\begin{aligned} T'(t) &= \frac{6(t^2+1) - 6t(2t)}{(t^2+1)^2} \\ &= \frac{6-6t^2}{(t^2+1)^2} \\ T'(t) &= 0 \\ \frac{6-6t^2}{(t^2+1)^2} &= 0 \\ 6-6t^2 &= 0 \\ t &= 1 \end{aligned}$$

Note we ignore the negative option since the question involves the interval $[0, \infty)$.
From part a) we know that $T(1) = 101.6$

d) 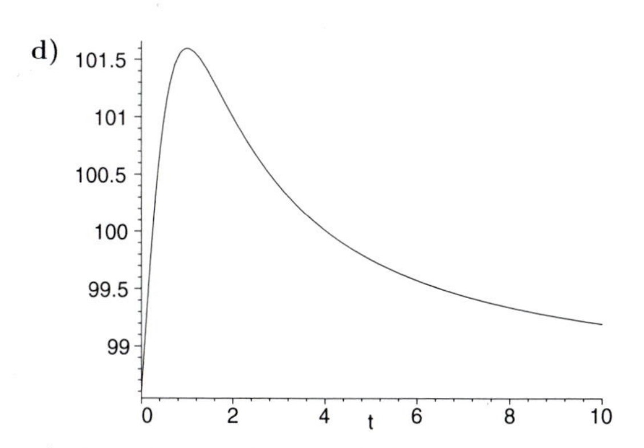

e) According to this model the temperature does not return to 98.6° since that temperature is reaches only when t approaches infinity.

51. a) $E(9) = 9 \cdot \frac{4}{9} = 4.00$

$E(8) = 9 \cdot \frac{4}{8} = 4.50$

$E(7) = 9 \cdot \frac{4}{7} \approx 5.14$

$E(6) = 9 \cdot \frac{4}{6} = 6.00$

$E(5) = 9 \cdot \frac{4}{5} = 7.20$

$E(4) = 9 \cdot \frac{4}{4} = 9.00$

$E(3) = 9 \cdot \frac{4}{3} = 12.00$

$E(2) = 9 \cdot \frac{4}{2} = 18.00$

$E(1) = 9 \cdot \frac{4}{1} = 36.00$

$E\left(\frac{2}{3}\right) = 9 \cdot \frac{4}{\frac{2}{3}} = 9 \cdot \left(4 \cdot \frac{3}{2}\right) = 54.00$

$E\left(\frac{1}{3}\right) = 9 \cdot \frac{4}{\frac{1}{3}} = 9 \cdot \left(4 \cdot \frac{3}{1}\right) = 108.00$

We complete the table.

Innings pitched (i)	Earned-run average (E)
9	4.00
8	4.50
7	5.14
6	6.00
5	7.20
4	9.00
3	12.00
2	18.00
1	36.00
$\frac{2}{3}$	54.00
$\frac{1}{3}$	108.00

b) As i approaches 0 from the right, the values of $E(i)$ increase without bound, so $\lim_{i \to 0^+} E(i) = \infty$.

c) Since $\lim_{i \to 0^+} E(i) = \infty$, the earned run average would be ∞.

53. Asymptotes can be thought of as "limiting lines" for graphs of functions. The graphs and limits on pages 201, 202, and 203 of the text illustrate vertical, horizontal, and oblique asymptotes.

55.

$$\begin{aligned}\lim_{x \to 5} \frac{x^2 - 6x + 5}{x^2 - 3x - 10} &= \lim_{x \to 5} \frac{(x-1)(x-5)}{(x+2)(x-5)} \\ &= \lim_{x \to 5} \frac{(x-1)}{(x+2)} \\ &= \frac{5-1}{5+2} \\ &= \frac{4}{7}\end{aligned}$$

57. We divide the numerator and the denominator by x^2, the highest power of x in the denominator.

$$\lim_{x \to \infty} \frac{-6x^3 + 7x}{2x^2 - 3x - 10} = \lim_{x \to \infty} \frac{-6x + \frac{7}{x}}{2 - \frac{3}{x} - \frac{10}{x^2}} = \frac{\lim_{x \to \infty} -6x + 0}{2 - 0 - 0} = -\infty$$

(The numerator increases without bound negatively while the denominator approaches 2.)

59. $\lim_{x \to 1} \frac{x^3 - 1}{x^2 - 1} = \lim_{x \to 1} \frac{(x-1)(x^2 + x + 1)}{(x-1)(x+1)} =$
$\lim_{x \to 1} \frac{x^2 + x + 1}{x + 1} = \frac{1 + 1 + 1}{1 + 1} = \frac{3}{2}$

61.

$$\begin{aligned}\lim_{x \to -\infty} \frac{2x^4 + x}{x + 1} &= \lim_{x \to -\infty} \frac{2x^3 + 1}{1 + \frac{1}{x}} \\ &= \frac{\lim_{x \to -\infty} 2x^3 + 1}{1 + 0} \\ &= -\infty\end{aligned}$$

63. Undefined

65. Since the numerator is bounded by ± 1 and the denominator grows indefinitely, then

$$\lim_{x \to \infty} \frac{\cos x}{x} = 0$$

see figure below

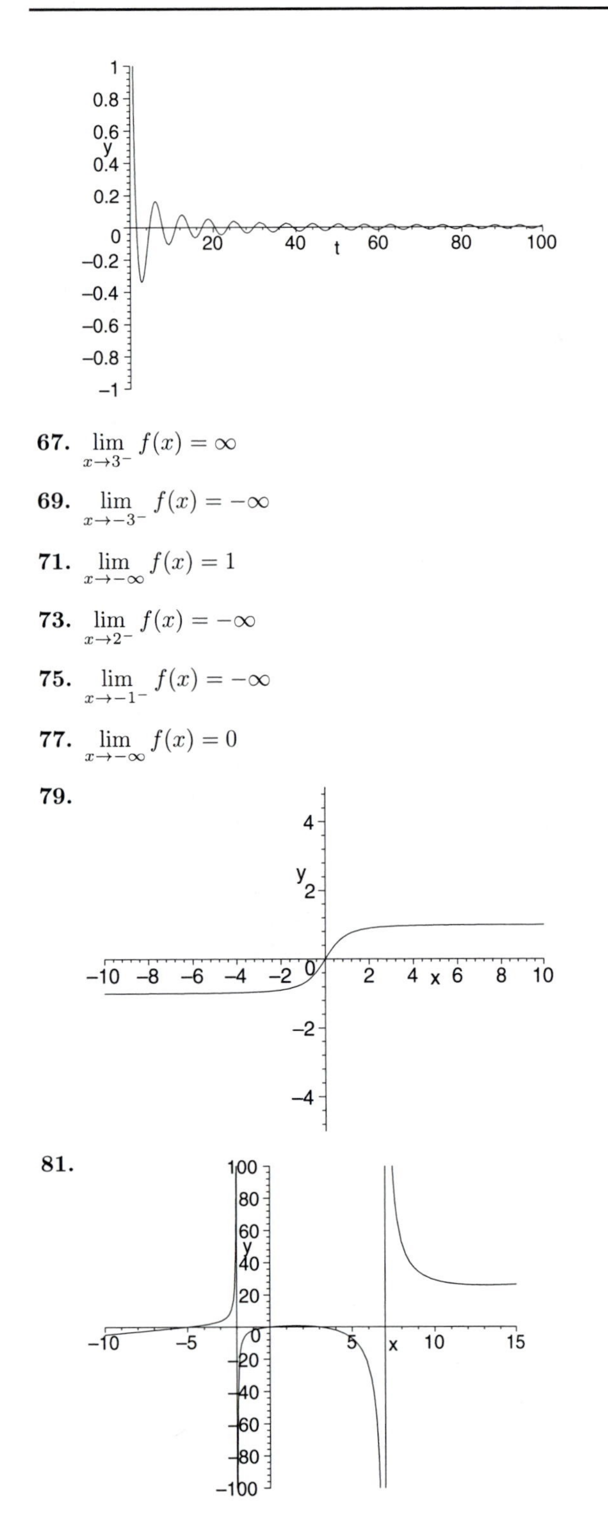

67. $\lim_{x \to 3^-} f(x) = \infty$

69. $\lim_{x \to -3^-} f(x) = -\infty$

71. $\lim_{x \to -\infty} f(x) = 1$

73. $\lim_{x \to 2^-} f(x) = -\infty$

75. $\lim_{x \to -1^-} f(x) = -\infty$

77. $\lim_{x \to -\infty} f(x) = 0$

79.

81.

83.

85.

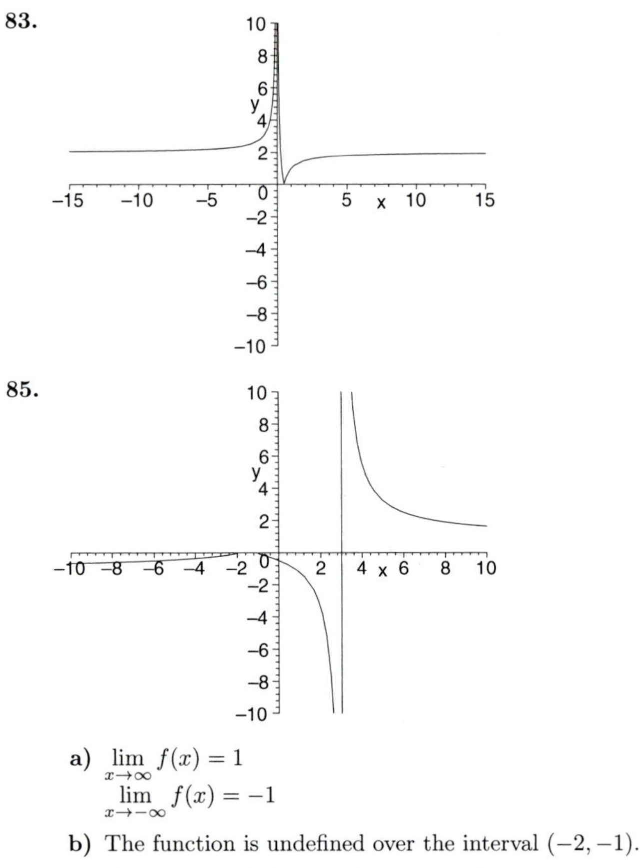

a) $\lim_{x \to \infty} f(x) = 1$
$\lim_{x \to -\infty} f(x) = -1$

b) The function is undefined over the interval $(-2, -1)$.

c) The domain of the function is given by

$$x < -2, \ -1 < x < 3, \text{ and } x > 3$$

The function is undefined on the interval $(-2, -1)$ since in that interval the radicant is negative.

d) $\lim_{x \to -2^+} f(x) = 0$
$\lim_{x \to -2^-} f(x) = 0$

Exercise Set 3.4

1. **a)** $x = 85$

b) $x = 0$

c) $y = 150$

d) $y = 210$

3. $f(x) = x^3 - x^2 - x + 2$

$$\begin{aligned}
f'(x) &= 3x^2 - 2x - 1 = (3x + 1)(x - 1) \\
f'(x) &= 0 \\
(3x + 1)(x + 1) &= 0 \\
x &= \frac{-1}{3} \quad \text{not in the interval} \\
x &= 1 \\
f(1) &= 1 - 1 - 1 + 2 \\
&= 1
\end{aligned}$$

Next, we check the endpoints of the interval
$f(0) = 0 - 0 - 0 + 2 = 2$
$f(2) = 8 - 4 - 2 + 2 = 4$
Thus we have an absolute maximim at $(2, 4)$ and an absolute minimum at $(1, 1)$.

5. $f(x) = 3x - 2$ $f'(x) = 3$, which means the function does not have critical values. Next, we check the endpoints of the interval
$f(-1) = -3 - 2 = -5$
$f(1) = 3 - 2 = 1$
Thus we have an absolute maximim at $(1, 1)$ and an absolute minimum at $(-1, -5)$.

7. $f(x) = 3 - 2x - 5x^2$

$$\begin{aligned} f'(x) &= -2 - 10x \\ f'(x) &= 0 \\ -2 - 10x &= 0 \\ x &= \frac{-1}{5} \\ f\left(\frac{-1}{5}\right) &= 3 + \frac{2}{5} - \frac{1}{5} \\ &= \frac{16}{5} \end{aligned}$$

Next, we check the endpoints of the interval
$f(-3) = 3 + 6 - 45 = -36$
$f(3) = 3 - 6 - 45 = -48$
Thus, we have an absolute maximum at $\left(\frac{-1}{5}, \frac{16}{5}\right)$ and an absolute minimum at $(3, -48)$.

9. $f(x) = 1 - x^3$

$$\begin{aligned} f'(x) &= -3x^2 \\ f'(x) &= 0 \\ -3x^2 &= 0 \\ x &= 0 \\ f(0) &= 1 - 0 \\ &= 1 \end{aligned}$$

Next, we check the endpoints of the interval
$f(-8) = 1 - (-512) = 513$
$f(8) = 1 - 512 = -511$
Thus, we have an absolute maximum at $(-8, 513)$ and an absolute minimum at $(8, -511)$.

11. $f(x) = 12 + 9x - 3x^2 - x^3$

$$\begin{aligned} f'(x) &= 9 - 6x - 3x^2 \\ &= 3(3 + x)(1 - x) \\ f'(x) &= 0 \\ 3(3 + x)(1 - x) &= 0 \\ x &= -3 \\ x &= 1 \\ f(-3) &= 12 - 27 - 27 + 27 = -15 \\ f(1) &= 12 + 9 - 3 - 1 = 17 \end{aligned}$$

Note that the critical values are the same as the endpoints of the interval.
Thus, we have an absolute maximum at $(1, 17)$ and an absolute minimum at $(-3, -15)$.

13. $f(x) = x^4 - 2x^3$

$$\begin{aligned} f'(x) &= 4x^3 - 6x^2 \\ &= 2x^2(2x - 3) \\ f'(x) &= 0 \\ 2x^2(2x - 3) &= 0 \\ x &= 0 \\ x &= \frac{3}{2} \\ f(0) &= 0 - 0 = 0 \\ f\left(\frac{3}{2}\right) &= \frac{81}{16} - \frac{27}{4} = -\frac{27}{16} \end{aligned}$$

Next, we check the endpoints of the interval
$f(-2) = 16 - 2(-8) = 32$
$f(2) = 16 - 2(8) = 0$
Thus, we have an absolute maximum at $(-2, 32)$ and an absolute minimum at $\left(\frac{3}{2}, -\frac{27}{16}\right)$.

15. $f(x) = x^4 - 2x^2 + 5$

$$\begin{aligned} f'(x) &= 4x^3 - 4x \\ &= 4x(x^2 - 1) \\ &= 4x(x - 1)(x + 1) \\ f'(x) &= 0 \\ 4x(x - 1)(x + 1) &= 0 \\ x &= 0 \\ x &= 1 \\ x &= -1 \\ f(0) &= 0 - 0 + 5 = 5 \\ f(1) &= 1 - 2 + 5 = 4 \\ f(-1) &= 1 - 2 + 5 = 4 \end{aligned}$$

Next, we check the endpoints of the interval
$f(-2) = 16 - 8 + 5 = 13$
$f(2) = 16 - 8 + 5 = 13$
Thus, we have an absolute maximum at $(-2, 13)$ and at $(-2, 13)$ and an absolute minimum at $(-1, 4)$ and at $(1, 4)$.

17. $f(x) = (x + 3)^{2/3} - 5$
$f'(x) = \frac{2}{3}(x + 3)^{-1/3}$, which is undefined at $x = -3$ (the only critical value).
$f'(-3) = 0 - 5 = -5$
Next, we check the endpoints of the interval
$f(-4) = 1 - 5 = -4$
$f(5) = 4 - 5 = -1$
Thus, we have an absolute maximum at $(5, -1)$ and an absolute minimum at $(-3, -5)$.

19. $f(x) = x + \frac{1}{x}$

$$\begin{aligned} f'(x) &= 1 - \frac{1}{x^2} \\ f'(x) &= 0 \\ 1 - \frac{1}{x^2} &= 0 \\ x^2 - 1 &= 0 \end{aligned}$$

$$\begin{aligned} x &= -1 \text{ not acceptable} \\ x &= 1 \\ f(1) &= 1+1=2 \end{aligned}$$

Next, we check the endpoint of the interval
$f(20) = 20 + 0.05 = 20.05$
Thus, we have an absolute maximum at $(20, 20.05)$ and an absolute minimum at $(1, 2)$.

21. $f(x) = \dfrac{x^2}{x^2+1}$

$$\begin{aligned} f'(x) &= \frac{2x(x^2+1) - 2x(x^2)}{(x^2+1)^2} \\ &= \frac{2x}{(x^2+1)^2} \\ f'(x) &= 0 \\ \frac{2x}{(x^2+1)^2} &= 0 \\ x &= 0 \\ f(0) &= \frac{0}{1} = 0 \end{aligned}$$

Next, we check the endpoints of the interval
$f(-2) = \dfrac{4}{4+1} = 0.8$
$f(2) = \dfrac{4}{4+1} = 0.8$
Thus, we have an absolute maximum at $(-2, 0.8)$ and at $(2, 0.8)$ and an absolute minimum at $(0, 0)$.

23. $f(x) = (x+1)^{1/3}$
$f'(x) = \frac{1}{3}(x+1)^{-2/3}$, which is undefined at $x = -1$ (the only critical value).
$f(-1) = 0$
Next, we check the endpoints of the interval
$f(-2) = (-1)^{-1/3} = -1$
$f(26) = (27)^{1/3} = 3$
Thus, we have an absolute maximum at $(26, 3)$ and an absolute minimum at $(-2, -1)$.

25. $f(x) = \dfrac{x+2}{x^2+5}$

$$\begin{aligned} f'(x) &= \frac{(x^2+5) - 2x(x+2)}{(x^2+5)^2} \\ &= \frac{5 - 4x - x^2}{(x^2+5)^2} \\ &= \frac{(5+x)(1-x)}{(x^2+5)^2} \\ f'(x) &= 0 \\ \frac{(5+x)(1-x)}{(x^2+5)^2} &= 0 \\ x &= -5 \\ x &= 1 \\ f(-5) &= \frac{-3}{25+5} = -0.1 \\ f(1) &= \frac{3}{1+5} = 0.5 \end{aligned}$$

Next, we check the endpoints of the interval
$f(-6) = \frac{-4}{36+5} = -0.0976$
$f(6) = \frac{8}{36+5} = 0.19512$
Thus, we have an absolute maximum at $(1, 0.5)$ and an absolute minimum at $(-5, -0.1)$.

27. $f(x) = x(x - x^2)^{1/2}$

$$\begin{aligned} f'(x) &= (x-x^2)^{1/2} + \frac{x(1-2x)}{2(x-x^2)^{1/2}} \\ f'(x) &= 0 \\ &\rightarrow \\ 2(x-x^2) &= x(2x-1) \\ 4x^2 - 3x &= 0 \\ x &= 0 \\ x &= \frac{3}{4} \\ f(0) &= 0 \\ f\left(\frac{3}{4}\right) &= \frac{3}{4}\left(\frac{3}{4} - \frac{9}{16}\right)^{1/2} = 0.32476 \end{aligned}$$

Next, we check the end point of the interval
$f(1) = 1(0) = 0$
Thus, we have an absolute maximum at $\left(\dfrac{3}{4}, 0.32476\right)$, and an absolute minimum at $(0, 0)$ and at $(1, 0)$.

29. $f(x) = x(x+3)^{1/2}$

$$\begin{aligned} f'(x) &= (x+3)^{1/2} + \frac{x}{2(x+3)^{1/2}} \\ f'(x) &= 0 \\ &\rightarrow \\ 2(x+3) &= -x \\ 3x + 6 &= 0 \\ x &= -2 \\ f(-2) &= -2(-2+3)^{1/2} = -2 \end{aligned}$$

Next, we check the end points of the interval
$f(-3) = -3(0) = 0$
$f(6) = 6(9)^{1/2} = 18$
Thus, we have an absolute maximum at $(6, 18)$, and an absolute minimum at $(-2, -2)$.

31. $f(x) = x + 2\ sin\ x$

$$\begin{aligned} f'(x) &= 1 + 2\ cos\ x \\ f'(x) &= 0 \\ 1 + 2\ cos\ x &= 0 \\ cos\ x &= \frac{-1}{2} \\ x &= \frac{2\pi}{3} \\ x &= \frac{4\pi}{3} \\ f\left(\frac{2\pi}{3}\right) &= \frac{2\pi}{3} + 2\ sin\left(\frac{2\pi}{3}\right) = 3.8264 \end{aligned}$$

$$f\left(\frac{4\pi}{3}\right) = \frac{4\pi}{3} + 2\,sin\left(\frac{4\pi}{3}\right) = 2.4567$$

Next, we check the endpoints of the interval
$f(0) = 0 + 0 = 0$
$f(2\pi) = 2\pi + 0 = 2\pi$
Thus, we have an absolute maximum at $(2\pi, 2\pi)$ and an absolute minimum at $(0, 0)$.

33. $f(x) = \dfrac{sin\ x}{2 + sin\ x}$

$$\begin{aligned} f'(x) &= \frac{(2+sin\ x)cos\ x - sin\ x\ cos\ x}{(2+sin\ x)^2} \\ &= \frac{2\ cos\ x}{(2+sin\ x)^2} \\ f'(x) &= 0 \\ \frac{2\ cos\ x}{(2+sin\ x)^2} &= 0 \\ 2\ cos\ x &= 0 \\ x &= \frac{\pi}{2} \\ x &= \frac{3\pi}{3} \\ f\left(\frac{\pi}{2}\right) &= \frac{1}{2+1} = \frac{1}{3} \\ f\left(\frac{3\pi}{2}\right) &= \frac{-1}{2-1} = -1 \end{aligned}$$

Next, we check the endpoints of the interval
$f(0) = \frac{0}{2} = 0$
$f(2\pi) = \frac{0}{2} = 0$
Thus, we have an absolute maximum at $\left(\frac{\pi}{2}, \frac{1}{3}\right)$ and an absolute minimum at $\left(\frac{3\pi}{2}, -1\right)$.

35. $f(x) = \dfrac{sin\ x}{(1 + sin\ x)^2}$

$$\begin{aligned} \frac{(1+sin\ x)^2 cos\ x - 2\ sin\ x(1+sin\ x)cos\ x}{(1+sin\ x)^4} &= f'(x) \\ \frac{cos\ x}{(1+sin\ x)^4} &= \\ f'(x) &= 0 \\ \frac{cos\ x}{(1+sin\ x)^4} &= 0 \\ cos\ x &= 0 \\ x &= \frac{\pi}{2} \\ f\left(\frac{\pi}{2}\right) &= \frac{1}{(1+1)^2} = \frac{1}{4} \end{aligned}$$

Note that $f'(x)$ is undefined at $x = -\frac{\pi}{2}$ but the value does not belong to the given interval.
Next, we check the endpoints of the interval
$f(0) = \frac{0}{1} = 0$
$f(\pi) = \frac{0}{1} = 0$
Thus, we have an absolute maximum at $(\left(\frac{\pi}{2}, \frac{1}{4}\right)$ and an absolute minimum at $(0, 0)$ and at $(\pi, 0)$.

37. $f(x) = 2x - tan\ x$

$$\begin{aligned} f'(x) &= 2 - sec^2\ x \\ &= 2 - \frac{1}{cos^2\ x} \\ f'(x) &= 0 \\ 2 - \frac{1}{cos^2\ x} &= 0 \\ cos^2\ x &= \frac{1}{2} \\ cos\ x &= \frac{-1}{\sqrt{2}} \\ cos\ x &= \frac{1}{\sqrt{2}} \\ x &= \frac{\pi}{4} \\ f\left(\frac{\pi}{4}\right) &= \frac{\pi}{2} - 1 = 0.5708 \end{aligned}$$

Next, we check the endpoints of the interval
$f(0) = 0 - 0 = 0$
$f\left(\frac{\pi}{3}\right) = \frac{2\pi}{3} - 1.73205 = 0.36234$
Thus, we have an absolute maximum at $\left(\frac{\pi}{4}, 0.5708\right)$ and an absolute minimum at $(0, 0)$.

39. $f(x) = 3\ sin\ x - 2\ sin^3\ x$

$$\begin{aligned} f'(x) &= 3\ cos\ x - 6\ sin^2\ x\ cos\ x \\ &= 3\ cos\ x(1 - 2\ sin^2\ x) \\ f'(x) &= 0 \\ 3\ cos\ x(1 - 2\ sin^2\ x) &= 0 \\ cos\ x &= 0 \\ &\text{or} \\ 1 - 2\ sin^2\ x &= 0 \\ sin^2\ x &= \frac{1}{2} \\ sin\ x &= \pm\frac{1}{\sqrt{2}} \\ x &= \frac{\pi}{2} \\ x &= \frac{3\pi}{2} \\ x &= \frac{\pi}{4} \\ x &= \frac{3\pi}{4} \\ x &= \frac{5\pi}{4} \\ x &= \frac{7\pi}{4} \\ f\left(\frac{\pi}{2}\right) &= 3 - 2 = 1 \\ f\left(\frac{3\pi}{2}\right) &= -3 + 2 = -1 \\ f\left(\frac{\pi}{4}\right) &= = 2.1213 - 0.7071 = 1.4142 \end{aligned}$$

$$f\left(\frac{3\pi}{4}\right) = = 2.1213 - 0.7071 = 1.4142$$
$$f\left(\frac{5\pi}{4}\right) = = -2.1213 + 0.7071 = -1.4142$$
$$f\left(\frac{7\pi}{4}\right) = = -2.1213 + 0.7071 = -1.4142$$

Next, we check the endpoints of the interval
$f(0) = 0 - 0 = 0$
$f(2\pi) = 0 - 0 = 0$
Thus, we have an absolute maximum at $\left(\frac{\pi}{4}, 1.4142\right)$ and at $\left(\frac{3\pi}{4}, 1.4142\right)$ and an absolute minimum at $\left(\frac{5\pi}{4}, -1.4142\right)$ and at $\left(\frac{7\pi}{4}, -1.4142\right)$.

41. $f(x) = x - \frac{4}{3}x^3$

$$\begin{aligned} f'(x) &= 1 - x^2 \\ f'(x) &= 0 \\ 1 - x^2 &= 0 \\ x &= -1 \text{ not acceptable} \\ x &= 1 \\ f(1) &= 1 - \frac{4}{3} = -\frac{1}{3} \end{aligned}$$

$f''(x) = -2x$
$f''(1) = -2 < 0$

The function has an absolute maximum of $\frac{-1}{3}$, no absolute minimum.

43. $f(x) = -0.001x^2 + 4.8x - 60$

$$\begin{aligned} f'(x) &= -0.002x + 4.8 \\ f'(x) &= 0 \\ -0.002x + 4.8 &= 0 \\ x &= 2400 \\ f(2400) &= -5760 + 11520 - 60 = 5700 \\ f''(x) &= -0.002 < 0 \end{aligned}$$

The fnction has an absolute maximum of 5700, no absolute minimum.

45. $f(x) = 2x + \frac{72}{x}$

$$\begin{aligned} f'(x) &= 2 - \frac{72}{x^2} \\ f'(x) &= 0 \\ 2 - \frac{72}{x^2} &= 0 \\ x^2 &= 36 \\ x &= 6 \\ f(6) &= 12 + 12 = 24 \\ f''(x) &= \frac{72}{x^3} \\ f''(6) &= \frac{1}{3} > 0 \end{aligned}$$

The function has an absolute minimum of 24, no absolute maximum.

47. $f(x) = x^2 + \frac{432}{x}$

$$\begin{aligned} f'(x) &= 2x - \frac{432}{x^2} \\ f'(x) &= 0 \\ 2x - \frac{432}{x^2} &= 0 \\ x^3 &= 216 \\ x &= 6 \\ f(6) &= 36 + 72 = 108 \\ f''(x) &= 2 + \frac{432}{x^3} \\ f''(6) &= 2 + 2 = 4 > 0 \end{aligned}$$

The function has an absolute minimum of 108, no absolute maximum.

49. $f(x) = (x + 1)^3$

$$\begin{aligned} f'(x) &= 3(x+1)^2 \\ f'(x) &= 0 \\ 3(x+1)^2 &= 0 \\ x &= -1 \\ f(-1) &= 0 \\ f''(x) &= 6(x+1) \\ f''(-1) &= 0 \end{aligned}$$

The function has no absolute maximum or minimum.

51. $f(x) = 2x - 3$
The function is linear, which means that on the interval $(-\infty, \infty)$ the function has no absolute maximum or minimum.

53. $f(x) = x^{2/3}$

$$f'(x) = \frac{2}{3}x^{-1/3}$$

$f''(x) = -\frac{2}{9}x^{-4/3}$ $f'(x)$ and $f''(x)$ are undefined at $x = 0$

$$\begin{aligned} f(-1) &= (-1)^{2/3} = 1 \\ f(0) &= (0)^{2/3} = 0 \\ f(1) &= (1)^{2/3} = 1 \\ f''(1) &= \frac{-2}{9} \\ f''(-1) &= \frac{-2}{9} \end{aligned}$$

The function has an absolute maximum of 1 and an absolute minimum of 0.

55. $f(x) = \frac{1}{3}x^3 - x + \frac{2}{3}$
The function grows indefinitely over $(-\infty, \infty)$, which means it has no absolute maximum or minimum.

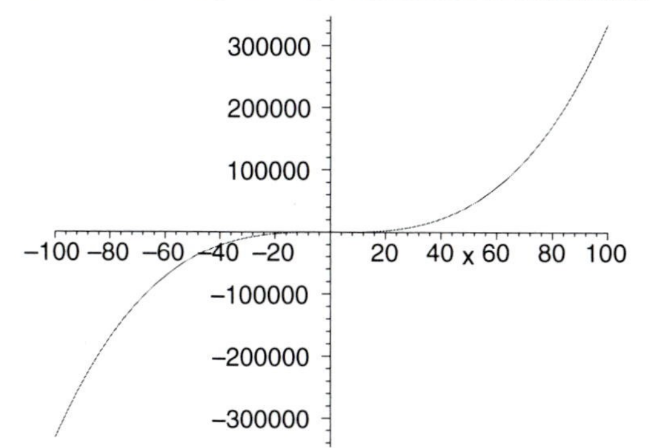

57. $f(x) = x^4 - 2x^2$

$$\begin{aligned} f'(x) &= 4x^3 - 4x \\ &= 4x(x^2-1) \\ f'(x) &= 0 \\ 4x(x^2-1) &= 0 \\ x &= 0 \\ x &= \pm 1 \\ f(-1) &= 1 - 2 = -1 \\ f(0) &= 0 - 0 = 0 \\ f(1) &= 1 - 2 = -1 \end{aligned}$$

The function has an absolute minimum of -1 and no absolute maximum.

59. $f(x) = \tan x + \cot x$

$$\begin{aligned} f'(x) &= \sec^2 x - \csc^2 x \\ &= \frac{\sin^2 x - \cos^2 x}{\cos^2 x \sin^2 x} \\ f'(x) &= 0 \\ \frac{\sin^2 x - \cos^2 x}{\cos^2 x \sin^2 x} &= 0 \\ \sin^2 x - \cos^2 x &= 0 \\ \sin x &= \pm \cos x \\ x &= \frac{\pi}{4} \\ f\left(\frac{\pi}{4}\right) &= 1 + 1 = 2 \end{aligned}$$

The function is undefined at $x = 0$ and $x = \frac{\pi}{2}$
$f''(x) = 2\sec^2 x \tan x + 2\csc^2 x \cot x$
$f''\left(\frac{\pi}{4}\right) = 1 + 1 = 2 > 0$ The function has an absolute minimum of 2 and no absolute maximum.

61. $f(x) = \frac{1}{\sin x + \cos x}$

$$\begin{aligned} f'(x) &= \frac{\cos x - \sin x}{(\sin x + \cos x)^2} \\ f'(x) &= 0 \\ \frac{\cos x - \sin x}{(\sin x + \cos x)^2} &= 0 \\ \cos x &= \sin x \\ x &= \frac{\pi}{4} \\ f\left(\frac{\pi}{4}\right) &= \frac{1}{\frac{1}{\sqrt{2}} + \frac{1}{\sqrt{2}}} \\ &= \frac{\sqrt{2}}{2} \\ f''(x) &= \frac{2\sin x \cos x - 1}{(\sin x + \cos x)^4} \\ f''\left(\frac{\pi}{4}\right) &= \frac{1}{(\frac{1}{\sqrt{2}} + \frac{1}{\sqrt{2}})^4} > 0 \end{aligned}$$

The function has an absolute minimum of $\frac{\sqrt{2}}{2}$ and no absolute maximum.

63. $f(x) = \frac{1}{x - 2\sin x}$

$$\begin{aligned} f'(x) &= \frac{2\cos x - 1}{(x - 2\sin x)^2} \\ f'(x) &= 0 \\ \frac{2\cos x - 1}{(x - 2\sin x)^2} &= 0 \\ \cos x &= \frac{1}{2} \\ x &= \frac{\pi}{3} \\ f\left(\frac{\pi}{3}\right) &= -1.46 \\ f''(x) &= \frac{-2\sin x(x - 2\sin x)}{(x - 2\sin x)^3} - \\ &\quad \frac{2(2\cos x - 1)(1 - \cos x)}{(x - 2\sin x)^3} \\ f''\left(\frac{\pi}{3}\right) &= -3.693 < 0 \end{aligned}$$

The function has an absolute maximum of -1.46 and no absolute minimum.

65. $f(x) = 2\csc x + \cot x$

$$\begin{aligned} f'(x) &= -2\csc x \cot x - \csc^2 x \\ f'(x) &= 0 \\ -2\csc x \cot x - \csc^2 x &= 0 \\ x &= \frac{4\pi}{3} \\ f\left(\frac{4\pi}{3}\right) &= -1.7321 \\ f''(x) &= -2\csc x\cot^2 x + 2\csc^3 x + \\ &\quad 2\csc^2 x \cot x \\ f''\left(\frac{4\pi}{3}\right) &= -0.7698 < 0 \end{aligned}$$

The function has an absolute maximum of -1.7321 and no absolute minimum.

67. $f(x) = tan\ x - 2\ sec\ x$

$$\begin{aligned} f'(x) &= sec^2\ x - 2\ sec\ x\ tan\ x \\ f'(x) &= 0 \\ sec^2\ x - 2\ sec\ x\ tan\ x &= 0 \\ x &= \frac{5\pi}{6} \\ f\left(\frac{5\pi}{6}\right) &= 1.7321 \\ f''(x) &= 2\ tan\ x\ sec^2\ x - 2\ sec^3\ x - \\ & \quad 2\ sec\ x\ tan^2\ x \\ f\left(\frac{5\pi}{6}\right) &= 2.309 > 0 \end{aligned}$$

The function has an absolute minimum of 1.7321 and no absolute maximum.

69. $f(x) = \dfrac{1}{1 - 2\ sin\ x}$

$$\begin{aligned} f'(x) &= \frac{2\ cos\ x}{(1-2\ sin\ x)^2} \\ f'(x) &= 0 \\ cos\ x &= 0 \\ x &= \frac{\pi}{2} \\ f\left(\frac{\pi}{2}\right) &= -1 \\ f''(x) &= \frac{8\ cos^2\ x}{(1-2\ sin\ x)^3} - \frac{2\ sin\ x}{(1-2\ sin\ x)^2} \\ f''\left(\frac{\pi}{2}\right) &= -2 < 0 \end{aligned}$$

The function has an absolute maximum of -1 and no absolute minimum.

71. $y = -6.1x^2 + 752x + 22620$

$$\begin{aligned} y' &= -12.2x + 752 \\ -12.2x + 752 &= 0 \\ x &= \frac{752}{12.2} = 61.64 \\ y'' &= -12.2 < 0 \end{aligned}$$

The maximum number of accidents occurs at $x = 61.64$ mph.

73. $r(x) = 104.5x^2 - 1501.5x + 6016$

$$\begin{aligned} r'(x) &= 209x - 1501.5 \\ 209x - 1501.5 &= 0 \\ x &= \frac{1501.5}{209} = 7.18 \\ r(7.18) &= 104.5(7.18)^2 - 1501.5(7.18) + 6016 \\ &= 15434.75 \\ r''(x) &= 209 > 0 \end{aligned}$$

The death rate is minimized at 7.18 hours of sleep per night.

75. a) $D(h) = 0.139443\ \sqrt{h} - 0.238382\ h^{0.3964}$

b) $D(h) = 0.139443\ \sqrt{h} - 0.238382\ h^{0.3964}$

$$\begin{aligned} D'(h) &= \frac{0.697215}{\sqrt{h}} - \frac{0.0944946248}{h^{0.6036}} \\ \frac{0.697215}{\sqrt{h}} - \frac{0.0944946248}{h^{0.6036}} &= 0 \\ h^{1.1036} &= 0.0658830698 \\ h &= 18.816 \\ D(18.816) &= 0.139443\ \sqrt{18.816} - \\ & \quad 0.0944946\ (18.816)^{0.3964} \\ &= -0.1581 \end{aligned}$$

We check the endpoints of the interval
$D(0) = 0.139443\ \sqrt{0} - 0.0944946(0)^{0.3964} = 0$
$D(215) = 0.139443\ \sqrt{215} - 0.0944946(215)^{0.3964} = 0.0408$

The function has an absolute maximum of 0.0408 and an absolute minimum of -0.1581.

77. Left to the student (answers vary).

79. a) $cos\ \theta_0 = \dfrac{\text{Adj}}{\text{Hyp}} = \frac{1}{\sqrt{3}}$

b) $sin\ \theta_0 = \dfrac{\text{Opp}}{\text{Hyp}} = \frac{\sqrt{2}}{\sqrt{3}}$

c) From Example 4

$$\begin{aligned} S''(\theta) &= \frac{3}{2}a^2\left[\frac{(\sqrt{3}sin(\theta))(sin^2(\theta))}{sin^4(\theta)}\right] - \\ & \quad \frac{3}{2}a^2\left[\frac{(1-\sqrt{3}cos(\theta))(2sin(\theta)cos(\theta))}{sin^4(\theta)}\right] \\ S''(\theta_0) &= \frac{3}{2}a^2\left[\frac{\sqrt{3}(\frac{\sqrt{2}}{\sqrt{3}})(\frac{\sqrt{2}}{\sqrt{3}})^2}{(\frac{\sqrt{2}}{\sqrt{3}})^4}\right] - \\ & \quad \frac{3}{2}a^2\left[\frac{(1-\sqrt{3}\frac{1}{\sqrt{3}})2(\frac{\sqrt{2}}{\sqrt{3}})(\frac{1}{\sqrt{3}})}{(\frac{\sqrt{2}}{\sqrt{3}})^4}\right] \\ &= \frac{3}{2}a^2\left[\frac{\frac{2\sqrt{2}}{3}}{\frac{4}{9}} - \frac{0}{\frac{4}{9}}\right] \\ &= \frac{9\sqrt{2}a^2}{4} \end{aligned}$$

81. Absolute maximum at $x = 4$ and absolute minimum at $x = 2$

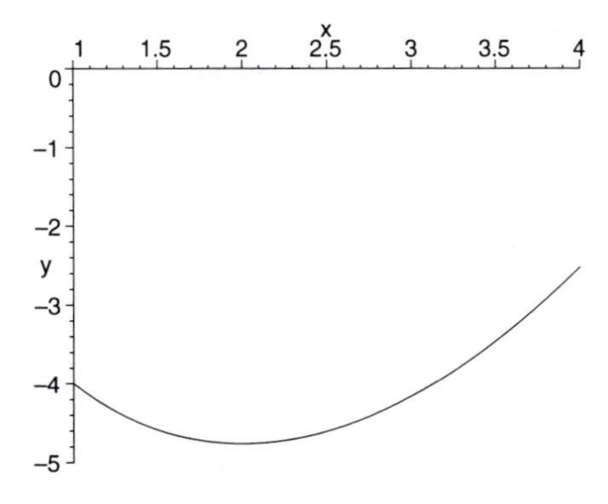

83. No absolute maximum and no absolute minimum

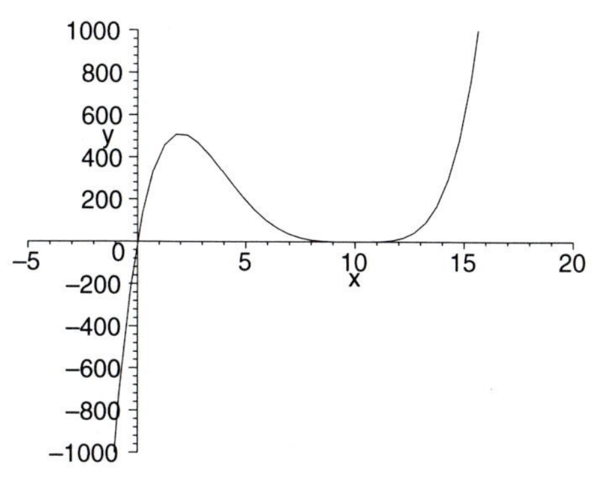

Exercise Set 3.5

1. $Q = xy$ and $x + y = 50$

$$\begin{aligned} Q &= x(50 - x) \\ &= 50x - x^2 \\ Q' &= 50 - 2x \\ 50 - 2x &= 0 \\ x &= 25 \\ y &= 50 - 25 = 25 \end{aligned}$$

The two numbers are $x = 25$ and $y = 25$

3. There cannot be a minimum product since the there is only one critical value for the function and the second derivative is positive for values of x.

5. $Q = xy$ and $y - x = 4$

$$\begin{aligned} Q &= x(x + 4) \\ &= x^2 + 4x \\ Q' &= 2x + 4 \\ 2x + 4 &= 0 \\ x &= -2 \\ y &= -2 + 4 = 2 \end{aligned}$$

The two number are $x = -2$ and $y = 2$

7. $Q = xy^2$ and $x + y^2 = 1$

$$\begin{aligned} Q &= x(1 - x) \\ &= x - x^2 \\ Q' &= 1 - 2x \\ 1 - 2x &= 0 \\ x &= \frac{1}{2} \\ y^2 &= 1 - \frac{1}{2} = \frac{1}{2} \\ y &= \frac{1}{\sqrt{2}} \end{aligned}$$

When $x = \frac{1}{2}$ and $y = \frac{1}{\sqrt{2}}$

$$\begin{aligned} Q &= \frac{1}{2}\left(\frac{1}{\sqrt{2}}\right)^2 \\ &= \frac{1}{4} \end{aligned}$$

9. $Q = 2x^2 + 3y^2$ and $x + y = 5$

$$\begin{aligned} Q &= 2x^2 + 3(5 - x)^2 \\ &= 2x^2 + 75 - 30x + 3x^2 \\ &= 5x^2 - 30x + 75 \\ Q' &= 10x - 30 \\ 10x - 30 &= 0 \\ x &= 3 \\ y &= 5 - 3 = 2 \end{aligned}$$

When $x = 3$ and $y = 2$

$$\begin{aligned} Q &= 2(3)^2 + 3(2)^2 \\ &= 18 + 12 \\ &= 30 \end{aligned}$$

11. $Q = x^2 + y^2$ and $x + y = 20$

$$\begin{aligned} Q &= x^2 + (20 - x)^2 \\ &= x^2 + 400 - 40x + x^2 \\ &= 2x^2 - 40x + 400 \\ Q' &= 4x - 40 \\ 4x - 40 &= 0 \\ x &= 10 \\ y &= 20 - 10 = 10 \end{aligned}$$

When $x = 10$ and $y = 10$

$$\begin{aligned} Q &= (10)^2 + (10)^2 \\ &= 100 + 100 \\ &= 200 \end{aligned}$$

13. $Q = xy$ and $\frac{4}{3}x^2 + y = 16$

$$\begin{aligned} Q &= x(16 - \frac{4}{3}x^2) \\ &= 16x - \frac{4}{3}x^3 \\ Q' &= 16 - 4x^2 \\ &= 4(4 - x^2) \\ 4(4 - x^2) &= 0 \\ x &= 2 \\ y &= 16 - \frac{4}{3}(4) = \frac{32}{3} \\ \text{and} & \end{aligned}$$

When $x = 2$ and $y = \frac{32}{3}$

$$\begin{aligned} Q &= 2\left(\frac{32}{3}\right) \\ &= \frac{64}{3} \end{aligned}$$

15. $Q = \sqrt{x} + \sqrt{y}$ and $x + y = 1$

$$\begin{aligned} Q &= \sqrt{x} + \sqrt{1-x} \\ &= x^{1/2} + (1-x)^{1/2} \\ Q' &= \frac{1}{2\sqrt{x}} - \frac{1}{2\sqrt{1-x}} \\ \frac{1}{2\sqrt{x}} - \frac{1}{2\sqrt{1-x}} &= 0 \\ \sqrt{x} &= \sqrt{1-x} \\ x &= 1-x \\ x &= \frac{1}{2} \\ y &= 1 - \frac{1}{2} = \frac{1}{2} \end{aligned}$$

When $x = \frac{1}{2}$ and $y = \frac{1}{2}$

$$\begin{aligned} Q &= \sqrt{\frac{1}{2}} + \sqrt{\frac{1}{2}} \\ &= \frac{2}{\sqrt{2}} \\ &= \sqrt{2} \end{aligned}$$

17. $A = lw$ and $2l + w = 20$

$$\begin{aligned} A &= l(20-2l) \\ &= 20l - 2l^2 \\ A' &= 20 - 4l \\ 20 - 4l &= 0 \\ l &= 5 \\ w &= 20 - 2(5) = 10 \end{aligned}$$

When $l = 5$ and $w = 10$

$$\begin{aligned} A &= 5(10) \\ &= 50 \end{aligned}$$

The rectangular fence is 5 yards by 10 yards with a maximum area of 50 squared yards.

19. $A = lw$ and $2l + 2w = 54 \to l + w = 27$

$$\begin{aligned} A &= l(27-l) \\ &= 27l - l^2 \\ A' &= 27 - 2l \\ 27 - 2l &= 0 \\ l &= 13.5 \\ w &= 27 - 13.5 = 13.5 \end{aligned}$$

When $l = 13.5$ and $w = 13.5$

$$\begin{aligned} A &= 13.5(13.5) \\ &= 182.25 \end{aligned}$$

The room is 13.5 feet by 13.5 feet yards with a maximum area of 182.25 squared feet.

21. Length: $l = 30 - 2x$, Width: $30 - 2x$, and $Height : h = x$

$$\begin{aligned} V &= (30-2x)(30-2x)x \\ &= 900x - 120x^2 + 4x^3 \\ V' &= 900 - 240x + 12x^2 \\ &= 12(15-x)(5-x) \\ 12(15-x)(5-x) &= 0 \\ x &= 15 \text{ not acceptable} \\ x &= 5 \\ l &= 30 - 2(5) = 20 \\ w &= 30 - 2(5) = 20 \\ h &= 5 \\ V &= 20(20)5 \\ &= 2000 \end{aligned}$$

The box has a length and width of 20 inches and a height of 5 inches with a maximum volume of 2000 squared inches.

23. Length: x, Width: x, and Height: y
$SA = x^2 + 4xy$ and $x^2y = 62.5 \to y = \frac{62.5}{x^2}$

$$\begin{aligned} SA &= x^2 + 4x\left(\frac{62.5}{x^2}\right) \\ &= x^2 + \frac{250}{x} \\ SA' &= 2x - \frac{250}{x^2} \\ 2x - \frac{250}{x^2} &= 0 \\ 2x^3 &= 250 \\ x &= 5 \\ y &= \frac{62.5}{5^2} = 2.5 \end{aligned}$$

When $x = 5$ and $y = 2.5$

$$\begin{aligned} SA &= 5^2 + 4(5)(2.5) \\ &= 75 \end{aligned}$$

The sqaure based box has dimensions $5 \times 5 \times 2.5$ inches and a minimum surface area of 75 squared inches.

25. $A = \frac{1}{2}(2\ sin\frac{\theta}{2})(cos\frac{\theta}{2})$

$$\begin{aligned} A &= sin\ \frac{\theta}{2}\ cos\ \frac{\theta}{2} \\ A' &= -sin^2\ \theta + cos^2\ \theta \\ -sin^2\ \frac{\theta}{2} + cos^2\ \frac{\theta}{2} &= 0 \\ sin\ \frac{\theta}{2} &= cos\ \frac{\theta}{2} \\ \frac{\theta}{2} &= \frac{\pi}{4} \\ \theta &= \frac{\pi}{2} \\ A &= sin\frac{\pi}{4}\ cos\frac{\pi}{4} \\ &= \frac{1}{2} \end{aligned}$$

The angle that maximizes the triangle is $\frac{\pi}{2}$ and the maximum area is $\frac{1}{2}$ square unit of length.

27. Price of ticket: x
Number of people at the game: $N = -10000x + 130000$

$$\begin{aligned} x(-10000x + 130000) + 1.50(-10000x + 130000) &= R \\ -10000x^2 + 130000x - 15000x + 19500 &= \\ -10000x^2 + 115000x + 19500 &= \\ -20000x + 115000 &= R' \\ -20000x + 115000 &= 0 \\ x &= 5.75 \\ -10000(5.75) + 130000 &= N \\ 72500 &= \end{aligned}$$

The maximum revenue occurs when the price of the ticket is \$5.75 and 72500 people attend the game.

29. Number of trees per acre: N
Number of bushels per tree: $B = -N + 50$

$$\begin{aligned} Y &= N(-N + 50) \\ &= -N^2 + 50N \\ Y' &= -2N + 50 \\ -2N + 50 &= 0 \\ N &= 25 \end{aligned}$$

The farmer should plant 25 trees per acre to maximize bushel yields per tree.

31. $x^2 y = 320 \rightarrow y = \frac{320}{x^2}$

$$\begin{aligned} C &= 0.15x^2 + 0.10x^2 + 0.025(4xy) \\ &= 0.25x^2 + 0.1x\left(\frac{320}{x^2}\right) \\ &= 0.25x^2 + \frac{32}{x} \\ C' &= 0.5x - \frac{32}{x^2} \\ 0.5x - \frac{32}{x^2} &= 0 \\ x^3 &= 64 \\ x &= 4 \\ y &= \frac{320}{16} \\ &= 20 \end{aligned}$$

The dimensions of the box that minimize the cost are $4 \times 4 \times 20$ feet.

33. Amount invested: x, Rate: r and $x = kr$

$$\begin{aligned} P &= 0.18kr - kr^2 \\ P' &= 0.18k - 2kr \\ 0.18k - 2kr &= 0 \\ r &= 0.09 \end{aligned}$$

Interest rate to maximize profit should be 9%

35. Price: p, Percentage: r, Ordered pair (p, r)
$(25, 2.13)$ and $(26, 2.09)$

a)

$$\begin{aligned} r - 2.13 &= -0.04(p - 25) \\ r &= -0.04p + 3.13 \end{aligned}$$

b) $R = p(100r)$, the 100 was needed to get the number of people with personalized license plates.

$$\begin{aligned} R &= p(-4p + 313) \\ &= -4p^2 + 313p \\ R' &= -8p + 313 \\ -8p + 313 &= 0 \\ p &= 39.13 \end{aligned}$$

The price that maximizes revenue is \$39.13.

37. Length: y, gerth: 4x, $4x + y = 84$

$$\begin{aligned} V &= x^2 y \\ &= x^2(84 - 4x) \\ &= 84x^2 - 4x^3 \\ V' &= 168x - 12x^2 \\ 168x - 12x^2 &= 0 \\ 12x(14 - x) &= 0 \\ x &= 0 \text{ not acceptable} \\ x &= 14 \\ y &= 84 - 4(14) = 28 \end{aligned}$$

The maximum box has dimensions $14 \times 14 \times 28$ inches.

39. $2y + 2x + \pi x = 24$

$$\begin{aligned} A &= 2xy + \frac{\pi}{2}x \\ &= 2x(-x - \frac{\pi}{2}x + 12) + \frac{\pi}{2}x^2 \\ &= -2x^2 - \frac{\pi}{2}x^2 + 24x \\ A' &= -4x - \pi x + 24 \\ -4x - \pi x + 24 &= 0 \\ x(4 + \pi) &= 24 \\ x &= \frac{24}{4 + \pi} \\ y &= -\frac{24}{4 + \pi} - \frac{\pi}{2}\left(\frac{24}{4 + \pi}\right) + 12 \\ &= \frac{24}{4 + \pi} \end{aligned}$$

41. Let x be the number

$$\begin{aligned} S &= \frac{1}{x} + 5x^2 \\ S' &= -\frac{1}{x^2} + 10x \\ -\frac{1}{x^2} + 10x &= 0 \\ x^3 &= \frac{1}{10} \\ x &= \sqrt[3]{\frac{1}{10}} \end{aligned}$$

43.

$$\begin{aligned} S &= \pi x^2 + (24-x)^2 \\ &= \pi x^2 + 576 - 48x + x^2 \\ &= (\pi+1)x^2 - 48x + 576 \\ S' &= 2(\pi+1)x - 48 \\ 2(\pi+1)x - 48 &= 0 \\ x &= \frac{24}{\pi+1} \\ 24 - x &= 24 - \frac{24}{\pi+1} \\ &= \frac{24\pi}{\pi+1} \end{aligned}$$

There is no maximum if the string is cut. A maximum could occur if the string is not cut and the whole string is used to make the circle.

45. left to the student

47. $Q = x^3 + 2y^3$ and $x + y = 1$

$$\begin{aligned} Q &= (1-y)^3 + 2y^3 \\ &= 1 - 3y + 3y^2 - y^3 + 2y^3 \\ &= y^3 + 3y^2 - 3y + 1 \\ Q' &= 3y^2 + 6y - 3 \\ 3y^2 + 6y - 3 &= 0 \\ y &= \frac{-6 \pm \sqrt{36+36}}{6} \\ &= -1 + \sqrt{2} \\ x &= 1 - (-1 \pm +\sqrt{2}) \\ &= 2 - \sqrt{2} \end{aligned}$$

When $x = 2 - \sqrt{2}$ and $y = -1 + \sqrt{2}$

$$\begin{aligned} Q &= (2-\sqrt{2})^3 + 2(-1+\sqrt{2})^3 \\ &= 8 - 12\sqrt{2} + 12 - 2\sqrt{2} + \\ &\quad 2(-1 + 3\sqrt{2} - 6 + 2\sqrt{2}) \\ &= 6 - 4\sqrt{2} \end{aligned}$$

Exercise Set 3.6

1. $f(x) = x^2$, $a = 3$
$f'(x) = 2x$

$$\begin{aligned} L(x) &= f(3) + f'(3)(x-3) \\ &= 9 + 6(x-3) \\ &= 6x - 9 \end{aligned}$$

3. $f(x) = \frac{1}{x}$, $a = 4$
$f'(x) = \frac{-1}{x^2}$

$$\begin{aligned} L(x) &= f(4) + f'(4)(x-4) \\ &= \frac{1}{4} - \frac{1}{16}(x-4) \\ &= -\frac{1}{16}x + \frac{1}{2} \end{aligned}$$

5. $f(x) = x^{3/2}$, $a = 4$
$f'(x) = \frac{3}{2}x^{1/2}$

$$\begin{aligned} L(x) &= f(4) + f'(4)(x-4) \\ &= 8 + 3(x-4) \\ &= 3x - 4 \end{aligned}$$

7. $f(x) = \cos x$, $a = 0$
$f'(x) = -\sin x$

$$\begin{aligned} L(x) &= f(0) + f'(0)(x-0) \\ &= 1 + 0(x-0) \\ &= 1 \end{aligned}$$

9. $f(x) = x \cos x$, $a = 0$
$f'(x) = -x \sin x + \cos x$

$$\begin{aligned} L(x) &= f(0) + f'(0)(x-0) \\ &= 0 + 1(x-0) \\ &= x \end{aligned}$$

11. $f(x) = \sqrt{x}$, $a = 16$
$f'(x) = \frac{1}{2\sqrt{x}}$

$$\begin{aligned} L(19) &= f(16) + f'(16)(19-16) \\ &= 4 + \frac{1}{8}(3) \\ &= 4.375 \end{aligned}$$

13. $f(x) = \sqrt{x}$, $a = 100$
$f'(x) = \frac{1}{2\sqrt{x}}$

$$\begin{aligned} L(99.1) &= f(100) + f'(100)(99.1-100) \\ &= 10 + \frac{1}{20}(-0.9) \\ &= 9.955 \end{aligned}$$

15. $f(x) = \sqrt[3]{x}$, $a = 8$
$f'(x) = \frac{1}{3\sqrt[3]{x^2}}$

$$\begin{aligned} L(10) &= f(8) + f'(8)(10-8) \\ &= 2 + \frac{1}{12}(2) \\ &= 2.16667 \end{aligned}$$

17. $f(x) = \sqrt{x}$, $a = 100$
$f'(x) = \frac{1}{2\sqrt{x}}$

$$\begin{aligned} L(97) &= f(100) + f'(100)(97-100) \\ &= 10 + \frac{1}{20}(-3) \\ &= 9.85 \end{aligned}$$

19. $f(x) = \sin x$, $a = 0$
$f'(x) = \cos x$

$$\begin{aligned} L(0.1) &= f(0) + f'(0)(0.1-0) \\ &= 0 + 1(0.1) \\ &= 0.1 \end{aligned}$$

21. $f(x) = tan\ x,\ a = 0$
$f'(x) = sec^2\ x$

$$\begin{aligned} L(-0.04) &= f(0) + f'(0)(-0.04 - 0) \\ &= 0 + 1(-0.04 - 0) \\ &= -0.04 \end{aligned}$$

23. $f(x) = \frac{1}{3}x^2 - x + 1over3$
$f'(x) = \frac{2}{3}x - 1$

$$\begin{aligned} x_{n+1} &= x_n - \frac{f(x_n)}{f'(x_n)} \\ x_1 &= 0 \\ x_2 &= 0 - \frac{0.33333}{-1} = 0.33333 \\ x_3 &= 0.33333 - \frac{0.03704}{-0.77778} = 0.38095 \\ x_4 &= 0.38095 - \frac{0.00076}{-0.74603} = 0.38197 \\ x_5 &= 0.38197 - \frac{0}{-7.454} = 0.38197 \end{aligned}$$

$x = 0.38197$

25. $f(x) = x^3 - 3x + 3$
$f'(x) = 3x^2 - 3$

$$\begin{aligned} x_{n+1} &= x_n - \frac{f(x_n)}{f'(x_n)} \\ x_1 &= -3 \\ x_2 &= -3 - \frac{-15}{24} = -2.375 \\ x_3 &= -2.375 - \frac{-3.27148}{13.92188} = -2.14001 \\ x_4 &= -2.14001 - \frac{-0.38045}{10.73892} = -2.10458 \\ x_5 &= -2.10458 - \frac{-0.00799}{10.28777} = -2.10381 \\ x_6 &= -2.10381 - \frac{-0.00007}{10.27805} = -2.10380 \\ x_7 &= -2.10380 - \frac{-0.00004}{10.27805} = -2.10380 \end{aligned}$$

$x = -2.10380$

27. $f(x) = x\sqrt{x+1} - 4$
$f'(x) = \frac{x}{\sqrt{x+1}} + \sqrt{x+1}$

$$\begin{aligned} x_{n+1} &= x_n - \frac{f(x_n)}{f'(x_n)} \\ x_1 &= 2 \\ x_2 &= 2 - \frac{-0.5359}{2.5774} = 2.2079 \\ x_3 &= 2.2079 - \frac{-0.0455}{2.6677} = 2.2250 \\ x_4 &= 2.2250 - \frac{-0.0043}{2.675} = 2.2266 \\ x_5 &= 2.2266 - \frac{-0.0004}{2.6757} = 2.2267 \\ x_6 &= 2.2267 - \frac{-0.0002}{2.6757} = 2.2268 \\ x_7 &= 2.2268 - \frac{0}{2.6757} = 2.2268 \end{aligned}$$

$x = 2.2268$

29. $f(x) = cos\ 2x - x$
$f'(x) = -2\ sin\ 2x - 1$

$$\begin{aligned} x_{n+1} &= x_n - \frac{f(x_n)}{f'(x_n)} \\ x_1 &= 0 \\ x_2 &= 0 - \frac{1}{-1} = 1 \\ x_3 &= 1 - \frac{-1.416}{-2.819} = 0.49769 \\ x_4 &= 0.49769 - \frac{0.04648}{-2.678} = 0.51505 \\ x_5 &= 0.51505 - \frac{-0.0003}{-2.715} = 0.51494 \\ x_6 &= 0.51494 - \frac{0}{-2.714} = 0.51494 \end{aligned}$$

$x = 0.51494$

31. $f(x) = sin\ x - cos\ x + x$
$f'(x) = cos\ x + sin\ x + 1$

$$\begin{aligned} x_{n+1} &= x_n - \frac{f(x_n)}{f'(x_n)} \\ x_1 &= 0 \\ x_2 &= 0 - \frac{-1}{2} = 0.5 \\ x_3 &= 0.5 - \frac{0.10184}{2.357} = 0.45679 \\ x_4 &= 0.45679 - \frac{0.00039}{2.3385} = 0.45663 \\ x_5 &= 0.45663 - \frac{0}{2.3385} = 0.45663 \end{aligned}$$

$x = 0.45663$

33. Initial guess $x = -1.5$ leads to solution $x = -1.142495$
Initial guess $x = 0.25$ leads to solution $x = 0.176245$
Initial guess $x = 4.9$ leads to solution $x = 4.96625$

35. Initial guess $x = -3$ leads to solution $x = -2.86640$
Initial guess $x = -0.5$ leads to solution $x = -0.56682$
Initial guess $x = 0.5$ leads to solution $x = 0.40865$

37. Use Newton's method on

$$0.05x^2 - 0.3x^3 - 0.0001$$

Initial guess $x = 0$ leads to solution $x = 0.05452$
Initial guess $x = 0.1$ leads to solution $x = 0.15229$
The two dosages are $x = 0.055$ and $x = 0.15$ cubic centimeters.

39. Use Newton's method on

$$-6.85 + 1.82t - 0.0596t^2 + 0.000758t^3$$

Initial guess $x = 4$ leads to solution $x = 4.34880$
The age at which the median weight of boys is 15 pounds is $t = 4.35$ months.

41. Use Newton's method on

$$-0.000775x^3 + 0.0696x^2 - 0.209x - 35.32$$

Initial guess $x = 30$ leads to solution $x = 29.93332$
The rate is 40 per 100000 in the end of 1959.

43. Use Newton's method on

$$-0.000054x^4 + 0.0067x^3 - 0.0997x^2 - 0.84x - 300.25$$

Initial guess $x = 60$ leads to solution $x = 57.09821$
Initial guess $x = 95$ leads to solution $x = 97.50401$
At the age of 57 years and 97.5 years 300 out of 100000 woemn will have breast cancer.

45. $f(r) = r^{15} - 0.99r^{14} - 0.0858\left[1 - \left(\frac{0.99}{r}\right)^{36}\right]$

$f'(r) = 15r^{14} - 13.86r^{13} + \left(\frac{2.151081148}{r^{37}}\right)$

$$\begin{aligned}
x_{n+1} &= x_n - \frac{f(x_n)}{f'(x_n)} \\
x_1 &= 1.1 \\
x_2 &= 1.1 - \frac{0.33386}{9.0507} = 1.0631 \\
x_3 &= 1.0631 - \frac{0.09303}{4.4002} = 1.042 \\
x_4 &= 1.0420 - \frac{0.02021}{2.5499} = 1.0340 \\
x_5 &= 1.0340 - \frac{0.00248}{1.9274} = 1.03276 \\
x_6 &= 1.03276 - \frac{0}{1.8279} = 1.03276
\end{aligned}$$

$r = 1.03276$

47. $f(r) = r^4 - 0.98r^3 - 0.1764\left[1 - \left(\frac{0.98}{r}\right)^{15}\right]$

$f'(r) = 4r^3 - 2.94r^2 - \frac{1.954253846}{r^{16}}$

After 7 iterations of Newton's method we reach $r = 1.08846$

49. **a)**

$$\begin{aligned}
v &= \frac{77000 \cdot 100 \cdot sec\,\frac{4\pi}{9}}{4000000} \\
&= 11.086\ \frac{cm}{s}
\end{aligned}$$

b) $f(t) = 1.925\ sec\ t$, $f'(t) = 11.086\ sec\ t\ tan\ t$

$$\begin{aligned}
L(0.01) &= f(\frac{4\pi}{9}) + f'(\frac{4\pi}{9})(0.01) \\
&= 11.7147
\end{aligned}$$

Difference in measurement $= 11.7147 - 11.086 = 0.6287\ \frac{cm}{s}$

c) The difference in measurement is more sensitive to angle measurement when the frequency f gets smaller.

51. **a)** $x_1 = 3$
$x_2 = 3 - \frac{-3}{-1} = 0$
$x_3 = 0 - \frac{0}{8} = 0$
$x = 0$

b) The tangent line at $x_1 = 3$ (which intersects x_{n+1}) intersected $x = 0$ instead of the closer solutions.

53. For $x = 0.04609$, $f = 259.674\ Hz$
For $x = 0.34184$, $f = 1925.96\ Hz$
For $x = 0.52360$, $f = 2950.01\ Hz$

55. For $x = 0.11602$, $f = 653.67\ Hz$
For $x = 0.32838$, $f = 1850.12\ Hz$
For $x = 0.42340$, $f = 2385.47\ Hz$

Exercise Set 3.7

1. $xy - x + 2y = 3$

$$\begin{aligned}
x\frac{dy}{dx} + y - 1 + 2\frac{dy}{dx} &= 0 \\
(x+2)\frac{dy}{dx} &= 1 - y \\
\frac{dy}{dx} &= \frac{1-y}{x+2}
\end{aligned}$$

For $\left(-5, \frac{2}{3}\right)$

$$\begin{aligned}
\frac{dy}{dx} &= \frac{1 - \frac{2}{3}}{-5+2} \\
&= \frac{1}{9}
\end{aligned}$$

3. $x^2 + y^2 = 1$

$$\begin{aligned}
2x + 2y\frac{dy}{dx} &= 0 \\
\frac{dy}{dx} &= \frac{-x}{y}
\end{aligned}$$

For $\left(\frac{1}{2}, \frac{\sqrt{3}}{2}\right)$

$$\begin{aligned}
\frac{dy}{dx} &= -\frac{\frac{1}{2}}{\frac{\sqrt{3}}{2}} \\
&= -\frac{1}{\sqrt{3}}
\end{aligned}$$

5. $x^2y - 2x^3 - y^3 + 1 = 0$

$$\begin{aligned}
x^2\frac{dy}{dx} + 2xy - 6x^2 - 3y^2\frac{dy}{dx} &= 0 \\
(x^2 - 3y^2)\frac{dy}{dx} &= 6x^2 - 2xy \\
\frac{dy}{dx} &= \frac{6x^2 - 2xy}{x^2 - 3y^2}
\end{aligned}$$

For $(2, -3)$

$$\begin{aligned}
\frac{dy}{dx} &= \frac{6(2)^2 - 2(2)(-3)}{(2)^2 - 3(-3)^2} \\
&= -\frac{36}{23}
\end{aligned}$$

7. $sin\ y + x^2 = cos\ y$

$$\begin{aligned} \frac{dy}{dx} cos\ y + 2x &= -\frac{dy}{dx} sin\ x \\ (cos\ y + sin\ y)\frac{dy}{dx} &= -2x \\ \frac{dy}{dx} &= \frac{-2x}{cos\ y + sin\ y} \end{aligned}$$

For $(1, 2\pi)$

$$\begin{aligned} \frac{dy}{dx} &= \frac{-2(1)}{cos\ 2\pi + sin\ 2\pi} \\ &= -2 \end{aligned}$$

9. $x\ sin\ x = y(1 + cos\ y)$

$$\begin{aligned} x\ cos\ x + sin\ x &= y(-sin\ y)\frac{dy}{dx} + (1 + cos\ y)\frac{dy}{dx} \\ x\ cos\ x + sin\ x &= (-y\ sin\ y + cos\ y + 1)\frac{dy}{dx} \\ \frac{dy}{dx} &= \frac{x\ cos\ x + sin\ x}{-y\ sin\ y + cos\ y + 1} \end{aligned}$$

For $\left(\frac{\pi}{2}, \frac{\pi}{3}\right)$

$$\begin{aligned} \frac{dy}{dx} &= \frac{\frac{\pi}{2} cos\ \frac{\pi}{2} + sin\ \frac{\pi}{2}}{-\frac{\pi}{3}\ sin\ \frac{\pi}{3} + cos\ \frac{\pi}{3} + 1} \\ &= \frac{6}{9 - \pi\sqrt{3}} \end{aligned}$$

11. $2xy + 3 = 0$

$$\begin{aligned} 2x\frac{dy}{dx} + 2y &= 0 \\ \frac{dy}{dx} &= \frac{-y}{x} \end{aligned}$$

13. $x^2 - y^2 = 16$

$$\begin{aligned} 2x - 2y\frac{dy}{dx} &= 0 \\ \frac{dy}{dx} &= \frac{x}{y} \end{aligned}$$

15. $y^5 = x^3$

$$\begin{aligned} 5y^4\frac{dy}{dx} &= 3x^2 \\ \frac{dy}{dx} &= \frac{3x^2}{5y^4} \end{aligned}$$

17. $x^2y^3 + x^3y^4 = 11$

$$\begin{aligned} 3x^2y^2\frac{dy}{dx} + 2xy^3 + 4x^3y^3\frac{dy}{dx} + 3x^2y^4 &= 0 \\ (3x^2y^2 + 4x^3y^3)\frac{dy}{dx} &= -2xy^3 - 3x^2y^4 \\ \frac{dy}{dx} &= \frac{-2xy^3 - 3x^2y^4}{3x^2y^2 + 4x^3y^3} \end{aligned}$$

19. $\sqrt{x} + \sqrt{y} = 1$

$$\begin{aligned} \frac{1}{2\sqrt{x}} + \frac{1}{2\sqrt{y}}\frac{dy}{dx} &= 0 \\ \frac{dy}{dx} &= -\frac{\sqrt{y}}{\sqrt{x}} \end{aligned}$$

21. $y^3 = \dfrac{x-1}{x+1}$

$$\begin{aligned} 3y^2\frac{dy}{dx} &= \frac{x + 1 - (x-1)}{(x+1)^2} \\ \frac{dy}{dx} &= \frac{2}{3(x+1)^2y^2} \end{aligned}$$

23. $x^{3/2} + y^{2/3} = 1$

$$\begin{aligned} \frac{3}{2}x^{1/2} + \frac{2}{3y^{1/3}}\frac{dy}{dx} &= 0 \\ \frac{dy}{dx} &= -\frac{\sqrt{x}}{\sqrt[3]{y}} \end{aligned}$$

25. $\dfrac{x^2y + xy + 1}{2x + y} = 1 \rightarrow x^2y + xy + 1 = 2x + y$

$$\begin{aligned} x^2\frac{dy}{dx} + 2xy + x\frac{dy}{dx} + y &= 2 + \frac{dy}{dx} \\ (x^2 + x - 1)\frac{dy}{dx} &= 2 - 2xy - y \\ \frac{dy}{dx} &= \frac{2 - 2xy - y}{x^2 + x - 1} \end{aligned}$$

27. $4\ sin\ x\ \ cos\ y = 3$

$$\begin{aligned} -4\ sin\ x\ sin\ y\frac{dy}{dx} + 4\ cos\ x\ cos\ y &= 0 \\ \frac{dy}{dx} &= \frac{cos\ x\ cos\ y}{sin\ x\ sin\ y} \\ &= cot\ x\ cot\ y \end{aligned}$$

29. $x + y = sin(\sqrt{y - x})$

$$\begin{aligned} 1 + \frac{dy}{dx} &= cos(\sqrt{y-x})\ \frac{\frac{dy}{dx} - 1}{2\sqrt{y-x}} \\ 2\sqrt{y-x} + 2\sqrt{y-x}\frac{dy}{dx} &= cos(\sqrt{y-x})\frac{dy}{dx} - \\ &\quad cos(\sqrt{y-x}) \\ (2\sqrt{y-x} - cos(\sqrt{y-x}))\frac{dy}{dx} &= -2\sqrt{y-x} + cos(\sqrt{y-x}) \\ \frac{dy}{dx} &= -\frac{2\sqrt{y-x} + cos(\sqrt{y-x})}{2\sqrt{y-x} - cos(\sqrt{y-x})} \end{aligned}$$

31. $A^3 + B^3 = 9$ When $A = 2$, $B = \sqrt[3]{9-8} = 1$

$$\begin{aligned} 3A^2\frac{dA}{dt} + 3B^2\frac{dB}{dt} &= 0 \\ 3(2)^2\frac{dA}{dt} + 3(1)^2(3) &= 0 \\ \frac{dA}{dt} &= \frac{-9}{12} \\ &= -\frac{3}{4} \end{aligned}$$

33. $V = \frac{4}{3}\pi r^3$

$$\begin{aligned}\frac{dV}{dt} &= 4\,\pi\,r^2\frac{dr}{dt}\\ &= 4\,\pi(1.2)^2(0.03)\\ &= 0.54287\ \frac{cm}{day}\end{aligned}$$

35. $V = \frac{p}{4Lv}\,(R^2 - r^2)$

a)

$$\begin{aligned}\frac{dV}{dt} &= \frac{2Rp}{4Lv}\frac{dR}{dt}\\ &= \frac{2R(100)}{4(1)(0.05)}\ \frac{dR}{dt}\\ &= 1000R\ \frac{dR}{dt}\end{aligned}$$

b)

$$\begin{aligned}\frac{dV}{dt} &= 1000R\ \frac{dR}{dt}\\ &= 1000(0.0075)(-0.0015)\\ &= -0.01125\ \frac{mm^3}{min}\end{aligned}$$

37. $S = \frac{\sqrt{hw}}{60}$

$$\begin{aligned}\frac{dS}{dt} &= \frac{h}{120\sqrt{hw}}\ \frac{dw}{dt}\\ &= \frac{180}{120\sqrt{(180)(85)}}(-4)\\ &= -0.0485\ \frac{m^2}{month}\end{aligned}$$

39. $D^2 = x^2 + y^2$ After one hour, $x = 25$, $y = 60$
$D = \sqrt{25^2 + 60^2} = 65$

$$\begin{aligned}2D\frac{dD}{dt} &= 2x\frac{dx}{dt} + 2y\frac{dy}{dt}\\ \frac{dD}{dt} &= \frac{x\frac{dx}{dt} + y\frac{dy}{dt}}{D}\\ &= \frac{25(25) + 60(60)}{65}\\ &= 65\ \ mph\end{aligned}$$

41. $tan\theta = \frac{h}{100}$

$$\begin{aligned}sec^2\,\theta\ \frac{d\theta}{dt} &= \frac{1}{100}\frac{dh}{dt}\\ \frac{dh}{dt} &= 100\ sec^2\,\theta\ \frac{d\theta}{dt}\\ &= 100\ sec^2(\frac{\pi}{6})(0.1)\\ &= 13.3333\ \frac{m}{min}\end{aligned}$$

43. **a)** $R^2 = x^2 + 9$
When $R = 10$, $x = \sqrt{100 - 9} = \sqrt{91}$

$$\begin{aligned}2R\ \frac{dR}{dt} &= 2x\ \frac{dx}{dt}\\ \frac{dx}{dt} &= \frac{R}{x}\frac{dR}{dt}\\ &= \frac{10}{\sqrt{91}}(2)\\ &= 2.0966\ \frac{ft}{sec}\end{aligned}$$

b) $sin\ \theta = \frac{3}{R}$
When $R = 10$, $\theta = sin^{-1}(\frac{3}{10}) = 0.30469\ \ rad$

$$\begin{aligned}cos\ \theta\frac{d\theta}{dt} &= -\frac{3}{R^2}\frac{dR}{dt}\\ \frac{d\theta}{dt} &= -\frac{3}{R^2\ cos\ \theta}\ \frac{dR}{dt}\\ &= -\frac{3}{100\ cos(0.30469)}(2)\\ &= -\frac{3}{5\ \sqrt{91}}\\ &= -0.0629\ \frac{rad}{sec}\end{aligned}$$

45. $xy + x - 2y = 4$

$$x\frac{dy}{dx} + y + 1 - 2\frac{dy}{dx} = 0$$

$$\begin{aligned}(x-2)\frac{dy}{dx} &= -1 - y\\ \frac{dy}{dx} &= \frac{y+1}{2-x}\\ \frac{d^2y}{dx^2} &= \frac{(2-x)(\frac{dy}{dx}) + (y+1)}{(2-x)^2}\\ &= \frac{(2-x)\frac{y+1}{2-x} + y + 1}{(2-x)^2}\\ &= \frac{(y+1) + (y+1)}{(2-x)^2}\\ &= \frac{2y+2}{(2-x)^2}\end{aligned}$$

47. $x^2 - y^2 = 5$

$$\begin{aligned}2x - 2y\frac{dy}{dx} &= 0\\ \frac{dy}{dx} &= \frac{x}{y}\\ \frac{d^2y}{dx^2} &= \frac{y - x\frac{dy}{dx}}{y^2}\\ &= \frac{y - \frac{x^2}{y}}{y^2}\\ &= \frac{y^2 - x^2}{y^3}\end{aligned}$$

49. Left to the student (answers vary)

51. **53.** **55.**

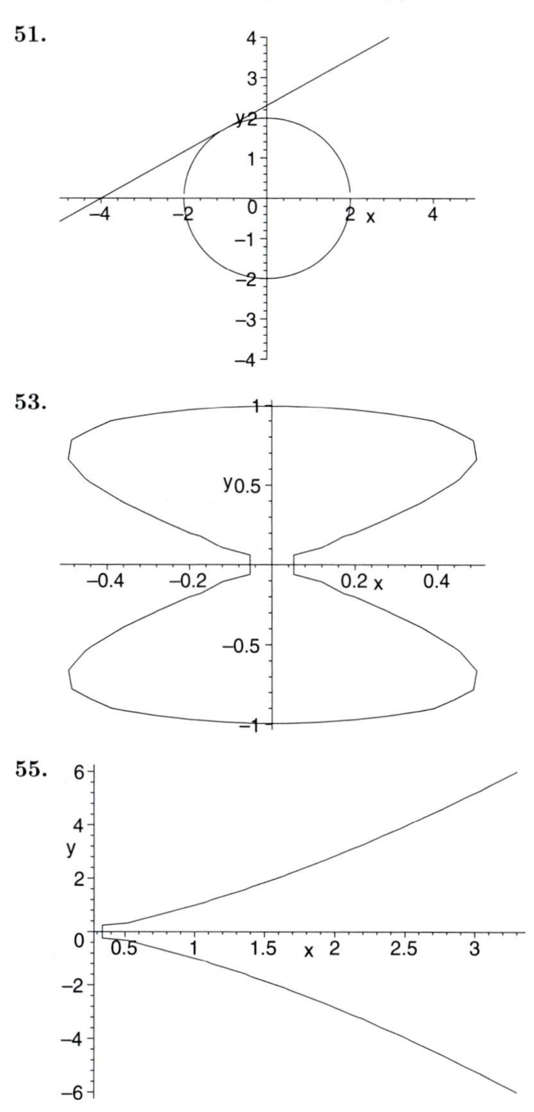

Chapter 4

Exponential and Logarithmic Functions

Exercise Set 4.1

1. Graph: $y = 4^x$

First we find some function values.

Note: For

$x = -2,\ y = 4^{-2} = \frac{1}{4^2} = \frac{1}{16} = 0.0625$

$x = -1,\ y = 4^{-1} = \frac{1}{4} = 0.25$

$x = 0,\quad y = 4^0 = 1$

$x = 1,\quad y = 4^1 = 4$

$x = 2,\quad y = 4^2 = 16$

x	y
-2	0.0625
-1	0.25
0	1
1	4
2	16

Plot these points and connect them with a smooth curve.

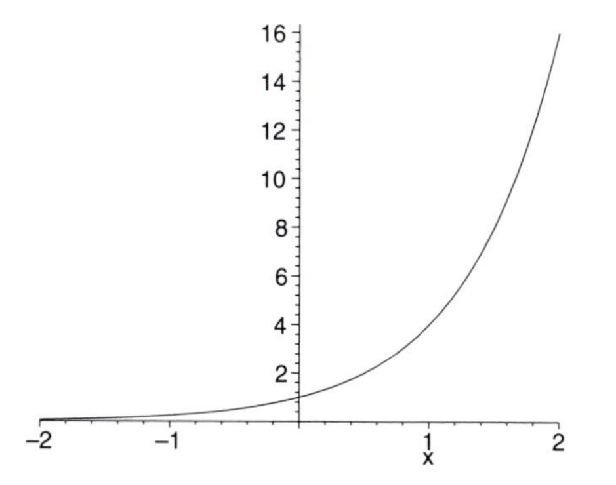

3. Graph: $y = (0.4)^x$

First we find some function values.

Note: For

$x = -2,\ y = (0.4)^{-2} = \frac{1}{(0.4)^2} = 6.25$

$x = -1,\ y = (0.4)^{-1} = \frac{1}{0.4} = 2.5$

$x = 0,\quad y = (0.4)^0 = 1$

$x = 1,\quad y = (0.4)^1 = 0.4$

$x = 2,\quad y = (0.4)^2 = 0.16$

x	y
-2	6.25
-1	2.5
0	1
1	0.4
2	0.16

Plot these points and connect them with a smooth curve.

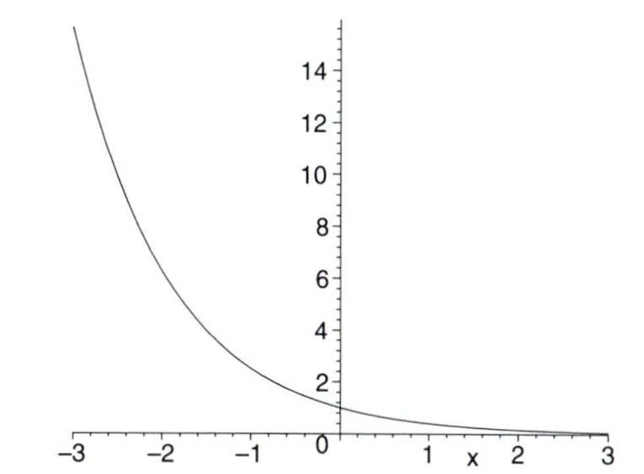

5. Graph: $x = 4^y$

First we find some function values.

Note: For

$y = -2,\ x = 4^{-2} = \frac{1}{4^2} = \frac{1}{16}$

$y = -1,\ x = 4^{-1} = \frac{1}{4}$

$y = 0,\quad x = 4^0 = 1$

$y = 1,\quad x = 4^1 = 4$

$y = 2,\quad x = 4^2 = 16$

x	y
$\frac{1}{16}$	-2
$\frac{1}{4}$	-1
1	0
4	1
16	2

Plot these points and connect them with a smooth curve.

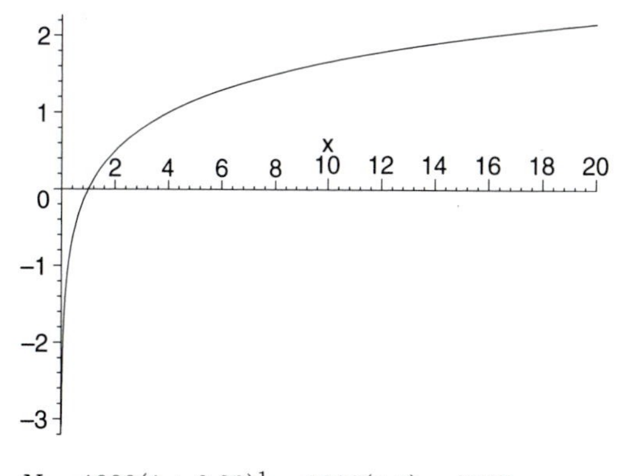

7. $N = 1000(1 + 0.20)^1 = 1000(1.2) = 1200$
$N = 1000(1 + 0.20)^2 = 1000(1.44) = 1440$
$N = 1000(1 + 0.20)^5 = 1000(2.488) = 2488$

9. $N = 286000000(1 + 0.021)^1 = 286000000(1.021) = 291006000$
$N = 286000000(1 + 0.021)^2 = 286000000(1.042441) = 298138126$
$N = 286000000(1 + 0.021)^5 = 286000000(1.109503586) = 3173188026$

11. $f(x) = e^{3x}$

$f'(x) = 3e^{3x} \qquad \left[\frac{d}{dx}e^{f(x)} = f'(x)e^{f(x)}\right]$

13. $f(x) = 5e^{-2x}$

$f'(x) = 5 \cdot \underbrace{(-2)e^{-2x}}$ $\left[\frac{d}{dx}[c \cdot f(x)] = c \cdot f'(x)\right]$

$\left[\frac{d}{dx}e^{f(x)} = f'(x)e^{f(x)}\right]$

$= -10e^{-2x}$

15. $f(x) = 3 - e^{-x}$

$f'(x) = 0 - \underbrace{(-1)e^{-x}}$ $\left[\frac{d}{dx}e^{f(x)} = f'(x)e^{f(x)}\right]$

$= e^{-x}$

17. $f(x) = -7e^{x}$

$f'(x) = -7e^{x}$ $\left[\frac{d}{dx} - 7e^x = -7 \cdot \frac{d}{dx}e^x\right]$

$\left[\frac{d}{dx}e^x = e^x\right]$

19. $f(x) = \frac{1}{2}e^{2x}$

$f'(x) = \frac{1}{2} \cdot \underbrace{2e^{2x}}$ $\left[\frac{d}{dx}[c \cdot f(x)] = c \cdot f'(x)\right]$

$\left[\frac{d}{dx}e^{f(x)} = f'(x)e^{f(x)}\right]$

$= e^{2x}$

21. $f(x) = x^4e^x$

$f'(x) = x^4 \cdot e^x + 4x^3 \cdot e^x$ Using the Product Rule

$= x^3e^x(x+4)$

23. $f(x) = (x^2+3x-9)e^x$

$f'(x) = (x^2+3x-9)e^x + (2x+3)e^x$ Using the Product Rule

$= (x^2+3x-9+2x+3)e^x$

$= (x^2+5x-6)e^x$

25. $f(x) = (\sin x)e^x$

$f'(x) = (\sin x)e^x + (\cos x)e^x$

27. $f(x) = \frac{e^x}{x^4}$

$f'(x) = \frac{x^4 \cdot e^x - 4x^3 \cdot e^x}{x^8}$ Using the Quotient Rule

$= \frac{x^3e^x(x-4)}{x^3 \cdot x^5}$ Factoring both numerator and denominator

$= \frac{x^3}{x^3} \cdot \frac{e^x(x-4)}{x^5}$

$= \frac{e^x(x-4)}{x^5}$ Simplifying

29. $f(x) = e^{-x^2+7x}$

$f'(x) = (-2x+7)e^{-x^2+7x}$ $\left[\frac{d}{dx}e^{f(x)} = f'(x)e^{f(x)}\right]$

$[f(x) = -x^2+7x,$
$f'(x) = -2x+7]$

31. $f(x) = e^{-x^2/2}$

$= e^{(-1/2)x^2}$

$f'(x) = \left(-\frac{1}{2} \cdot 2x\right)e^{(-1/2)x^2}$

$= -xe^{-x^2/2}$

33. $y = e^{\sqrt{x-7}}$

$= e^{(x-7)^{1/2}}$

$\frac{dy}{dx} = \frac{1}{2}(x-7)^{-1/2} \cdot e^{(x-7)^{1/2}}$

$= \frac{e^{(x-7)^{1/2}}}{2(x-7)^{1/2}}$

$= \frac{e^{\sqrt{x-7}}}{2\sqrt{x-7}}$

35. $y = \sqrt{e^x - 1}$

$= (e^x-1)^{1/2}$

$\frac{dy}{dx} = \frac{1}{2}(e^x-1)^{-1/2} \cdot e^x$ [Extended Power Rule; $\frac{d}{dx}(e^x-1) = e^x - 0 = e^x$]

$= \frac{e^x}{2\sqrt{e^x-1}}$

37. $y = \tan(e^x+1)$

$\frac{dy}{dx} = \sec^2(e^x+1) \cdot e^x$

$= e^x \sec^2(e^x+1)$

39. $y = e^{\tan x}$

$\frac{dy}{dx} = e^{\tan x} \cdot \sec^2 x$

41. $y = (2x + \cos x)e^{3x+1}$

$\frac{dy}{dx} = (2x+\cos x) \cdot e^{3x+1}(1) + e^{3x+1}(2 - \sin x)$

$= e^{3x+1}(2x + \cos x - \sin x + 2)$

43. $y = xe^{-2x} + e^{-x} + x^3$

$\frac{dy}{dx} = x \cdot (-2) \cdot e^{-2x} + 1 \cdot e^{-2x} + (-1) \cdot e^{-x} + 3x^2$

$= -2xe^{-2x} + e^{-2x} - e^{-x} + 3x^2$

$= (1-2x)e^{-2x} - e^{-x} + 3x^2$

45. $y = 1 - e^{-x}$

$\frac{dy}{dx} = 0 - (-1)e^{-x}$

$= e^{-x}$

47. $y = 1 - e^{-kx}$

$\frac{dy}{dx} = 0 - (-k)e^{-kx}$

$= ke^{-kx}$

49. $y = (e^{3x}+1)^5$

$\frac{dy}{dx} = 5(e^{3x}+1)^4 \cdot 3e^{3x}$

$= 15e^{3x}(e^{3x}+1)^4$

51. $y = \dfrac{e^{3t} - e^{7t}}{e^{4t}}$

$= \dfrac{e^{3t}(1 - e^{4t})}{e^{3t} \cdot e^{t}}$ Factoring

$= \dfrac{1 - e^{4t}}{e^{t}}$ Simplifying

$\dfrac{dy}{dt} = \dfrac{e^t(-4e^{4t}) - e^t(1 - e^{4t})}{(e^t)^2}$ Using the Quotient Rule

$= \dfrac{e^t(-4e^{4t} - 1 + e^{4t})}{e^t \cdot e^t}$

$= \dfrac{-3e^{4t} - 1}{e^t}$

$= -3e^{3t} - e^{-t}$

53. $y = \dfrac{e^x}{x^2 + 1}$

$\dfrac{dy}{dx} = \dfrac{(x^2+1)e^x - 2x \cdot e^x}{(x^2+1)^2}$

$= \dfrac{e^x(x^2 - 2x + 1)}{(x^2+1)^2}$

$= \dfrac{e^x(x-1)^2}{(x^2+1)^2}$

55. $f(x) = e^{\sqrt{x}} + \sqrt{e^x}$

$= e^{x^{1/2}} + e^{x/2}$

$f'(x) = \dfrac{1}{2}x^{-1/2}e^{x^{1/2}} + \dfrac{1}{2}e^{x/2}$

$= \dfrac{e^{\sqrt{x}}}{2\sqrt{x}} + \dfrac{\sqrt{e^x}}{2}$

57. $f(x) = e^{x/2} \cdot \sqrt{x-1}$

$= e^{x/2} \cdot (x-1)^{1/2}$

$f'(x) = e^{x/2} \cdot \dfrac{1}{2}(x-1)^{-1/2} + \dfrac{1}{2}e^{x/2} \cdot (x-1)^{1/2}$

$= \dfrac{1}{2}e^{x/2}\left((x-1)^{-1/2} + (x-1)^{1/2}\right)$

$= \dfrac{1}{2}e^{x/2}\left(\dfrac{1}{\sqrt{x-1}} + \sqrt{x-1}\right)$

$= \dfrac{1}{2}e^{x/2}\left(\dfrac{1}{\sqrt{x-1}} + \sqrt{x-1} \cdot \dfrac{\sqrt{x-1}}{\sqrt{x-1}}\right)$

$= \dfrac{1}{2}e^{x/2}\left(\dfrac{1}{\sqrt{x-1}} + \dfrac{x-1}{\sqrt{x-1}}\right)$

$= \dfrac{1}{2}e^{x/2}\left(\dfrac{x}{\sqrt{x-1}}\right)$

59. $f(x) = \dfrac{e^x - e^{-x}}{e^x + e^{-x}}$

$f'(x) = \dfrac{(e^x+e^{-x})(e^x+e^{-x}) - (e^x-e^{-x})(e^x-e^{-x})}{(e^x+e^{-x})^2}$

$= \dfrac{(e^{2x}+e^0+e^0+e^{-2x}) - (e^{2x}-e^0-e^0+e^{-2x})}{(e^x+e^{-x})^2}$

$= \dfrac{e^{2x}+1+1+e^{-2x}-e^{2x}+1+1-e^{-2x}}{(e^x+e^{-x})^2}$

$= \dfrac{4}{(e^x+e^{-x})^2}$

61. Graph: $f(x) = e^{2x}$

Using a calculator we first find some function values.

Note: For

$x = -2,\ f(-2) = e^{2(-2)} = e^{-4} = 0.0183$

$x = -1,\ f(-1) = e^{2(-1)} = e^{-2} = 0.1353$

$x = \ \ 0,\ f(0) = e^{2\cdot 0} = e^0 = 1$

$x = \ \ 1,\ f(1) = e^{2\cdot 1} = e^2 = 7.3891$

$x = \ \ 2,\ f(2) = e^{2\cdot 2} = e^4 = 54.598$

x	$f(x)$
−2	0.0183
−1	0.1353
0	1
1	7.3891
2	54.598

Plot these points and connect them with a smooth curve.

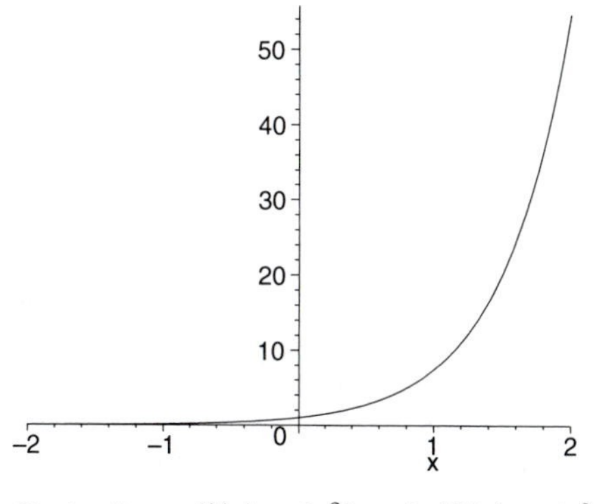

Derivatives. $f'(x) = 2e^{2x}$ and $f''(x) = 4e^{2x}$.

Critical points of f. Since $f'(x) > 0$ for all real numbers x, we know that the derivative exists for all real numbers and there is no solution of the equation $f'(x) = 0$. There are no critical points and therefore no maximum or minimum values.

Increasing. Since $f'(x) > 0$ for all real numbers x, the function f is increasing over the entire real line, $(-\infty, \infty)$.

Inflection points. Since $f''(x) > 0$ for all real numbers x, the equation $f''(x) = 0$ has no solution and there are no points of inflection.

Concavity. Since $f''(x) > 0$ for all real numbers x, the function f' is increasing and the graph is concave up over the entire real line.

63. Graph: $f(x) = e^{-2x}$

Using a calculator we first find some function values.

Note: For

$x = -2,\ f(-2) = e^{-2(-2)} = e^4 = 54.598$

$x = -1,\ f(-1) = e^{-2(-1)} = e^2 = 7.3891$

$x = \ \ 0,\ f(0) = e^{-2\cdot 0} = e^0 = 1$

$x = \ \ 1,\ f(1) = e^{-2\cdot 1} = e^{-2} = 0.1353$

$x = \ \ 2,\ f(2) = e^{-2\cdot 2} = e^{-4} = 0.0183$

x	$f(x)$
-2	54.598
-1	7.3891
0	1
1	0.1353
2	0.0183

Plot these points and connect them with a smooth curve.

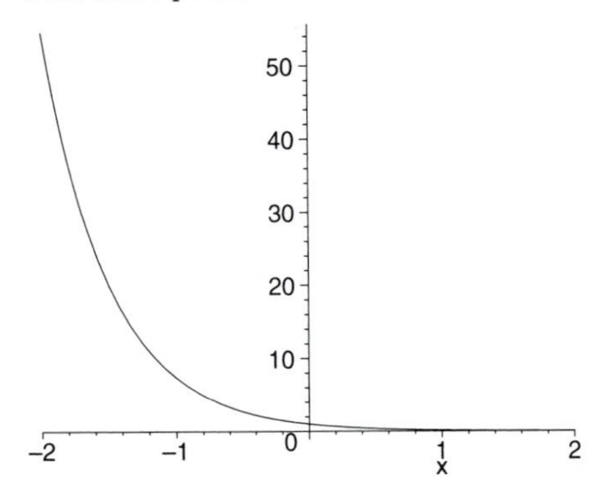

Derivatives. $f'(x) = -2e^{-2x}$ and $f''(x) = 4e^{-2x}$.

Critical points of f. Since $f'(x) < 0$ for all real numbers x, we know that the derivative exists for all real numbers and there is no solution of the equation $f'(x) = 0$. There are no critical points and therefore no maximum or minimum values.

Decreasing. Since $f'(x) < 0$ for all real numbers x, the function f is decreasing over the entire real line, $(-\infty, \infty)$.

Inflection points. Since $f''(x) > 0$ for all real numbers x, the equation $f''(x) = 0$ has no solution and there are no points of inflection.

Concavity. Since $f''(x) > 0$ for all real numbers x, the function f' is increasing and the graph is concave up over the entire real line.

65. Graph: $f(x) = 3 - e^{-x}$, for nonnegative values of x.

Using a calculator we first find some function values.

Note: For

$x = 0$, $f(0) = 3 - e^{-0} = 3 - 1 = 2$

$x = 1$, $f(1) = 3 - e^{-1} = 3 - 0.3679 = 2.6321$

$x = 2$, $f(2) = 3 - e^{-2} = 3 - 0.1353 = 2.8647$

$x = 3$, $f(3) = 3 - e^{-3} = 3 - 0.0498 = 2.9502$

$x = 4$, $f(4) = 3 - e^{-4} = 3 - 0.0183 - 2.9817$

$x = 6$, $f(6) = 3 - e^{-6} = 3 - 0.0025 = 2.9975$

x	$f(x)$
0	2
1	2.6321
2	2.8647
3	2.9502
4	2.9817
6	2.9975

Plot these points and connect them with a smooth curve.

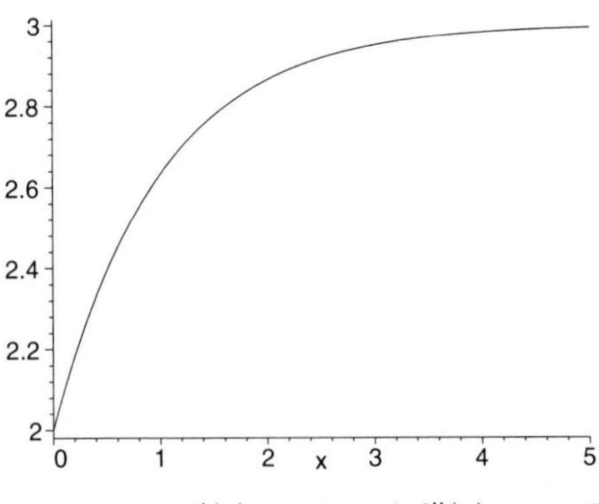

Derivatives. $f'(x) = e^{-x}$ and $f''(x) = -e^{-x}$.

Critical points of f. Since $f'(x) > 0$ for all real numbers x, we know that the derivative exists for all real numbers and there is no solution of the equation $f'(x) = 0$. There are no critical points and therefore no maximum or minimum values.

Increasing. Since $f'(x) > 0$ for all real numbers x, the function f is increasing over the entire real line, $(-\infty, \infty)$.

Inflection points. Since $f''(x) < 0$ for all real numbers x, the equation $f''(x) = 0$ has no solution and there are no points of inflection.

Concavity. Since $f''(x) < 0$ for all real numbers x, the function f' is decreasing and the graph is concave down over the entire real line.

67. - 71. Left to the student

73. We first find the slope of the tangent line at $(0, 1)$, $f'(0)$:

$$f(x) = e^x$$
$$f'(x) = e^x$$
$$f'(0) = e^0 = 1$$

Then we find the equation of the line with slope 1 and containing the point $(0, 1)$:

$$y - y_1 = m(x - x_1) \quad \text{Point-slope equation}$$
$$y - 1 = 1(x - 0)$$
$$y - 1 = x$$
$$y = x + 1$$

75. Left to the student

77. $C(t) = 10t^2e^{-t}$

a) $C(0) = 10 \cdot 0^2 \cdot e^{-0} = 0$ ppm

$$C(1) = 10 \cdot 1^2 \cdot e^{-1}$$
$$\approx 10(0.367879)$$
$$\approx 3.7 \text{ ppm}$$

$$\begin{aligned} C(2) &= 10 \cdot 2^2 \cdot e^{-2} \\ &\approx 40(0.135335) \\ &\approx 5.4 \text{ ppm} \end{aligned}$$

$$\begin{aligned} C(3) &= 10 \cdot 3^2 \cdot e^{-3} \\ &\approx 90(0.049787) \\ &\approx 4.48 \text{ ppm} \end{aligned}$$

$$\begin{aligned} C(10) &= 10 \cdot 10^2 \cdot e^{-10} \\ &\approx 1000(0.000045) \\ &\approx 0.05 \text{ ppm} \end{aligned}$$

b) We plot the points $(0,0)$, $(1,3.7)$, $(2,5.4)$, $(3,4.48)$, and $(10,0.05)$ and other points as needed. Then we connect the points with a smooth curve.

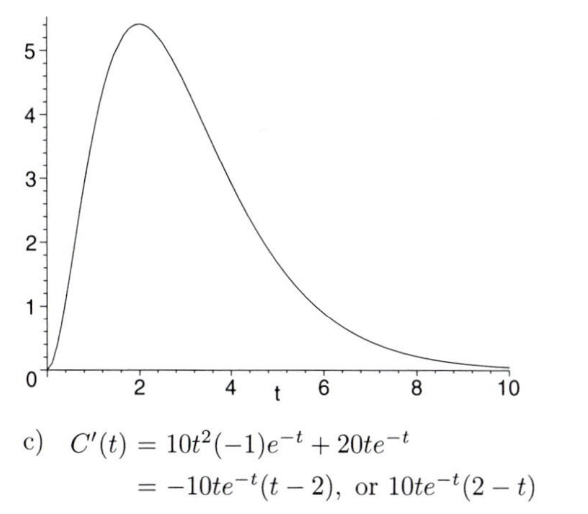

c) $C'(t) = 10t^2(-1)e^{-t} + 20te^{-t}$

$= -10te^{-t}(t-2)$, or $10te^{-t}(2-t)$

d) We find the maximum value of $C(t)$ on $[0, \infty)$.

$C'(t)$ exists for all t in $[0, \infty)$. We solve $C'(t) = 0$.

$10te^{-t}(2-t) = 0$

$t(2-t) = 0 \quad (10e^t \neq 0)$

$t = 0$ or $2 - t = 0$

$t = 0$ or $2 = t$

Since the function has two critical points and the interval is not closed, we must examine the graph to find the maximum value. We see that the maximum value of the concentration is about 5.4 ppm. It occurs at $t = 2$ hr.

e) The derivative represents the rate of change of the concentration of the medication with respect to the time t.

79. a) $r(T) = 0.153\left(e^{0.141(T-9.5)} - e^{4.2159-0.153(39.4-T)}\right)$
$= 0.153\left(e^{0.141T-1.3395} - e^{0.153T-1.8123}\right)$

b)

$$\begin{aligned} r'(t) &= 0.153 \cdot \left(e^{0.141T-1.3395}\right)(0.141) - \\ &\quad 0.153 \cdot \left(e^{0.153T-1.8123}\right)(0.153) \\ &= 0.021573\left(e^{0.141T-1.3395}\right) - \\ &\quad 0.023409\left(e^{0.153T-1.8123}\right) \end{aligned}$$

81. $y = x^{3/2}\, e^{3x^3+2x-1}$

$$\begin{aligned} \frac{dy}{dx} &= x^{3/2}[e^{3x^3+2x-1}(9x^2+2)] + e^{3x^3+2x-1}(\frac{3}{2}x^{1/2}) \\ &= \frac{1}{2}\sqrt{x}\, e^{3x^3+2x-1}(2x(9x^2+2)+3) \\ &= \frac{1}{2}\sqrt{x}\, e^{3x^3+2x-1}(18x^3+4x^2+3) \end{aligned}$$

83. $y = x^{1/2} + (e^x)^{1/2} + (xe^x)^{1/2}$

$$\begin{aligned} \frac{dy}{dx} &= \frac{1}{2\sqrt{x}} + \frac{1}{2\sqrt{e^x}} \cdot e^x + \\ &\quad \frac{1}{2\sqrt{xe^x}} \cdot (xe^x + e^x) \\ &= \frac{1}{2}\left(\frac{xe^x + e^x}{\sqrt{xe^x}} + \sqrt{e^x} + \frac{1}{\sqrt{x}}\right) \end{aligned}$$

85. $y = sin(cos\ e^x)$

$$\begin{aligned} \frac{dy}{dx} &= cos(cos\ e^x) \cdot -sin\ e^x \cdot e^x \\ &= -e^x\ sin(e^x)\ cos(cos\ e^x) \end{aligned}$$

87. $y = 1 + e^{1+e^{1+e^x}}$

$$\begin{aligned} \frac{dy}{dx} &= e^{1+e^{1+e^x}} \cdot e^{1+e^x} \cdot e^x \\ &= e^{x+1+e^x+1+e^{1+e^x}} \\ &= e^{x+2+e^x+e^{1+e^x}} \end{aligned}$$

89. $f(t) = (1+t)^{1/t}$

$$\begin{aligned} f(1) &= 2^1 = 2 \\ f(0.5) &= 1.5^2 = 2.25 \\ f(0.2) &= 1.2^5 = 2.48832 \\ f(0.1) &= 1.1^{10} = 2.59374 \\ f(0.001) &= 1.001^{1000} = 2.71692 \end{aligned}$$

91. $f(x) = x^2e^{-x}$

$$\begin{aligned} f'(x) &= x^2(-e^{-x}) + 2x(e^{-x}) \\ &= e^{-x}(2x - x^2) \end{aligned}$$

Solve $f'(x) = 0$

$$\begin{aligned} e^{-x}(2x - x^2) &= 0 \\ 2x - x^2 &= 0 \\ x(2-x) &= 0 \\ x &= 0 \\ \text{and} & \\ x &= 2 \end{aligned}$$

We use test values to determine the nature of the critical values we found.
$f'(1) = e^{-1}(2-1) = e^{-1} > 0$
$f'(3) = e^{-3}(6-9) = -3e^{-3} < 0$
Therefore, the function has a maximum at $x = 2$. We find $f(2)$

$$\begin{aligned} f(2) &= 2^2 \cdot e^{-2} \\ &= \frac{4}{e^2} \end{aligned}$$

The maximum occurs at $(2, \frac{4}{e^2})$

93. $F(t) = e^{-(9/(t-15)+0.56/(35-t))}$

a) $\lim_{t\to 15^+} F(t) = 0$

b) $\lim_{t\to 35^-} F(t) = 0$

c)

$$\begin{aligned} F'(t) &= e^{-(9/(t-15)+0.56/(35-t))}\cdot \\ &\quad -(-9/(t-15)^2 + 0.56/(35-t)^2 \end{aligned}$$

Solve for $F'(t) = 0$

$$\begin{aligned} \left(\frac{9}{(t-15)^2} - \frac{0.56}{(35-t)^2}\right) &= 0 \\ \frac{9}{(t-15)^2} &= \frac{0.56}{(35-t)^2} \\ 9(35-t)^2 - 0.56(t-15)^2 &= 0 \\ 11025 - 630t + 9t^2 - 0.56t^2 + 16.8t - 126 &= 0 \\ 8.44t^2 - 613.2t + 10899 &= 0 \end{aligned}$$

Using the quadratic formula we get: $x = 31.0071$. The other zero of the function falls outside the interval $15 < t < 35$. Using test values to determine the sign of the first derivative on either side of $x = 31.0071$ gives $F'(20) > 0$ and $F'(33) < 0$. Therefore $F(t)$ has a maximum at $x = 31.0071$. Find $F(31.0071)$

$$\begin{aligned} F(31.0071) &= e^{-(9/16.0071+0.56/3.9983)} \\ &= e^{-0.702499766} \\ &= 0.49535 \end{aligned}$$

95. $D(t) = 34.4 - \frac{30.48}{1+29.44e^{-0.072t}}$

a) Find $D(110)$ and $D(150)$

$$\begin{aligned} D(110) &= 34.4 - \frac{30.48}{1+29.44e^{-0.072(110)}} \\ &= 34.4 - 30.157 \\ &= 4.253 \\ D(150) &= 34.4 - \frac{30.48}{1+29.44e^{-0.072(150)}} \\ &= 34.4 - 30.462 \\ &= 3.938 \end{aligned}$$

b) $\lim t \to \infty D(t) = 3.92$ Which is to say that the annual death rate in Mexico will not go below 3.92 per 1000 citizens.

97. Rewrite as follows $f(x) = 3.2e^{1.07x} - 5$, which gives $f'(x) = 3.424e^{1.07x}$, and now try to find the zero of $f(x)$ using Newton's method starting with a guess of $x_n = 1.5$

$$\begin{aligned} x_{n+1} &= x_n - f(x_n)/f'(x_n) \\ &= 1.5 - \frac{10.929}{17.044} = 0.8588 \\ &= 0.8588 - \frac{3.021}{8.5825} = 0.5068 \\ &= 0.5068 - \frac{0.50373}{5.889} = 0.4213 \\ &= 0.4213 - \frac{0.02257}{5.3742} = 0.4171 \\ &= 0.4171 - \frac{0.000049}{5.3501} = 0.417091 \\ &= 0.417091 - \frac{0.0000013}{5.35} = 0.417091 \end{aligned}$$

99. $P = 1 - \frac{1}{1+e^{-0.055-0.083T}}$

$$\begin{aligned} P(20) &= 1 - \frac{1}{1+e^{-0.055-0.083(20)}} \\ &= 1 - 0.84748 \\ &= 0.15252 \\ P(22) &= 1 - \frac{1}{1+e^{-0.055-0.083(22)}} \\ &= 1 - 0.86773 \\ &= 0.13227 \end{aligned}$$

The probability is declining by $0.15252 - 0.13227 = 0.02025$ per hour

101. The student use the power rule on an exponent that is not a constant, there is the error in the students work. The correct answer is $\frac{d}{dx}e^x = e^x$

103. Left to the student

105. Relative minimum at $(0, 0)$ and a relative maximum at $(2, 0.5413)$

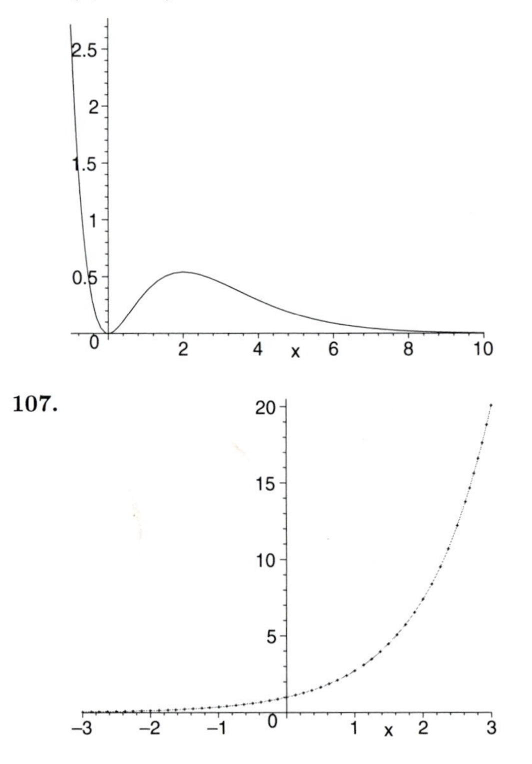

109.

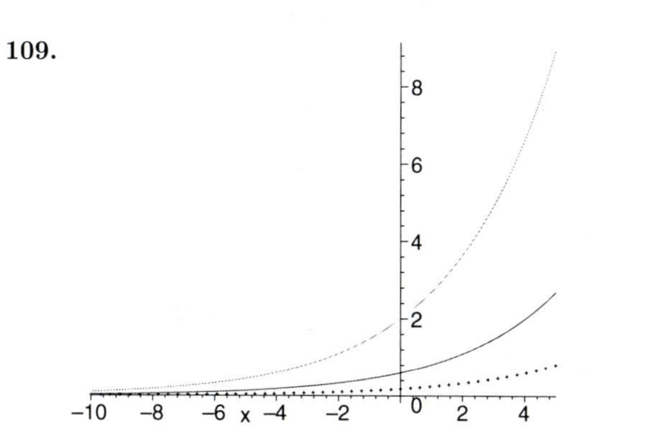

Exercise Set 4.2

1. $\log_2 8 = 3$ Logarithmic equation

$2^3 = 8$ Exponential equation; 2 is the base, 3 is the exponent

3. $\log_8 2 = \frac{1}{3}$ Logarithmic equation

$8^{1/3} = 2$ Exponential equation; 8 is the base, 1/3 is the exponent

5. $\log_a K = J$ Logarithmic equation

$a^J = K$ Exponential equation; a is the base, J is the exponent

7. $-\log_{10} h = p$ Logarithmic equation

$\log_{10} h = -p$ Multiplying by -1

$10^{-p} = h$ Exponential equation; 10 is the base, $-p$ is the exponent.

9. $e^M = b$ Exponential equation; e is the base, M is the exponent

$\log_e b = M$ Logarithmic equation

or $\ln\, b = M$ $\ln\, b$ is the abbreviation for $\log_e b$

11. $10^2 = 100$ Exponential equation; 10 is the base, 2 is the exponent

$\log_{10} 100 = 2$ Logarithmic equation

13. $10^{-1} = 0.1$ Exponential equation; 10 is the base, -1 is the exponent

$\log_{10} 0.1 = -1$ Logarithmic equation

15. $M^p = V$ Exponential equation; M is the base, p is the exponent

$\log_M V = p$ Logarithmic equation

17. $\log_b 15 = \log_b 3 \cdot 5$

$= \log_b 3 + \log_b 5$ (P1)

$= 1.099 + 1.609$

$= 2.708$

19. $\log_b \frac{1}{5} = \log_b 1 - \log_b 5$ (P2)

$= 0 - \log_b 5$ (P6)

$= -\log_b 5$

$= -1.609$

21. $\log_b 5b = \log_b 5 + \log_b b$ (P1)

$= 1.609 + 1$ (P4)

$= 2.609$

23. $\ln 20 = \ln 4 \cdot 5$

$= \ln 4 + \ln 5$ (P1)

$= 1.3863 + 1.6094$

$= 2.9957$

25. $\ln \frac{1}{4} = \ln 1 - \ln 4$ (P2)

$= 0 - 1.3863$ (P6)

$= -1.3863$

27. $\ln \sqrt{e^8} = \ln e^{8/2}$

$= \ln e^4$

$= 4$ (P5)

29. $\ln 3927 = 8.275631$ Using a calculator and rounding to six decimal places

31. $\ln 0.0182 = -4.006334$

33. $\ln 8100 = 8.999619$

35. $e^t = 100$

$\ln e^t = \ln 100$ Taking the natural logarithm on both sides

$t = \ln 100$ (P5)

$t = 4.605170$ Using a calculator

$t \approx 4.6$

37. $e^t = 60$

$\ln e^t = \ln 60$ Taking the natural logarithm on both sides

$t = \ln 60$ (P5)

$t = 4.094345$ Using a calculator

$t \approx 4.1$

39. $e^{-t} = 0.1$

$\ln e^{-t} = \ln 0.1$ Taking the natural logarithm on both sides

$-t = \ln 0.1$ (P5)

$t = -\ln 0.1$

$t = -(-2.302585)$ Using a calculator

$t = 2.302585$

$t \approx 2.3$

41. $e^{-0.02t} = 0.06$

$\ln e^{-0.02t} = \ln 0.06$ Taking the natural logarithm on both sides

$-0.02t = \ln 0.06$ (P5)

$t = \frac{\ln 0.06}{-0.02}$

$t = \frac{-2.813411}{-0.02}$ Using a calculator

$t \approx 141$

43. $y = -6\ln x$

$\frac{dy}{dx} = -6 \cdot \frac{1}{x}$ $\left[\frac{d}{dx}[c \cdot f(x)] = c \cdot f'(x)\right]$

$= -\frac{6}{x}$

45. $y = x^4 \ln x - \frac{1}{2}x^2$

$\frac{dy}{dx} = \underbrace{x^4 \cdot \frac{1}{x} + 4x^3 \cdot \ln x} - \frac{1}{2} \cdot 2x$ ← Product Rule on $x^4 \ln x$

$= x^3 + 4x^3 \ln x - x$

$= x^3(1 + 4\ln x) - x$

47. $y = \frac{\ln x}{x^4}$

$\frac{dy}{dx} = \frac{x^4 \cdot \frac{1}{x} - 4x^3 \cdot \ln x}{x^8}$ Using the Quotient Rule

$= \frac{x^3 - 4x^3 \ln x}{x^8}$

$= \frac{x^3(1 - 4\ln x)}{x^3 \cdot x^5}$ Factoring both numerator and denominator

$= \frac{x^3}{x^3} \cdot \frac{1 - 4\ln x}{x^5}$

$= \frac{1 - 4\ln x}{x^5}$ Simplifying

49. $y = \ln \frac{x}{4}$

$y = \ln x - \ln 4$ (P2)

$\frac{dy}{dx} = \frac{1}{x} - 0$

$= \frac{1}{x}$

51. $y = \ln \cos x$

$\frac{dy}{dx} = \frac{-\sin x}{\cos x}$

$= -\tan x$

53. $f(x) = \ln(\ln 4x)$

$f'(x) = 4 \cdot \frac{1}{4x} \cdot \frac{1}{\ln 4x}$ $\left[\frac{d}{dx} \ln g(x) = g'(x) \cdot \frac{1}{g(x)}\right]$

$\left[\frac{d}{dx} \ln 4x = 4 \cdot \frac{1}{4x}\right]$

$= \frac{1}{x} \cdot \frac{1}{\ln 4x}$

$= \frac{1}{x \ln 4x}$

55. $f(x) = \ln\left(\frac{x^2 - 7}{x}\right)$

$f'(x) = \frac{x \cdot 2x - 1 \cdot (x^2 - 7)}{x^2} \cdot \frac{1}{\frac{x^2 - 7}{x}}$

$\left[\frac{d}{dx} \ln g(x) = g'(x) \cdot \frac{1}{g(x)}\right]$

(Using Quotient Rule to find $\frac{d}{dx}\frac{x^2 - 7}{x}$)

$= \frac{2x^2 - x^2 + 7}{x^2} \cdot \frac{x}{x^2 - 7}$ $\left[\frac{1}{\frac{x^2 - 7}{x}} = \frac{x}{x^2 - 7}\right]$

$= \frac{x^2 + 7}{x^2} \cdot \frac{x}{x^2 - 7}$

$= \frac{x}{x} \cdot \frac{x^2 + 7}{x(x^2 - 7)}$

$= \frac{x^2 + 7}{x(x^2 - 7)}$

We could also have done this another way using P2:

$f(x) = \ln\left(\frac{x^2 - 7}{x}\right)$

$= \ln(x^2 - 7) - \ln x$

$f'(x) = 2x \cdot \frac{1}{x^2 - 7} - \frac{1}{x}$

$= \frac{2x}{x^2 - 7} \cdot \frac{x}{x} - \frac{1}{x} \cdot \frac{x^2 - 7}{x^2 - 7}$ Multiplying by forms of 1

$= \frac{2x^2}{x(x^2 - 7)} - \frac{x^2 - 7}{x(x^2 - 7)}$

$= \frac{2x^2 - x^2 + 7}{x(x^2 - 7)}$

$= \frac{x^2 + 7}{x(x^2 - 7)}$

57. $f(x) = e^x \ln x$

$f'(x) = e^x \cdot \frac{1}{x} + e^x \cdot \ln x$ Using the Product Rule

$= e^x\left(\frac{1}{x} + \ln x\right)$

59. $f(x) = e^x \sec x$

$f'(x) = e^x \cdot \sec x \tan x + e^x \sec x$

$= e^x \sec x(\tan x + 1)$

61. $f(x) = \ln(e^x + 1)$

$f'(x) = e^x \cdot \frac{1}{e^x + 1}$ $\left[\frac{d}{dx}(e^x + 1) = e^x\right]$

$= \frac{e^x}{e^x + 1}$

63. $f(x) = (\ln x)^2$

$f'(x) = 2(\ln x)^1 \cdot \frac{1}{x}$ Extended Power Rule

$= \frac{2\ln x}{x}$

65. $y = (\ln x)^{-4}$

$$\frac{dy}{dx} = -4(\ln x)^{-5} \cdot \frac{1}{x} = \frac{-4(\ln x)^{-5}}{x}$$

67. $f(t) = \ln(t^3+1)^5$

$$f'(t) = 5(t^3+1)^4 \cdot 3t^2 \cdot \frac{1}{(t^3+1)^5} = \frac{15t^2}{t^3+1} \quad \text{Simplifying}$$

69. $f(x) = [\ln(x+5)]^4$

$$f'(x) = 4[\ln(x+5)]^3 \cdot 1 \cdot \frac{1}{x+5} = \frac{4[\ln(x+5)]^3}{x+5}$$

71. $f(t) = \ln[(t^3+3)(t^2-1)]$

$$f'(t) = [(t^3+3)(2t)+(3t^2)(t^2-1)] \cdot \frac{1}{(t^3+3)(t^2-1)} = \frac{2t^4+6t+3t^4-3t^2}{(t^3+3)(t^2-1)} = \frac{5t^4-3t^2+6t}{(t^3+3)(t^2-1)}$$

73. $y = \ln \frac{x^5}{(8x+5)^2}$

$y = \ln x^5 - \ln(8x+5)^2$ Property 2

$$\frac{dy}{dx} = 5x^4 \cdot \frac{1}{x^5} - 2(8x+5) \cdot 8 \cdot \frac{1}{(8x+5)^2} = \frac{5}{x} - \frac{16}{8x+5}$$

$$= \frac{5}{x} \cdot \frac{8x+5}{8x+5} - \frac{16}{8x+5} \cdot \frac{x}{x} \quad \text{Multiplying by forms of 1}$$

$$= \frac{40x+25-16x}{x(8x+5)} = \frac{24x+25}{8x^2+5x}$$

75. $y = \frac{\ln\, sin\, x}{sin\, x}$

$$\frac{dy}{dx} = \frac{sin\, x\left(\frac{1}{sin\, x} \cdot cos\, x\right) - \ln\, sin\, x \cdot cos\, x}{sin^2\, x} = \frac{cos\, x(1-\ln\, sin\, x)}{sin^2\, x}$$

77. $y = \frac{x^{n+1}}{n+1}\left(\ln x - \frac{1}{n+1}\right)$

$$\frac{dy}{dx} = \frac{x^{n+1}}{n+1}\left(\frac{1}{x} - 0\right) + x^n\left(\ln x - \frac{1}{n+1}\right)$$

$$\left[\frac{d}{dx}\frac{x^{n+1}}{n+1} = \frac{1}{n+1} \cdot (n+1)x^n = x^n\right]$$

$$= \frac{x^n}{n+1} + x^n \ln x - \frac{x^n}{n+1} = x^n \ln x$$

79. $y = \ln\left(t+\sqrt{1+t^2}\right) = \ln[t+(1+t^2)^{1/2}]$

$$\frac{dy}{dx} = \left[1+\frac{1}{2}(1+t^2)^{-1/2} \cdot 2t\right] \cdot \frac{1}{t+(1+t^2)^{1/2}}$$

$$= \left(1+\frac{t}{(1+t^2)^{1/2}}\right) \cdot \frac{1}{t+(1+t^2)^{1/2}}$$

$$= \left(\frac{(1+t^2)^{1/2}}{(1+t^2)^{1/2}} + \frac{t}{(1+t^2)^{1/2}}\right) \cdot \frac{1}{t+(1+t^2)^{1/2}}$$

$$= \frac{t+(1+t^2)^{1/2}}{(1+t^2)^{1/2}} \cdot \frac{1}{t+(1+t^2)^{1/2}} = \frac{1}{(1+t^2)^{1/2}} = \frac{1}{\sqrt{1+t^2}}$$

81. $y = sin(\ln\, x)$

$$\frac{dy}{dx} = cos(\ln\, x) \cdot \frac{1}{x} = \frac{cos(\ln\, x)}{x}$$

83. $y = (sin\, x)\ln\,(tan\, x)$

$$\frac{dy}{dx} = (sin\, x)\left(\frac{sec^2\, x}{tan\, x}\right) + (cos\, x)\ \ln\,(tan\, x) = cos\, x\ \ln\,(tan\, x) + sec\, x$$

85. $y = \ln\,(sec\, 2x + tan\, 2x)$

$$\frac{dy}{dx} = \frac{1}{sec\, 2x + tan\, 2x} \cdot (2sec\, 2x\ tan\, 2x + 2\ sec^2\, 2x) = \frac{2\ sec\, 2x(sec\, 2x + tan\, 2x)}{sec\, 2x + tan\, 2x} = 2\ sec\, 2x$$

87. $y = \ln\left(\frac{e^x - e^{-x}}{e^x + e^{-x}}\right)$

$$\frac{dy}{dx} = \frac{e^x+e^{-x}}{e^x-e^{-x}} \cdot \frac{(e^x+e^{-x})(e^x+e^{-x})}{(e^x+e^{-x})^2} - \frac{e^x+e^{-x}}{e^x-e^{-x}} \cdot \frac{(e^x-e^{-x})(e^x-e^{-x})}{(e^x+e^{-x})^2}$$

$$= \frac{e^{2x}+1+1+e^{-2x}-e^{2x}+1+1-e^{-2x}}{(e^x-e^{-x})(e^x+e^{-x})} = \frac{4}{e^{2x}-e^{-2x}} = \frac{4e^{2x}}{e^{4x}-1}$$

89. Using the point $(10, 164)$ and $(70, 3045)$ we get

$$164 = A(10)^c$$

$$\rightarrow$$

$$A = \frac{164}{10^c}$$

$$3045 = A(70)^c$$

$$\rightarrow$$

$$\begin{aligned}
3045 &= \frac{164}{10^c} \cdot 70^c \\
\frac{3045}{164} &= (7)^c \\
c &= \log_7\left(\frac{3045}{164}\right) \\
&= 1.501297 \\
\rightarrow & \\
A &= \frac{164}{10^{1.501297}} \\
&= 5.170666
\end{aligned}$$

Therefore, $y = 5.17x^{1.50}$

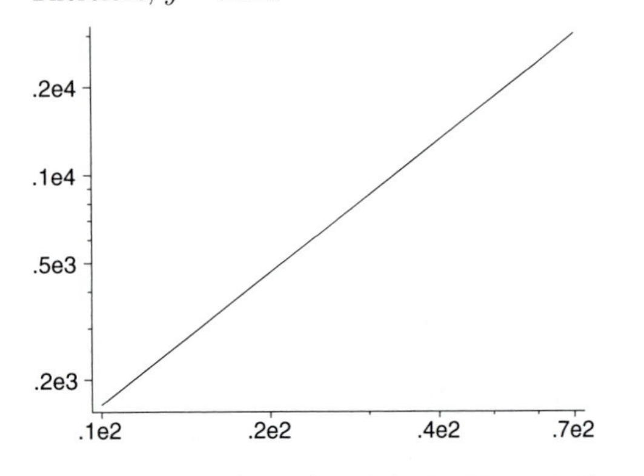

91. Using the point $(30, 25)$ and $(1000000000, 250)$ we get

$$\begin{aligned}
250 &= A1000000000^c \\
\rightarrow & \\
A &= \frac{250}{1000000000^c} \\
25 &= A30^c \\
\rightarrow & \\
25 &= \frac{250}{1000000000^c} \cdot 30^c \\
\frac{25}{250} &= \left(\frac{3}{100000000}\right)^c \\
c &= \frac{\log(\frac{1}{10})}{\log(\frac{3}{100000000})} \\
&= 0.133 \\
\rightarrow & \\
A &= \frac{250}{1000000000^{0.133}} \\
&= 15.907
\end{aligned}$$

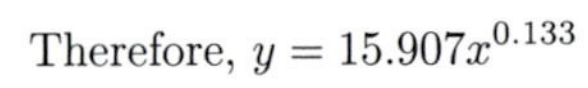
Therefore, $y = 15.907x^{0.133}$

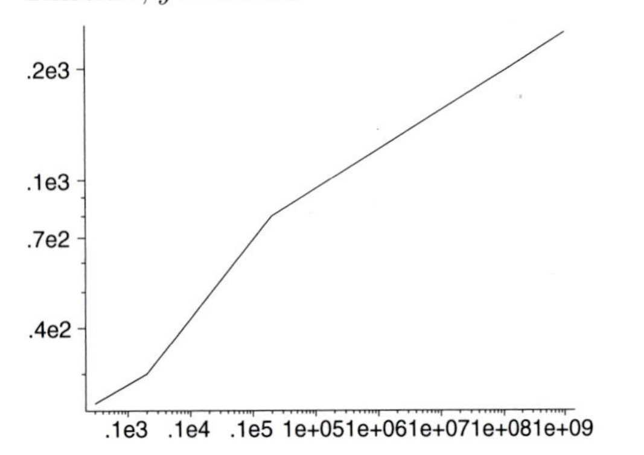

93. Using the point $(25000, 1079)$ and $(500000, 29)$ we get

$$\begin{aligned}
1079 &= A25000^c \\
\rightarrow & \\
A &= \frac{1079}{25000^c} \\
29 &= A500000^c \\
\rightarrow & \\
29 &= \frac{1079}{25000^c} \cdot 500000^c \\
\frac{29}{1079} &= (20)^c \\
c &= \frac{\log(\frac{29}{1079})}{\log(20)} \\
&= -1.207 \\
\rightarrow & \\
A &= \frac{1079}{25000^{-1.207}} \\
&= 219449722.1
\end{aligned}$$

Therefore, $y = 219449722.1x^{-1.207}$

.1e4
.5e3
.1e3
.5e2
.5e5 1e+05 2e+05 5e+05

95. a) $P(t) = 100(1 - e^{-0.2t})$

$P(1) = 100(1 - e^{-0.2\cdot 1})$ Substituting 1 for t

$= 100(1 - e^{-0.2})$

$= 100(0.181269)$ Using a calculator

$\approx 18.1\%$

$P(6) = 100(1 - e^{-0.2\cdot 6})$

$= 100(1 - e^{-1.2})$

$= 100(0.698806)$

$\approx 69.9\%$

b) $P(t) = 100(1 - e^{-0.2t})$

$P'(t) = 100[-(-0.2)e^{-0.2t}]$

$= 20e^{-0.2t}$

c)
$$P(t) = 100(1 - e^{-0.2t})$$
$$90 = 100(1 - e^{-0.2t}) \quad \text{Replacing } P(t) \text{ by } 90$$
$$0.9 = 1 - e^{-0.2t}$$
$$-0.1 = -e^{-0.2t} \quad \text{Adding } -1$$
$$0.1 = e^{-0.2t} \quad \text{Multiplying by } -1$$
$$\ln 0.1 = \ln e^{-0.2t} \quad \text{Taking the natural logarithm on both sides}$$
$$\ln 0.1 = -0.2t$$
$$\frac{\ln 0.1}{-0.2} = t$$
$$\frac{-2.302585}{-0.2} = t \quad \text{Using a calculator}$$
$$11.5 \approx t$$

Thus it will take approximately 11.5 months for 90% of the doctors to become aware of the new medicine.

d) $\lim\limits_{t\to\infty} P(t) = \lim\limits_{t\to\infty} 100(1 - e^{-0.2t}) = 100.$

This indicates that 100% of doctors will eventually accept the new medicine.

97. a) $S(t) = 68 - 20\ln(t+1),\ t \geq 0$
$$S(0) = 68 - 20\ln(0+1) \quad \text{Substituting 0 for } t$$
$$= 68 - 20\ln 1$$
$$= 68 - 20 \cdot 0$$
$$= 68 - 0$$
$$= 68$$

Thus the average score when they initially took the test was 68%.

b)
$$S(4) = 68 - 20\ln(4+1) \quad \text{Substituting 4 for } t$$
$$= 68 - 20\ln 5$$
$$= 68 - 20(1.609438) \quad \text{Using a calculator}$$
$$= 68 - 32.18876$$
$$\approx 36\%$$

c)
$$S(24) = 68 - 20\ln(24+1) \quad \text{Substituting 24 for } t$$
$$= 68 - 20\ln 25$$
$$= 68 - 20(3.218876) \quad \text{Using a calculator}$$
$$= 68 - 64.37752$$
$$\approx 3.6\%$$

d) First we reword the question:

3.6 (the average score after 24 months) is what percent of 68 (the average score when $t = 0$).

Then we translate and solve:
$$3.6 = x \cdot 68$$
$$\frac{3.6}{68} = x$$
$$0.052941 = x$$
$$5\% \approx x$$

e) $S(t) = 68 - 20\ln(t+1),\ t \geq 0$
$$S'(t) = 0 - 20 \cdot 1 \cdot \frac{1}{t+1}$$
$$= -\frac{20}{t+1}$$

f) $S'(t) < 0$ for all $t \geq 0$. Thus $S(t)$ is a decreasing function and has a maximum value of 68% when $t = 0$.

g) $\lim\limits_{t\to\infty} S(t) = \lim\limits_{t\to\infty} 68 - 20\ \ln\ (t+1) = -\infty.$

Clearly, the score cannot be less than 0, but this limit indicates that eventually everything will be forgotten.

99. $v(p) = 0.37\ln p + 0.05$ p in thousands, v in ft per sec

a)
$$v(531) = 0.37\ln 531 + 0.05 \quad \text{Substituting 531 for } p$$
$$= 0.37(6.274762) + 0.05$$
$$= 2.321662 + 0.05$$
$$= 2.371662$$
$$\approx 2.37 \text{ ft/sec}$$

b)
$$v(7900) = 0.37\ln 7900 + 0.05$$
$$= 0.37(8.974618) + 0.05$$
$$= 3.320609 + 0.05$$
$$= 3.370609$$
$$\approx 3.37 \text{ ft/sec}$$

c)
$$v'(p) = 0.37 \cdot \frac{1}{p} + 0$$
$$= \frac{0.37}{p}$$

d) $v'(p)$ is the acceleration of the walker.

101. $f(x) = \ln[\ln\ x]^3$
$$f'(x) = \frac{1}{[\ln\ x]^3} \cdot 3\ [\ln\ x]^2 \cdot \frac{1}{x}$$
$$= \frac{3}{x\ \ln\ x}$$

103. Using L'Hospital's Rule (taking the derivative of the numerator and denominator then applying the limit)
$$\lim_{h\to 0} \frac{\ln(1+h)}{h} = \lim_{h\to 0} \frac{\frac{1}{(1+h)}}{1}$$
$$= \lim_{h\to 0} \frac{1}{(1+h)}$$
$$= \frac{1}{1+0}$$
$$= 1$$

105.
$$P = P_0 e^{-kt}$$
$$\frac{P}{P_0} = e^{kt}$$
$$\ln\left(\frac{P}{P_0}\right) = kt\ \ln(e)$$

$$\begin{aligned}\ln\left(\frac{P}{P_0}\right) &= kt\\ \frac{\ln\left(\frac{P}{P_0}\right)}{k} &= t\end{aligned}$$

107. Left to the student

109. **a)** $r(T) = 0.124\left(e^{0.129(T-9.5)} - e^{4.128-0.144(41.5-T)}\right)$

$r(T) = 0.124\left(e^{0.129T-1.2255} - e^{0.144T-1.848}\right)$

b) $\frac{dr}{dT} = 0.124(0.129e^{0.129T-1.2255} - 0.144e^{0.144T-1.848})$

$\frac{dr}{dT} = 0.015996e^{0.129T-1.2255} - 0.017856e^{0.144T-1.848})$

c)

$$\begin{aligned}\frac{dr}{dT} &= 0\\ 0.015996e^{0.129T-1.2255} &= 0.017856e^{0.144T-1.848}\\ \left(\frac{0.015996}{0.017856}\right) &= e^{0.144-1.848-(0.129T-1.2255)}\\ 0.890844286 &= e^{0.015T-0.6225}\\ \ln(0.890844286) &= 0.015T-0.6225\\ T &= \frac{\ln(0.890844286)+0.6225}{0.015}\\ &= 34.17\end{aligned}$$

111.

$$\begin{aligned}A\left(be^{b(T-T_L)} - ce^{b(T_U-T_L)-c(T_U-T)}\right) &= 0\\ ce^{b(T_U-T_L)-c(T_U-T)} &= be^{b(T-T_L)}\\ e^{b(T_U-T_L)-c(T_U-T)-b(T-T_L)} &= \frac{b}{c}\\ e^{b(T_U)-c(T_U)-cT-bT} &= \frac{b}{c}\\ b(T_U)-c(T_U)-cT-bT &= \ln(\frac{b}{c})\\ T_U(b-c)-T(b-c) &= \ln(\frac{b}{c})\\ \ln(\frac{b}{c})-T_U(b-c) &= -T(b-c)\\ T_U-\frac{1}{b-c}\ln(\frac{b}{c}) &= T\end{aligned}$$

Plugging the values of the constants left to the student.

113.

a - b)

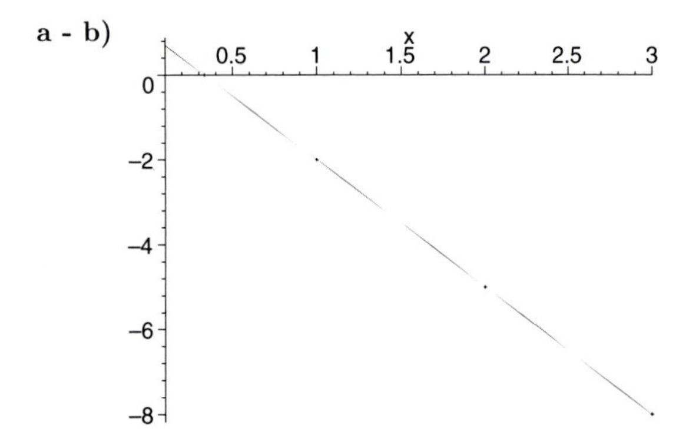

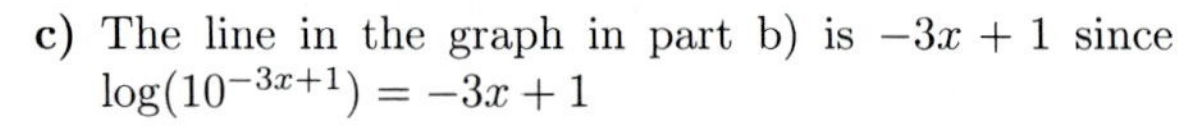
c) The line in the graph in part b) is $-3x+1$ since $\log(10^{-3x+1}) = -3x+1$

115. **a)**

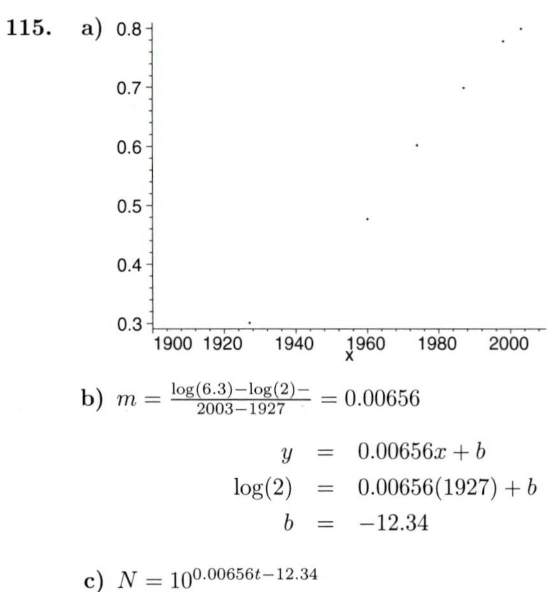

b) $m = \frac{\log(6.3)-\log(2)-}{2003-1927} = 0.00656$

$$\begin{aligned}y &= 0.00656x+b\\ \log(2) &= 0.00656(1927)+b\\ b &= -12.34\end{aligned}$$

c) $N = 10^{0.00656t-12.34}$

117. The statement means that the graph of the tangent line of the function $y = \ln(4x)$ is given by the graph of $y = \frac{1}{x}$

119. $\sqrt[e]{e} = e^{1/e} = 1.44467$ which is larger than $\sqrt[x]{x} = x^{1/x}$ for all $x > 0$

121. $\lim_{x\to\infty} \ln x = \infty$

123.

125.

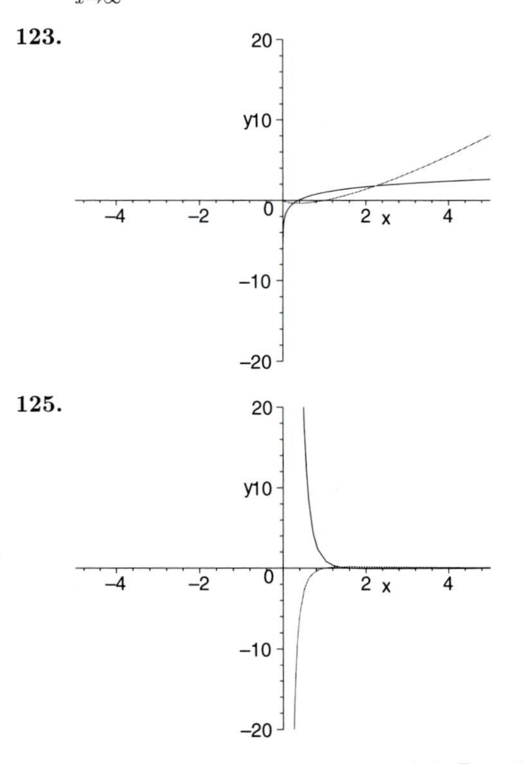

127. The function has a minimum at $(1/\sqrt{e}, -1/2e)$

129. $y = 21.5x^{0.364}$

131. $y = 1316788.481x^{-0.958}$

133. $x = 203.19$

135. $x = 204.41$

137. $x = 83.82$ and $x = 22444.11$

139. $x = 85.68$ and$x = 9503.27$

Exercise Set 4.3

1. The solution of $\frac{dQ}{dt} = kQ$ is $Q(t) = ce^{kt}$, where t is the time. At $t = 0$, we have some "initial" population $Q(0)$ that we will represent by Q_0. Thus $Q_0 = Q(0) = ce^{k \cdot 0} = ce^0 = c \cdot 1 = c$.

Thus, $Q_0 = c$, so we can express $Q(t)$ as $Q(t) = Q_0\, e^{kt}$.

3. The solution of $\frac{dy}{dt} = 2y$ is
$y = c\, e^{2t}$, $y = 5$
when $t = 0$ gives

$$\begin{aligned} 5 &= c\, e^{2(0)} \\ 5 &= c \cdot 1 \\ 5 &= c \end{aligned}$$

Therefore, $y = 5\, e^{2t}$

5. **a)** $P = 1000\, e^{0.033t}$

b) when $t = 30$

$$\begin{aligned} P &= 1000\, e^{0.033(30)} \\ &= 1000\, e^{0.99} \\ &= 2691 \end{aligned}$$

c) when $t = 60$

$$\begin{aligned} P &= 1000\, e^{0.033(60)} \\ &= 1000\, e^{1.98} \\ &= 7243 \end{aligned}$$

d) when $t = 1440$

$$\begin{aligned} P &= 1000\, e^{0.033(1440)} \\ &= 1000\, e^{47.52} \\ &= 4.34 \times 10^{23} \end{aligned}$$

e) The generation time is

$$\begin{aligned} T &= \frac{\ln(2)}{k} \\ &= \frac{\ln(2)}{0.033} \\ &= 21 \end{aligned}$$

7. **a)** $k = \frac{\ln(2)}{T}$

$$\begin{aligned} k &= \frac{\ln(2)}{0.47} \\ &= 1.4748 \end{aligned}$$

b) $P = 200\, e^{1.4748t}$ for $t = 3$

$$\begin{aligned} P &= 200\, e^{1.4748(3)} \\ &= 200\, e^{4.4244} \\ &= 16692 \end{aligned}$$

c) For $t = 24$

$$\begin{aligned} P &= 200\, e^{1.4748(24)} \\ &= 200\, e^{35.3952} \\ &= 4.7 \times 10^{17} \end{aligned}$$

d) Find t when $P = 3(200) = 600$

$$\begin{aligned} 600 &= 200\, e^{1.4748t} \\ 3 &= e^{1.4748t} \\ \ln(3) &= 1.4748t \\ \frac{\ln(3)}{1.4748} &= t \\ 0.7449 &= t \end{aligned}$$

9. Find P_o for $P = 20000$, $k = \frac{\ln(2)}{40}$, and $t = 120$

$$\begin{aligned} P &= P_o\, e^{kt} \\ 20000 &= P_o\, e^{\frac{\ln(2)}{40}(120)} \\ 20000 &= P_o e^{3\ln(2)} = 8P_o \\ \frac{20000}{8} &= P_o \\ 2500 &= P_o \end{aligned}$$

11. **a)** $P = c\, e^{0.009t}$, when $t = 0$, $P = 281$ gives

$$\begin{aligned} 281 &= c\, e^{0.009(0)} \\ 281 &= c \cdot 1 \\ c &= 281 \end{aligned}$$

Thus, $P = 281\, e^{0.009t}$

b) Find P when $t = 15$

$$\begin{aligned} P &= 281\, e^{0.009(15)} \\ &= 281\, e^{0.135} \\ &= 321.6 \text{ million} \end{aligned}$$

c) $T = \frac{\ln(2)}{0.009} = 77.02$ years

13. The balance grows at the rate given by
$$\frac{dP}{dt} = 0.065P.$$

a) $P(t) = P_0\, e^{0.065t}$

b) $P(1) = 1000e^{0.065 \cdot 1}$ Substituting 1000 for P_0 and 1 for t

$= 1000e^{0.065}$

$= 1000(1.067159)$

≈ 1067.16

The balance after 1 year is $1067.16.

$$\begin{aligned} P(2) &= 1000e^{0.065\cdot 2} && \text{Substituting 1000 for } P_0 \text{ and 2 for } t \\ &= 1000e^{0.13} \\ &= 1000(1.138828) \\ &\approx 1138.83 \end{aligned}$$

The balance after 2 years is \$1138.83.

c) $$\begin{aligned} T &= \frac{\ln 2}{k} \\ &= \frac{0.693147}{0.065} && \text{Substituting 0.693147 for } \ln 2 \text{ and 0.065 for } k \\ &\approx 10.7 \end{aligned}$$

An investment of \$1000 will double itself in 10.7 years.

15. $$\begin{aligned} k &= \frac{\ln 2}{T} \\ &= \frac{0.693147}{10} && \text{Substituting 0.693147 for } \ln 2 \text{ and 10 for } T \\ &= 0.0693147 \\ &\approx 6.9\% \end{aligned}$$

The annual interest rate is 6.9%.

17. $T = \dfrac{\ln 2}{k} = \dfrac{\ln 2}{0.035} \approx 19.8$ yr

19. $k = \dfrac{\ln 2}{T} = \dfrac{\ln 2}{6.931} \approx 0.10 \approx 10\%$

21. $T = \dfrac{\ln 2}{k} = \dfrac{\ln 2}{0.02794} \approx 24.8$ yr

23. $$\begin{aligned} R(b) &= e^{21.4b} && \text{See Example 7} \\ 80 &= e^{21.4b} && \text{Substituting 80 for } R(b) \\ \ln 80 &= \ln e^{21.4b} \\ \ln 80 &= 21.4b \\ \frac{\ln 80}{21.4} &= b \\ 0.20 &\approx b && \text{Rounding to the nearest hundredth} \end{aligned}$$

Thus when the blood alcohol level is 0.20%, the risk of having an accident is 80%.

25. a) $$\begin{aligned} P(t) &= P_0\, e^{kt} \\ 216,000,000 &= 2,508,000\, e^{k\cdot 200} \\ \frac{216,000,000}{2,508,000} &= e^{200k} \\ \ln \frac{216,000}{2508} &= \ln\, e^{200k} \\ \frac{\ln \frac{216,000}{2508}}{200} &= k \\ 0.022 &\approx k \end{aligned}$$

The growth rate was approximately 2.2%.

b) It is reasonable to assume that the population of the United States grew exponentially between 1776 and 1976 rather than linearly or in some other pattern.

27. $(0, 47432)$ and $(40, 432976)$

a) From the point $(0, 47432)$ we get

$$\begin{aligned} y &= a \cdot b^t \\ 47432 &= a \cdot b^0 \\ 47432 &= a \end{aligned}$$

From the point $(40, 432976)$ we get

$$\begin{aligned} 432976 &= 47432 \cdot b^{40} \\ \frac{432976}{47432} &= b^{40} \\ \ln\left(\frac{432976}{47432}\right) &= 40\ln(b) \\ \frac{\ln\left(\frac{432976}{47432}\right)}{40} &= \ln(b) \\ 0.0552846 &= \ln(b) \end{aligned}$$

So, $y = 47432\, e^{0.0552846t}$

b) When $t = 50$

$$\begin{aligned} y &= 47432\, e^{0.0552846(50)} \\ &= 752595 \end{aligned}$$

c) When $y = 1000000$

$$\begin{aligned} 1000000 &= 47432\, e^{0.0552846t} \\ \frac{1000000}{47432} &= e^{0.0552846t} \\ \ln\left(\frac{1000000}{47432}\right) &= 0.0552846t \\ \frac{\ln\left(\frac{1000000}{47432}\right)}{0.0552846} &= t \\ 55.14 &= t \end{aligned}$$

Which corresponds to the year 2015

d) $T = \dfrac{\ln(2)}{0.0552846} = 12.54$ years

e) As expected the answers in this exercise are close to those of Exercise 26.

29. $(0, 85)$ and $(4, 109)$

a) From the point $(0, 85)$ we get

$$\begin{aligned} y &= a \cdot b^t \\ 85 &= a \cdot b^0 \\ 85 &= a \end{aligned}$$

From the point $(4, 109)$ we get

$$\begin{aligned} 109 &= 85 \cdot b^4 \\ \frac{109}{85} &= b^4 \\ \ln\left(\frac{109}{85}\right) &= 4\ln(b) \\ \frac{\ln\left(\frac{109}{85}\right)}{4} &= \ln(b) \\ 0.062174 &= \ln(b) \end{aligned}$$

So, $y = 85\, e^{0.062174t}$

b) When $t = 5$

$$\begin{aligned} y &= 85\ e^{0.062174(5)} \\ &= 116 \end{aligned}$$

c) When $y = 1000000$

$$\begin{aligned} 1000 &= 85\ e^{0.062174t} \\ \frac{1000}{85} &= e^{0.062174t} \\ \ln\left(\frac{1000}{85}\right) &= 0.062174t \\ \frac{\ln\left(\frac{1000}{85}\right)}{0.062174} &= t \\ 39.65 &= t \end{aligned}$$

d) $T = \dfrac{\ln(2)}{0.062174} = 11.15$ years

e) As expected the answers in this exercise are close to those of Exercise 28.

31. a) Using a graphing calculator $P(x) = 14.88 \cdot 0.9996^t$

b) $\ln(0.9996) = -0.0004$ So, $P(x) = 14.88\ e^{-0.0004t}$

c)

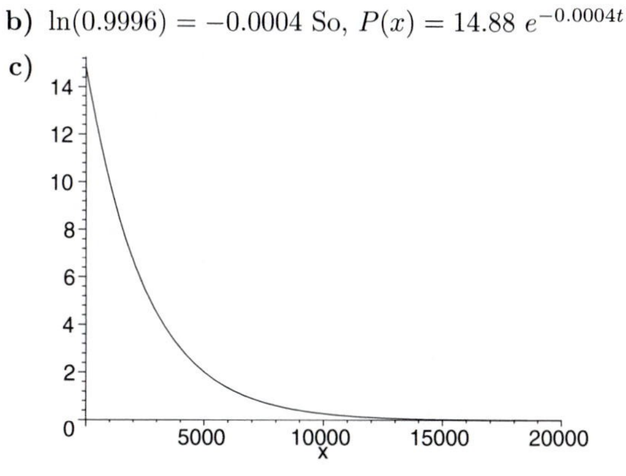

d) $P'(x) = 14.88 \cdot -0.0004\, e^{-0.0004t} = -0.00595\ e^{-0.0004t}$

e) When x gets closer to sea level the slope of the tangent line at $x = 0$ is about -0.005

f) $P(A(x)) = x = A(P(x))$ since the exponential function and the logrithmic functions are inverses of each other.

33. $G(x) = \dfrac{100}{1 + 43.3\ e^{-0.0425x}}$

a) When $x = 100$

$$\begin{aligned} G(100) &= \frac{100}{1 + 43.3\ e^{-0.0425(100)}} \\ &= \frac{100}{1 + 43.3\ e^{-4.25}} \\ &= 61.82\% \end{aligned}$$

When $x = 150$

$$\begin{aligned} G(150) &= \frac{100}{1 + 43.3\ e^{-0.0425(150)}} \\ &= \frac{100}{1 + 43.3\ e^{-6.375}} \\ &= 91.13\% \end{aligned}$$

When $x = 300$

$$\begin{aligned} G(300) &= \frac{100}{1 + 43.3\ e^{-0.0425(300)}} \\ &= \frac{100}{1 + 43.3\ e^{-12.75}} \\ &= 99.99\% \end{aligned}$$

b)

$$\begin{aligned} G'(x) &= 100 \cdot -1(1 + 43.3\ e^{-0.0425x})^{-2} \times \\ &\quad (43.3 \cdot -0.0425\ e^{-0.0425x}) \\ &= \frac{184.025\ e^{-0.0425x}}{(1 + 43.3\ e^{-0.0425x})^2} \end{aligned}$$

c)

$$G''(x) = \frac{\left(-184.025\ e^{-0.0425x}\right)\left[0.0425 - 1.84025\ e^{-0.0425x}\right]}{(1 + 43.3\ e^{-0.0425x})^3}$$

The possible inflection point occurs when $G''(x) = 0$ or undefined, which means the only possible inflection points occurs at

$$\begin{aligned} 0.0425 - 1.84025\ e^{-0.0425x} &= 0 \\ e^{-0.0425x} &= \frac{0.0425}{1.84025} \\ -0.0425x &= \ln\left(\frac{0.0425}{1.84025}\right) \\ x &= \frac{\ln\left(\frac{0.0425}{1.84025}\right)}{-0.0425} \\ &= 88.66 \end{aligned}$$

d)

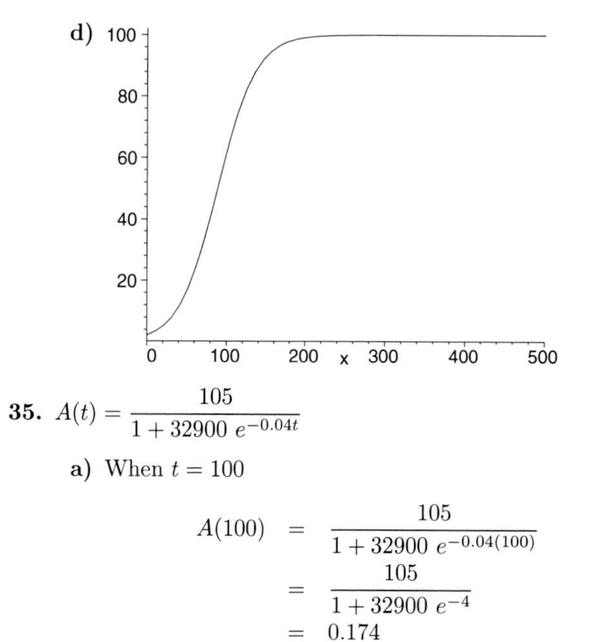

35. $A(t) = \dfrac{105}{1 + 32900\ e^{-0.04t}}$

a) When $t = 100$

$$\begin{aligned} A(100) &= \frac{105}{1 + 32900\ e^{-0.04(100)}} \\ &= \frac{105}{1 + 32900\ e^{-4}} \\ &= 0.174 \end{aligned}$$

When $t = 150$

$$\begin{aligned} A(150) &= \frac{105}{1 + 32900\ e^{-0.04(150)}} \\ &= \frac{105}{1 + 32900\ e^{-6}} \\ &= 1.272 \end{aligned}$$

When $t = 200$

$$\begin{aligned} A(200) &= \frac{105}{1 + 32900\, e^{-0.04(200)}} \\ &= \frac{105}{1 + 32900\, e^{-8}} \\ &= 8.723 \end{aligned}$$

b)

$$\begin{aligned} A'(x) &= 105 \cdot -1(1 + 32900\, e^{-0.04t})^{-2} \times \\ &\quad (32900 \cdot -0.04\, e^{-0.04t}) \\ &= \frac{138180\, e^{-0.04t}}{(1 + 32900\, e^{-0.04t})^2} \end{aligned}$$

c)

$$\begin{aligned} A''(t) &= \frac{\left(e^{-0.04t}\right)\left[1 + 32900\, e^{-0.04t}\right]}{(1 + 32900\, e^{-0.04t})^3} \times \\ &\quad \left[-5527.2\left(1 + 32900\, e^{-0.04t}\right) + 363689760\right] \end{aligned}$$

The possible inflection point occurs when $A''(t) = 0$ or undefined, which means the possible inflection points occurs at

$$\begin{aligned} 0 &= -5527.2\left(1 + 32900\, e^{-0.04t}\right) + \\ &\quad 363689760 \\ 1 + 32900\, e^{-0.04t} &= \frac{363689760}{5527.2} \\ \frac{\frac{363689760}{5527.2} - 1}{32900} &= e^{-0.04t} \\ \ln\left(\frac{\frac{363689760}{5527.2} - 1}{32900}\right) &= -0.04t \\ \frac{\ln\left(\frac{\frac{363689760}{5527.2} - 1}{32900}\right)}{-0.04} &= t \\ t &= -17.33 \end{aligned}$$

not an acceptable answer

or

$$\begin{aligned} 1 + 32900\, e^{-0.04t} &= 0 \\ e^{-0.04t} &= \frac{1}{32900} \\ -0.04t &= \ln\left(\frac{1}{32900}\right) \\ t &= \frac{\ln\left(\frac{1}{32900}\right)}{-0.04} \\ &= 260.031 \end{aligned}$$

We find $A(260.031)$

$$\begin{aligned} A(200) &= \frac{105}{1 + 32900\, e^{-0.04(260.031)}} \\ &= \frac{105}{1 + 32900\, e^{-10.40124}} \\ &= 52.5 \end{aligned}$$

d)

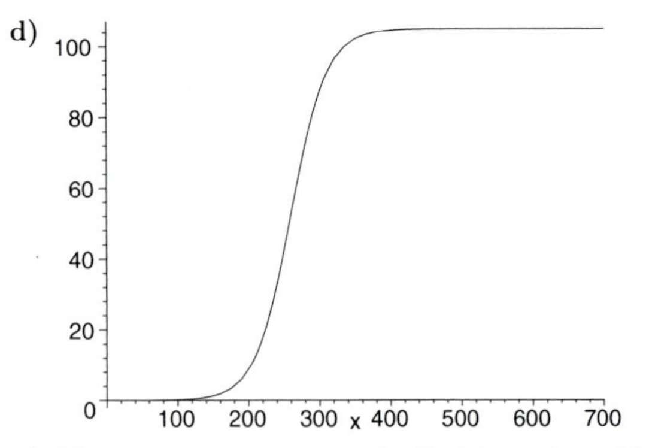

e) The maximum occurs at the limiting value which is 105

37. $P(t) = 100\%(1 - e^{-0.4t})$

a) $P(0) = 100\%(1 - e^{-0.4(0)}) = 100\%(1 - e^0) = 100\%(1 - 1) = 100\%(0) = 0\%$

$P(1) = 100\%(1 - e^{-0.4(1)}) = 100\%(1 - e^{-0.4}) \approx 33.0\%$

$P(2) = 100\%(1 - e^{-0.4(2)}) = 100\%(1 - e^{-0.8}) \approx 55.1\%$

$P(3) = 100\%(1 - e^{-0.4(3)}) = 100\%(1 - e^{-1.2}) \approx 69.9\%$

$P(5) = 100\%(1 - e^{-0.4(5)}) = 100\%(1 - e^{-2}) \approx 86.5\%$

$P(12) = 100\%(1 - e^{-0.4(12)}) = 100\%(1 - e^{-4.8}) \approx 99.2\%$

$P(16) = 100\%(1 - e^{-0.4(16)}) = 100\%(1 - e^{-6.4}) \approx 99.8\%$

b)
$$\begin{aligned} P'(t) &= 100\%[-(-0.4)\, e^{-0.4t}] \\ &= 100\%(0.4)\, e^{-0.4t} \\ &= 0.4\, e^{-0.4t} \qquad (100\% = 1) \end{aligned}$$

c) The derivative $P'(t)$ exists for all real numbers. The equation $P'(t) = 0$ has no solution. Thus, the function has no critical points and hence no relative extrema. $P'(t) > 0$ for all real numbers, so $P(t)$ is increasing on $[0, \infty)$. $P''(t) = -0.16\, e^{-0.4t}$, so $P''(t) < 0$ for all real numbers and hence is concave down on $[0, \infty)$.

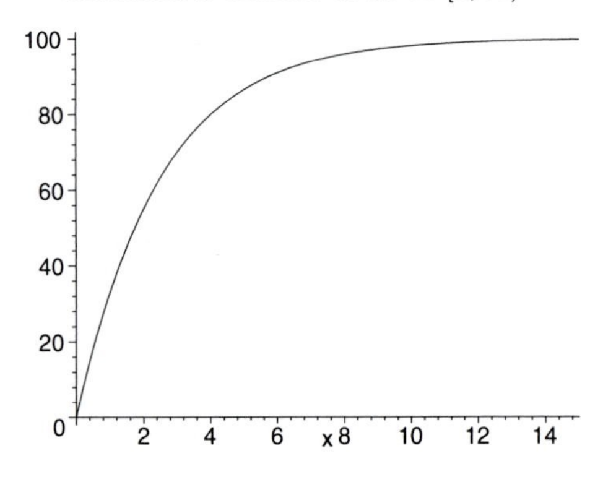

39. $i = e^{0.073} - 1 = 0.0757 = 7.57\%$

41. $k = \ln(0.0924 + 1) = 0.0884 = 8.84\%$

43.

$$\begin{aligned} 3P_0 &= P_0 e^{kt} \\ 3 &= e^{kt} \\ \frac{\ln(3)}{k} &= t \end{aligned}$$

45. Answers vary

47. $2 = e^{24k}$

$$k = \frac{ln(2)}{24} = 0.0289$$

49. $k = \dfrac{\ln(2)}{T}$ and $T = \dfrac{\ln(2)}{k}$

k	1%	2%	4.621%	6.931%	14%
T	69.315	34.657	15	10	4.951

The graph of $T = \ln\ 2/k$ is not linear since the independent variable appears in the denominator

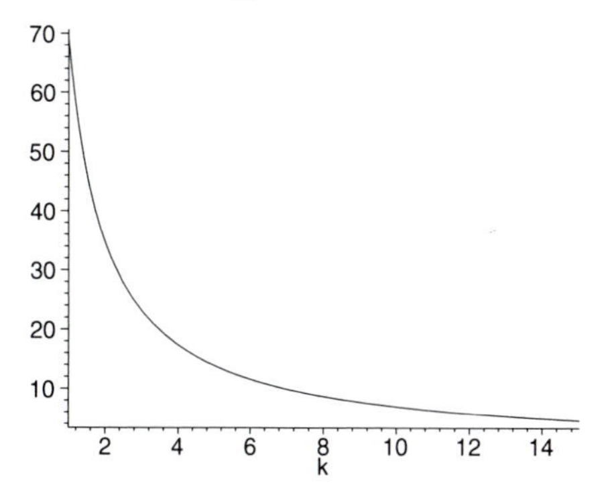

51. The rule of 69 is used to approximate the doubling time (generation time) of growing things. The name comes from the approximation to $\ln(2) = 0.6931$ which is used to find the doubling time

Exercise Set 4.4

1. (d)

3. (a)

5. (c)

7. $k = \dfrac{\ln 2}{T} = \dfrac{\ln 2}{22} \approx 0.032 \approx 3.2\%$ per yr

9. $k = \dfrac{\ln 2}{T} = \dfrac{\ln 2}{25} \approx 0.028 \approx 2.8\%$ per year

11. $k = \dfrac{\ln 2}{T} = \dfrac{\ln 2}{23,105} \approx 0.00003 \approx 0.003\%$ per yr

13. $P(t) = P_0\ e^{-kt}$

$P(20) = 1000\ e^{-0.231(20)}$ Substituting 1000 for P_0, 0.231 for k and 20 for t

$= 1000\ e^{-4.62}$

≈ 9.9

Thus 9.9 grams of polonium will remain after 20 minutes.

15. $N(t) = N_0\ e^{-0.0001205t}$ See Example 3(b).

If a piece of wood has lost 90% of its carbon-14 from an initial amount P_0, then 10% P_0 is the amount present. To find the age of the wood, we solve the following equation for t:

$10\%\ P_0 = P_0\ e^{-0.0001205t}$

Substituting 10% P_0 for $P(t)$

$0.1 = e^{-0.0001205t}$

$\ln\ 0.1 = \ln\ e^{-0.0001205t}$

$\ln\ 0.1 = -0.0001205t$

$-2.302585 = -0.0001205t$ Using a calculator

$\dfrac{-2.302585}{-0.0001205} = t$

$19,109 \approx t$

Thus, the piece of wood is about 19,109 years old.

17. If an artifact has lost 60% of its carbon-14 from an initial amount P_0, then 40% P_0 is the amount present. To find the age t we solve the following equation for t.

$40\% P_0 = P_0\ e^{-0.0001205t}$

Substituting 40% P_0 for $P(t)$ (See Example 3(b).)

$0.4 = e^{-0.0001205t}$

$\ln\ 0.4 = \ln\ e^{-0.0001205t}$

$\ln\ 0.4 = -0.0001205t$

$-0.916291 = -0.0001205t$ Using a calculator

$\dfrac{-0.916291}{-0.0001205} = t$

$7604 \approx t$

The artifact is about 7604 years old.

19. For carbon-14 the decay rate is 0.0001205

The amount of carbon-14 that remains is

$$\begin{aligned} N &= N_0\ e^{-0.0001205 \cdot 10000} \\ &= N_0\ e^{-1.205} \\ &= 0.2997\ N_0 \end{aligned}$$

Amount lost is $100 - 29.97 = 70.03\%$

21. $k = \dfrac{\ln(1/2)}{1.3 \times 10^9} = -5.332 \times 10^{-10}$

$$\begin{aligned} 0.0884N_0 &= N_0\ e^{-5.332\times 10^{-10}t} \\ 0.0884 &= e^{-5.332\times 10^{-10}t} \\ \ln(0.0884) &= -5.332 \times 10^{-10}t \\ t &= \frac{\ln(0.0884)}{-5.332 \times 10^{-10}} \\ &= 4.55 \times 10^9 \text{ years} \end{aligned}$$

23.

$$\begin{aligned} 0.653N_0 &= N_0\, e^{-5.332\times 10^{-10}t} \\ 0.653 &= e^{-5.332\times 10^{-10}t} \\ \ln(0.653) &= -5.332\times 10^{-10}t \\ t &= \frac{\ln(0.653)}{-5.332\times 10^{-10}} \\ &= 7.99\times 10^{8} \text{ years} \end{aligned}$$

25. a) When A decomposes at a rate proportional to the amount of A present, we know that

$\frac{dA}{dt} = -kA.$

The solution of this equation is $A = A_0\, e^{-kt}$.

b) We first find k. The half-life of A is 3 hr.

$k = \frac{\ln 2}{T}$

$k = \frac{0.693147}{3}$ Substituting 0.693147 for ln 2 and 3 for T

≈ 0.23, or 23%

We now substitute 8 for A_0, 1 for A, and 0.23 for k and solve for t.

$A = A_0\, e^{-kt}$

$1 = 8\, e^{-0.23t}$ Substituting

$\frac{1}{8} = e^{-0.23t}$

$0.125 = e^{-0.23t}$

$\ln 0.125 = \ln e^{-0.23t}$

$-2.079442 = -0.23t$ Using a calculator

$\frac{-2.079442}{-0.23} = t$

$9 \approx t$

After 9 hr there will be 1 gram left.

27. a) $W = W_0\, e^{-0.008t}$

$k = 0.008$, or 0.8%

The starving animal loses 0.8% of its weight each day.

b) $W = W_0\, e^{-0.008t}$

$W = W_0\, e^{-0.008(30)}$ Substituting 30 for t

$W = W_0\, e^{-0.24}$

$W = 0.786628\, W_0$ Using a calculator

$W \approx 78.7\%\, W_0$

Thus, after 30 days, 78.7% of the initial weight remains.

29. a) $I = I_0\, e^{-\mu x}$

$I = I_0\, e^{-1.4(1)}$ Substituting 1.4 for μ and 1 for x

$I = I_0\, e^{-1.4}$

$I = I_0(0.246597)$ Using a calculator

$I \approx 25\%\, I_0$

$I = I_0\, e^{-1.4(2)}$ Substituting 1.4 for μ and 2 for x

$I = I_0\, e^{-2.8}$

$I = I_0(0.060810)$ Using a calculator

$I \approx 6.1\%\, I_0$

$I = I_0\, e^{-1.4(3)}$ Substituting 1.4 for μ and 3 for x

$I = I_0\, e^{-4.2}$

$I = I_0(0.014996)$ Using a calculator

$I \approx 1.5\%\, I_0$

b) $I = I_0\, e^{-\mu x}$

$I = I_0\, e^{-1.4(10)}$ Substituting 1.4 for μ and 10 for x

$I = I_0\, e^{-14}$

$I = I_0(0.00000083)$ Using a calculator

$I \approx 0.00008\%\, I_0$

31. a) $T(t) = ae^{-kt} + C$ Newton's law of Cooling

At $t = 0$, $T = 100°$. We solve the following equation for a.

$100 = ae^{-k\cdot 0} + 75$ Substituting 100 for T, 0 for t, and 75 for C

$25 = ae^0$

$25 = a$ $(e^0 = 1)$

The value of the constant is 25°.

Thus, $T(t) = 25\, e^{-kt} + 75$.

b) Now we find k using the fact that at $t = 10$, $T = 90°$.

$T(t) = 25\, e^{-kt} + 75$

$90 = 25\, e^{-k\cdot 10} + 75$ Substituting 90 for T and 10 for t

$15 = 25\, e^{-10k}$

$\frac{15}{25} = e^{-10k}$

$0.6 = e^{-10k}$

$\ln 0.6 = e^{-10k}$

$\ln 0.6 = -10k$

$\frac{\ln 0.6}{-10} = k$

$\frac{-0.510826}{-10} = k$

$0.05 \approx k$

Thus, $T(t) = 25\, e^{-0.05t} + 75$.

c) $T(t) = 25\, e^{-0.05t} + 75$

$T(20) = 25\, e^{-0.05(20)} + 75$ Substituting 20 for t

$= 25\, e^{-1} + 75$

$= 25(0.367879) + 75$ Using a calculator

$= 9.196975 + 75$

≈ 84.2

The temperature after 20 minutes is 84.2°.

d)
$$T(t) = 25\,e^{-0.05t} + 75$$
$$80 = 25\,e^{-0.05t} + 75 \quad \text{Substituting 80 for } T$$
$$5 = 25\,e^{-0.05t}$$
$$\frac{5}{25} = e^{-0.05t}$$
$$0.2 = e^{-0.05t}$$
$$\ln\,0.2 = \ln\,e^{-0.05t}$$
$$\ln\,0.2 = -0.05t$$
$$\frac{\ln\,0.02}{-0.05} = t$$
$$\frac{-1.609438}{-0.05} = t$$
$$32 \approx t$$

It takes 32 minutes for the liquid to cool to 80°.

e) $T'(t) = -1.25e^{-0.05t}$; the rate of change of the temperature is $-1.25e^{-0.05t}$ degrees per minute.

33. We first find a in the equation $T(t) = ae^{-kt} + C$.

Assuming the temperature of the body was normal when the murder occurred, we have $T = 98.6°$at $t = 0$. Thus

$98.6° = ae^{-k\cdot 0} + 60°$ (Room temperature is 60°.)

so

$a = 38.6°$.

Thus T is given by $T(t) = 38.6\,e^{-kt} + 60$.

We want to find the number of hours N since the murder was committed. To find N we must first determine k. From the two temperature readings, we have

$$85.9 = 38.6\,e^{-kN} + 60, \text{ or } 25.9 = 38.6\,e^{-kN}$$
$$83.4 = 38.6\,e^{-k(N+1)} + 60, \text{ or } 23.4 = 38.6\,e^{-k(N+1)}$$

Dividing the first equation by the second, we get

$$\frac{25.9}{23.4} = \frac{38.6\,e^{-kN}}{38.6\,e^{-k(N+1)}} = e^{-kN+k(N+1)} = e^k.$$

We solve this equation for k:

$$\ln\frac{25.9}{23.4} = \ln\,e^k$$
$$\ln\,1.106838 = k$$
$$0.10 \approx k$$

Now we substitute 0.10 for k in the equation $25.9 = 38.6\,e^{-kN}$ and solve for N.

$$25.9 = 38.6\,e^{-0.10N}$$
$$\frac{25.9}{38.6} = e^{-0.10N}$$
$$\ln\frac{25.9}{38.6} = \ln\,e^{-0.10N}$$
$$\ln\,0.670984 = -0.10N$$
$$\frac{-0.399009}{-0.10} = N$$
$$4\text{ hr} \approx N$$

The coroner arrived at 11 P.M., so the murder was committed about 7 P.M.

35. **a)** Let 1995 correspond to $t = 0$

$$48.8 = 51.9\,e^{6k}$$
$$\ln\left(\frac{48.8}{51.9}\right) = 6k$$
$$\frac{\ln\left(\frac{48.8}{51.9}\right)}{6} = k$$
$$0.010265 = k$$

Thus, $N = 51.9\,e^{-0.010265t}$

b) In 2015, $t = 20$

$$N = 51.9\,e^{-0.010265\cdot 20}$$
$$= 51.9 \cdot 0.8144029589$$
$$= 42.27$$

c)

$$1 = 51.9\,e^{-0.010265t}$$
$$\frac{1}{51.9} = e^{-0.010265t}$$
$$\ln\left(\frac{1}{51.9}\right) = -0.010265t$$
$$\frac{\ln\left(\frac{1}{51.9}\right)}{-0.010265} = t$$
$$384.7 = t$$

The population of Ukraine will be 1 in the year 2380.

37. a)
$$P(t) = 50\,e^{-0.004t}$$
$$P(375) = 50\,e^{-0.004(375)} \quad \text{Substituting 375 for } t$$
$$= 50\,e^{-1.5}$$
$$= 50(0.223130)$$
$$\approx 11$$

After 375 days, 11 watts will be available.

b)
$$T = \frac{\ln\,2}{k}$$
$$= \frac{0.693147}{0.004} \quad \text{Substituting 0.693147 for } \ln\,2 \text{ and 0.004 for } k$$
$$\approx 173$$

The half-life of the power supply is 173 days.

c)
$$P(t) = 50\,e^{-0.004t}$$
$$10 = 50\,e^{-0.004t} \quad \text{Substituting 10 for } P(t)$$
$$\frac{10}{50} = e^{-0.004t}$$
$$0.2 = e^{-0.004t}$$
$$\ln\,0.2 = \ln\,e^{-0.004t}$$
$$\ln\,0.2 = -0.004t$$
$$\frac{\ln\,0.2}{-0.004} = t$$
$$\frac{-1.609438}{-0.004} = t \quad \text{Using a calculator}$$
$$402 \approx t$$

The satellite can stay in operation 402 days.

d) When $t = 0$,

$P = 50\,e^{-0.004(0)}$ Substituting 0 for t

$= 50\,e^0$

$= 50 \cdot 1$

$= 50$

At the beginning the power output was 50 watts.

e) $P'(t) = -0.2e^{-0.004t}$; the power is changing at a rate of $-0.2e^{-0.004t}$ watts per day.

39. a) $y = 84.94353992 - 0.5412834098 \ln x$

b) When $x = 8$, $y = 83.8\%$;

when $x = 10$, $y = 83.7\%$;

when $x = 24$, $y = 83.2\%$;

when $x = 36$, $y = 83.09\%$

c)

$$82 = 84.94353992 - 0.5412834098 \ln x$$
$$-2.94353992 = -0.5412834098 \ln x$$
$$5.438075261 \approx \ln x$$
$$x \approx e^{5.438075261}$$
$$x \approx 230$$

The test scores will fall below 82% after about 230 months.

d) $y' = -\dfrac{0.5412834098}{x}$; the test scores are changing at the rate of $-\dfrac{0.5412834098}{x}$ percent per month.

41. **a)** Let 1980 correspond to $t = 0$

$$48 = 52\,e^{20k}$$
$$\ln\left(\frac{48}{52}\right) = 20k$$
$$\frac{\ln\left(\frac{48}{52}\right)}{20} = k$$
$$-0.004002 = k$$

Thus, $N = 52\,e^{-0.004002t}$

b) In 2010, $t = 30$

$$N = 52\,e^{-0.004002 \cdot 30}$$
$$= 52 \cdot 0.8868636211$$
$$= 46.12$$

c)

$$10 = 52\,e^{-0.004002t}$$
$$\frac{10}{52} = e^{-0.004002t}$$
$$\ln\left(\frac{10}{52}\right) = -0.004002t$$
$$\frac{\ln\left(\frac{10}{52}\right)}{-0.004002} = t$$
$$411.9 = t$$

The population of Russia will be 100 million in the year 2392.

Exercise Set 4.5

1. $5^4 = e^{4 \cdot \ln 5}$ Theorem 13: $a^x = e^{x \cdot \ln a}$

$\approx e^{4(1.609438)}$ Using a calculator

$\approx e^{6.4378}$

3. $3.4^{10} = e^{10 \cdot \ln 3.4}$ Theorem 13: $a^x = e^{x \cdot \ln a}$

$\approx e^{10(1.223775)}$ Using a calculator

$\approx e^{12.238}$

5. $4^k = e^{k \cdot \ln 4}$ Theorem 13: $a^x = e^{x \cdot \ln a}$

7. $8^{kT} = e^{kT \cdot \ln 8}$ Theorem 13: $a^x = e^{x \cdot \ln a}$

9. $y = 6^x$

$\dfrac{dy}{dx} = (\ln 6)6^x$ Theorem 14: $\dfrac{dy}{dx}a^x = (\ln a)a^x$

11. $f(x) = 10^x$

$f'(x) = (\ln 10)10^x$ Theorem 14

13. $f(x) = x(6.2)^x$

$f'(x) = x\left[\dfrac{d}{dx}(6.2)^x\right] + \left[\dfrac{d}{dx}x\right](6.2)^x$ Product Rule

$= x \cdot \underbrace{(\ln 6.2)(6.2)^x}_{\text{Theorem 14}} + 1 \cdot (6.2)^x$

$= (6.2)^x[x \ln 6.2 + 1]$

15. $y = x^3 10^x$

$\dfrac{dy}{dx} = x^3 \cdot \underbrace{(\ln 10)10^x}_{\text{Theorem 14}} + 3x^2 \cdot 10^x$ Product Rule

$= 10^x x^2(x \ln 10 + 3)$

17. $y = \log_4 x$

$\dfrac{dy}{dx} = \dfrac{1}{\ln 4} \cdot \dfrac{1}{x}$ Theorem 16: $\dfrac{d}{dx}\log_a x = \dfrac{1}{\ln a} \cdot \dfrac{1}{x}$

19. $f(x) = 2 \log x$

$f'(x) = 2 \cdot \dfrac{d}{dx}\log x$

$= 2 \cdot \dfrac{1}{\ln 10} \cdot \dfrac{1}{x}$ Theorem 16 ($\log x = \log_{10} x$)

$= \dfrac{2}{\ln 10} \cdot \dfrac{1}{x}$

21. $f(x) = \log \dfrac{x}{3}$

$f(x) = \log x - \log 3$ (P2)

$f'(x) = \dfrac{1}{\ln 10} \cdot \dfrac{1}{x} - 0$ Theorem 16 ($\log x = \log_{10} x$)

$= \dfrac{1}{\ln 10} \cdot \dfrac{1}{x}$

23. $y = x^3 \log_8 x$

$\frac{dy}{dx} = \underbrace{x^3\left(\frac{1}{\ln 8}\cdot\frac{1}{x}\right)}_{\text{Theorem 16}} + 3x^2 \cdot \log_8 x$ Product Rule

$= x^2 \cdot \frac{1}{\ln 8} + 3x^2 \cdot \log_8 x$

$= x^2\left(\frac{1}{\ln 8} + 3\log_8 x\right)$

25. $y = csc\, x \log_2 x$

$\frac{dy}{dx} = csc\, x \cdot \frac{1}{\ln 2}\cdot\frac{1}{x} - csc\, x\, cot\, x \cdot \frac{\ln x}{\ln 2}$

$= \frac{csc\, x - x\, csc\, x\, cot\, x\, ln\, x}{x \ln 2}$

27. $y = \log_{10} sin\, x$

$\frac{dy}{dx} = \frac{1}{\ln 10}\cdot\frac{1}{sin\, x}\cdot cos\, x$

$= \frac{1}{\ln 10}\cdot cot\, x$

29. $y = \log_x 3$

$y = \frac{\ln 3}{\ln x}$

$\frac{dy}{dx} = -\ln 3 \cdot \frac{-1}{\ln^2 x}\cdot\frac{1}{x}$

$= \frac{-\ln 3}{\ln^2 x}$

31. $g(x) = (\log_x 10)(\log_{10} x)$

$g'(x) = \log_x 10 \cdot \frac{1}{x \ln 10} + \log_{10} x \cdot \left(\ln 10 \cdot \frac{-1}{\ln^2 x}\cdot\frac{1}{x}\right)$

$= \frac{\ln 10}{\ln x}\cdot\frac{1}{x \ln 10} + \frac{\ln x}{\ln 10}\cdot\left(\ln 10 \cdot \frac{-1}{\ln^2 x}\cdot\frac{1}{x}\right)$

$= 0$

33. a) $N(t) = 250,000\left(\frac{1}{4}\right)^t$

$N'(t) = 250,000\frac{d}{dx}\left(\frac{1}{4}\right)^t$

$= 250,000 \cdot \left(\ln \frac{1}{4}\right)\left(\frac{1}{4}\right)^t$ Theorem 14

$= 250,000(\ln 1 - \ln 4)\left(\frac{1}{4}\right)^t$ (P2)

$= -250,000(\ln 4)\left(\frac{1}{4}\right)^t$ $(\ln 1 = 0)$

b) The rate of change of the number of cans still in use, when 250,000 cans are initially distributed, is $-250,000(\ln 4)\left(\frac{1}{4}\right)^t$ cans per year.

35. $R = \log \frac{I}{I_0}$

$R = \log \frac{10^5 \cdot I_0}{I_0}$ Substituting $10^5 \cdot I_0$ for I

$= \log 10^5$

$= 5$ (P5)

The magnitude on the Richter scale is 5.

37. a) $I = I_0 \cdot 10^R$

$I = I_0 \cdot 10^7$ Substituting 7 for R

$= 10^7 \cdot I_0$

b) $I = I_0 \cdot 10^R$

$I = I_0 \cdot 10^8$ Substituting 8 for R

$= 10^8 \cdot I_0$

c) The intensity in (b) is 10 times that in (a).

$10^8\, I_0 = 10 \cdot 10^7\, I_0$

d) $I = I_0\, 10^R$

$\frac{dI}{dR} = I_0 \cdot \frac{d}{dR}\, 10^R$ I_0 is a constant

$= I_0 \cdot (\ln 10)10^R$ Theorem 14

$= (I_0 \cdot \ln 10)10^R$

e) The intensity is changing at a rate of $(I_0 \cdot \ln 10)^R$.

39. a) $R = \log \frac{I}{I_0}$

$R = \log I - \log I_0$ (P2)

$\frac{dR}{dI} = \frac{1}{\ln 10}\cdot\frac{1}{I} - 0$ Theorem 16; I_0 is a constant.

$= \frac{1}{\ln 10}\cdot\frac{1}{I}$

b) The magnitude is changing at a rate of $\frac{1}{\ln 10}\cdot\frac{1}{I}$.

41. a) $y = m \log x + b$

$\frac{dy}{dx} = m \cdot \frac{d}{dx}\log x + 0$ m and b are constants

$= m\left(\frac{1}{\ln 10}\cdot\frac{1}{x}\right)$ Theorem 16

$= \frac{m}{\ln 10}\cdot\frac{1}{x}$

b) The response is changing at a rate of $\frac{m}{\ln 10}\cdot\frac{1}{x}$.

43. **a)** When $t = 0$ $[OH^-] = x = 10^{-7}$ moles/liter

b) When $t = 0$ $[H^+] = \frac{10^{-14}}{10^{-7}} = 10^{-7}$ moles/liter

c) When $t = 0$ $pH = -\log\left(10^{-7}\right) = 7$

d)

$$\begin{aligned} [H^+][OH^-] &= 10^{-14} \\ \log\left([H^+][OH^-]\right) &= \log(10^{-14} \\ \log\left([H^+][0.002t + 10^{-7}]\right) &= -14 \\ \log[H^+] + \log\left(0.002t + 10^{-7}\right) &= -14 \\ 14 + \log\left(0.002t + 10^{-7}\right) &= -\log[H^+] \end{aligned}$$

Thus, $pH = 14 + \log\left(0.002t + 10^{-7}\right)$

e) $\frac{d}{dt}\,[OH^-] = \frac{0.002}{\ln 10\ (0.002t + 10^{-7}}$

f) The pH is changing most rapidly at $t = 0$ which corresponds to a pH of 7

45. $y = x^x,\ x > 0$

$y = e^{x \ln x}$ Theorem 13: $a^x = e^{x \ln a}$

$\frac{dy}{dx} = \left(x \cdot \frac{1}{x} + 1 \cdot \ln x\right) e^{x \ln x}$

$= (1 + \ln x)x^x$ Substituting x^x for $e^{x \ln x}$

47. $f(x) = x^{e^x},\ x > 0$

$f(x) = e^{e^x \ln x}$ Theorem 13: $a^x = e^{x \ln a}$

$f'(x) = \left(e^x \cdot \frac{1}{x} + e^x \ln x\right) e^{e^x \ln x}$

$= e^x \left(\frac{1}{x} + \ln x\right) x^{e^x}$ Substituting x^{e^x} for $e^{e^x \ln x}$

$= e^x\, x^{e^x} \left(\ln x + \frac{1}{x}\right)$

49. $y = \log_a f(x),\ f(x) > 0$

$a^y = f(x)$ Exponential equation

$e^{y \ln a} = f(x)$ Theorem 13

Differential implicitly to find dy/dx.

$$\frac{d}{dx} e^{y \ln a} = \frac{d}{dx} f(x)$$

$$\frac{dy}{dx} \cdot \ln a \cdot e^{y \ln a} = f'(x)$$

$\frac{dy}{dx} \cdot \ln a \cdot f(x) = f'(x)$ Substituting $f(x)$ for $e^{y \ln a}$

$$\frac{dy}{dx} = \frac{1}{\ln a} \cdot \frac{f'(x)}{f(x)}$$

51. Since a^x can be written as $e^{x \cdot \ln a}$, we can find the derivative of $f(x) = a^x$ using the rule for differentiating an exponential function, base e.

Chapter 5

Integration

Exercise Set 5.1

1. $\int x^6\,dx$

$= \frac{x^{6+1}}{6+1} + C \quad \left(\int x^r\,dx = \frac{r^{r+1}}{r+1} + C\right)$

$= \frac{x^7}{7} + C$

3. $\int 2\,dx$

$= 2x + C$ (For k a constant, $\int k\,dx = kx + C$.)

5. $\int x^{1/4}\,dx$

$= \frac{x^{1/4+1}}{\frac{1}{4}+1} + C \quad \left(\int x^r\,dx = \frac{x^{r+1}}{r+1} + C\right)$

$= \frac{x^{5/4}}{\frac{5}{4}} + C$

$= \frac{4}{5}x^{5/4} + C$

7. $\int (x^2 + x - 1)\,dx$

$= \int x^2\,dx + \int x\,dx - \int 1\,dx$

The integral of a sum is the sum of the integrals.

$= \frac{x^3}{3} + \frac{x^2}{2} - x + C$ ← DON'T FORGET THE C!

$\left[\int x^r\,dx = \frac{x^{r+1}}{r+1} + C\right]$

(For k a constant, $\int k\,dx = kx + C$.)

9. $\int (t^2 - 2t + 3)\,dt$

$= \int t^2\,dt - \int 2t\,dt + \int 3\,dt$

The integral of a sum is the sum of the integrals.

$= \frac{t^3}{3} - 2\cdot\frac{t^2}{2} + 3t + C$

$\left[\int x^r\,dx = \frac{x^{r+1}}{r+1} + C\right]$

(For k a constant, $\int k\,dx = kx + C$.)

$= \frac{t^3}{3} + t^2 + 3t + C$

11. $\int 5\,e^{8x}\,dx$

$= \frac{5}{8}\,e^{8x} + C \quad \left(\int be^{ax}\,dx = \frac{b}{a}\,e^{ax} + C\right)$

13. $\int (w^3 - w^{8/7})\,dw$

$= \int x^3\,dx - \int x^{8/7}\,dx$

The integral of a sum is the sum of the integrals.

$= \frac{w^4}{4} - \frac{w^{8/7+1}}{\frac{8}{7}+1} + C \quad \left[\int x^r\,dx = \frac{x^{r+1}}{r+1} + C\right]$

$= \frac{w^4}{4} - \frac{w^{15/7}}{\frac{15}{7}} + C$

$= \frac{w^4}{4} - \frac{7}{15}w^{15/7} + C$

15. $\int \frac{1000}{r}\,dr$

$= 1000\int \frac{1}{r}\,dr$ The integral of a constant times a function is the constant times the integral.

$= 1000\ln r + C$ $\int \frac{1}{x}\,dx = \ln x + C,\ x > 0$; we generally consider $r > 0$

17. $\int \frac{dx}{x^2} = \int \frac{1}{x^2}\,dx = \int x^{-2}\,dx$

$= \frac{x^{-2+1}}{-2+1} + C \quad \left[\int x^r\,dx = \frac{x^{r+1}}{r+1} + C\right]$

$= \frac{x^{-1}}{-1} + C$

$= -x^{-1} + C$, or $-\frac{1}{x} + C$

19. $\int \sqrt{s}\,ds = \int s^{1/2}\,ds$

$= \frac{s^{1/2+1}}{\frac{1}{2}+1} + C$

$= \frac{s^{3/2}}{\frac{3}{2}} + C$

$= \frac{2}{3}s^{3/2} + C$

21. $\int \frac{-6}{\sqrt[3]{x^2}}\,dx = \int \frac{-6}{x^{2/3}}\,dx = \int -6x^{-2/3}\,dx$

$$= -6 \int x^{-2/3}\,dx$$
$$= -6 \cdot \frac{x^{-2/3+1}}{-\frac{2}{3}+1} + C$$
$$= -6 \cdot \frac{x^{1/3}}{\frac{1}{3}} + C$$
$$= -6 \cdot 3x^{1/3} + C$$
$$= -18x^{1/3} + C$$

23. $\int 8\,e^{-2x}\,dx = \frac{8}{-2}\,e^{-2x} + C$

$$= -4\,e^{-2x} + C$$

25. $\int \left(x^2 - \frac{3}{2}\sqrt{x} + x^{-4/3}\right) dx$

$$= \int x^2\,dx - \frac{3}{2}\int x^{1/2}\,dx + \int x^{-4/3}\,dx$$
$$= \frac{x^{2+1}}{2+1} - \frac{3}{2}\cdot\frac{x^{1/2+1}}{\frac{1}{2}+1} + \frac{x^{-4/3+1}}{-\frac{4}{3}+1} + C$$
$$= \frac{x^3}{3} - \frac{3}{2}\cdot\frac{x^{3/2}}{\frac{3}{2}} + \frac{x^{-1/3}}{-\frac{1}{3}} + C$$
$$= \frac{x^3}{3} - x^{3/2} - 3x^{-1/3} + C$$

27. $\int 5\,sin\,2\theta\,d\theta = \frac{-5}{2}\,cos\,2\theta + C$

29. $\int \left(5\,sin\,5x - 4\,cos\,2x\right) dx$

$$= -cos\,5x - 2\,sin\,2x + C$$

31. $\int 3\,sec^2\,3x\,dx = tan\,3x + C$

33. $\int \frac{1}{3}\,sec\,\frac{x}{9}\,tan\,\frac{x}{9}\,dx = 3\,sec\,\frac{x}{9} + C$

35. $\int (sec\,x + tan\,x)sec\,x\,dx = \int (sec^2\,x + sec\,x\,tan\,x)\,dx$

$$= tan\,x + sec\,x + C$$

37. $\int \left[\frac{1}{t} + \frac{1}{t^2} - \frac{1}{e^t}\right] dt = \ln t - \frac{1}{t} + \frac{1}{e^t} + C$

39. Find the function f such that

$f'(x) = x - 3$ and $f(2) = 9$.

We first find $f(x)$ by integrating.

$$f(x) = \int (x-3)\,dx$$
$$= \int x\,dx - \int 3\,dx$$
$$= \frac{x^2}{2} - 3x + C$$

The condition $f(2) = 9$ allows us to find C.

$$f(x) = \frac{x^2}{2} - 3x + C$$
$$f(2) = \frac{2^2}{2} - 3\cdot 2 + C = 9 \quad \text{Substituting 2 for } x \text{ and 9 for } f(2)$$
$$2 - 6 + C = 9$$
$$C = 13$$

Thus, $f(x) = \frac{x^2}{2} - 3x + 13$.

41. Find the function f such that

$f'(x) = x^2 - 4$ and $f(0) = 7$.

We first find $f(x)$ by integrating.

$$f(x) = \int (x^2 - 4)\,dx$$
$$= \int x^2\,dx - \int 4\,dx$$
$$= \frac{x^3}{3} - 4x + C$$

The condition $f(0) = 7$ allows us to find C.

$$f(x) = \frac{x^3}{3} - 4x + C$$
$$f(0) = \frac{0^3}{3} - 4\cdot 0 + C = 7 \quad \text{Substituting 0 for } x \text{ and 7 for } f(0)$$

Solving for C we get $C = 7$.

Thus, $f(x) = \frac{x^3}{3} - 4x + 7$.

43. $f(x) = \int 2\,cos\,3x\,dx = \frac{2}{3}\,sin\,3x + C$

$$f(0) = \frac{2}{3}\,sin\,(0) + C = 1$$
$$C = 1$$
$$f(x) = \frac{2}{3}\,sin\,3x + 1$$

45. $f(x) = \int 5\,e^{2x}\,dx = \frac{5}{2}\,e^{2x} + C$

$$f(0) = \frac{5}{2} + C = -10$$
$$C = \frac{-25}{2}$$
$$f(x) = \frac{5}{2}\,e^{2x} - \frac{25}{2}$$

47. $f(t) = \int 157t + 1000\,dt = \frac{157}{2}t^2 + 1000t + C$

$$f(0) = 0 + 0 + C$$
$$156239 = C$$
$$f(t) = \frac{157}{2}t^2 + 1000t + 156239$$
$$f(2) = \frac{157}{2}(2)^2 + 1000(2) + 156239$$
$$= 158553$$

49. $v(t) = 3t^2, \quad s(0) = 4$

We find $s(t)$ by integrating $v(t)$.

$$\begin{aligned} s(t) &= \int v(t)\,dt \\ &= \int 3t^2\,dt \\ &= 3 \cdot \frac{t^3}{3} + C \\ &= t^3 + C \end{aligned}$$

The condition $s(0) = 4$ allows us to find C.

$s(0) = 0^3 + C = 4$ Substituting 0 for t and 4 for $s(0)$

Solving for C, we get $C = 4$.

Thus, $s(t) = t^3 + 4$.

51. $a(t) = 4t, \quad v(0) = 20$

We find $v(t)$ by integrating $a(t)$.

$$\begin{aligned} v(t) &= \int a(t)\,dt \\ &= \int 4t\,dt \\ &= 4 \cdot \frac{t^2}{2} + C \\ &= 2t^2 + C \end{aligned}$$

The condition $v(0) = 20$ allows us to find C.

$v(0) = 2 \cdot 0^2 + C = 20$ Substituting 0 for t and 20 for $v(0)$

Solving for C, we get $C = 20$.

Thus, $v(t) = 2t^2 + 20$.

53. $a(t) = -2t + 6, \quad v(0) = 6 \quad$ and $s(0) = 10$

We find $v(t)$ by integrating $a(t)$.

$$\begin{aligned} v(t) &= \int a(t)\,dt \\ &= \int (-2t + 6)\,dt \\ &= -t^2 + 6t + C_1 \end{aligned}$$

The condition $v(0) = 6$ allows us to find C_1.

$v(0) = -0^2 + 6 \cdot 0 + C_1 = 6$ Substituting 0 for t and 6 for $v(0)$

Solving for C_1, we get $C_1 = 6$.

Thus, $v(t) = -t^2 + 6t + 6$.

We find $s(t)$ by integrating $v(t)$.

$$\begin{aligned} s(t) &= \int v(t)\,dt \\ &= \int (-t^2 + 6t + 6)\,dt \\ &= -\frac{t^3}{3} + 3t^2 + 6t + C_2 \end{aligned}$$

The condition $s(0) = 10$ allows us to find C_2.

$s(0) = -\frac{0^3}{3} + 3 \cdot 0^2 + 6 \cdot 0 + C_2$ Substituting 0 for t and 10 for $s(0)$

Solving for C_2, we get $C_2 = 10$.

Thus, $s(t) = -\frac{1}{3}t^3 + 3t^2 + 6t + 10$.

55.

$$\begin{aligned} a(t) &= -32 \text{ ft/sec}^2 \\ v(0) &= \text{initial velocity} = v_0 \\ s(0) &= \text{initial height} = s_0 \end{aligned}$$

We find $v(t)$ by integrating $a(t)$.

$$\begin{aligned} v(t) &= \int a(t)\,dt \\ &= \int (-32)\,dt \\ &= -32t + C_1 \end{aligned}$$

The condition $v(0) = v_0$ allows us to find C_1.

$v(0) = -32 \cdot 0 + C_1 = v_0$ Substituting 0 for t and v_0 for $v(0)$

$C_1 = v_0$

Thus, $v(t) = -32t + v_0$.

We find $s(t)$ by integrating $v(t)$.

$$\begin{aligned} s(t) &= \int v(t)\,dt \\ &= \int (-32t + v_0)\,dt \\ &= -16t^2 + v_0 t + C_2 \quad v_0 \text{ is constant} \end{aligned}$$

The condition $s(0) = s_0$ allows us to find C_2.

$s(0) = -16 \cdot 0^2 + v_0 \cdot 0 + C_2 = s_0$ Substituting 0 for t and s_0 for $s(0)$

$C_2 = s_0$

Thus, $s(t) = -16t^2 + v_0 t + s_0$.

57. $a(t) = k$ Constant acceleration

$v(t) = \int a(t)\,dt = \int k\,dt = kt \quad (v(0) = 0;$ thus $C = 0.)$

$$s(t) = \int v(t)\,dt = \int kt\,dt = k \cdot \frac{t^2}{2} = \frac{1}{2}kt^2$$

$(s(0) = 0;$ thus $C = 0.)$

We know that

$$a(t) = k = \frac{60 \text{ mph}}{\frac{1}{2} \text{ min}}$$

and that

$$t = \frac{1}{2} \text{ min.}$$

Thus

$$\begin{aligned} s(t) &= \frac{1}{2}kt^2 \\ s\left(\frac{1}{2}\text{ min}\right) &= \frac{1}{2} \cdot \frac{60 \text{ mph}}{\frac{1}{2} \text{ min}} \cdot \left(\frac{1}{2}\text{ min}\right)^2 \\ &= \frac{1}{2} \cdot \frac{60 \text{ mi}}{\text{hr}} \cdot \frac{1}{2} \text{ min} \\ &= \frac{1}{2} \cdot \frac{60 \text{ mi}}{\text{hr}} \cdot \frac{1}{120} \text{ hr} \\ &= \frac{60}{240} \text{ mi} \\ &= \frac{1}{4} \text{ mi} \end{aligned}$$

The car travels $\frac{1}{4}$ mi during that time.

59. $a(t) = -68.5$

$v(t) = \int a(t)\,dt = \int -68.5\,dt = -68.5t + C$

(v(0)=132)

$132 = 0 + C \to C = 102.7$

$s(t) = \int v(t)\,dt = \int -68.5t + 132\,dt$

$= -34.25t^2 + 132t + C$

(s(0)=0; *thus* $C = 0$)

$s(t) = -34.25t^2 + 132t$

The time it takes to go from $132 ft/sec$ to 0 is

$$\begin{aligned} -68.5t + 132 &= 0 \\ 68.5t &= 132 \\ t &= \frac{132}{68.5} \\ &= 1.927 \end{aligned}$$

Thus

$s(1.927) = -34.25(1.927)^2 + 132(1.927)$

$= 127.182$

The car travels almost 127.2 feet during that time.

61. $M'(t) = 0.2t - 0.003t^2$

a) We integrate to find $M(t)$.

$M(t) = \int(0.2t - 0.003t^2)\,dt$

$= 0.1t^2 - 0.001t^3 + C$

We use $M(0) = 0$ to find C.

$M(0) = 0.1(0)^2 - 0.001(0)^3 + C = 0$

$C = 0$

$M(t) = 0.1t^2 - 0.001t^3$

b) $M(8) = 0.1(8)^2 - 0.001(8)^3$

$= 6.4 - 0.512$

$= 5.888$

≈ 6 words

63. $N'(t) = 38.2e^{0.0376t}$

a)

$$\begin{aligned} N(t) &= \int 38.2e^{0.0376t}\,dt \\ &= 1015.957e^{0.0376t} + C \\ 1622 &= 1015.957 + C \\ 606.043 &= C \\ N(t) &= 1015.957e^{0.0376t} + 606.043 \end{aligned}$$

b)

$$\begin{aligned} N(30) &= 1015.957e^{0.0376(30)} + 606.043 \\ &= 3742 \end{aligned}$$

c)

$$\begin{aligned} N(60) &= 1015.957e^{0.0376(60)} + 606.043 \\ &= 10294 \end{aligned}$$

65. $f'(t) = t^{\sqrt{3}}$

$$\begin{aligned} f(t) &= \int t^{\sqrt{3}}\,dt \\ &= \frac{t^{\sqrt{3}+1}}{\sqrt{3}+1} + C \\ 8 &= 0 + C \\ 8 &= C \\ f(t) &= \frac{t^{\sqrt{3}+1}}{\sqrt{3}+1} + 8 \end{aligned}$$

67.

$$\begin{aligned} \int (x-1)^2x^3\,dx &= \int (x^5 - 2x^4 + x^3)\,dx \\ &= \frac{x^6}{6} - \frac{2}{5}x^5 + \frac{x^4}{4} + C \end{aligned}$$

69.

$$\begin{aligned} \int \frac{(t+3)^2}{\sqrt{t}}dt &= \int (t^{3/2} + 6t^{1/2} + 9t^{-1/2})\,dt \\ &= \frac{2}{5}t^{5/2} + 4t^{3/2} + 18t^{1/2} + C \end{aligned}$$

71.

$$\begin{aligned} \int (t+1)^3\,dt &= \int (t^3 + 3t^2 + 3t + 1)\,dt \\ &= \frac{t^4}{4} + t^3 + \frac{3}{2}t^2 + t + C \end{aligned}$$

73.

$$\begin{aligned} \int be^{ax}\,dx &= b\int e^{ax}\,dx \\ &= \frac{b}{a}e^{ax} + C \end{aligned}$$

75.

$$\begin{aligned} \int \sqrt[3]{64x^4}\,dx &= \sqrt[3]{64}\int x^{4/3}\,dx \\ &= 4 \cdot \frac{3}{7}x^{7/3} + C \\ &= \frac{12}{7}x^{7/3} + C \end{aligned}$$

77.

$$\begin{aligned} \int \frac{t^3+8}{t+2}dt &= \int \frac{(t+2)(t^2-2t+4)}{(t+2)}dt \\ &= \int (t^2 - 2t + 4)dt \\ &= \frac{t^3}{3} - t^2 + 4t + C \end{aligned}$$

79.

$$\begin{aligned}\int (\cos^3 x + \cos x \sin^2 x)\,dx &= \int [\cos x(\cos^2 x + \sin^2 x)]\,dx \\ &= \int \cos x\,dx \\ &= \sin x + C\end{aligned}$$

81.

$$\begin{aligned}\int \cot^2 2x\,dx &= \int (\csc^2 2x - 1)\,dx \\ &= -\frac{\cot 2x}{2} - x + C\end{aligned}$$

83. Answers could vary. The antiderivative of a function represents the area under the curve of that function.

Exercise Set 5.2

1. a) $f(x) = \dfrac{1}{x^2}$

In the drawing in the text the interval $[1, 7]$ has been divided into 6 subintervals, each having width $1 \left(\Delta x = \dfrac{7-1}{6} = 1\right)$.

The heights of the rectangles shown are

$f(1) = \dfrac{1}{1^2} = 1$

$f(2) = \dfrac{1}{2^2} = \dfrac{1}{4} = 0.2500$

$f(3) = \dfrac{1}{3^2} = \dfrac{1}{9} \approx 0.1111$

$f(4) = \dfrac{1}{4^2} = \dfrac{1}{16} = 0.0625$

$f(5) = \dfrac{1}{5^2} = \dfrac{1}{25} = 0.0400$

$f(6) = \dfrac{1}{6^2} = \dfrac{1}{36} \approx 0.0278$

The area of the region under the curve over $[1, 7]$ is approximately the sum of the areas of the 6 rectangles.

Area of each rectangle:

1st rectangle: $1 \cdot 1 = 1$

$[f(1) = 1 \text{ and } \Delta x = 1]$

2nd rectangle: $0.2500 \cdot 1 = 0.2500$

$[f(2) = 0.2500 \text{ and } \Delta x = 1]$

3rd rectangle: $0.1111 \cdot 1 = 0.1111$

4th rectangle: $0.0625 \cdot 1 = 0.0625$

5th rectangle: $0.0400 \cdot 1 = 0.0400$

6th rectangle: $0.0278 \cdot 1 = 0.0278$

The total area is $1 + 0.2500 + 0.1111 + 0.0625 + 0.0400 + 0.0278$, or 1.4914.

b) $f(x) = \dfrac{1}{x^2}$

The interval $[1, 7]$ has been divided into 12 subintervals, each having width 0.5 $\left(\Delta x = \dfrac{7-1}{12} = \dfrac{6}{12} = 0.5\right)$. The heights of six of the rectangles were computed in part (a). The others are computed below:

$f(1.5) = \dfrac{1}{(1.5)^2} = \dfrac{1}{2.25} \approx 0.4444$

$f(2.5) = \dfrac{1}{(2.5)^2} = \dfrac{1}{6.25} \approx 0.1600$

$f(3.5) = \dfrac{1}{(3.5)^2} = \dfrac{1}{12.25} \approx 0.0816$

$f(4.5) = \dfrac{1}{(4.5)^2} = \dfrac{1}{20.25} \approx 0.0494$

$f(5.5) = \dfrac{1}{(5.5)^2} = \dfrac{1}{30.25} \approx 0.0331$

$f(6.5) = \dfrac{1}{(6.5)^2} = \dfrac{1}{42.25} \approx 0.0237$

The area of the region under the curve over $[1, 7]$ is approximately the sum of the areas of the 12 rectangles.

Area of each rectangle:

1st rectangle: $1(0.5) = 0.5$

$[f(1) = 1 \text{ and } \Delta x = 0.5]$

2nd rectangle: $0.4444(0.5) = 0.2222$

$[f(1.5) \approx 0.4444 \text{ and } \Delta x = 0.5]$

3rd rectangle: $0.2500(0.5) = 0.1250$

4th rectangle: $0.1600(0.5) = 0.0800$

5th rectangle: $0.1111(0.5) \approx 0.0556$

6th rectangle: $0.0816(0.5) = 0.0408$

7th rectangle: $0.0625(0.5) \approx 0.0313$

8th rectangle: $0.0494(0.5) = 0.0247$

9th rectangle: $0.0400(0.5) = 0.0200$

10th rectangle: $0.0331(0.5) \approx 0.0166$

11th rectangle: $0.0278(0.5) = 0.0139$

12th rectangle: $0.0237(0.5) \approx 0.0119$

The total area is $0.5 + 0.2222 + 0.1250 + 0.0800 + 0.0556 + 0.0408 + 0.0313 + 0.0247 + 0.0200 + 0.0166 + 0.0139 + 0.0119$, or 1.1420. (Answers may vary slightly depending on when rounding was done.)

3. The shaded region represents an antiderivative. It also represents velocity, the antiderivative of acceleration.

5. The shaded region represents an antiderivative. It also represents total energy used in time t.

7. The shaded region represents an antiderivative. It also represents the amount of the drug in the blood.

9. The shaded region represents an antiderivative. It also represents the number of words memorized in time t.

11. $\Delta x = \frac{2-0}{4} = \frac{1}{2}$

$$\begin{aligned} \int_0^2 x^2\,dx &= f(1/2)(1/2) + f(1)(1/2) + f(3/2)(1/2) + \\ &\quad f(2)(1/2) \\ &= (1/4)(1/2) + (1)(1/2) + (9/4)(1/2) + (4)(1/2) \\ &= 1/8 + 1/2 + 9/8 + 2 \\ &= 3.75 \end{aligned}$$

13. $\Delta x = \frac{5-4}{6} = \frac{1}{6}$

$$\begin{aligned} \int_4^5 x\,dx &= f(25/6)(1/6) + f(26/6)(1/6) + f(27/6)(1/6) + \\ &\quad f(28/6)(1/6) + f(29/6)(1/6) + f(5)(1/6) \\ &= (25/6)(1/6) + (26/6)(1/6) + (27/6)(1/6) + \\ &\quad (28/6)(1/6) + (29/6)(1/6) + (5)(1/6) \\ &= 25/36 + 26/36 + 27/36 + 28/36 + 29/36 + 5/6 \\ &= 4.58333 \end{aligned}$$

15. $\Delta x = \frac{\pi-0}{4} = \frac{\pi}{4}$

$$\begin{aligned} \int_0^\pi sin\ x\,dx &= sin(\pi/4)(\pi/4) + sin(2\pi/4)(\pi/4) + \\ &\quad sin(3\pi/4)(\pi/4) + sin(\pi)(\pi/4) \\ &= (1/\sqrt{2})(\pi/4) + (1)(\pi/4) + (1/\sqrt{2})(\pi/4) + 0 \\ &= 1.89612 \end{aligned}$$

17. a) $\int_a^b f(x)\,dx = 0$, because there is the same area above the x-axis as below. That is, the area is $A - A$, or 0.

b) $\int_a^b f(x)\,dx < 0$, because there is more area below the x-axis than above. The area is $A - 2A$, or $-A$.

19. $P'(t) = 200e^{-t}$

$$\begin{aligned} \Delta t &= \frac{2-0}{6} = \frac{1}{3} \\ \int_0^2 P'(t)\,dt &= P'(1/3)(1/3) + P'(2/3)(1/3) + P'(1)(1/3) + \\ &\quad P'(4/3)(1/3) + P'(5/3)(1/3) + P'(2)(1/3) \\ &= 143.31(1/3) + 102.68(1/3) + 73.58(1/3) + \\ &\quad 52.72(1/3) + 37.78(1/3) + 27.07(1/3) \\ &= 145.71333 \approx 146 \end{aligned}$$

21. $P'(t) = -500(20 - t)$

$$\begin{aligned} \Delta t &= \frac{20-0}{5} = 4 \\ \int_0^{20} P'(t)\,dt &= P'(4)(4) + P'(8)4 + P'(12)(4) + \\ &\quad P'(16)(4) + P'(20)(4) \\ &= (-8000)(4) + (-6000)(4) + \\ &\quad (-4000)(4) + (-2000)(4) + 0 \\ &= -80000 \end{aligned}$$

23. $v(t) = 3t^2 + 2t$

$$\begin{aligned} \Delta t &= \frac{5-1}{4} = 1 \\ \int_1^5 v(t)\,dt &= v(1)(1) + v(2)(1) + v(3)(1) + v(4)(1) \\ &= 5 + 16 + 33 + 56 \\ &= 110 \end{aligned}$$

25. $f(x) = x$. Since we are integrating over $[0, 2]$ the length of the subintervals is given by

$$\Delta x = \frac{2-0}{n} = \frac{2}{n}$$

Now, we need an expression for x_i:
$x_0 = 0$
$x_1 = 0 + 2/n = 2/n$
$x_2 = 2/n + 2/n = 4/n = 2(2/n)$
$x_3 = 2(2/n) + 2/n = 3(2/n)$
In general, we can write $x_i = i \cdot (\frac{2}{n})$

$$\begin{aligned} \sum_{i=1}^n f(x_i)\Delta x &= \sum_{i=1}^n f(i \cdot \frac{2}{n})\frac{2}{n} \\ &= \sum_{i=1}^n i \cdot \frac{2}{n} \cdot \frac{2}{n} \\ &= \sum_{i=1}^n i \cdot \frac{4}{n^2} \\ &= \frac{4}{n^2}\sum_{i=1}^n i \\ &= \frac{4}{n^2} \cdot \frac{n(n+1)}{2} \\ &= \frac{2(n+1)}{n} \\ \int_0^2 x dx &= \lim_{n\to\infty} \sum_{i=1}^n f(x_i)\Delta x \\ &= \lim_{n\to\infty} \frac{2(n+1)}{n} \\ &= \lim_{n\to\infty} 2 + \frac{2}{n} \\ &= 2 \end{aligned}$$

27. $f(x) = 3x^2$. Since we are integrating over $[0, 1]$ the length of the subintervals is given by

$$\Delta x = \frac{1-0}{n} = \frac{1}{n}$$

Now, we need an expression for x_i:
$x_0 = 0$
$x_1 = 0 + 1/n = 1/n$
$x_2 = 1/n + 1/n = 1/n = 2(1/n)$
$x_3 = 2(1/n) + 1/n = 3(1/n)$
In general, we can write $x_i = i \cdot (\frac{1}{n})$

$$\sum_{i=1}^n f(x_i)\Delta x = \sum_{i=1}^n f(i \cdot \frac{1}{n})\frac{1}{n}$$

$$\begin{aligned}
&= \sum_{i=1}^{n} i^2 \cdot 3(\frac{1}{n})^2 \cdot \frac{1}{n} \\
&= \sum_{i=1}^{n} i^2 \cdot \frac{3}{n^3} \\
&= \frac{3}{n^3} \sum_{i=1}^{n} i^2 \\
&= \frac{3}{n^3} \cdot \frac{n(n+1)(2n+1)}{6} \\
&= \frac{2n^3+3n^2+n}{2n^3}
\end{aligned}$$

$$\begin{aligned}
\int_0^1 3x^2 dx &= \lim_{n\to\infty} \sum_{i=1}^{n} f(x_i)\Delta x \\
&= \lim_{n\to\infty} \frac{2n^3+3n^2+n}{2n^3} \\
&= \lim_{n\to\infty} 1 + \frac{3}{2n} + \frac{1}{2n^2} \\
&= 1
\end{aligned}$$

29. $f(x) = x^3$. Since we are integrating over $[0,4]$ the length of the subintervals is given by

$$\Delta x = \frac{4-0}{n} = \frac{4}{n}$$

Now, we need an expression for x_i:
$x_0 = 0$
$x_1 = 0 + 4/n = 4/n$
$x_2 = 4/n + 4/n = 8/n = 2(4/n)$
$x_3 = 2(4/n) + 4/n = 3(4/n)$
In general, we can write $x_i = i \cdot (\frac{4}{n})$

$$\begin{aligned}
\sum_{i=1}^{n} f(x_i)\Delta x &= \sum_{i=1}^{n} f(i \cdot \frac{4}{n})\frac{4}{n} \\
&= \sum_{i=1}^{n} i^3 \cdot (\frac{4}{n})^3 \cdot \frac{4}{n} \\
&= \sum_{i=1}^{n} i^3 \cdot \frac{256}{n^4} \\
&= \frac{256}{n^4} \sum_{i=1}^{n} i^3 \\
&= \frac{256}{n^4} \cdot \frac{n2(n+1)^2}{4} \\
&= \frac{64(n^4+2n^3+n^2)}{n^4}
\end{aligned}$$

$$\begin{aligned}
\int_0^4 x^3 dx &= \lim_{n\to\infty} \sum_{i=1}^{n} f(x_i)\Delta x \\
&= \lim_{n\to\infty} \frac{64(n^4+2n^3+n^2)}{n^4} \\
&= \lim_{n\to\infty} 64 + \frac{128}{n} + \frac{64}{n^2} \\
&= 64
\end{aligned}$$

31. $f(x) = x^2$. Since we are integrating over $[1,3]$ the length of the subintervals is given by

$$\Delta x = \frac{3-1}{n} = \frac{2}{n}$$

Now, we need an expression for x_i:
$x_0 = 1$
$x_1 = 1 + 2/n$
$x_2 = 1 + 2/n + 2/n = 1 + 2(2/n)$
$x_3 = 1 + 2(2/n) + 2/n = 1 + 3(2/n)$
In general, we can write $x_i = 1 + i(\frac{2}{n})$

$$\begin{aligned}
\sum_{i=1}^{n} f(x_i)\Delta x &= \sum_{i=1}^{n} f(1+\frac{2i}{n})\frac{2}{n} \\
&= \sum_{i=1}^{n} (1+\frac{2i}{n})^2 \cdot \frac{2}{n} \\
&= \sum_{i=1}^{n} \frac{2}{n}\left[1 + \frac{4i}{n} + \frac{4i^2}{n^2}\right] \\
&= \frac{2}{n}\left[\sum_{i=1}^{n} 1 + \sum_{i=1}^{n} \frac{4i}{n} + \sum_{i=1}^{n} \frac{4i^2}{n^2}\right] \\
&= \frac{2}{n}\left[n + \frac{4}{n}\sum_{i=1}^{n} i + \frac{4}{n^2}\sum_{i=1}^{n} i^2\right] \\
&= \frac{2}{n}\left[n + \frac{4}{n} \cdot \frac{n(n+1)}{2}\right] + \\
&\quad \frac{2}{n}\left[\frac{4}{n^2} \cdot \frac{n(n+1)(2n+1)}{6}\right] \\
&= 2 + \frac{8}{n^2} \cdot \frac{n^2+n}{2} + \frac{8}{n^3} \cdot \frac{2n^3+3n^2+n}{6} \\
&= 2 + 4 + \frac{4}{n} + \frac{8}{3} + \frac{4}{n} + \frac{4}{3n^2} \\
&= \frac{26}{3} + \frac{8}{n} + \frac{4}{3n^2}
\end{aligned}$$

$$\begin{aligned}
\int_1^3 x^2 dx &= \lim_{n\to\infty} \sum_{i=1}^{n} f(x_i)\Delta x \\
&= \lim_{n\to\infty} \left[\frac{26}{3} + \frac{8}{n} + \frac{4}{3n^2}\right] \\
&= \frac{26}{3}
\end{aligned}$$

33. $\int_0^2 (x + x^2)dx = \int_0^2 x\ dx + \int_0^2 x^2\ dx$
The first integral was evaluated in Exercise 25, and yielded a value of 2. The second integral is now computed: $\Delta x = \frac{2-0}{n} = \frac{2}{n}$
$x_0 = 0$
$x_1 = 0 + 2/n = 2/n$
$x_2 = 2/n + 2/n = 2(2/n)$
$x_3 = 2(2/n) + 2/n = 3(2/n)$
In general, we can write $x_i = i \cdot \frac{2}{n}$

$$\begin{aligned}
\sum_{i=1}^{n} f(x_i)\Delta x &= \sum_{i=1}^{n} f(i \cdot \frac{2}{n})\frac{2}{n} \\
&= \sum_{i=1}^{n} i^2 \cdot (\frac{2}{n})^2 \frac{2}{n}
\end{aligned}$$

$$= \sum_{i=1}^{n} i^2 \cdot \frac{8}{n^3}$$
$$= \frac{8}{n^3}\sum_{i=1}^{n} i^2$$
$$= \frac{8}{n^3} \cdot \frac{n(n+1)(2n+1)}{6}$$
$$= \frac{4}{3}\frac{2n^3+3n^2+n}{n^3}$$
$$= \frac{4}{3}\left[2+\frac{3}{n}+\frac{1}{n^2}\right]$$
$$= \frac{8}{3}+\frac{4}{n}+\frac{4}{3n^2}$$
$$\int_0^2 x^2\,dx = \lim_{n\to\infty} \frac{8}{3}+\frac{4}{n}+\frac{4}{3n^2}$$
$$= \frac{8}{3}$$

Thus,

$$\int_0^2 (x+x^2)\,dx = 2+\frac{8}{3} = \frac{14}{3}$$

35. - 41. Left to the student

43. $\int_0^{\pi} sin\ x\ dx = 2$

45. $\int_0^4 \sqrt{x}\ dx = 5.33334$

47. $\int_2^4 \ln(x)\ dx = 2.15888$

Exercise Set 5.3

1. Find the area under the curve $y = 4$ on the interval $[1, 3]$.

$A(x) = \int 4\,dx$
$= 4x + C$

Since we know that $A(1) = 0$ (there is no area above the number 1), we can substitute for x and $A(x)$ to determine C.

$A(1) = 4 \cdot 1 + C = 0$ Substituting 1 for x and 0 for $A(1)$

Solving for C we get:

$4 + C = 0$
$C = -4$

Thus, $A(x) = 4x - 4$.

Then the area on the interval $[1, 3]$ is $A(3)$.

$A(3) = 4 \cdot 3 - 4$ Substituting 3 for x
$= 12 - 4$
$= 8$

3. Find the area under the curve $y = 2x$ on the interval $[1, 3]$.

$A(x) = \int 2x\,dx$
$= x^2 + C$

Since we know that $A(1) = 0$ (there is no area above the number 1), we can substitute for x and $A(x)$ to determine C.

$A(1) = 1^2 + C = 0$ Substituting 1 for x and 0 for $A(1)$

Solving for C we get:

$1 + C = 0$
$C = -1$

Thus, $A(x) = x^2 - 1$.

Then the area on the interval $[1, 3]$ is $A(3)$.

$A(3) = 3^2 - 1$ Substituting 3 for x
$= 9 - 1$
$= 8$

5. Find the area under the curve $y = x^2$ on the interval $[0, 5]$.

$A(x) = \int x^2\,dx$
$= \frac{x^3}{3} + C$

Since we know that $A(0) = 0$ (there is no area above the number 0), we can substitute for x and $A(x)$ to determine C.

$A(0) = \frac{0^3}{3} + C = 0$ Substituting 0 for x and 0 for $A(0)$

Solving for C, we get $C = 0$:

Thus, $A(x) = \frac{x^3}{3}$.

Then the area on the interval $[0, 5]$ is $A(5)$.

$A(5) = \frac{5^3}{3}$ Substituting 5 for x
$= \frac{125}{3}$, or $41\frac{2}{3}$

7. Find the area under the curve $y = x^3$ on the interval $[0, 1]$.

$A(x) = \int x^3\,dx$
$= \frac{x^4}{4} + C$

Since we know that $A(0) = 0$, we can substitute for x and $A(x)$ to determine C.

$A(0) = \frac{0^4}{4} + C = 0$ Substituting 0 for x and 0 for $A(0)$

Solving for C, we get $C = 0$.

Thus, $A(x) = \frac{x^4}{4}$.

Then the area on the interval $[0, 1]$ is $A(1)$.

$A(1) = \frac{1^4}{4}$ Substituting 1 for x
$= \frac{1}{4}$

9. Find the area under the curve $y = 4 - x^2$ on the interval $[-2, 2]$.

$A(x) = \int (4 - x^2)\,dx$
$= 4x - \frac{x^3}{3} + C$

Since we know that $A(-2) = 0$, (there is no area above the number -2),we can substitute for x and $A(x)$ to determine C.

$$A(-2) = 4(-2) - \frac{(-2)^3}{3} + C = 0$$

Substituting -2 for x and 0 for $A(-2)$

Solving for C, we get:

$$-8 + \frac{8}{3} + C = 0$$

$$-\frac{24}{8} + \frac{8}{3} + C = 0$$

$$-\frac{16}{3} + C = 0$$

$$C = \frac{16}{3}$$

Thus, $A(x) = 4x - \frac{x^3}{3} + \frac{16}{3}$.

The area on the interval $[-2, 2]$ is $A(2)$.

$$A(2) = 4 \cdot 2 - \frac{2^3}{3} + \frac{16}{3} \quad \text{Substituting 2 for } x$$

$$= 8 - \frac{8}{3} + \frac{16}{3}$$

$$= \frac{24}{3} - \frac{8}{3} + \frac{16}{3}$$

$$= \frac{32}{3}, \text{ or } 10\frac{2}{3}$$

11. Find the area under the curve $y = e^x$ on the interval $[0, 3]$.

$$A(x) = \int e^x\,dx$$

$$= e^x + C$$

Since we know that $A(0) = 0$, (there is no area above the number 0),we can substitute for x and $A(x)$ to determine C.

$$A(0) = e^0 + C = 0 \quad \text{Substituting 0 for } x \text{ and 0 for } A(0)$$

$$1 + C = 0 \quad (e^0 = 1)$$

$$C = -1$$

Thus, $A(x) = e^x - 1$.

The area on the interval $[0, 3]$ is $A(3)$.

$$A(3) = e^3 - 1$$

$$= 20.085537 - 1 \quad \text{Using a calculator}$$

$$\approx 19.086$$

13. Find the area under the curve $y = \frac{3}{x}$ on the interval $[1, 6]$.

$$A(x) = \int \frac{3}{x}\,dx = 3\int \frac{1}{x}\,dx$$

$$= 3\ln x + C$$

Since we know that $A(1) = 0$, (there is no area above the number 1), we can substitute for x and $A(x)$ to determine C.

$$A(1) = 3\ln 1 + C = 0 \quad \text{Substituting 1 for } x \text{ and 0 for } A(1)$$

$$3 \cdot 0 + C = 0 \quad (\ln 1 = 0)$$

$$C = 0$$

Thus, $A(x) = 3\ln x$.

The area on the interval $[1, 6]$ is $A(6)$.

$$A(6) = 3\ln 6 \quad \text{Substituting 6 for } x$$

$$\approx 5.375 \quad \text{Using a calculator}$$

15. $\int_0^{1.5} (x - x^2)\,dx$

$$= \left[\frac{x^2}{2} - \frac{x^3}{3}\right]_0^{1.5}$$

$$= \left(\frac{(1.5)^2}{2} - \frac{(1.5)^3}{3}\right) - \left(\frac{0^2}{2} - \frac{0^3}{3}\right)$$

Substituting 0 for x

Substituting 1.5 for x

$$= \left(\frac{2.25}{2} - \frac{3.375}{3}\right) - (0 - 0)$$

$$= 1.125 - 1.125$$

$$= 0$$

The area above the x-axis is equal to the area below the x-axis.

17. $\int_0^{3\pi/2} \cos x\,dx$

$$= \left[\sin x\right]_0^{3\pi/2}$$

$$= \sin(3\pi/2) - \sin(0)$$

$$= -1 - 0$$

$$= -1$$

This means that the area between $[0, \frac{\pi}{2}]$ (area above the x-axis) is less than that area between $[\frac{\pi}{2}, \frac{3\pi}{2}]$ (area below the x-axis) by 1 unit of area.

19. - 35.] Left to the student

37. $\int_a^b e^t\,dt$

$$= [e^t]_a^b$$

$$= e^b - e^a$$

39. $\int_a^b 3t^2\,dt$

$$= \left[3 \cdot \frac{t^3}{3}\right]_a^b$$

$$= [t^3]_a^b$$

$$= b^3 - a^3$$

41. $\int_1^e \left(x + \frac{1}{x}\right)dx$

$$= \left[\frac{x^2}{2} + \ln x\right]_1^e$$

$$= \left(\frac{e^2}{2} + \ln e\right) - \left(\frac{1^2}{2} + \ln 1\right)$$

$$= \frac{e^2}{2} + 1 - \frac{1}{2} \quad (\ln e = 1,\ \ln 1 = 0)$$

$$= \frac{e^2}{2} + \frac{1}{2}$$

43. $\int_0^{\pi/6} \frac{5}{2} \, sin\, 2x \, dx$

$= \left[-5\,\frac{cos\, 2x}{4} \right]_0^{\pi/6}$

$= -5\, cos(\pi/6) - (-5cos(0))$

$= -\frac{5}{8} - (-\frac{5}{4})$

$= \frac{5}{8}$

45. $\int_{-4}^{1} \frac{10}{17} t^3 \, dt$

$= \frac{10}{17} \int_{-4}^{1} t^3 \, dt$

$= \frac{10}{17} \left[\frac{t^4}{4} \right]_{-4}^{1}$

$= \frac{10}{17} \left(\frac{1^4}{4} - \frac{(-4)^4}{4} \right)$

$= \frac{10}{17} \left(\frac{1}{4} - 64 \right)$

$= \frac{10}{17} \cdot \left(-\frac{255}{4} \right)$

$= -\frac{2550}{68}$

$= -\frac{1275}{34}$

47. Find the area under $y = x^3$ on $[0, 2]$.

$\int_0^2 x^3 \, dx$

$= \left[\frac{x^4}{4} \right]_0^2$

$= \frac{2^4}{4} - \frac{0^4}{4}$

$= 4 - 0$

$= 4$

49. Find the area under $y = x^2 + x + 1$ on $[2, 3]$.

$\int_2^3 (x^2 + x + 1) \, dx$

$= \left[\frac{x^3}{3} + \frac{x^2}{2} + x \right]_2^3$

$= \left(\frac{3^3}{3} + \frac{3^2}{2} + 3 \right) - \left(\frac{2^3}{3} + \frac{2^2}{2} + 2 \right)$

$= \left(9 + \frac{9}{2} + 3 \right) - \left(\frac{8}{3} + 2 + 2 \right)$

$= 12 + \frac{9}{2} - \frac{8}{3} - 4$

$= 8 + \frac{27}{6} - \frac{16}{6}$

$= 8 + \frac{11}{6}$

$= 8 + 1\frac{5}{6}$

$= 9\frac{5}{6}$

51. Find the area under $y = 5 - x^2$ on $[-1, 2]$.

$\int_{-1}^2 (5 - x^2) \, dx$

$= \left[5x - \frac{x^3}{3} \right]_{-1}^2$

$= \left(5 \cdot 2 - \frac{2^3}{3} \right) - \left[5(-1) - \frac{(-1)^3}{3} \right]$

$= \left(10 - \frac{8}{3} \right) - \left(-5 + \frac{1}{3} \right)$

$= \left(\frac{30}{3} - \frac{8}{3} \right) - \left(-\frac{15}{3} + \frac{1}{3} \right)$

$= \frac{22}{3} - \left(-\frac{14}{3} \right)$

$= \frac{22}{3} + \frac{14}{3}$

$= \frac{36}{3}$

$= 12$

53. Find the area under $y = e^x$ on $[-1, 5]$.

$\int_{-1}^5 e^x \, dx$

$= [e^x]_{-1}^5$

$= e^5 - e^{-1}$

$= e^5 - \frac{1}{e}$

55.

$$\begin{aligned} \int_2^3 \frac{x^2 - 1}{x - 1} dx &= \int_2^3 \frac{(x-1)(x+1)}{x-1} dx \\ &= \int_2^3 (x+1) \, dx \\ &= \left[\frac{1}{2} x^2 + x \right]_2^3 \\ &= \left[\frac{9}{2} + 3 \right] - [2 + 2] \\ &= \frac{7}{2} \end{aligned}$$

57.

$$\begin{aligned} \int_4^{16} (x-1)\sqrt{x} \, dx &= \int_4^{16} (x^{3/2} - x^{1/2}) \, dx \\ &= \left[\frac{2}{5} x^{5/2} - \frac{2}{3} x^{3/2} \right]_4^{16} \\ &= \left[\frac{2}{5}(16)^{5/2} - \frac{2}{3}(16)^{3/2} \right] - \\ & \quad \left[\frac{2}{5}(4)^{5/2} - \frac{2}{3}(4)^{3/2} \right] \\ &= \frac{2048}{5} - \frac{128}{3} - \frac{64}{5} + \frac{16}{3} \\ &= \frac{5392}{15} \end{aligned}$$

59.

$$\begin{aligned}\int_1^8 \frac{\sqrt[3]{x^2}-1}{\sqrt[3]{x}}dx &= \int_1^8 (x^{1/3}-x^{-1/3})\ dx\\ &= \left[\frac{3}{4}x^{4/3}-\frac{3}{2}x^{2/3}\right]_1^8\\ &= \left[\frac{3}{4}8^{4/3}-\frac{3}{2}8^{2/3}\right]-\\ &\quad \left[\frac{3}{4}1^{4/3}-\frac{3}{2}1^{2/3}\right]\\ &= 12-6-\frac{3}{4}+\frac{3}{2}\\ &= \frac{27}{4}\end{aligned}$$

61.

$$\begin{aligned}\int_1^2 (4x+3)(5x-2)dx &= \int_1^2 (20x^2+7x-6)\ dx\\ &= \left[\frac{20}{3}x^3+\frac{7}{2}x^2-6x\right]_1^2\\ &= \frac{160}{3}+14-12-\frac{20}{3}-\frac{7}{2}+6\\ &= \frac{307}{6}\end{aligned}$$

63.

$$\begin{aligned}\int_0^1 (t+1)^3 dt &= \int_0^1 (t^3+3t^2+3t+1)dt\\ &= \left[\frac{t^4}{4}+t^3+\frac{3t^2}{2}+t\right]_0^1\\ &= \frac{1}{4}+1+\frac{3}{2}+1-0\\ &= \frac{15}{4}\end{aligned}$$

65.

$$\begin{aligned}\int_1^3 \frac{t^5-t}{t^3}dt &= \int_1^3 (t^2-t^{-2})dt\\ &= \left[\frac{t^3}{3}+t^{-1}\right]_1^3\\ &= 9+\frac{1}{3}-\frac{1}{3}-1\\ &= 8\end{aligned}$$

67.

$$\begin{aligned}\int_3^5 \frac{x^2-4}{x-2}dx &= \int_3^5 \frac{(x-2)(x+2)}{x-2}dx\\ &= \int_3^5 (x+2)dx\\ &= \left[\frac{x^2}{2}+2x\right]_3^5\\ &= \frac{25}{2}+10-\frac{9}{2}-6\\ &= 12\end{aligned}$$

69. The average is given by

$$\begin{aligned}\frac{1}{1-(-1)}\int_{-1}^1 2x^3 dx &= \frac{1}{2}\int_{-1}^1 2x^3 dx\\ &= \frac{1}{2}\cdot\left[\frac{1}{2}x^4\right]_{-1}^1\\ &= \frac{1}{4}(1^4-(-1)^4)\\ &= 0\end{aligned}$$

71. The average is given by

$$\begin{aligned}\frac{1}{1-0}\int_0^1 e^x dx &= \int_0^1 e^x dx\\ &= \left[e^x\right]_0^1\\ &= e-1\end{aligned}$$

73. The average is given by

$$\begin{aligned}\frac{1}{2-0}\int_0^2 (x^2-x+1)dx &= \frac{1}{2}\int_0^2 (x^2-x+1)dx\\ &= \frac{1}{2}\left[\frac{x^3}{3}-\frac{x^2}{2}+x\right]_0^2\\ &= \frac{1}{2}\left(\frac{8}{3}-2+2-0\right)\\ &= \frac{4}{3}\end{aligned}$$

75. The average is given by

$$\begin{aligned}\frac{1}{6-2}\int_0^2 (3x+1)dx &= \frac{1}{4}\int_2^6 (3x+1)dx\\ &= \frac{1}{4}\left[\frac{3x^2}{2}+x\right]_2^6\\ &= \frac{1}{4}(54+6-6-2)\\ &= 13\end{aligned}$$

77. The average is given by

$$\begin{aligned}\frac{1}{1-0}\int_0^1 x^n dx &= \int_0^1 x^n dx\\ &= \left[\frac{x^{n+1}}{n+1}\right]_0^1\\ &= \frac{1}{n+1}-0\\ &= \frac{1}{n+1}\end{aligned}$$

79. The distance is given by

$$\begin{aligned}\int_1^5 (3t^2+2t)dt &= \left[t^3+t^2\right]_1^5\\ &= 125+25-1-1\\ &= 148\end{aligned}$$

81. The population increase is given by

$$\begin{aligned}\int_0^2 200e^{-t}dt &= \left[-200e^{-t}\right]_0^2 \\ &= -200e^{-2}+200 \\ &= 200\left(1-\frac{1}{e^2}\right)\end{aligned}$$

83. The population decrease is given by

$$\begin{aligned}\int_0^{10}(-500(20-t)dt &= -500\left[20t-\frac{t^2}{2}\right]_0^{10} \\ &= -500(200-50) \\ &= -75000\end{aligned}$$

85. The work done is given by

$$\begin{aligned}&\int_2^{11} 71.3x-4.15x^2+0.434x^3\,dx \\ =& \left[35.65x^2-1.383x^3+.1085x^4\right]_2^{11} \\ =& [4061-133.27] \\ =& 3927.73 \ N\cdot mm\end{aligned}$$

87. **a)** The initial dosage is 42.03 $\mu g/ml$

b) The aberage amoount is given by

$$\begin{aligned}&\frac{1}{120-10}\int_{10}^{120} 42.03e^{-0.01050t}\,dt \\ =& \frac{1}{110}\left[\frac{42.03}{-0.01050}e^{-0.01050t}\right]_{10}^{120} \\ =& \frac{[-1135-(-3604)]}{110} \\ =& 22.45 \ \mu g/ml\end{aligned}$$

89. **a)**

$$\begin{aligned}\int_0^b 2\pi r\,dr &= 2\pi\left[\frac{r^2}{2}\right]_0^b \\ &= 2\pi\frac{b^2}{2} \\ &= \pi b^2\end{aligned}$$

b) The circle could be thought of as a rectangle with length πb and width b

91. The lower limit of the integral, $x=1$, was not evaluated

Exercise Set 5.4

1. First graph the system of equations and shade the region bounded by the graphs.

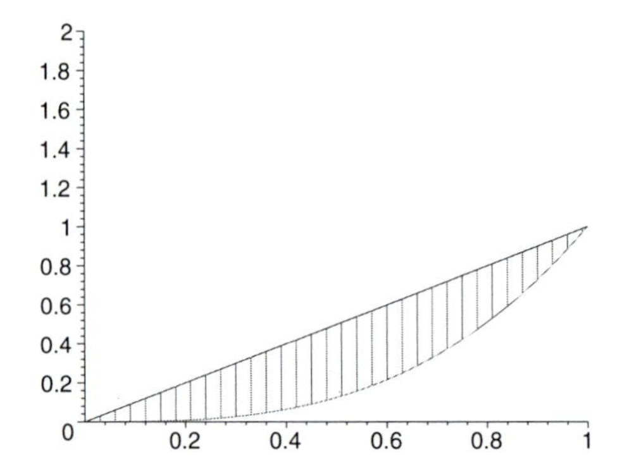

Here the boundaries are easily determined by looking at the graph. Note which is the upper graph. Here it is $x \geq x^3$ over the interval $[0,1]$.

Compute the area as follows:

$$\begin{aligned}&\int_0^1 (x-x^3)\,dx \\ =& \left[\frac{x^2}{2}-\frac{x^4}{4}\right]_0^1 \\ =& \left(\frac{1^2}{2}-\frac{1^4}{4}\right)-\left(\frac{0^2}{2}-\frac{0^4}{4}\right) \\ =& \frac{1}{2}-\frac{1}{4} \\ =& \frac{1}{4}\end{aligned}$$

3. First graph the system of equations and shade the region bounded by the graphs.

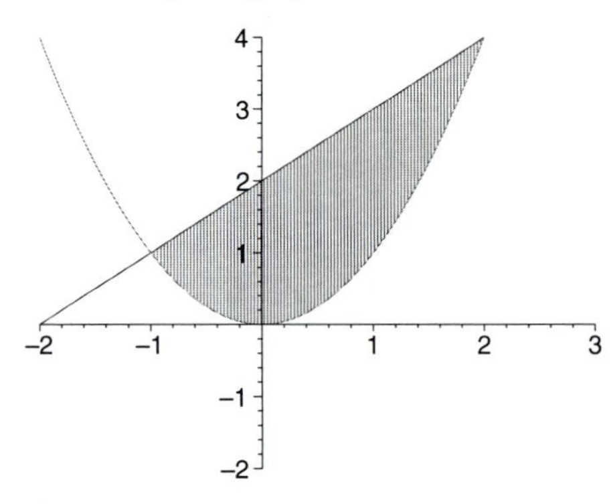

Here the boundaries are easily determined by the graph. Note which is the upper graph. Here it is $(x+2) \geq x^2$ over the interval $[-1,2]$.

Compute the area as follows:

$$\int_{-1}^{2}[(x+2)-x^2]\,dx$$
$$=\int_{-1}^{2}(-x^2+x+2)\,dx$$
$$=\left[-\frac{x^3}{3}+\frac{x^2}{2}+2x\right]_{-1}^{2}$$
$$=\left[-\frac{2^3}{3}+\frac{2^2}{2}+2\cdot 2\right]-\left[-\frac{(-1)^3}{3}+\frac{(-1)^2}{2}+2(-1)\right]$$
$$=\left(-\frac{8}{3}+2+4\right)-\left(\frac{1}{3}+\frac{1}{2}-2\right)$$
$$=-\frac{8}{3}+2+4-\frac{1}{3}-\frac{1}{2}+2$$
$$=-\frac{9}{3}-\frac{1}{2}+2+4+2=4\frac{1}{2},\text{ or }\frac{9}{2}$$

5. First graph the system of equations and shade the region bounded by the graphs.

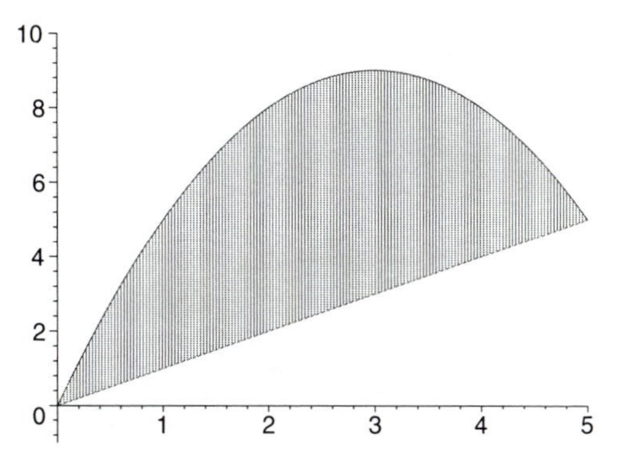

Here the boundaries are easily determined by the graph. Note which is the upper graph. Here it is $(6x-x^2)\geq x$ over the interval $[0,5]$.

Compute the area as follows:

$$\int_{0}^{5}[(6x-x^2)-x]\,dx$$
$$=\int_{0}^{5}(-x^2+5x)\,dx$$
$$=\left[-\frac{x^3}{3}+5\cdot\frac{x^2}{2}\right]_{0}^{5}$$
$$=\left(-\frac{5^3}{3}+5\cdot\frac{5^2}{2}\right)-\left(-\frac{0^3}{3}+5\cdot\frac{0^2}{2}\right)$$
$$=\left(-\frac{125}{3}+\frac{125}{2}\right)-0$$
$$=-\frac{250}{6}+\frac{375}{6}$$
$$=\frac{125}{6}$$

7. First graph the system of equations and shade the region bounded by the graphs.

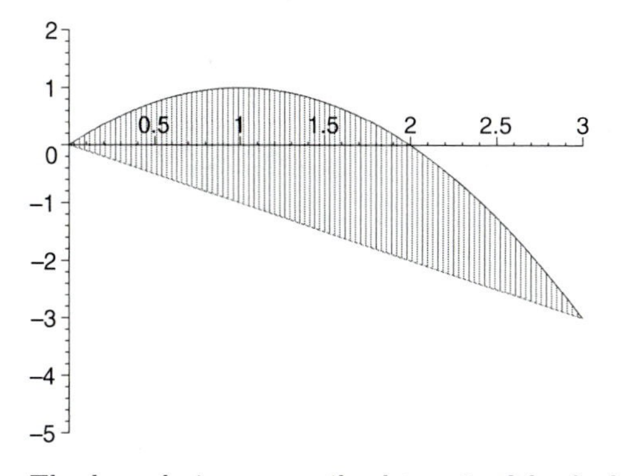

The boundaries are easily determined by looking at the graph. Note which is the upper graph over the shaded region. Here it is $(2x-x^2)\geq -x$ over the interval $[0,3]$.

Compute the area as follows:

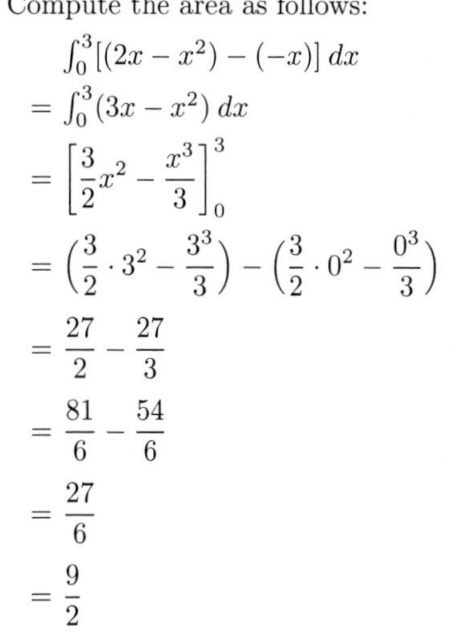

$$\int_{0}^{3}[(2x-x^2)-(-x)]\,dx$$
$$=\int_{0}^{3}(3x-x^2)\,dx$$
$$=\left[\frac{3}{2}x^2-\frac{x^3}{3}\right]_{0}^{3}$$
$$=\left(\frac{3}{2}\cdot 3^2-\frac{3^3}{3}\right)-\left(\frac{3}{2}\cdot 0^2-\frac{0^3}{3}\right)$$
$$=\frac{27}{2}-\frac{27}{3}$$
$$=\frac{81}{6}-\frac{54}{6}$$
$$=\frac{27}{6}$$
$$=\frac{9}{2}$$

9. First graph the system of equations and shade the region bounded by the graphs.

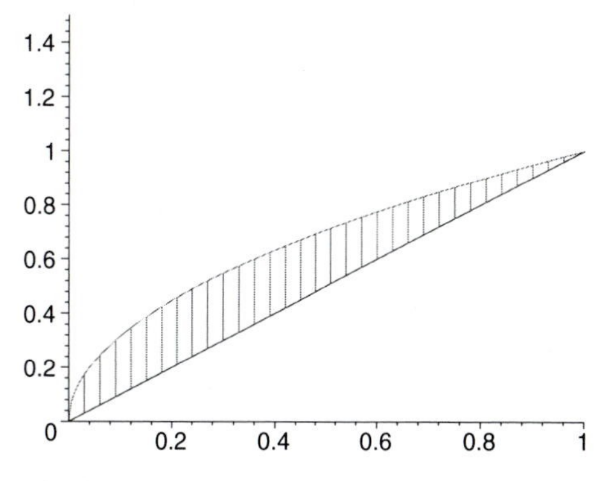

The boundaries are easily determined by looking at the graph. Note which is the upper graph over the shaded region. Here it is $\sqrt[4]{x}\geq x$ over the interval $[0,1]$.

Compute the area as follows:

$\int_0^1 (\sqrt[4]{x} - x)\, dx$

$= \int_0^1 (x^{1/4} - x)\, dx$

$= \left[\frac{4}{5}x^{5/4} - \frac{1}{2}x^2\right]_0^1$

$= \left(\frac{4}{5} \cdot 1^{5/4} - \frac{1}{2} \cdot 1^2\right) - \left(\frac{4}{5} \cdot 0^{5/4} - \frac{1}{2} \cdot 0^2\right)$

$= \frac{4}{5} - \frac{1}{2} - 0$

$= \frac{8}{10} - \frac{5}{10}$

$= \frac{3}{10}$

11. Graph the system of equations and shade the region bounded by the graphs.

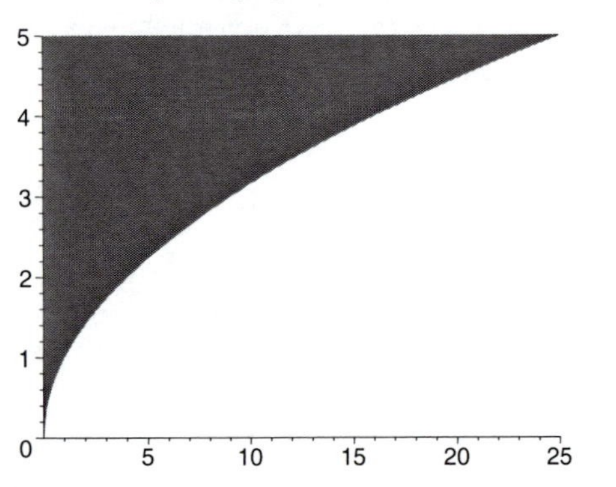

The boundaries are easily determined by looking at the graph. Here $5 \geq \sqrt{x}$ over the interval $[0, 25]$.

Compute the area as follows:

$\int_0^{25} (5 - \sqrt{x})\, dx$

$= \int_0^{25} (5 - x^{1/2})\, dx$

$= \left[5x - \frac{x^{3/2}}{3/2}\right]_0^{25}$

$= \left[5x - \frac{2}{3}x^{3/2}\right]_0^{25}$

$= \left(5 \cdot 25 - \frac{2}{3} \cdot 25^{3/2}\right) - \left(5 \cdot 0 - \frac{2}{3} \cdot 0^{3/2}\right)$

$= 125 - \frac{250}{3} - 0 \qquad [25^{3/2} = (5^2)^{3/2} = 5^3 = 125]$

$= \frac{375}{3} - \frac{250}{3}$

$= \frac{125}{3}$, or $41\frac{2}{3}$

13. First graph the system of equations and shade the region bounded by the graphs.

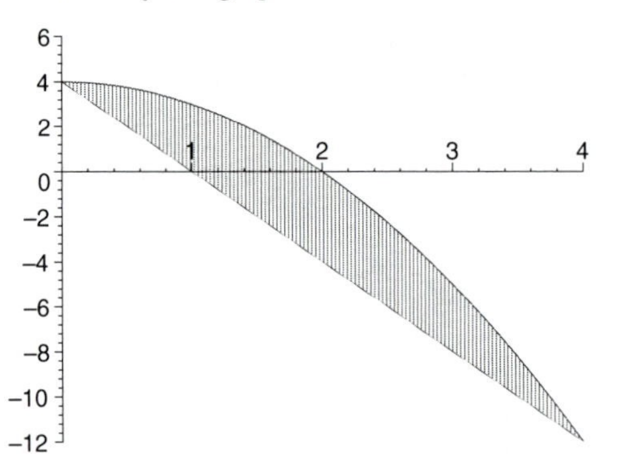

Then determine the first coordinates of possible points of intersection by solving a system of equations as follows. At the points of intersection, $y = 4 - x^2$ and $y = 4 - 4x$, so

$4 - x^2 = 4 - 4x$

$0 = x^2 - 4x$

$0 = x(x - 4)$

$x = 0$ *or* $x = 4$

Thus the interval with which we are concerned is $[0, 4]$. Note that $4 - x^2 \geq 4 - 4x$ over the interval $[0, 4]$.

Compute the area as follows:

$\int_0^4 [(4 - x^2) - (4 - 4x)]\, dx$

$= \int_0^4 (-x^2 + 4x)\, dx$

$= \left[-\frac{x^3}{3} + 2x^2\right]_0^4$

$= \left(-\frac{4^3}{3} + 2 \cdot 4^2\right) - \left(-\frac{0^3}{3} + 2 \cdot 0^2\right)$

$= -\frac{64}{3} + 32$

$= -\frac{64}{3} + \frac{96}{3}$

$= \frac{32}{3}$

15. First graph the system of equations and shade the region bounded by the graphs.

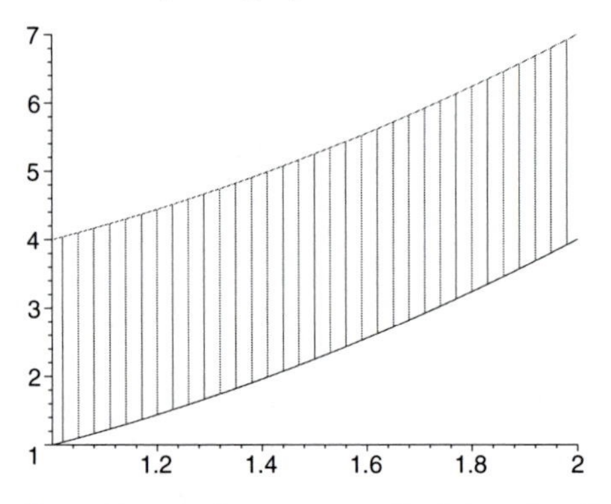

From the graph we can easily determine the interval with which we are concerned. Here $x^2 + 3 \geq x^2$ over the interval $[1, 2]$.

Compute the area as follows:

$\int_1^2 [(x^2+3) - x^2]\, dx$

$= \int_1^2 3\, dx$

$= \big[3x\big]_1^2$

$= 3 \cdot 2 - 3 \cdot 1$

$= 6 - 3$

$= 3$

17. First graph the system of equations and shade the region bounded by the graphs.

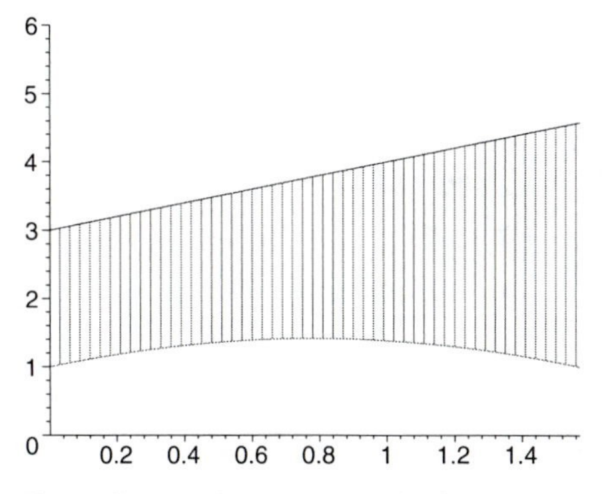

From the graph we can easily determine the interval with which we are concerned. Here $x + 3 \geq \sin x + \cos x$ over the interval $[0, \frac{\pi}{2}]$.

Compute the area as follows:

$\int_0^{\frac{\pi}{2}} [x + 3 - \sin x - \cos x]\, dx$

$= \left[\frac{1}{2}x^2 + 3x + \cos x - \sin x\right]_0^{\frac{\pi}{2}}$

$= \left[\frac{\pi^2}{8} + \frac{3\pi}{2} - 1\right] - [1]$

$= \frac{\pi^2}{8} + \frac{3\pi}{2} - 2$

19. First graph the system of equations and shade the region bounded by the graphs.

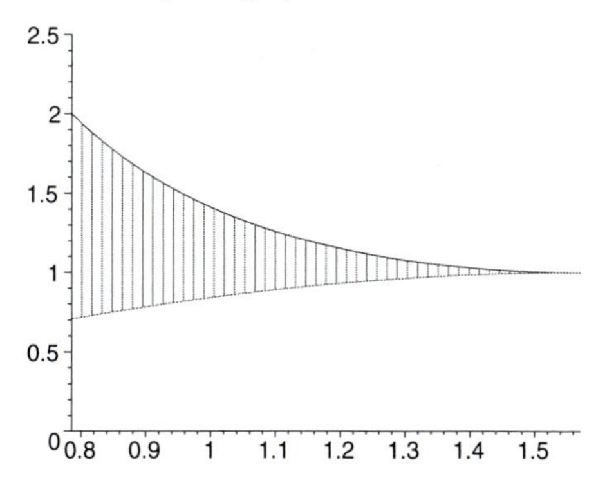

From the graph we can easily determine the interval with which we are concerned. Here $x + 3 \geq \sin x + \cos x$ over the interval $[\frac{\pi}{4}, \frac{\pi}{2}]$.

Compute the area as follows:

$\int_{\frac{\pi}{4}}^{\frac{\pi}{2}} [\csc^2 x - \sin x]\, dx$

$= \left[-\cot x + \cos x\right]_{\frac{\pi}{4}}^{\frac{\pi}{2}}$

$= [0] - \left[-1 + \frac{1}{\sqrt{2}}\right]$

$= 1 - \frac{1}{\sqrt{2}}$

21. First graph the system of equations and shade the region bounded by the graphs.

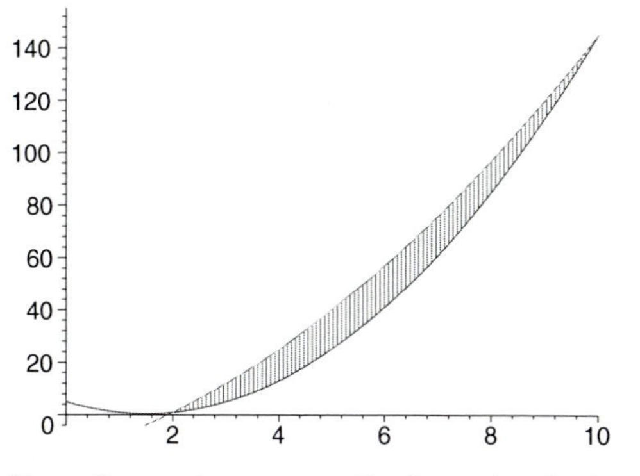

From the graph we can easily determine the interval with which we are concerned. Here $x^2 + 6x - 15 \geq 2x^2 - 6x + 5$ over the interval $[2, 10]$.

Compute the area as follows:

$\int_2^{10} [x^2 + 6x - 15 - 2x^2 + 6x - 5]\, dx$

$= \int_2^{10} [-x^2 + 12x - 20]\, dx$

$= \left[-\frac{1}{3}x^3 + 6x^2 - 20x\right]_2^{10}$

$= \left[-\frac{1000}{3} + 600 - 200\right] - \left[-\frac{8}{3} + 24 - 40\right]$

$= \frac{256}{3}$

23. $f(x) \geq g(x)$ on $[-5, -1]$, and $g(x) \geq f(x)$ on $[-1, 3]$. We use two integrals to find the total area.

$\int_{-5}^{-1} [(x^3 + 3x^2 - 9x - 12) - (4x + 3)]\, dx +$

$\int_{-1}^{3} [(4x + 3) - (x^3 + 3x^2 - 9x - 12)]\, dx$

$= \int_{-5}^{-1} (x^3 + 3x^2 - 13x - 15)\, dx +$

$\int_{-1}^{3} (-x^3 - 3x^2 + 13x + 15)\, dx$

$= \left[\frac{x^4}{4} + x^3 - \frac{13x^2}{2} - 15x\right]_{-5}^{-1} +$

$\left[-\frac{x^4}{4} - x^3 + \frac{13x^2}{2} + 15x\right]_{-1}^{3}$

$= \left[\frac{(-1)^4}{4} + (-1)^3 - \frac{13(-1)^2}{2} - 15(-1)\right] -$

$\left[\frac{(-5)^4}{4} + (-5)^3 - \frac{13(-5)^2}{2} - 15(-5)\right] +$

$\left[-\frac{3^4}{4} - 3^3 + \frac{13(3)^2}{2} + 15 \cdot 3\right] -$

$\left[-\frac{(-1)^4}{4} - (-1)^3 + \frac{13(-1)^2}{2} + 15(-1)\right]$

$= \left(\frac{1}{4} - 1 - \frac{13}{2} + 15\right) - \left(\frac{625}{4} - 125 - \frac{325}{2} + 75\right) +$

$\left(-\frac{81}{4} - 27 + \frac{117}{2} + 45\right) - \left(-\frac{1}{4} + 1 + \frac{13}{2} - 15\right)$

$= \frac{31}{4} + \frac{225}{4} + \frac{225}{4} + \frac{31}{4}$

$= 128$

25. $f(x) \geq g(x)$ on $[1, 4]$. We find the area.

$$\int_1^4 [(4x - x^2) - (x^2 - 6x + 8)]\, dx$$
$$= \int_1^4 (-2x^2 + 10x - 8)\, dx$$
$$= \left[-\frac{2x^3}{3} + 5x^2 - 8x \right]_1^4$$
$$= \Big(-\frac{2 \cdot 4^3}{3} + 5 \cdot 4^2 - 8 \cdot 4 \Big) - \Big(-\frac{2 \cdot 1^3}{3} + 5 \cdot 1^2 - 8 \cdot 1 \Big)$$
$$= \Big(-\frac{128}{3} + 80 - 32 \Big) - \Big(-\frac{2}{3} + 5 - 8 \Big)$$
$$= \frac{16}{3} + \frac{11}{3} = \frac{27}{3}$$
$$= 9$$

27. Find the area under

$$f(x) = \begin{cases} 4 - x^2, & \text{if } x < 0 \\ 4, & \text{if } x \geq 0 \end{cases} \quad \text{on } [-2, 3]$$

We have to break the integral into two parts in order to complete this problem

$$\int_{-2}^3 f(x)\, dx$$
$$= \int_{-2}^0 f(x)\, dx + \int_0^3 f(x)\, dx$$
$$= \int_{-2}^0 (4 - x^2)\, dx + \int_0^3 4\, dx$$
$$= \left[4x - \frac{x^3}{3} \right]_{-2}^0 + [4x]_0^3$$
$$= \left\{ \left[4 \cdot 0 - \frac{0^3}{3} \right] - \left[4(-2) - \frac{(-2)^3}{3} \right] \right\} + (4 \cdot 3 - 4 \cdot 0)$$
$$= (0 - 0) - \Big(-8 + \frac{8}{3} \Big) + (12 - 0)$$
$$= -\Big(-\frac{24}{3} + \frac{8}{3} \Big) + 12$$
$$= -\Big(-\frac{16}{3} \Big) + 12$$
$$= \frac{16}{3} + \frac{36}{3}$$
$$= \frac{52}{3}, \text{ or } 17\frac{1}{3}$$

29. First graph the system of equations and shade the region bounded by the graph.

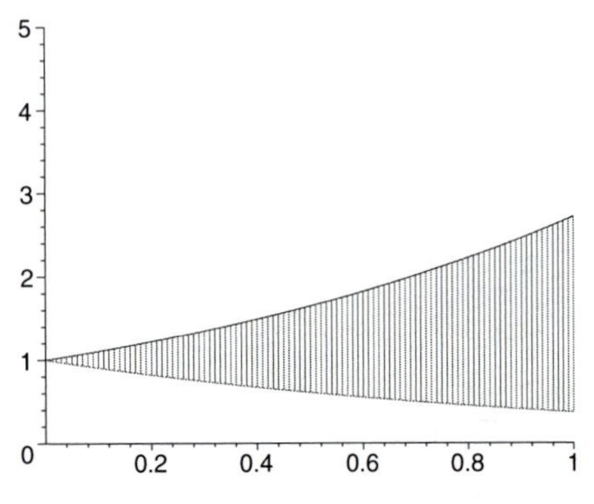

From the graph we can easily determine the interval with which we are concerned. Here $e^x \geq e^{-x}$ over the interval $[0, 1]$.

Compute the area as follows:

$$\int_0^1 (e^x - e^{-x})\, dx$$
$$= \left[e^x + e^{-x} \right]_0^1$$
$$= (e^1 + e^{-1}) - (e^0 + e^{-0})$$
$$= \Big(e + \frac{1}{e} \Big) - (1 + 1)$$
$$= e + \frac{1}{e} - 2$$
$$= \frac{e^2 - 2e + 1}{e}$$
$$= \frac{(e-1)^2}{e} \approx 1.086$$

31. First graph the system of equations and shade the region bounded by the graph.

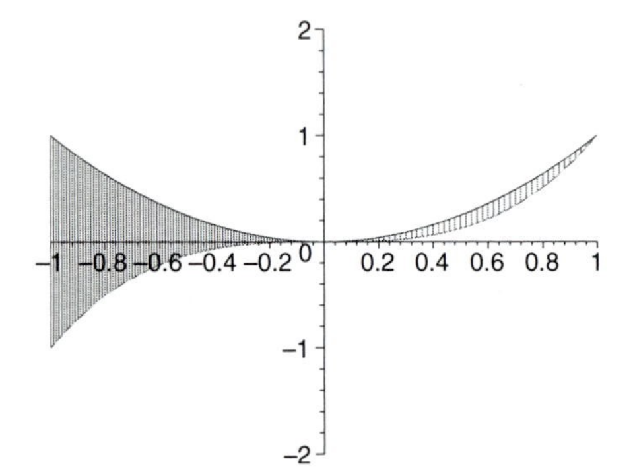

From the graph we can easily determine the interval with which we are concerned. Here $x^2 \geq x^3$ over the interval $[-1, 1]$.

Compute the area as follows:

$$\int_{-1}^1 (x^2 - x^3)\, dx$$
$$= \left[\frac{x^3}{3} - \frac{x^4}{4} \right]_{-1}^1$$
$$= \Big(\frac{1^3}{3} - \frac{1^4}{4} \Big) - \Big(\frac{(-1)^3}{3} - \frac{(-1)^4}{4} \Big)$$
$$= \Big(\frac{1}{3} - \frac{1}{4} \Big) - \Big(-\frac{1}{3} - \frac{1}{4} \Big)$$
$$= \frac{1}{3} - \frac{1}{4} + \frac{1}{3} + \frac{1}{4} = \frac{2}{3}$$

33. On $[0, 11]$, $f(x) \geq g(x)$.

$$\int_0^{11} [f(x) - g(x)]\, dx$$
$$= \int_0^{11} [0.3x + 6.15x^2 - 0.552x^3]\, dx$$
$$= \left[0.15x^2 + \frac{14.45}{3} x^3 - 0.138x^4 \right]_0^{11}$$
$$= [726.24 - 0]$$
$$= 726.24 \;\; N \cdot mm$$

35. $T(t) = 25 + 3e^{-t} - 20(t - 0.5)^2$, $f(t) = 8$

$$\int_0^1 [T(t) - f(t)]\, dt$$

$$= \int_0^1 [17 + 3e^{-t} - 20(t-0.5)^2]\, dt$$

$$= \left[17t - 3e^{-t} - \frac{20}{3}(t-0.5)^3\right]_0^1$$

$$= [17 - \tfrac{3}{e} - \tfrac{5}{6} - [-3 + \tfrac{5}{6}]$$

$= 17.23$ degree days

37.

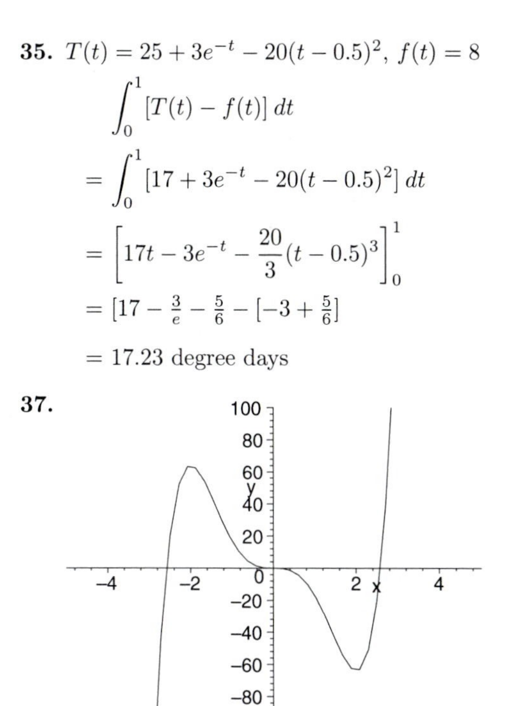

From the graph we see that the first coordinate of the relative maximum is $x = -2$ and the first coordinate of the relative minimum is $x = 2$

$$\int_{-2}^0 [3x^5 - 20x^3 - 0]\, dx + \int_0^2 [0 - 3x^5 + 20x^3]\, dx$$

$$= \left[\frac{1}{2}x^6 - 5x^4\right]_{-2}^0 + \left[-\frac{1}{2}x^6 + 5x^4\right]_0^2$$

$$= [-32 + 80] - [32 - 80]$$

$$= 96$$

39. $V = \dfrac{p}{4Lv}(R^2 - r^2)$

$$\begin{aligned} Q &= \int_0^R [2\pi \cdot \frac{p}{4Lv}(R^2 - r^2)r]\, dr \\ &= \frac{\pi p}{2Lv}\int_0^R [R^2 r - r^3]\, dr \\ &= \frac{\pi p}{2Lv}\left[\frac{R^2}{2}r^2 - \frac{1}{4}r^4\right]_0^R \\ &= \frac{\pi p}{2Lv}\left[\frac{R^4}{2} - \frac{1}{4}R^4\right] \\ &= \frac{\pi p R^4}{8Lv} \end{aligned}$$

41. The area bounded in the graph is 24.961

43. The area bounded in the graph is 416.708

45. a)

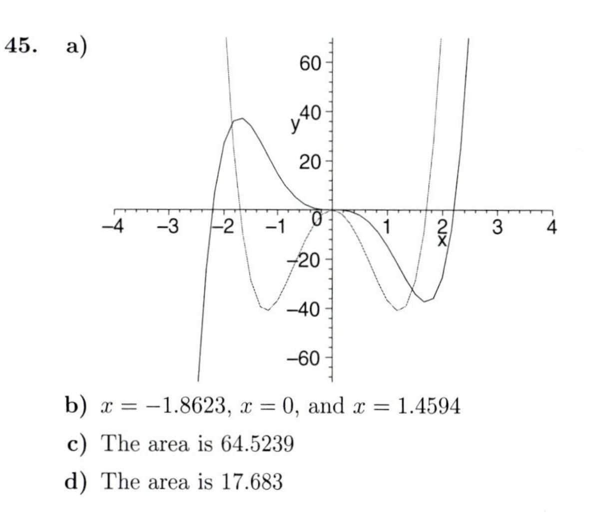

b) $x = -1.8623$, $x = 0$, and $x = 1.4594$

c) The area is 64.5239

d) The area is 17.683

Exercise Set 5.5

1. $\displaystyle\int \frac{3x^2\, dx}{7 + x^3}$

Let $u = 7 + x^3$, then $du = 3x^2\, dx$.

$= \displaystyle\int \frac{du}{u}$ Substituting u for $7 + x^3$ and du for $3x^2\, dx$

$= \displaystyle\int \frac{1}{u}\, du$

$= \ln u + C$ Using Formula C

$= \ln(7 + x^3) + C$

3. $\int e^{4x}\, dx$

Let $u = 4x$, then $du = 4\, dx$.

We do not have $4\, dx$. We only have dx and need to supply a 4. We do this by multiplying by $\frac{1}{4} \cdot 4$ as follows.

$\displaystyle\frac{1}{4} \cdot 4\int e^{4x}\, dx$ Multiplying by 1

$= \displaystyle\frac{1}{4}\int 4e^{4x}\, dx$

$= \displaystyle\frac{1}{4}\int e^{4x}(4\, dx)$

$= \displaystyle\frac{1}{4}\int e^u\, du$ Substituting u for $4x$ and du for $4\, dx$

$= \displaystyle\frac{1}{4}e^u + C$ Using Formula B

$= \displaystyle\frac{1}{4}e^{4x} + C$

5. $\int e^{x/2}\, dx = \int e^{(1/2)x}\, dx$

Let $u = \frac{1}{2}x$, then $du = \frac{1}{2}\, dx$.

We do not have $\frac{1}{2}\, dx$. We only have dx and need to supply a $\frac{1}{2}$ by multiplying by $2 \cdot \frac{1}{2}$ as follows.

$2 \cdot \frac{1}{2} \int e^{x/2}\, dx$ Multiplying by 1

$= 2 \int \frac{1}{2} e^{x/2}\, dx$

$= 2 \int e^{x/2} \left(\frac{1}{2}\, dx\right)$

$= 2 \int e^u\, du$ Substituting u for $x/2$ and du for $\frac{1}{2}\, dx$

$= 2\, e^u + C$ Using Formula B

$= 2\, e^{x/2} + C$

7. $\int x^3\, e^{x^4}\, dx$

Let $u = x^4$, then $du = 4x^3\, dx$.

We do not have $4x^3\, dx$. We only have $x^3\, dx$ and need to supply a 4. We do this by multiplying by $\frac{1}{4} \cdot 4$ as follows.

$\frac{1}{4} \cdot 4 \int x^3\, e^{x^4}\, dx$ Multiplying by 1

$= \frac{1}{4} \int 4x^3\, e^{x^4}\, dx$

$= \frac{1}{4} \int e^{x^4} (4x^3\, dx)$

$= \frac{1}{4} \int e^u\, du$ Substituting u for x^4 and du for $4x^3\, dx$

$= \frac{1}{4} \cdot e^u + C$ Using Formula B

$= \frac{1}{4} e^{x^4} + C$

9. $\int t^2\, e^{-t^3}\, dt$

Let $u = -t^3$, then $du = -3t^2\, dt$.

We do not have $-3t^2\, dt$. We only have $t^2\, dt$. We need to supply a -3 by multiplying by $-\frac{1}{3} \cdot (-3)$ as follows.

$-\frac{1}{3} \cdot (-3) \int t^2\, e^{-t^3}\, dt$ Multiplying by 1

$= -\frac{1}{3} \int -3t^2\, e^{-t^3}\, dt$

$= -\frac{1}{3} \int e^{-t^3} (-3t^2\, dt)$

$= -\frac{1}{3} \int e^u\, du$ Substituting u for $-t^3$ and du for $-3t^2\, dt$

$= -\frac{1}{3} e^u + C$ Using Formula B

$= -\frac{1}{3} e^{-t^3} + C$

11. $\int \frac{\ln 4x\, dx}{x}$

Let $u = \ln 4x$, then $du = \frac{1}{x}\, dx$.

$= \int \ln 4x \left(\frac{1}{x}\, dx\right)$

$= \int u\, du$ Substituting u for $\ln 4x$ and du for $\frac{1}{x}\, dx$

$= \frac{u^2}{2} + C$ Using Formula A

$= \frac{(\ln 4x)^2}{2} + C$

13. $\int \frac{dx}{1+x}$

Let $u = 1 + x$, then $du = dx$.

$= \int \frac{du}{u}$ Substituting u for $1 + x$ and du for dx

$= \int \frac{1}{u}\, du$

$= \ln u + C$ Using Formula C

$= \ln (1 + x) + C$

15. $\int 3\, sin(4x + 2)\, dx$

Let $u = 4x + 2$, then $du = 4\, dx$.

$= \frac{3}{4} \int sin(u)\, du$ Substituting u for $4x + 2$ and $\frac{1}{4} du$ for dx

$= -\frac{3}{4} cos(u) + C$

$= -\frac{3}{4} cos(4x + 2) + C$

17. $\int csc(2x + 3)\, cot(2x + 3)\, dx$

Let $u = 2x + 3$, then $du = 2\, dx$.

$= \frac{1}{2} \int csc(u)\, cot(u)\, du$ Substituting u for $2x + 3$ and $\frac{1}{2} du$ for dx

$= -csc(u) + C$

$= -csc(2x + 3) + C$

19. $\int \frac{dx}{4 - x}$

Let $u = 4 - x$, then $du = -dx$.

We do not have $-dx$. We only have dx and need to supply a -1 by multiplying by $-1 \cdot (-1)$ as follows.

$$-1 \cdot (-1) \int \frac{dx}{4-x} \quad \text{Multiplying by 1}$$

$$= -1 \int -1 \cdot \frac{dx}{4-x}$$

$$= -\int \frac{1}{4-x}(-dx)$$

$$= -\int \frac{1}{u}\,du \quad \text{Substituting } u \text{ for } 4-x \text{ and } du \text{ for } -dx$$

$$= -\ln u + C \quad \text{Using Formula C}$$

$$= -\ln(4-x) + C$$

21. $\int t^2(t^3-1)^7\,dt$

Let $u = t^3 - 1$, then $du = 3t^2\,dt$.

We do not have $3t^2\,dt$. We only have $t^2\,dt$. We need to supply a 3 by multiplying by $\frac{1}{3} \cdot 3$ as follows.

$$\frac{1}{3} \cdot 3 \int t^2(t^3-1)^7\,dt \quad \text{Multiplying by 1}$$

$$= \frac{1}{3} \int 3t^2(t^3-1)^7\,dt$$

$$= \frac{1}{3} \int (t^3-1)^7\,3t^2\,dt$$

$$= \frac{1}{3} \int u^7\,du \quad \text{Substituting } u \text{ for } t^3-1 \text{ and } du \text{ for } 3t^2\,dt$$

$$= \frac{1}{3} \cdot \frac{u^8}{8} + C \quad \text{Using Formula A}$$

$$= \frac{1}{24}(t^3-1)^8 + C$$

23. $\int x\, sin\, x^2\,dx$

Let $u = x^2$, then $du = 2x\,dx$.

$$= \frac{1}{2} \int sin\; u\,du$$

$$= -\frac{1}{2} cos\; u + C$$

$$= -\frac{1}{2} cos\; x^2 + C$$

25. $\int (x+1)\, sec^2(x^2+2x+3)\,dx$

Let $u = x^2 + 2x + 3$, then $du = 2x + 2\,dx$.

$$= \frac{1}{2} \int sec^2\; u\,du$$

$$= \frac{1}{2} tan(u) + C$$

$$= \frac{1}{2} tan(x^2+2x+3) + C$$

27. $\int \frac{e^x\,dx}{4+e^x}$

Let $u = 4 + e^x$, then $du = e^x\,dx$.

$$= \int \frac{du}{u} \quad \text{Substituting } u \text{ for } 4+e^x \text{ and } du \text{ for } e^x\,dx$$

$$= \int \frac{1}{u}\,du$$

$$= \ln u + C \quad \text{Using Formula C}$$

$$= \ln(4+e^x) + C$$

29. $\int \frac{\ln x^2}{x}\,dx$

Let $u = \ln x^2$, then $du = \left(2x \cdot \frac{1}{x^2}\right) dx = \frac{2}{x}\,dx$.

We do not have $\frac{2}{x}\,dx$. We only have $\frac{1}{x}\,dx$ and need to supply a 2 by multiplying by $\frac{1}{2} \cdot 2$ as follows.

$$\frac{1}{2} \cdot 2 \int \frac{\ln x^2}{x}\,dx \quad \text{Multiplying by 1}$$

$$= \frac{1}{2} \int 2 \cdot \frac{\ln x^2}{x}\,dx$$

$$= \frac{1}{2} \int \ln x^2 \cdot \frac{2}{x}\,dx$$

$$= \frac{1}{2} \int u\,du \quad \text{Substituting } u \text{ for } \ln x^2 \text{ and } du \text{ for } \frac{2}{x}\,dx$$

$$= \frac{1}{2} \cdot \frac{u^2}{2} + C \quad \text{Using Formula A}$$

$$= \frac{u^2}{4} + C$$

$$= \frac{1}{4}(\ln x^2)^2 + C,$$

$$\text{or } \frac{1}{4}(2\ln x)^2 + C = \frac{1}{4} \cdot 4(\ln x)^2 + C = (\ln x)^2 + C$$

31. $\int \frac{dx}{x \ln x}$

Let $u = \ln x$, then $du = \frac{1}{x}\,dx$.

$$= \int \frac{1}{\ln x}\left(\frac{1}{x}\,dx\right)$$

$$= \int \frac{1}{u}\,du \quad \text{Substituting } u \text{ for } \ln x \text{ and } du \text{ for } \frac{1}{x}\,dx$$

$$= \ln u + C \quad \text{Using Formula C}$$

$$= \ln(\ln x) + C$$

33. $\int \sqrt{ax+b}\,dx$, or $\int (ax+b)^{1/2}\,dx$

Let $u = ax + b$, then $du = a\,dx$.

We do not have $a\,dx$. We only have dx and need to supply an a by multiplying by $\frac{1}{a} \cdot a$ as follows.

$$\frac{1}{a}\cdot a\int\sqrt{ax+b}\,dx \quad \text{Multiplying by 1}$$
$$=\frac{1}{a}\int a\sqrt{ax+b}\,dx$$
$$=\frac{1}{a}\int\sqrt{ax+b}\,(a\,dx)$$
$$=\frac{1}{a}\int\sqrt{u}\,du \quad \text{Substituting } u \text{ for } ax+b \text{ and } du \text{ for } a\,dx$$
$$=\frac{1}{a}\int u^{1/2}\,du$$
$$=\frac{1}{a}\cdot\frac{u^{3/2}}{\frac{3}{2}}+C \quad \text{Using Formula A}$$
$$=\frac{2}{3a}\cdot u^{3/2}+C$$
$$=\frac{2}{3a}(ax+b)^{3/2}+C$$

35. $\int b\,e^{ax}\,dx$

$= b\int e^{ax}\,dx$

Let $u = ax$, then $du = a\,dx$.

We do not have $a\,dx$. We only have dx and need to supply an a by multiplying by $\frac{1}{a}\cdot a$ as follows.

$$=b\cdot\frac{1}{a}\cdot a\int e^{ax}\,dx$$
$$=\frac{b}{a}\int a\,e^{ax}\,dx$$
$$=\frac{b}{a}\int e^{ax}\,(a\,dx)$$
$$=\frac{b}{a}\int e^{u}\,du \quad \text{Substituting } u \text{ for } ax \text{ and } du \text{ for } a\,dx$$
$$=\frac{b}{a}e^{u}+C \quad \text{Using Formula B}$$
$$=\frac{b}{a}e^{ax}+C$$

37. $\int a\,sin(bx+c)\,dx$

Let $u = bx+c$, then $du = b\,dx$.

$$=\frac{1}{b}\int sin\,u\,du$$
$$=-\frac{1}{b}\,cos(u)+D$$
$$=-\frac{1}{b}\,cos(bx+c)+D$$

39. $\int\frac{3x^2\,dx}{(1+x^3)^5}$

Let $u = 1+x^3$, then $du = 3x^2\,dx$.

$$=\int\frac{1}{(1+x^3)^5}\cdot 3x^2\,dx$$
$$=\int\frac{1}{u^5}\,du \quad \text{Substituting } u \text{ for } 1+x^3 \text{ and } du \text{ for } 3x^2\,dx$$
$$=\int u^{-5}\,du$$
$$=\frac{u^{-4}}{-4}+C \quad \text{Using Formula A}$$
$$=-\frac{1}{4u^4}+C$$
$$=-\frac{1}{4(1+x^3)^4}+C$$

41. $\int cot\,x\,dx = \int\frac{cos\,x}{sin\,x}\,dx$

Let $u = sin\,x$, then $du = cos\,x\,dx$.

$$=\int\frac{du}{u}$$
$$=\ln u + C$$
$$=\ln\,(sin\,x)+C$$

43. $\int 5t\sqrt{1-4t^2}\,dt$

Let $u = 1-4t^2$, then $du = -8t\,dt$.

$$=-\frac{5}{8}\int u^{1/2}\,du$$
$$=-\frac{5}{12}u^{3/2}+C$$
$$=-\frac{5}{12}(1-4t^2)^{3/2}+C$$

45. $\int e^t\,sin\,e^t\,dt$

Let $u = \sqrt{w}$, then $du = \frac{dw}{2\sqrt{w}}$.

$$=\frac{1}{2}\int e^u\,du$$
$$=\frac{1}{2}e^u+C$$
$$=\frac{1}{2}e^{\sqrt{w}}+C$$

47. $\int sin^2x\,cos\,x\,dx$

Let $u = sin\,x$, then $du = cos\,x\,dx$.

$$=\int u^2\,du$$
$$=\frac{1}{3}u^3+C$$
$$=\frac{1}{3}sin^3x+C$$

49. $\int e^t \, sin \; e^t \, dt$

Let $u = -sin \; t$, then $du = -cos \; t \, dt$.

$= -\int e^u \, du$

$= -e^u + C$

$= -e^{sin \; t} + C$

51. $\int r^2 \, sin(3r^3 + 7) \, dr$

Let $u = 3r^3$, then $du = 9r^2 \, dr$.

$= \frac{1}{9}\int sin \; u \, du$

$= -\frac{1}{9} cos \; u + C$

$= -\frac{1}{9} cos(3r^3 + 7) + C$

53. $\int_0^1 2x \, e^{x^2} \, dx$

First find the indefinite integral.

$\int 2x \, e^{x^2} \, dx$

Let $u = x^2$, then $du = 2x \, dx$.

$= \int e^{x^2} (2x \, dx)$

$= \int e^u \, du$ Substituting u for x^2 and du for $2x \, dx$

$= e^u + C$

$= e^{x^2} + C$

Then evaluate the definite integral on $[0, 1]$.

$\int_0^1 2x \, e^{x^2} \, dx$

$= \left[e^{x^2}\right]_0^1$

$= e^{1^2} - e^{0^2}$

$= e - 1$

55. $\int_0^1 x(x^2 + 1)^5 \, dx$

First find the indefinite integral.

$\int x(x^2 + 1)^5 \, dx$

Let $u = x^2 + 1$, then $du = 2x \, dx$.

We only have $x \, dx$ and need to supply a 2 by multiplying by $\frac{1}{2} \cdot 2$.

$\frac{1}{2} \cdot 2 \int x(x^2 + 1)^5 \, dx$ Multiplying by 1

$= \frac{1}{2}\int 2x(x^2 + 1)^5 \, dx$

$= \frac{1}{2}\int (x^2 + 1)^5 \cdot 2x \, dx$

$= \frac{1}{2}\int u^5 \, du$ Substituting u for $x^2 + 1$ and du for $2x \, dx$

$= \frac{1}{2} \cdot \frac{u^6}{6} + C$Using Formula A

$= \frac{(x^2 + 1)^6}{12} + C$

Then evaluate the definite integral on $[0, 1]$.

$\int_0^1 x(x^2 + 1)^5 \, dx$

$= \left[\frac{(x^2 + 1)^6}{12}\right]_0^1$

$= \frac{(1^2 + 1)^6}{12} - \frac{(0^2 + 1)^6}{12}$

$= \frac{64}{12} - \frac{1}{12}$

$= \frac{63}{12}$

$= \frac{21}{4}$

57. $\int_0^4 \frac{dt}{1 + t}$

First find the indefinite integral.

$\int \frac{dt}{1 + t}$

Let $u = 1 + t$, then $du = dt$.

$= \int \frac{du}{u}$ Substituting u for $1 + t$ and du for dt

$= \int \frac{1}{u} \, du$

$= \ln \, u + C$ Using Formula C

$= \ln \, (1 + t) + C$

Then evaluate the definite integral on $[0, 4]$.

$\int_0^4 \frac{dt}{1 + t}$

$= \left[\ln \, (1 + t)\right]_0^4$

$= \ln \, (1 + 4) - \ln \, (1 + 0)$

$= \ln \, 5 - \ln \, 1$

$= \ln \, 5 - 0$

$= \ln \, 5$

59. $\int_1^4 \frac{2x + 1}{x^2 + x - 1} \, dx$

First find the indefinite integral.

$\int \frac{2x + 1}{x^2 + x - 1} \, dx$

Let $u = x^2 + x = 1$, then $du = (2x + 1) \, dx$.

$= \int \frac{1}{x^2 + x - 1} (2x + 1) \, dx$

$= \int \frac{1}{u} \, du$ Substituting u for $x^2 + x - 1$ and du for $(2x + 1) \, dx$

$= \ln \, u + C$

$= \ln \, (x^2 + x - 1) + C$

Then evaluate the definite integral on $[1, 4]$.

$$\int_1^4 \frac{2x+1}{x^2+x-1}\,dx$$
$$= \left[\ln(x^2+x-1)\right]_1^4$$
$$= \ln(4^4+4-1) - \ln(1^2+1-1)$$
$$= \ln 19 - \ln 1$$
$$= \ln 19 \qquad (\ln 1 = 0)$$

61. $\int_0^b e^{-x}\,dx$

First find the indefinite integral.

$\int e^{-x}\,dx$

Let $u = -x$, then $du = -dx$.

We only have dx and need to supply a -1 by multiplying by $-1 \cdot (-1)$.

$$-1 \cdot (-1) \int e^{-x}\,dx$$
$$= -\int -e^{-x}\,dx$$
$$= -\int e^{-x}\,(-dx)$$
$$= -\int e^u\,du \qquad \text{Substituting } u \text{ for } -x \text{ and } du \text{ for } -dx$$
$$= -e^u + C \qquad \text{Using Formula B}$$
$$= -e^{-x} + C$$

Then evaluate the definite integral on $[0, b]$.

$$\int_0^b e^{-x}\,dx$$
$$= \left[-e^{-x}\right]_0^b$$
$$= (-e^{-b}) - (-e^{-0})$$
$$= -e^{-b} + e^0$$
$$= -e^{-b} + 1$$
$$= 1 - \frac{1}{e^b}$$

63. $\int_0^b m\,e^{-mx}\,dx$

First find the indefinite integral.

$\int m\,e^{-mx}\,dx \qquad m$ is a constant

Let $u = -mx$, then $du = -m\,dx$.

We only have $m\,dx$ and need to supply a -1 by multiplying by $-1 \cdot (-1)$.

$$-1 \cdot (-1) \int m\,e^{-mx}\,dx$$
$$= -\int -m\,e^{-mx}\,dx$$
$$= -\int e^{-mx}\,(-m\,dx)$$
$$= -\int e^u\,du \qquad \text{Substituting } u \text{ for } -mx \text{ and } du \text{ for } -m\,dx$$
$$= -e^u + C \qquad \text{Using Formula B}$$
$$= -e^{-mx} + C$$

Then evaluate the definite integral on $[0, b]$.

$$\int_0^b me^{-mx}\,dx$$
$$= \left[-e^{-mx}\right]_0^b$$
$$= (-e^{-mb}) - (-e^{-m\cdot 0})$$
$$= -e^{-mb} + 1 \qquad (e^{-m\cdot 0} = e^0 = 1)$$
$$= 1 - e^{-mb}$$
$$= 1 - \frac{1}{e^{mb}}$$

65. $\int_0^4 (x-6)^2\,dx$

First find the indefinite integral.

$\int (x-6)^2\,dx$

Let $u = x - 6$, then $du = dx$.

$$= \int u^2\,du \qquad \text{Substituting } u \text{ for } x-6 \text{ and } du \text{ for } dx$$
$$= \frac{u^3}{3} + C$$
$$= \frac{(x-6)^3}{3} + C$$

Then evaluate the definite integral on $[0, 4]$.

$$\int_0^4 (x-6)^2\,dx$$
$$= \left[\frac{(x-6)^3}{3}\right]_0^4$$
$$= \frac{(4-6)^3}{3} - \frac{(0-6)^3}{3}$$
$$= -\frac{8}{3} - \left(-\frac{216}{3}\right)$$
$$= -\frac{8}{3} + \frac{216}{3}$$
$$= \frac{208}{3}$$

67. $\int_{-1/3}^0 \cos(\pi x + \pi/3)\,dx$

First find the indefinite integral.

$\int \cos(\pi x + \pi/3)^2\,dx$

Let $u = \pi x + \pi/3$, then $du = \pi dx$.

$$= \int \frac{1}{\pi} \cos u\,du \qquad \text{Substituting } u \text{ for } \pi x + \pi/3 \text{ and } \frac{1}{\pi}du \text{ for } dx$$
$$= \frac{1}{\pi}\sin u + C$$
$$= \frac{1}{\pi}\sin(\pi x + \pi/3) + C$$

Then evaluate the definite integral on $[-1/3, 0]$.

$$\int_{-1/3}^0 \cos(\pi x + \pi/3)\,dx$$
$$= \left[\frac{1}{\pi}\sin(\pi x + \pi/3)\right]_{-1/3}^0$$
$$= \frac{\sqrt{3}}{2\pi} - 0$$
$$= \frac{\sqrt{3}}{2\pi}$$

69. $\int_0^2 \frac{3x^2\,dx}{(1+x^3)^5}$

From Exercise 39 we know that the indefinite integral is

$$\int \frac{3x^2\,dx}{(1+x^3)^5} = -\frac{1}{4(1+x^3)^4} + C.$$

Now we evaluate the definite integral on $[0,2]$.

$$\int_0^2 \frac{3x^2\,dx}{(1+x^3)^5}$$
$$= \left[-\frac{1}{4(1+x^3)^4}\right]_0^2$$
$$= \left[-\frac{1}{4(1+2^3)^4}\right] - \left[-\frac{1}{4(1+0^3)^4}\right]$$
$$= -\frac{1}{26,244} + \frac{1}{4}$$
$$= \frac{6560}{26,244}$$
$$= \frac{1640}{6561}$$

71. $\int_0^{\sqrt{7}} 7x\sqrt[3]{1+x^2}\,dx = 7\int_0^{\sqrt{7}} x\sqrt[3]{1+x^2}\,dx$

First find the indefinite integral.

$7\int x\sqrt[3]{1+x^2}\,dx$

Let $u = 1+x^2$, then $du = 2x\,dx$.

$$= 7\cdot\frac{1}{2}\cdot 2\int x\sqrt[3]{1+x^2}\,dx$$
$$= 7\cdot\frac{1}{2}\int 2x\sqrt[3]{1+x^2}\,dx$$
$$= \frac{7}{2}\int \sqrt[3]{1+x^2}\,(2x\,dx)$$
$$= \frac{7}{2}\int \sqrt[3]{u}\,du$$
$$= \frac{7}{2}\int u^{1/3}\,du$$
$$= \frac{7}{2}\cdot\frac{u^{4/3}}{4/3} + C$$
$$= \frac{21}{8}u^{4/3} + C$$
$$= \frac{21}{8}(1+x^2)^{4/3} + C$$

Then evaluate the definite integral on $[0,\sqrt{7}]$.

$$7\int_0^{\sqrt{7}} x\sqrt[3]{1+x^2}\,dx$$
$$= \left[\frac{21}{8}(1+x^2)^{4/3}\right]_0^{\sqrt{7}}$$
$$= \frac{21}{8}(1+(\sqrt{7})^2)^{4/3} - \frac{21}{8}(1+0^2)^{4/3}$$
$$= \frac{21}{8}\cdot 8^{4/3} - \frac{21}{8}\cdot 1^{4/3}$$
$$= \frac{21}{8}\cdot 16 - \frac{21}{8}\cdot 1$$
$$= 42 - \frac{21}{8}$$
$$= \frac{315}{8}$$

73. **a)** $K\int_0^H (H-x)^{3/2}\,dx$

$$= \left[-\frac{2K}{5}(H-x)^{5/2}\right]_0^H$$
$$= 0 - \frac{-2K}{5}H^{5/2}$$
$$= \frac{2}{5}KH^{5/2}$$

b) $K\int_0^{H/2} (H-x)^{3/2}\,dx$

$$= \left[-\frac{2K}{5}(H-x)^{5/2}\right]_0^{H/2}$$
$$= \frac{-2K}{5}(H/2)^{5/2} - \frac{-2K}{5}H^{5/2}$$
$$= \frac{-\sqrt{2}KH^{5/2}}{20} + \frac{2KH^{5/2}}{5}$$
$$= \frac{1}{20}(8-\sqrt{2})KH^{5/2}$$

c) We divide the answers from the previous two parts

$$\frac{\frac{1}{20}(8-\sqrt{2})KH^{5/2}}{\frac{2}{5}KH^{5/2}}$$
$$= \frac{(8-\sqrt{2})}{20}\cdot\frac{5}{2}$$
$$= \frac{(8-\sqrt{2})}{8}$$
$$= 0.8232$$

d) Upper half proportion is given by

$$1 - \frac{(8-\sqrt{2})}{8} = \frac{\sqrt{2}}{8}$$
$$= 0.1768$$

75. We have to break the integral into two parts

$$\int_{-2}^0 [-x\sqrt{4-x^2} - 0]\,dx + \int_0^2 [0 - (-x\sqrt{4-x^2})]\,dx$$
$$= \left[\frac{1}{3}(4-x^2)^{3/2}\right]_{-2}^0 + \left[-\frac{1}{3}(4-x^2)^{3/2}\right]_0^2$$
$$= \frac{8}{3} - 0 + 0 - \frac{-8}{3}$$
$$= \frac{16}{3}$$

77. Using the hint

$$\int \frac{t^2+2t}{(t+1)^2}\,dt$$
$$= \int 1 - \frac{1}{(t+1)^2}\,dt$$
$$= \int 1 - (t+1)^{-2}\,dt$$
$$= t + \frac{1}{(t+1)} + C$$

79. Using the hint

$$\int \frac{x+3}{x+1}\,dx$$
$$= \int 1 + \frac{2}{(x+1)}\,dx$$
$$= x + 2\ln|x+1| + C$$

81. Let $u = \ln x$ then $du = \dfrac{dx}{x}$

$$\int \frac{dx}{x(\ln x)^n}$$
$$= \int u^{-n}\,du$$
$$= \frac{u^{-n+1}}{-n+1} + C$$
$$= \frac{(\ln x)^{-n+1}}{1-n} + C$$
$$= \frac{1}{(1-n)\,(\ln x)^{n-1}} + C$$

83. Let $u = \ln(\ln x)$ then $du = \dfrac{dx}{x\,\ln x}$

$$\int \frac{dx}{x\,\ln x\,[\ln(\ln x)]}$$
$$= \int \frac{du}{u}$$
$$= \ln u + C$$
$$= \ln|\ln(\ln x)| + C$$

85. Using the hint

$$\int sec\ x\,dx$$
$$= \int \frac{sec\ x\ tan\ x + sec^2 x}{tan\ x + sec\ x}\,dx$$

Let $u = tan\ x + sec\ x$ then $du = sec^2 x + sec\ x\ tan\ x\,dx$

$$= \int \frac{du}{u}$$
$$= \ln u + C$$
$$= \ln|sec\ x + tan\ x| + C$$

87. Left to the student

Exercise Set 5.6

1. $\int 5x\,e^{5x}\,dx = \int x(5e^{5x}\,dx)$

Let

$u = x$ and $dv = 5e^{5x}\,dx$.

Then $du = dx$ and $v = e^{5x}$.

$$\int \underset{u}{x}(\underset{dv}{5e^{5x}\,dx}) = \underset{u}{x}\cdot\underset{v}{e^{5x}} - \int \underset{v}{e^{5x}}\cdot\underset{du}{dx}$$

Using Theorem 7: $\int u\,dv = uv - \int v\,du$

$$= xe^{5x} - \frac{1}{5}e^{5x} + C$$

3. $\int x\ sin\ x\,dx$

Let

$u = x$ and $dv = sin\ x\,dx$.

Then $du = dx$ and $v = -cos\ x$

$$\begin{aligned}\int x\ sin\ x\,dx &= x\cdot -cos\ x + \int cos\ x\,dx\\ &= -x\ cos\ x + sin\ x + C\\ &= sin\ x - x\ cos\ x + C\end{aligned}$$

5. $\int xe^{2x}\,dx$

Let

$u = x$ and $dv = e^{2x}\,dx$.

Then $du = dx$ and $v = \dfrac{1}{2}e^{2x}$

$$\begin{aligned}\int xe^{2x}\,dx &= x\cdot\frac{1}{2}e^{2x} - \int \frac{1}{2}e^{2x}dx\\ &= \frac{1}{2}xe^{2x} - \frac{1}{4}e^{2x}\end{aligned}$$

7. $\int xe^{-2x}\,dx$

Let

$u = x$ and $dv = e^{-2x}\,dx$.

Then $du = dx$ and $v = -\dfrac{1}{2}e^{-2x}$.

$$\int x\ e^{-2x}\,dx = x\cdot\left(-\frac{1}{2}e^{-2x}\right) - \int\left(-\frac{1}{2}e^{-2x}\right)dx$$
$$= -\frac{1}{2}\,xe^{-2x} - \frac{-\frac{1}{2}}{-2}e^{-2x} + C$$
$$\left(\int be^{ax}\,dx = \frac{b}{a}e^{ax} + C\right)$$
$$= -\frac{1}{2}xe^{-2x} - \frac{1}{4}e^{-2x} + C$$

9. $\int x^2\ \ln\ x\,dx = \int(\ln\ x)\,x^2\,dx$

Let

$u = \ln\ x$ and $dv = x^2\,dx$.

Then $du = \dfrac{1}{x}\,dx$ and $v = \dfrac{x^3}{3}$.

$$\int \underset{u}{(\ln\ x)}\ \underset{dv}{x^2\,dx} = \underset{u}{\ln\ x}\cdot\underset{v}{\frac{x^3}{3}} - \int \underset{v}{\frac{x^3}{3}}\cdot\underset{du}{\frac{1}{x}\,dx}$$

Integration by Parts

$$= \frac{x^3}{3}\ln\ x - \frac{1}{3}\int x^2\,dx$$
$$= \frac{x^3}{3}\ln\ x - \frac{1}{3}\cdot\frac{x^3}{3} + C$$
$$= \frac{x^3}{3}\ln\ x - \frac{x^3}{9} + C$$

11. $\int x\ \ln\ x^2\,dx$

Let

$u = \ln\ x^2$ and $dv = x\,dx$.

Then $du = \dfrac{1}{x^2}\cdot 2x\,dx = \dfrac{2}{x}\,dx$ and $v = \dfrac{x^2}{2}$.

$$\int (\ln x^2)\, x\, dx = (\ln x^2)\cdot\frac{x^2}{2} - \int \frac{x^2}{2}\cdot\frac{2}{x}\, dx$$
$$= \frac{x^2}{2}\ln x^2 - \int x\, dx$$
$$= \frac{x^2}{2}\ln x^2 - \frac{x^2}{2} + C$$

13. $\int \ln(x+3)\, dx$

Let

$u = \ln(x+3)$ and $dv = dx$.

Then $du = \dfrac{1}{x+3}\, dx$ and $v = x+3$

$$\int \ln(x+3)\, dx = \ln(x+3)\cdot(x+3) - \int x+3\cdot\frac{dx}{x+3)}$$
$$= (x+3)\ln(x+3) - \int dx$$
$$= (x+3)\ln(x+3) - x + C$$

15. $\int (x+2)\ \ln x\, dx = \int (\ln x)(x+2)\, dx$

Let

$u = \ln x$ and $dv = (x+2)\, dx$.

Then

$du = \dfrac{1}{x}\, dx$ and $v = \dfrac{(x+2)^2}{2}$.

$u = \ln x$ and $dv = (x+2)\, dx$.

Then

$du = \dfrac{1}{x}\, dx$ and $v = \dfrac{(x+2)^2}{2}$.

$\int (\ln x)(x+2)\, dx$

$$= (\ln x)\cdot\frac{(x+2)^2}{2} - \int \frac{(x+2)^2}{2}\cdot\frac{1}{x}\, dx$$
$$= \frac{x^2+4x+4}{2}\ln x - \int \frac{x^2+4x+4}{2x}\, dx$$
$$= \frac{x^2+4x+4}{2}\ln x - \int \left(\frac{x}{2}+2+\frac{2}{x}\right) dx$$
$$= \frac{x^2+4x+4}{2}\ln x - \frac{1}{2}\int x\, dx - 2\int dx - 2\int \frac{1}{x}\, dx$$
$$= \frac{x^2+4x+4}{2}\ln x - \frac{1}{2}\cdot\frac{x^2}{2} - 2\cdot x - 2\cdot\ln x + C$$
$$= \frac{x^2+4x+4}{2}\ln x - \frac{1}{2}\cdot\frac{x^2}{2} - 2\cdot x - 2\cdot\ln x + C$$
$$= \left(\frac{x^2+4x+4}{2} - 2\right)\ln x - \frac{x^2}{4} - 2x + C$$
$$= \left(\frac{x^2}{2} + 2x\right)\ln x - \frac{x^2}{4} - 2x + C$$

17. $\int (x-1)\ \sin x\, dx$ Let $u = x-1$ and $dv = \sin x\, dx$.

Then $du = dx$ and $v = -\cos x$

$$\int (x-1)\ \sin x\, dx = (x-1)\cdot -\cos x + \int \cos x\, dx$$
$$= (1-x)\ \cos x + \sin x + C$$

19. $\int x\sqrt{x+2}\, dx$

Let

$u = x$ and $dv = \sqrt{x+2}\, dx = (x+2)^{1/2}\, dx$.

Then

$du = dx$ and $v = \dfrac{(x+2)^{3/2}}{3/2} = \dfrac{2}{3}(x+2)^{3/2}$.

$$\overset{u}{}\quad \overset{dv}{}$$
$$\int x\ \sqrt{x+2}\, dx$$
$$= x\cdot\frac{2}{3}(x+2)^{3/2} - \int \frac{2}{3}(x+2)^{3/2}\, dx$$
$$= \frac{2}{3}x(x+2)^{3/2} - \frac{2}{3}\int (x+2)^{3/2}\, dx$$
$$= \frac{2}{3}x(x+2)^{3/2} - \frac{2}{3}\cdot\frac{(x+2)^{5/2}}{5/2} + C$$
$$= \frac{2}{3}x(x+2)^{3/2} - \frac{4}{15}(x+2)^{5/2} + C$$

21. $\int x^3\ \ln 2x\, dx = \int (\ln 2x)(x^3\, dx)$

Let

$u = \ln 2x$ and $dv = x^3\, dx$.

Then $du = \dfrac{1}{x}\, dx$ and $v = \dfrac{x^4}{4}$.

$$\int (\ln 2x)\ (x^3\, dx) = (\ln 2x)\cdot\frac{x^4}{4} - \int \frac{x^4}{4}\cdot\frac{1}{x}\, dx$$
$$= \frac{x^4}{4}\ln 2x - \frac{1}{4}\int x^3\, dx$$
$$= \frac{x^4}{4}\ln 2x - \frac{1}{4}\cdot\frac{x^4}{4} + C$$
$$= \frac{x^4}{4}\ln 2x - \frac{x^4}{16} + C$$

23. $\int x^2\, e^x\, dx$

Let

$u = x^2$ and $dv = e^x\, dx$.

Then $du = 2x\, dx$ and $v = e^x$.

$$\underset{u}{}\ \underset{dv}{}\qquad \underset{u}{}\ \underset{v}{}\qquad \underset{v}{}\ \underset{du}{}$$
$$\int x^2\, e^x\, dx = x^2\, e^x - \int e^x\cdot 2x\, dx$$

Integration by Parts

$$= x^2\, e^x - \int 2xe^x\, dx$$

We evaluate $\int 2xe^x\, dx$ using the Integration by Parts formula.

$\int 2xe^x\, dx$

Let

$u = 2x$ and $dv = e^x\, dx$.

Then

$du = 2\, dx$ and $v = e^x$.

$$\int 2x\ e^x\, dx = 2x\cdot e^x - \int 2e^x\, dx$$
$$= 2xe^x - 2e^x + K$$

Thus,

$\int x^2\, e^x\, dx = x^2 e^x - (2xe^x - 2e^x + K)$

$\quad = x^2\, e^x - 2xe^x + 2e^x + C \quad (C = -K)$

Since we have an integral $\int f(x)g(x)\, dx$ where $f(x)$, or x^2, can be differentiated repeatedly to a derivative that is eventually 0 and $g(x)$, or e^x, can be integrated repeatedly easily, we can use tabular integration.

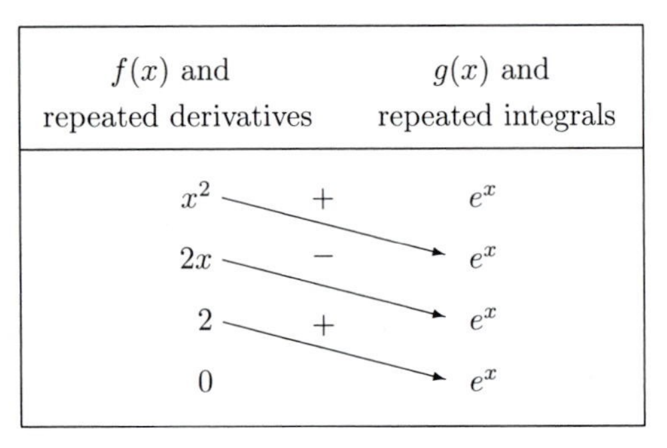

$f(x)$ and repeated derivatives		$g(x)$ and repeated integrals
x^2	+	e^x
$2x$	−	e^x
2	+	e^x
0		e^x

We add the products along the arrows, making the alternate sign changes.

$\int x^2\, e^x\, dx = x^2 e^x - 2xe^x + 2e^x + C$

25. $\int x^2\, sin\ 2x\, dx$

Let

$u = x^2$ and $dv = sin\ 2x\, dx$.

Then $du = 2x\, dx$ and $v = -\dfrac{1}{2}\, cos\ 2x$.

$$\int x^2\, sin\ 2x\, dx \;=\; x^2 \cdot -\frac{cos\ 2x}{2} + \frac{1}{2}\int cos\ 2x \cdot 2x\, dx$$

We evaluate $\int 2x\, cos\ 2x\, dx$ using the Integration by Parts formula

Let

$u = x$ and $dv = 2\, cos\ 2x\, dx$.

Then

$du = dx$ and $v = sin\ 2x$.

$$\int 2xcos\ 2x\, dx \;=\; x\, sin\ 2x - \int sin\ 2x\, dx$$
$$\;=\; x\, sin\ 2x + \frac{1}{2}cos\ 2x + K$$

Thus,

$\int x^2\, sin\ 2x\, dx = -\dfrac{1}{2}x^2\, cos\ 2x - \dfrac{1}{2}x\, sin\ 2x - \dfrac{1}{4}\, cos\ 2x - K$

Since we have an integral $\int f(x)g(x)\, dx$ where $f(x)$, or x^2, can be differentiated repeatedly to a derivative that is eventually 0 and $g(x)$, or e^x, can be integrated repeatedly easily, we can use tabular integration.

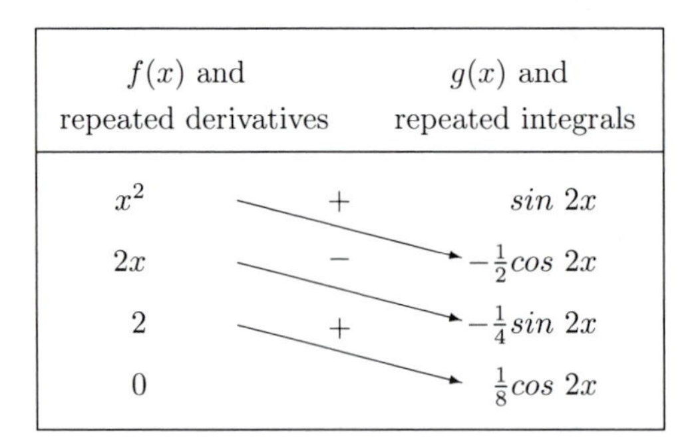

$f(x)$ and repeated derivatives		$g(x)$ and repeated integrals
x^2	+	$sin\ 2x$
$2x$	−	$-\frac{1}{2}cos\ 2x$
2	+	$-\frac{1}{4}sin\ 2x$
0		$\frac{1}{8}cos\ 2x$

We add the products along the arrows, making the alternate sign changes.

$\int x^2\, sin\ 2x\, dx = -\frac{1}{2}x^2\, cos\ 2x + \frac{1}{2}\, x\, sin\ 2x + \frac{1}{4}\, cos\ 2x + C$

27. $\int x^3\, e^{-2x}\, dx$

We will use tabular integration.

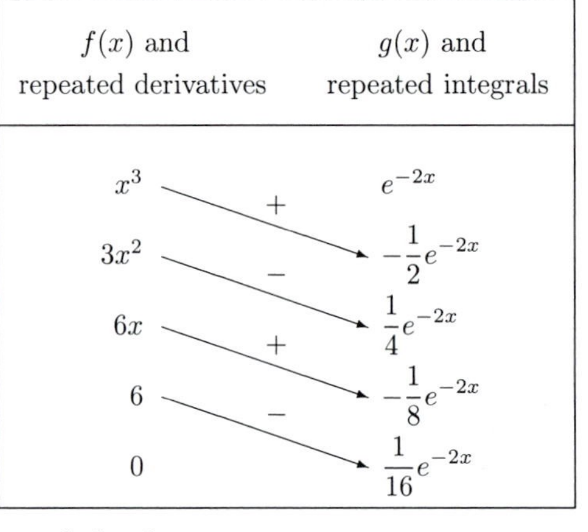

$f(x)$ and repeated derivatives		$g(x)$ and repeated integrals
x^3	+	e^{-2x}
$3x^2$	−	$-\frac{1}{2}e^{-2x}$
$6x$	+	$\frac{1}{4}e^{-2x}$
6	−	$-\frac{1}{8}e^{-2x}$
0		$\frac{1}{16}e^{-2x}$

$\int x^3\, e^{-2x}\, dx$

$= x^3\Big(-\dfrac{1}{2}e^{-2x}\Big) - 3x^2\Big(\dfrac{1}{4}e^{-2x}\Big) + 6x\Big(-\dfrac{1}{8}e^{-2x}\Big) -$

$\quad 6\Big(\dfrac{1}{16}e^{-2x}\Big) + C$

$= -\dfrac{1}{2}x^3\, e^{-2x} - \dfrac{3}{4}x^2\, e^{-2x} - \dfrac{3}{4}xe^{-2x} - \dfrac{3}{8}e^{-2x} + C$

$= e^{-2x}\Big(-\dfrac{1}{2}x^3 - \dfrac{3}{4}x^2 - \dfrac{3}{4}x - \dfrac{3}{8}\Big) + C$

29. $\int x\, sec^2 x\, dx$

Let

$u = x$ and $dv = sec^2 x\, dx$.

Then

$du = dx$ and $v = tan\ x$

$$\int x\, sec^2 x\, dx \;=\; x\, tan\ x - \int tan\ x\, dx$$
$$\;=\; x\, tan\ x + \ln|\, cos\ x\,| + C$$

31. $\int_1^2 x^2 \ln x\, dx$

In Exercise 9 above we found the indefinite integral.

$$\int x^2 \ln x\, dx = \frac{x^3}{3} \ln x - \frac{x^3}{9} + C$$

Evaluate the definite integral.

$$\begin{aligned}\int_1^2 x^2 \ln x\, dx &= \left[\frac{x^3}{3} \ln x - \frac{x^3}{9}\right]_1^2 \\ &= \left(\frac{2^3}{3} \ln 2 - \frac{2^3}{9}\right) - \left(\frac{1^3}{3} \ln 1 - \frac{1^3}{9}\right) \\ &= \left(\frac{8}{3} \ln 2 - \frac{8}{9}\right) - \left(\frac{1}{3} \ln 1 - \frac{1}{9}\right) \\ &= \frac{8}{3} \ln 2 - \frac{8}{9} + \frac{1}{9} \qquad (\ln 1 = 0) \\ &= \frac{8}{3} \ln 2 - \frac{7}{9}\end{aligned}$$

33. $\int_2^6 \ln(x+3)\, dx$

In Exercise 13 above we found the indefinite integral.

$\int \ln(x+3)\, dx = (x+3)\ln(x+3) - x + C$

Evaluate the definite integral.

$$\begin{aligned}&\int_2^6 \ln x(x+3)\, dx \\ &= \big[(x+3)\ln(x+3) - x\big]_2^6 \\ &= [(6+3)\ln(6+3) - 6] - [(2+3)\ln(2+3) - 2] \\ &= (9 \ln 9 - 6) - (5 \ln 5 - 2) \\ &= 9 \ln 9 - 6 - 5 \ln 5 + 2 \\ &= 9 \ln 9 - 5 \ln 5 - 4\end{aligned}$$

35. a) We first find the indefinite integral.

$\int xe^x\, dx$

Let

$u = x$ and $dv = e^x\, dx$.

Then

$du = dx$ and $v = e^x$.

$$\begin{aligned}\int xe^x\, dx &= xe^x - \int e^x\, dx \\ &= xe^x - e^x + C\end{aligned}$$

b) Evaluate the definite integral.

$$\begin{aligned}\int_0^1 xe^x\, dx &= \big[xe^x - e^x\big]_0^1 \\ &= (1 \cdot e^1 - e^1) - (0 \cdot e^0 - e^0) \\ &= (e - e) - (0 - 1) \\ &= 0 - (-1) \\ &= 1\end{aligned}$$

37. $\int_0^{5\pi/6} 3x \cos x\, dx$ Let

$u = 3x$ and $dv = \cos x\, dx$.

Then

$du = 3\, dx$ and $v = \sin x$

$$\begin{aligned}\int 3x \cos x\, dx &= 3x \sin x - \int 3 \sin x\, dx \\ &= 3x \sin x + 3 \cos x + C\end{aligned}$$

Thus,

$$\begin{aligned}\int_0^{5\pi/6} 3x \cos x\, dx &= \Big[3x \sin x + 3 \cos x\Big]_0^{5\pi/6} \\ &= \left[\frac{5\pi}{4} - \frac{3\sqrt{3}}{2}\right] - [0 + 3] \\ &= \frac{5\pi}{4} - 3 - \frac{3\sqrt{3}}{2}\end{aligned}$$

39. $M'(t) = 10t\sqrt{t+15}$

Let $u = 10t$, and $dv = \sqrt{t+15}\, dt$

Then $du = 10dt$ and $v = \frac{2}{3}(t+15)^{3/2}$

$$\begin{aligned}M(t) &= \int 10t\sqrt{t+15}\, dt \\ &= \frac{20}{3}\, t(t+15)^{3/2} - \frac{20}{3} \int (t+15)^{3/2}\, dt \\ &= \frac{20}{3}\, t(t+15)^{3/2} - \frac{8}{3}(t+15)^{5/2} + C \\ 150000 &= 0 - \frac{8}{3} \cdot (15)^{5/2} + C \\ C &= 150000 + \frac{8(15)^{5/2}}{3} \\ M(t) &= \frac{20}{3}\, t(t+15)^{3/2} - \frac{8}{3}(t+15)^{5/2} + 150000 + \frac{8(15)^{5/2}}{3} \\ M(10) &= \frac{20}{3}\, 10(25)^{3/2} - \frac{8}{3}(25)^{3/2} + 150000 + \frac{8(15)^{5/2}}{3} \\ &= \frac{25000}{3} - \frac{25000}{3} + 150000 + \frac{8(15)^{5/2}}{3} \\ &= 152324\end{aligned}$$

41. a) We first find the indefinite integral.

$\int 10te^{-t}\, dt = 10 \int te^{-t}\, dt$

Let

$u = t$ and $dv = e^{-t}\, dt$.

Then

$du = dt$ and $v = -e^{-t}$.

$$\begin{aligned}10 \int te^{-t}\, dt &= 10\big[t(-e^{-t}) - \int -e^{-t}\, dt\big] \\ &= 10(-te^{-t} + \int e^{-t}\, dt) \\ &= 10(-te^{-t} - e^{-t} + K) \\ &= -10te^{-t} - 10e^{-t} + C \quad (C = 10K)\end{aligned}$$

Then evaluate the definite integral.

$$\begin{aligned}\int_0^T te^{-t}\, dt &= \big[-10te^{-t} - 10e^{-t}\big]_0^T \\ &= (-10Te^{-T} - 10e^{-T}) - \\ &\qquad (-10 \cdot 0 \cdot e^{-0} - 10e^{-0}) \\ &= (-10Te^{-T} - 10e^{-T}) - (0 - 10) \\ &= -10Te^{-T} - 10e^{-T} + 10 \\ &= -10\big[e^{-T}(T+1) - 1\big], \text{ or} \\ &\quad 10\big[e^{-T}(-T-1) + 1\big]\end{aligned}$$

b) Substitute 4 for T.

$$\int_0^4 te^{-t}\,dt = -10\left[e^{-4}(4+1)-1\right]$$
$$= -50e^{-4}+10$$
$$\approx -50(0.018316)+10$$
$$\approx -0.915800+10$$
$$\approx 9.084$$

43. Left to the student

45. Let $u = \ln x$ and $dv = x^{1/2}$
Then $du = \dfrac{dx}{x}$ and $v = \dfrac{2}{3}x^{3/2}$

$$\int \sqrt{x}\ln x\,dx = \frac{2}{3}x^{3/2}\ln x - \frac{2}{3}\int x^{1/2}\,dx$$
$$= \frac{2}{3}x^{3/2}\ln x - \frac{4}{9}x^{3/2}+C$$

47. Let $u = te^t$ and $dv = \dfrac{1}{(t+1)^2}\,dt$
Then $du = (te^t+e^t)\,dt$ and $v = \dfrac{-1}{t+1}$

$$\int \frac{te^t}{(t+1)^2}\,dt = \frac{-te^t}{t+1} + \int \frac{te^t}{t+1}\,dt + \int \frac{e^t}{t+1}\,dt$$

For the first integral on the right hand side
Let $u = \dfrac{t}{t+1}$ and $dv = e^t\,dt$
Then $du = \dfrac{dt}{t+1}$ and $v = e^t$
For the second integral on the right hand side
Let $u = \dfrac{1}{t+1}$ and $dv = e^t\,dt$
Then $du = \dfrac{-dt}{(t+1)^2}$ and $v = e^t$ $\displaystyle\int \frac{te^t}{(t+1)^2}\,dt$

$$= \frac{-te^t}{t+1} + \int \frac{te^t}{t+1}\,dt + \int \frac{e^t}{t+1}\,dt$$
$$= \frac{-te^t}{t+1} + \frac{te^t}{t+1} - \int \frac{e^t}{(t+1)^2}\,dt + \frac{e^t}{t+1} + \int \frac{e^t}{(t+1)^2}\,dt$$
$$= \frac{e^t}{t+1}+C$$

49. Let $u = \ln x$ and $dv = x^{-1/2}$
Then $du = \dfrac{dx}{x}$ and $v = 2x^{\frac{1}{2}}$

$$\int \frac{\ln x}{x^{1/2}}\,dx = 2\sqrt{x}\,\ln x - \int 2x^{-1/2}\,dx$$
$$= 2\sqrt{x}\,\ln x - 4\sqrt{x}+C$$

51. Let $u = 13t^2-48$ and $dv = (4t+7)^{-1/5}$
Then $du = 26t\,dt$ and $v = \dfrac{5}{16}(4t+7)^{4/5}$

$$\int \frac{13t^2-48}{\sqrt[5]{4t+7}}\,dt = \frac{5}{16}(13t^2-48)(4t+7)^{4/5} - \int \frac{65}{8}t(4t+7)^{4/5}\,dt$$

Let $u = \dfrac{65}{8}t$ and $dv = (4t+7)^{4/5}\,dt$
Then $du = 26t\,dt$ and $v = \dfrac{5}{36}(4t+7)^{9/5}$ $\displaystyle\int \frac{13t^2-48}{\sqrt[5]{4t+7}}\,dt$

$$= \frac{5}{16}(13t^2-48)(4t+7)^{4/5} - \int \frac{65}{8}t(4t+7)^{4/5}\,dt$$
$$= \frac{5}{16}(13t^2-48)(4t+7)^{4/5} - \frac{325t(4t+7)^{9/4}}{288} - \frac{325}{288}\int (4t+7)^{9/5}\,dt$$
$$= \frac{5(13t^2-48)(4t+7)^{4/5}}{16} - \frac{325t(4t+7)^{9/4}}{288} - \frac{1625(4t+7)^{14/5}}{4032}+C$$

53. Left to the student

55. Left to the student

57. Left to the student

59. $\displaystyle\int_1^{10} x^5\ln x\,dx = 355986.43$

Exercise Set 5.7

1. $\int xe^{-3x}\,dx$

This integral fits Formula 6 in Table 1.

$$\int xe^{ax}\,dx = \frac{1}{a^2}\cdot e^{ax}(ax-1)+C$$

In our integral $a=-3$, so we have, by the formula,

$$\int xe^{-3x}\,dx = \frac{1}{(-3)^2}\cdot e^{-3x}(-3x-1)+C$$
$$= \frac{1}{9}e^{-3x}(-3x-1)+C$$
$$\text{or } -\frac{1}{9}e^{-3x}(3x+1)+C$$

3. $\int 5^x\,dx$

This integral fits Formula 11 in Table 1.

$$\int a^x\,dx = \frac{a^x}{\ln a}+C,\ a>0,\ a\neq 1$$

In our integral $a=5$, so we have, by the formula,

$$\int 5^x\,dx = \frac{5^x}{\ln 5}+C$$

5. $\displaystyle\int \frac{1}{16-x^2}\,dx$

This integral fits Formula 27 in Table 1.

$$\int \frac{1}{a^2-x^2}\,dx = \frac{1}{2a}\ln\left(\frac{a+x}{a-x}\right)+C$$

In our integral $a=4$, so we have, by the formula,

$$\int \frac{1}{16-x^2}\,dx = \int \frac{1}{4^2-x^2}\,dx$$
$$= \frac{1}{2\cdot 4}\ln\frac{4+x}{4-x}+C$$
$$= \frac{1}{8}\ln\frac{4+x}{4-x}+C$$

7. $\displaystyle\int \frac{x}{5-x}\,dx$

This integral fits Formula 30 in Table 1.

$$\int \frac{x}{ax+b}\,dx = \frac{b}{a^2} + \frac{x}{a} - \frac{b}{a^2}\ \ln(ax+b) + C$$

In our integral $a = -1$ and $b = 5$, so we have, by the formula,

$$\int \frac{x}{5-x}\,dx = \frac{5}{(-1)^2} + \frac{x}{(-1)} - \frac{5}{(-1)^2}\ \ln(-1\cdot x+5)+C$$
$$= 5 - x - 5\ \ln(5-x) + C$$

9. $\displaystyle\int \frac{1}{x(5-x)^2}\,dx$

This integral fits Formula 33 in Table 1.

$$\int \frac{1}{x(ax+b)^2}\,dx = \frac{1}{b(ax+b)} + \frac{1}{b^2}\ \ln\left(\frac{x}{ax+b}\right) + C$$

In our integral $a = -1$ and $b = 5$, so we have, by the formula,

$$\int \frac{1}{x(5-x)^2}\,dx = \int \frac{1}{x(-x+5)^2}\,dx$$
$$= \frac{1}{5(-x+5)} + \frac{1}{5^2}\ \ln\left(\frac{x}{-x+5}\right) + C$$
$$= \frac{1}{5(5-x)} + \frac{1}{25}\ \ln\left(\frac{x}{5-x}\right) + C$$

11. $\int \ln 3x\,dx$

$= \int (\ln 3 + \ln x)dx$

$= \int \ln 3\,dx + \int \ln x\,dx$

$= (\ln 3)\,x + \int \ln x\,dx$

The integral in the second term fits Formula 8 in Table 1.

$\int \ln x\,dx = x\ \ln x - x + C$

$\int \ln 3x\,dx = (\ln 3)x + \int \ln x\,dx$

$= (\ln 3)x + x \ln x - x + C$

13. $\int x^4\,e^{5x}\,dx$

This integral first Formula 7 in Table 1.

$$\int x^n\,e^{ax}\,dx = \frac{x^n\,e^{ax}}{a} - \frac{n}{a}\int x^{n-1}\,e^{ax}\,dx$$

In our integral $n = 4$ and $a = 5$, so we have, by the formula,

$\int x^4\,e^{5x}\,dx$

$$= \frac{x^4\,e^{5x}}{5} - \frac{4}{5}\int x^3\,e^{5x}\,dx$$

In the integral in the second term where $n = 3$ and $a = 5$, we again apply Formula 7.

$$= \frac{x^4\,e^{5x}}{5} - \frac{4}{5}\left[\frac{x^3\,e^{5x}}{5} - \frac{3}{5}\int x^2\,e^{5x}\,dx\right]$$

We continue to apply Formula 7.

$$= \frac{x^4\,e^{5x}}{5} - \frac{4}{25}x^3\,e^{5x} + \frac{12}{25}\left[\frac{x^2\,e^{5x}}{5} - \frac{2}{5}\int x\,e^{5x}\,dx\right]$$
$$= \frac{x^4\,e^{5x}}{5} - \frac{4}{25}x^3\,e^{5x} + \frac{12}{125}x^2\,e^{5x} - \frac{24}{125}\left[\frac{x\,e^{5x}}{5} - \frac{1}{5}\int x^0\,e^{5x}\,dx\right]$$
$$= \frac{x^4\,e^{5x}}{5} - \frac{4}{25}x^3\,e^{5x} + \frac{12}{125}x^2\,e^{5x} - \frac{24}{625}x\,e^{5x} + \frac{24}{625}\int e^{5x}\,dx$$

We now apply Formula 5, $\displaystyle\int e^{ax}\,dx = \frac{1}{a}\cdot e^{ax} + C.$

$$= \frac{x^4\,e^{5x}}{5} - \frac{4}{25}x^3\,e^{5x} + \frac{12}{125}x^2\,e^{5x} - \frac{24}{625}x\,e^{5x} + \frac{24}{3125}e^{5x} + C$$

15. $\int x^3\ sin\ x\ \ln x\,dx$

This integral fits Formula 22 in Table 1.

$$\int x^3\ sin\ x\,dx$$
$$= -x^3\ cos\ x + 3\int x^2\ cos\ x$$
$$= -x^3\ cos\ x + 3\left[x^2\ sin\ x - 2\int x\ sin\ x\right]$$
$$= -x^3\ cos\ x + 3x^2\ sin\ x - 6\left[-x\ cos\ x + sin\ x\right] + C$$
$$= -x^3\ cos\ x + 3x^2\ sin\ x + 6x\ cos\ x - 6\ sin\ x + C$$

17. $\int sec\ 2x\,dx$

This integral fits Formula 16 in Table 1.

$$\int sec\ 2x\,dx = \frac{1}{2}\ \ln|\ sec\ 2x + tan\ 2x\ | + C$$

19. $\int 2\ tan(2x+1)\,dx$. Let $u = 2x+1$ then $du = 2\,dx$

This integral fits Formula 14 in Table 1.

$$\int tan\ u\,du = -\ln|\ cos\ u\ | + C$$
$$= -\ln|\ cos(2x+1)\ | + C$$

21. $\displaystyle\int \frac{dx}{\sqrt{x^2+7}}$

This integral fits Formula 24 in Table 1.

$$\int \frac{1}{\sqrt{x^2+a^2}}\,dx = \ln(x + \sqrt{x^2+a^2}) + C$$

In our integral $a^2 = 7$, so we have, by the formula,

$$\int \frac{dx}{\sqrt{x^2+7}} = \ln(x + \sqrt{x^2+7}) + C$$

23. $\displaystyle\int \frac{10\,dx}{x(5-7x)^2} = 10\int \frac{1}{x(-7x+5)^2}\,dx$

This integral fits Formula 33 in Table 1.

$$\int \frac{1}{x(ax+b)^2}\,dx = \frac{1}{b(ax+b)} + \frac{1}{b^2}\ln\left(\frac{x}{ax+b}\right) + C$$

In our integral $a = -7$ and $b = 5$, so we have, by the formula,

$$= 10\int \frac{1}{x(-7x+5)^2}\,dx$$
$$= 10\left[\frac{1}{5(-7x+5)} + \frac{1}{5^2}\ln\left(\frac{x}{-7x+5}\right)\right] + C$$
$$= \frac{2}{5-7x} + \frac{2}{5}\ln\left(\frac{x}{5-7x}\right) + C$$

25. $\displaystyle\int \frac{-5}{4x^2-1}\,dx = -5\int \frac{1}{4x^2-1}\,dx$

This integral almost fits Formula 26 in Table 1.

$$\int \frac{1}{x^2-a^2}\,dx = \frac{1}{2a}\ln\left(\frac{x-a}{x+a}\right) + C$$

But the x^2 coefficient needs to be 1. We factor out 4 as follows. Then we apply Formula 26.

$$-5\int \frac{1}{4x^2-1}\,dx = -5\int \frac{1}{4\left(x^2-\frac{1}{4}\right)}\,dx$$
$$= -\frac{5}{4}\int \frac{1}{x^2-\frac{1}{4}}\,dx \quad \left(a^2 = \frac{1}{4},\ a = \frac{1}{2}\right)$$
$$= -\frac{5}{4}\left[\frac{1}{2\cdot\frac{1}{2}}\ln\left(\frac{x-\frac{1}{2}}{x+\frac{1}{2}}\right)\right] + C$$
$$= -\frac{5}{4}\ln\left(\frac{x-1/2}{x+1/2}\right) + C$$

27. $\int \sqrt{4m^2+16}\,dm$

This integral almost fits Formula 34 in Table 1.

$$\int \sqrt{x^2+a^2}\,dx$$
$$= \frac{1}{2}\left[x\sqrt{x^2+a^2} + a^2\ln\left(x+\sqrt{x^2+a^2}\right)\right] + C$$

But the x^2 coefficient needs to be 1. We factor out 4 as follows. Then we apply Formula 34.

$$\int \sqrt{4m^2+16}\,dm$$
$$= \int \sqrt{4(m^2+4)}\,dm$$
$$= 2\int \sqrt{m^2+4}\,dm$$
$$= 2\cdot\frac{1}{2}\left[m\sqrt{m^2+4} + 4\ln\left(m+\sqrt{m^2+4}\right)\right] + C$$
$$= m\sqrt{m^2+4} + 4\ln\left(m+\sqrt{m^2+4}\right) + C$$

29. $\displaystyle\int \frac{-5\ln x}{x^3}\,dx = -5\int x^{-3}\ln x\,dx$

This integral fits Formula 10 in Table 1.

$$\int x^n\ln x\,dx = x^{n+1}\left[\frac{\ln x}{n+1} - \frac{1}{(n+1)^2}\right] + C,\ n \neq -1$$

In our integral $n = -3$, so we have, by the formula,

$$-5\int x^{-3}\ln x\,dx$$
$$= -5\left[x^{-3+1}\left(\frac{\ln x}{-3+1} - \frac{1}{(-3+1)^2}\right)\right] + C$$
$$= -5\left[x^{-2}\left(\frac{\ln x}{-2} - \frac{1}{4}\right)\right] + C$$
$$= \frac{5\ln x}{2x^2} + \frac{5}{4x^2} + C$$

31. $\displaystyle\int \frac{e^x}{x^{-3}}\,dx = \int x^3 e^x\,dx$

This integral fits Formula 7 in Table 1.

$$\int x^n e^{ax}\,dx = \frac{x^n e^{ax}}{a} - \frac{n}{a}\int x^{n-1}e^{ax}\,dx$$

In our integral $n = 3$ and $a = 1$, so we have, by the formula,

$$\int x^3 e^x\,dx$$
$$= x^3 e^x - 3\int x^2 e^x\,dx$$

We continue to apply Formula 7.

$$= x^3 e^x - 3\left(x^2 e^x - 2\int x e^x\,dx\right) \quad (n = 2,\ a = 1)$$
$$= x^3 e^x - 3x^2 e^x + 6\int x e^x\,dx$$
$$= x^3 e^x - 3x^2 e^x + 6\left(x e^x - \int x^0 e^x\,dx\right) \quad (n = 1,\ a = 1)$$
$$= x^3 e^x - 3x^2 e^x + 6x e^x - 6\int e^x\,dx$$
$$= x^3 e^x - 3x^2 e^x + 6x e^x - 6e^x + C$$

33. $\int x\sqrt{1+2x}\,dx$

This integral fits Formula 35 in Table 1.

$$\int x\sqrt{a+bx}\,dx = \frac{2}{15b^2}(3bx-2a)(a+bx)^{3/2} + C$$

In our integral $a = 1$ and $b = 2$.

$$\int x\sqrt{1+2x}\,dx$$
$$= \frac{2}{15\cdot 2^2}(3\cdot 2x - 2\cdot 1)(1+2x)^{3/2} + C$$
$$= \frac{2}{60}(6x-2)(1+2x)^{3/2} + C$$
$$= \frac{1}{30}\cdot 2(3x-1)(1+2x)^{3/2} + C$$
$$= \frac{1}{15}(3x-1)(1+2x)^{3/2} + C$$

35. $\int (\ln x)^4\,dx$

$$= x\ln^4 x - 4x\ln^3 x + 12x\ln^3 x - 24x\ln x + 24x + C$$

37. $\displaystyle\int \frac{1}{x(x^2-1)}\,dx$

$$= \frac{1}{2}\ln|x^2-1| - \ln|x| + C$$

39. $\int x^4 \; sin\; 3x\, dx$

$$= -\frac{1}{3}\, x^4 \;cos\; 3x + \frac{4}{9}\, x^3 \;sin\; 3x + \frac{4}{9}\, x^2 \;cos\; 3x - \frac{8}{81}\;cos\; 3x - \frac{8}{27}\, x\; sin\; 3x + C$$

41. $\int e^{2x} \; sin\; 3x\, dx$

$$= \frac{-3}{13}\, e^{2x} \;cos\; 3x + \frac{2}{13}\, e^{2x}\; sin\; 3x + C$$

43. **a)** Trapezoid: 0.742984098

b) Simpson: 0.74685538

45. **a)** Trapezoid: 0.449975605

b) Simpson: 0.447138991

47. **a)** Trapezoid: 1.503577487

b) Simpson: 1.505472368

49. **a)** Trapezoid: 0.270958739

b) Simpson: 0.270918581

51. $p'(t) = \dfrac{1}{t(2+t)^2}$ The integral follows formula number 33 in Table 1 with $a = 1$ and $b = 2$

$$\begin{aligned} p(t) &= \int p'(t)\, dt \\ &= \int \frac{1}{t(2+t)^2}\, dt \\ &= \frac{1}{2(2+t)} + \frac{1}{4}\, \ln \left| \frac{t}{2+t} \right| + C \end{aligned}$$

Applying the initial condition $p(2) = 0.8267$

$$\begin{aligned} 0.8267 &= \frac{1}{2(2+2)} + \frac{1}{4}\, \ln \left| \frac{2}{2+2} \right| + C \\ 0.8267 &= 0.125 - 0.1733 + C \\ C &= 0.5284 \\ p(t) &= \frac{1}{2(2+t)} + \frac{1}{4}\, \ln \left| \frac{2}{2+t} \right| + 0.5284 \end{aligned}$$

53. Using a spreadsheet and following Example 6, the number of degree days between 9:00 P.M. July 5 and 9:00 P.M. July 7, 2003 are 17.

55. $\int \dfrac{x\, dx}{4x^2 - 12x + 9} = \int \dfrac{x\, dx}{(2x-3)^2)}$
Using formula number in Table 1 with $a = 2$ and $b = -3$

$$\begin{aligned} \int \frac{x\, dx}{(2x-3)} &= \frac{-3}{4(2x-3)} + \frac{1}{4}\, \ln \left| \frac{x}{2x-3} \right| + C \\ &= \frac{3}{4(3-2x)} + \frac{1}{4}\, \ln \left| \frac{x}{2x-3} \right| + C \end{aligned}$$

57. $\int e^x\, \sqrt{e^{2x}+1}\, dx$
Let $y = e^x$ then $dy = e^x\, dx$
Using formula number in Table 1 with $a = 1$

$$\begin{aligned} \int e^x\, \sqrt{e^2x+1}\, dx &= \int \sqrt{y^2+1}\, dy \\ &= \ln \left| y + \sqrt{y^2+1} \right| + C \\ &= \ln \left| e^x + \sqrt{e^{2x}+1} \right| + C \end{aligned}$$

59. Let $y = \ln x$ then $dy = \dfrac{dx}{x}$
Using formula number in Table 1 with $a = 7$

$$\begin{aligned} &\int \frac{\sqrt{(\ln x)^2 + 49}}{2x}\, dx \\ &= \frac{1}{2} \int \sqrt{y^2+49}\, dy \\ &= \frac{1}{2}\, \ln \left| y + \sqrt{y^2+49} \right| + C \\ &= \frac{1}{2}\, \ln \left| \ln x + \sqrt{(\ln x)^2+49} \right| + C \end{aligned}$$

61. **a)** $\int_{-h}^{h} ax^3 + bx^2 + cx + d\, dx$

$$\begin{aligned} &= \left[\frac{ax^4}{4} + \frac{bx^3}{3} + \frac{cx^2}{2} + dx \right]_{-h}^{h} \\ &= \frac{ah^4}{4} + \frac{bh^3}{3} + \frac{ch^2}{2} + dh - \left(\frac{ah^4}{4} - \frac{bh^3}{3} + \frac{ch^2}{2} - dh \right) \\ &= \frac{2bh^3}{3} + 2dh \end{aligned}$$

b) $f(h) = ah^3 + bh^2 + ch + d$
$f(0) = d$
$f(-h) = -ah^3 + bh^2 - ch + d$

c) $\int_{-h}^{h} ax^3 + bx^2 + cx + d\, dx$

$$\begin{aligned} &= \frac{2bh^3}{3} + 2dh \\ &= \frac{h}{3}\left[2bh^2 + 6d\right] \\ &= \frac{h}{3}\left[-ah^3 + bh^2 - ch + d + 4d + d + ch + bh^2 + ah^3\right] \\ &= \frac{h}{3}\left[f(-h) + 4f(0) + f(h)\right] \end{aligned}$$

d) Left to the student (answers Vary)

Exercise Set 5.8

1. Find the volume of the solid of revolution generated by rotating about the x-axis the region under the graph of

$y = x$

from $x = 0$ to $x = 1$.

$V = \int_a^b \pi\left[f(x)\right]^2 dx$ Volume of a solid of revolution

$V = \int_0^1 \pi\, x^2\, dx$ Substituting 0 for a, 1 for b, and x for $f(x)$

$= \left[\pi \cdot \frac{x^3}{3}\right]_0^1$

$= \frac{\pi}{3}\left[x^3\right]_0^1$

$= \frac{\pi}{3}(1^3 - 0^3)$

$= \frac{\pi}{3} \cdot 1$

$= \frac{\pi}{3}$

3. Find the volume of the solid of revolution generated by rotating about the x-axis the region under the graph of

$y = \sqrt{sin\ x}$

from $x = 0$ to $x = \frac{\pi}{2}$.

$V = \int_a^b \pi\left[f(x)\right]^2 dx$

$V = \int_0^{\pi/2} \pi\ sin\ x\ dx$

$= \pi\left[-\cos x\right]_0^{\pi/2}$

$= \pi\left[0 - (-1)\right]$

$= \pi$

5. Find the volume of the solid of revolution generated by rotating about the x-axis the region under the graph of

$y = e^x$

from $x = -2$ to $x = 5$.

$V = \int_a^b \pi\left[f(x)\right]^2 dx$ Volume of a solid of revolution

$V = \int_{-2}^5 \pi\left[e^x\right]^2 dx$ Substituting -2 for a, 5 for b, and e^x for $f(x)$

$= \int_{-2}^5 \pi\, e^{2x}\, dx$

$= \left[\pi \cdot \frac{1}{2}e^{2x}\right]_{-2}^5$

$= \frac{\pi}{2}\left[e^{2x}\right]_{-2}^5$

$= \frac{\pi}{2}\left(e^{2\cdot 5} - e^{2(-2)}\right)$

$= \frac{\pi}{2}(e^{10} - e^{-4})$

7. Find the volume of the solid of revolution generated by rotating about the x-axis the region under the graph of

$y = \frac{1}{x}$

from $x = 1$ to $x = 3$.

$V = \int_a^b \pi\left[f(x)\right]^2 dx$ Volume of a solid of revolution

$V = \int_1^3 \pi\left[\frac{1}{x}\right]^2 dx$ Substituting 1 for a, 3 for b, and $\frac{1}{x}$ for $f(x)$

$= \int_1^3 \pi \cdot \frac{1}{x^2}\, dx$

$= \int_1^3 \pi\, x^{-2}\, dx$

$= \left[\pi \frac{x^{-1}}{-1}\right]_1^3$

$= -\pi\left[\frac{1}{x}\right]_1^3$

$= -\pi\left(\frac{1}{3} - \frac{1}{1}\right)$

$= -\pi \cdot \left(-\frac{2}{3}\right)$

$= \frac{2}{3}\pi$

9. Find the volume of the solid of revolution generated by rotating about the x-axis the region under the graph of

$y = \frac{2}{\sqrt{x}}$

from $x = 1$ to $x = 3$.

$V = \int_a^b \pi\left[f(x)\right]^2 dx$ Volume of a solid of revolution

$V = \int_1^3 \pi\left[\frac{2}{\sqrt{x}}\right]^2 dx$ Substituting 1 for a, 3 for b, and $\frac{2}{\sqrt{x}}$ for $f(x)$

$= \int_1^3 \pi \cdot \frac{4}{x}\, dx$

$= 4\pi\left[\ln x\right]_1^3$

$= 4\pi(\ln 3 - \ln 1)$

$= 4\pi \ln 3$ $(\ln 1 = 0)$

11. Find the volume of the solid of revolution generated by rotating about the x-axis the region under the graph of

$y = 4$

from $x = 1$ to $x = 3$.

$V = \int_a^b \pi\left[f(x)\right]^2 dx$ Volume of a solid of revolution

$V = \int_1^3 \pi\left[4\right]^2 dx$ Substituting 1 for a, 3 for b, and 4 for $f(x)$

$= \int_1^3 16\pi\, dx$

$= 16\pi\left[x\right]_1^3$

$= 16\pi(3 - 1)$

$= 32\pi$

13. Find the volume of the solid of revolution generated by rotating about the x-axis the region under the graph of

$$y = x^2$$

from $x = 0$ to $x = 2$.

$V = \int_a^b \pi\left[f(x)\right]^2 dx$ Volume of a solid of revolution

$V = \int_0^2 \pi\left[x^2\right]^2 dx$ Substituting 0 for a, 2 for b, and x^2 for $f(x)$

$$= \int_0^2 \pi x^4\, dx$$

$$= \left[\pi \cdot \frac{x^5}{5}\right]_0^2$$

$$= \frac{\pi}{5}(2^5 - 0^5)$$

$$= \frac{32}{5}\pi$$

15. Find the volume of the solid of revolution generated by rotating about the x-axis the region under the graph of

$$y = \cos x$$

from $x = 0$ to $x = \frac{\pi}{2}$.

$$V = \int_a^b \pi\left[f(x)\right]^2 dx$$

$$V = \int_0^{\pi/2} \pi \cos^2 x\, dx$$

$$= \pi\left[\frac{1}{2}x + \frac{1}{2}\sin x \cos x\right]_0^{\pi/2}$$

$$= \frac{\pi}{2}\left[\frac{\pi}{2} + 0 - 0\right]$$

$$= \frac{\pi^2}{4}$$

17. Find the volume of the solid of revolution generated by rotating about the x-axis the region under the graph of

$$y = \tan x$$

from $x = 0$ to $x = \frac{\pi}{4}$.

$$V = \int_a^b \pi\left[f(x)\right]^2 dx$$

$$V = \int_0^{\pi/2} \pi \tan^2 x\, dx$$

$$= \pi\left[\tan x + x\right]_0^{\pi/4}$$

$$= \pi\left[1 + \frac{\pi}{4} - 0\right]$$

$$= \pi + \frac{\pi^2}{4}$$

19. Find the volume of the solid of revolution generated by rotating about the x-axis the region under the graph of

$$y = \sqrt{1+x}$$

from $x = 2$ to $x = 10$.

$V = \int_a^b \pi\left[f(x)\right]^2 dx$ Volume of a solid of revolution

$V = \int_2^{10} \pi\left(\sqrt{1+x}\right)^2 dx$ Substituting 2 for a, 10 for b, and $\sqrt{1+x}$ for $f(x)$

$$= \int_2^{10} \pi(1+x)\, dx$$

$$= \left[\pi \cdot \frac{(1+x)^2}{2}\right]_2^{10}$$

$$= \frac{\pi}{2}\left[(1+x)^2\right]_2^{10}$$

$$= \frac{\pi}{2}[(1+10)^2 - (1+2)^2]$$

$$= \frac{\pi}{2}(11^2 - 3^2)$$

$$= \frac{\pi}{2}(121 - 9)$$

$$= \frac{\pi}{2} \cdot 112$$

$$= 56\pi$$

21. Find the volume of the solid of revolution generated by rotating about the x-axis the region under the graph of

$$y = \sqrt{4 - x^2}$$

from $x = -2$ to $x = 2$.

$V = \int_a^b \pi\left[f(x)\right]^2 dx$ Volume of a solid of revolution

$V = \int_{-2}^{2} \pi\left(\sqrt{4-x^2}\right)^2 dx$ Substituting -2 for a, 2 for b, and $\sqrt{4-x^2}$ for $f(x)$

$$= \int_{-2}^{2} \pi(4 - x^2)\, dx$$

$$= \pi\left[4x - \frac{x^3}{3}\right]_{-2}^{2}$$

$$= \pi\left[\left(4 \cdot 2 - \frac{2^3}{3}\right) - \left(4 \cdot (-2) - \frac{(-2)^3}{3}\right)\right]$$

$$= \pi\left(8 - \frac{8}{3} + 8 - \frac{8}{3}\right)$$

$$= \pi\left(16 - \frac{16}{3}\right)$$

$$= \pi\left(\frac{48}{3} - \frac{16}{3}\right)$$

$$= \frac{32}{3}\pi$$

23. Find the volume of the solid with cross-sectional area

$$A(x) = \frac{1}{2}x^2$$

from $x = 0$ to $x = 6$.

$$V = \int_a^b A(x)\,dx$$
$$V = \int_0^6 \left(\frac{1}{2}x^2\right) dx$$
$$= \left[\frac{1}{6}x^3\right]_0^6$$
$$= \left(36 - 0\right)$$
$$= 36$$

25. Find the volume of the solid with cross-sectional area
$$A(x) = \frac{\sqrt{3}}{2}x^2$$
from $x = 0$ to $x = 9$.
$$V = \int_a^b A(x)\,dx$$
$$V = \int_0^9 \left(\frac{\sqrt{3}}{2}x^2\right) dx$$
$$= \left[\frac{\sqrt{3}}{6}x^3\right]_0^9$$
$$= \left(\frac{243\sqrt{3}}{2} - 0\right)$$
$$= \frac{243\sqrt{3}}{2}$$

27. $A(x) = x^4$
from $x = 0$ to $x = 4$.
$$V = \int_a^b A(x)\,dx$$
$$V = \int_0^4 \left(x^4\right) dx$$
$$= \left[\frac{1}{5}x^5\right]_0^4$$
$$= \frac{1024}{5} - 0$$
$$= \frac{1024}{5}$$

29. $A(x) = 2x^2$
from $x = 0$ to $x = 5$.
$$V = \int_a^b A(x)\,dx$$
$$V = \int_0^5 \left(2x^2\right) dx$$
$$= \left[\frac{2}{3}x^3\right]_0^5$$
$$= \frac{250}{3} - 0$$
$$= \frac{250}{3}$$

31. $10 = K \cdot 8 \to K = \frac{5}{4}$ (See Example 4 on page 410 as a reference)

$$A(x) = \frac{25}{16}x^2$$
from $x = 0$ to $x = 8$.
$$V = \int_a^b A(x)\,dx$$
$$V = \int_0^8 \left(\frac{25}{16}x^2\right) dx$$
$$= \left[\frac{25}{48}x^3\right]_0^8$$
$$= \frac{800}{3} - 0$$
$$= \frac{800}{3}$$

33. For $r = 1.5$ when $H = 75$ and $x = 0$, $K = 38.2285$
Thus, $r(x) = 0.05886(75 - x)^{3/4}$
$$V = \int_a^b \pi \left[r(x)\right]^2 dx$$
$$V = \int_0^{75} \pi \left(0.05886(75 - x)^{3/4}\right)^2 dx$$
$$= \int_0^{75} \pi \left(0.003464(75 - x)^{3/2}\right) dx$$
$$= -0.003464\pi \left[\frac{2}{5}(75 - x)^{5/2}\right]_0^{75}$$
$$= -0.003464\pi \left(0 - \frac{-2}{5}\,75^{5/2}\right)$$
$$= 212.058$$

35. Find the volume of the solid of revolution generated by rotating about the x-axis the region under the graph of
$$y = \sqrt{\ln x}$$
from $x = e$ to $x = e^3$.
$$V = \int_a^b \pi \left[f(x)\right]^2 dx$$
$$V = \int_e^{e^3} \pi \left[\sqrt{\ln x}\right]^2 dx$$
$$= \int_3^{e^3} \pi \ln x\,dx$$
$$= \pi \left[x \ln x - x\right]_e^{e^3}$$
$$= \pi[(e^3 \ln e^3 - e^3) - (e \ln e - e)]$$
$$= \pi[(3e^3 - e^3) - (e - e)]$$
$$= 2\pi e^3$$

37. **a)** The resulting solid of revolution is a cone with a base radius of r and a height of h

b)
$$V = \int_0^h \pi y^2\,dx$$
$$= \pi \int_0^h \frac{r^2}{h^2}x^2\,dx$$
$$= \frac{\pi r^2}{3h^2}\left[x^3\right]_0^h$$

$$= \frac{\pi r^2}{3h^2}\left[h^3 - 0\right]$$
$$= \frac{1}{3}\,\pi r^2\,h$$

Exercise Set 5.9

1. $\displaystyle\int_2^\infty \frac{dx}{x^2}$

$= \displaystyle\lim_{b\to\infty}\int_2^b x^{-2}\,dx$

$= \displaystyle\lim_{b\to\infty}\left[\frac{x^{-1}}{-1}\right]_2^b$

$= \displaystyle\lim_{b\to\infty}\left[-\frac{1}{x}\right]_2^b$

$= \displaystyle\lim_{b\to\infty}\left(-\frac{1}{b}-\left(-\frac{1}{2}\right)\right)$

$= \displaystyle\frac{1}{2} \quad \left(\text{As } b\to\infty,\ -\frac{1}{b}\to 0 \text{ and } -\frac{1}{b}+\frac{1}{2}\to\frac{1}{2}.\right)$

The limit does exist. Thus the improper integral is convergent.

3. $\displaystyle\int_4^\infty \frac{dx}{x} = \lim_{b\to\infty}\int_4^b \frac{1}{x}\,dx$

$= \displaystyle\lim_{b\to\infty}\left[\ln x\right]_4^b$

$= \displaystyle\lim_{b\to\infty}(\ln b - \ln 4)$

Note that $\ln b$ increases indefinitely as b increases. Therefore, the limit does not exist. If the limit does not exist, we say the improper integral is divergent.

5. $\displaystyle\int_{-\infty}^{-1} \frac{dt}{t^2}$

$= \displaystyle\lim_{b\to-\infty}\int_b^{-1} t^{-2}\,dt$

$= \displaystyle\lim_{b\to-\infty}\left[\frac{t^{-1}}{-1}\right]_b^{-1}$

$= \displaystyle\lim_{b\to-\infty}\left[-\frac{1}{t}\right]_b^{-1}$

$= \displaystyle\lim_{b\to-\infty}\left(1-\frac{1}{b}\right)$

$= \displaystyle 1 \quad \left(\text{As } b\to-\infty,\ -\frac{1}{b}\to 0 \text{ and } 1-\frac{1}{b}\to 1.\right)$

The limit does exist. Thus the improper integral is convergent.

7. $\displaystyle\int_0^\infty 4\,e^{-4x}\,dx = \lim_{b\to\infty}\int_0^b 4\,e^{-4x}\,dx$

$= \displaystyle\lim_{b\to\infty}\left[\frac{4}{-4}e^{-4x}\right]_0^b$

$= \displaystyle\lim_{b\to\infty}\left[-\,e^{-4x}\right]_0^b$

$= \displaystyle\lim_{b\to\infty}\left[-e^{-4b} - (-e^{-4\cdot 0})\right]$

$= \displaystyle\lim_{b\to\infty}\left(-\frac{1}{e^{4b}}+1\right)$

$= 1$

The integral is convergent.

9. $\displaystyle\int_0^\infty e^x\,dx = \lim_{b\to\infty}\int_0^b e^x\,dx$

$= \displaystyle\lim_{b\to\infty}\left[e^x\right]_0^b$

$= \displaystyle\lim_{b\to\infty}(e^b - e^0)$

$= \displaystyle\lim_{b\to\infty}(e^b - 1)$

As $b\to\infty$, $e^b\to\infty$. Thus the limit does not exist. The improper integral is divergent.

11. $\displaystyle\int_{-\infty}^\infty e^{2x}\,dx = \lim_{b\to\infty}\int_{-b}^b e^{2x}\,dx$

$= \displaystyle\lim_{b\to\infty}\left[\frac{1}{2}e^{2x}\right]_{-b}^b$

$= \displaystyle\lim_{b\to\infty}(e^b - e^{-b})$

As $b\to\pm\infty$, $e^b\to\infty$. Thus the limit does not exist. The improper integral is divergent.

13. $\displaystyle\int_{-\infty}^\infty \frac{t\,dt}{(1+t^2)^3} = \lim_{b\to\infty}\int_{-b}^b \frac{t\,dt}{(1+t^2)^3}$

$= \displaystyle\lim_{b\to\infty}\left[\frac{-1}{4(1+t^2)^2}\right]_{-b}^b$

$= \displaystyle\lim_{b\to\infty}[0]$

$= 0$

The integral is convergent.

15. $\displaystyle\int_{-\infty}^\infty 2x\,e^{-3x^2}\,dx = \lim_{b\to\infty}\int_{-b}^b 2xe^{-3x^2}\,dx$

$= \displaystyle\lim_{b\to\infty}\left[\frac{-1}{3}e^{-3x^2}\right]_{-b}^b$

$= \displaystyle\lim_{b\to\infty}\left(\frac{-1}{3}e^{-3b^2} - \frac{-1}{3}e^{-3b^2}\right)$

As $b\to\pm\infty$, $e^b\to\infty$. Thus the limit does not exist. The improper integral is divergent.

17. $\displaystyle\int_0^\infty 2t^2\,e^{-2t}\,dt = \lim_{b\to\infty}\int_0^b 2t^2\,e^{-2t}\,dt$

$= \displaystyle\lim_{b\to\infty}\left[t^2\,e^{-2t} - t\,e^{-2t} + \frac{1}{2}\,e^{-2t}\right]_0^b$

$= \displaystyle\lim_{b\to\infty}\left(-b^2\,e^{-2b} - b\,e^{-2b} - \frac{1}{2}\,e^{-2b} - \frac{-1}{2}\right)$

$= \displaystyle\frac{1}{2}$

The integral is convergent.

19. $$\int_{-\infty}^{1} 2x\, e^{3x}\, dx = \lim_{b\to-\infty} \int_{b}^{1} 2x\, e^{3x}\, dx$$
$$= \lim_{b\to-\infty} \left[\frac{2}{3}x\, e^{3x} - \frac{2}{9}\, e^{3x}\right]_b^1$$
$$= \lim_{b\to-\infty} \left(\frac{2}{3}e^3 - \frac{2}{9}e^3 - \frac{2}{3}b\, e^{3b} + \frac{2}{9}e^{3b}\right)$$
$$= \frac{4}{9}e^3$$

The integral is convergent.

21. $$\int_{1}^{\infty} \frac{dx}{x^{1/2}} = \lim_{b\to\infty} \int_{1}^{b} x^{-1/2}\, dx$$
$$= \lim_{b\to\infty} \left[2x^{1/2}\right]_1^b$$
$$= \lim_{b\to\infty} \left(2b^{1/2} - 2\right)$$
$$= \infty$$

The integral is divergent.

23. $$\int_{0}^{\infty} \frac{1 - \sin x}{(x + \cos x)^2}\, dx = \lim_{b\to\infty} \int_{0}^{b} (1 - \sin x)(x + \cos x)^{-2}\, dx$$
$$= \lim_{b\to\infty} \left[\frac{-1}{(x + \cos x)}\right]_0^b$$
$$= \lim_{b\to\infty} \left(\frac{-1}{b + \cos b} + 1\right)$$
$$= 1$$

The integral is convergent.

25. $$\int_{1}^{\infty} \frac{1}{x(x+1)^2}\, dx = \lim_{b\to\infty} \int_{1}^{b} \frac{1}{x^3 + 2x^2 + 3x}\, dx$$
$$= \lim_{b\to\infty} \left[\ln x + \frac{1}{x+1} - \ln(x+1)\right]_1^b$$
$$= \lim_{b\to\infty} \left[\ln\frac{x}{x+1} + \frac{1}{x+1}\right]_1^b$$
$$= \left(\ln 1 + 0 - \ln\frac{1}{2} - \frac{1}{2}\right)$$
$$= \ln 2 - \frac{1}{2}$$

The integral is convergent.

27. $$\int_{1}^{\infty} \frac{x}{\sqrt{x+2}}\, dx = \lim_{b\to\infty} \int_{1}^{b} \frac{x}{\sqrt{x+2}}\, dx$$
$$= \lim_{b\to\infty} \left[x - 2 \;- \ln(x+2)\right]_1^b$$
$$= \lim_{b\to\infty} \left[b - 2\ln(b+2) - 1 + 2\ln(3)\right]$$

The limit does not exist. The integral is divergent.

29. $$\int_{e}^{\infty} \frac{\ln x}{x}\, dx = \lim_{b\to\infty} \int_{e}^{b} \frac{\ln x}{x}\, dx$$
$$= \lim_{b\to\infty} \left[\ln^2 x\right]_e^b$$
$$= \lim_{b\to\infty} \left[\ln^2 b - 1\right]$$

The limit does not exist. The integral is divergent.

31. $$\int_{-\infty}^{\infty} 3xe^{-x^2/2}\, dx = \lim_{b\to\infty} \int_{-b}^{b} 3xe^{x^2/2}\, dx$$
$$= \lim_{b\to\infty} \left[-3\, e^{-x^2/2}\right]_{-b}^{b}$$
$$= \lim_{b\to\infty} \left[-3\, e^{-b^2/2} + 3\, e^{-b^2/2}\right]$$
$$= 0$$

The integral is convergent.

33. $\int_0^\infty m\, e^{-mx}\, dx,\ m > 0$
$$= \lim_{b\to\infty} \int_{0}^{b} m\, e^{-mx}\, dx$$
$$= \lim_{b\to\infty} \left[\frac{m}{-m}e^{-mx}\right]_0^b$$
$$= \lim_{b\to\infty} \left[-\, e^{-mx}\right]_0^b$$
$$= \lim_{b\to\infty} [-e^{-mb} - (-e^{-m\cdot 0})]$$
$$= \lim_{b\to\infty} \left(1 - \frac{1}{e^{mb}}\right)$$
$$= 1$$

The limit does exist. Thus the improper integral is convergent.

35. The area is given by
$$\int_{2}^{\infty} \frac{1}{x^2}\, dx = \lim_{b\to\infty} \int_{2}^{b} x^{-2}\, dx$$
$$= \lim_{b\to\infty} \left[-\, x^{-1}\right]_2^b$$
$$= \lim_{b\to\infty} \left[-\frac{1}{x}\right]_2^b$$
$$= \lim_{b\to\infty} \left(-\frac{1}{b} - \left(-\frac{1}{2}\right)\right)$$
$$= \frac{1}{2}$$

The area of the region is $\frac{1}{2}$.

37. The area is given by
$\int_0^\infty 2x\, e^{-x^2}\, dx$
$$= \lim_{b\to\infty} \int_{0}^{b} 2x\, e^{-x^2}\, dx$$
(We use the substitution $u = -x^2$ to integrate.)
$$= \lim_{b\to\infty} \left[-\, e^{-x^2}\right]_0^b$$
$$= \lim_{b\to\infty} [-e^{-b^2} - (-e^{-0^2})]$$
$$= \lim_{b\to\infty} \left(-\frac{1}{e^{b^2}} + 1\right)$$
$$= 1 \qquad \left(\text{As } b \to \infty,\ -\frac{1}{e^{b^2}} \to 0 \text{ and } -\frac{1}{e^{b^2}} + 1 \to 1.\right)$$

The area of the region is 1.

39. Note that 60.1 days $= \dfrac{60.1}{365}$ yr ≈ 0.164658.

a)
$$\begin{aligned}
\frac{1}{2}P_0 &= P_0\, e^{-k(0.164658)}\\
0.5 &= e^{-0.164658k}\\
\ln 0.5 &= \ln e^{-0.164658k}\\
\ln 0.5 &= -0.164658k\\
\frac{\ln 0.5}{-0.164658} &= k\\
4.20963 &\approx k
\end{aligned}$$

The decay rate is 420.963% per year.

b) The first month is $\dfrac{1}{12}$ yr.

$$\begin{aligned}
E &= \textstyle\int_0^{1/12} 10\, e^{-4.20963t}\, dt\\
&= \frac{10}{-4.29063}\left[e^{-4.20963t}\right]_0^{1/12}\\
&= \frac{10}{-4.20963}\left[e^{-4.20963(\frac{1}{12})} - e^{-4.20963(0)}\right]\\
&= \frac{10}{-4.20963}(e^{-0.3508025} - 1)\\
&\approx 0.702858 \text{ rems}
\end{aligned}$$

c)
$$\begin{aligned}
E &= \textstyle\int_0^{\infty} 10\, e^{-4.20963t}\, dt\\
&= \lim_{b\to\infty}\int_0^b 10\, e^{-4.20963t}\, dt\\
&= \frac{10}{4.20963} \qquad \left[\int_0^{\infty} Pe^{-kt}\, dt = \frac{P}{k}\right]\\
&\approx 2.37551 \text{ rems}
\end{aligned}$$

41.
$$\begin{aligned}
\frac{P}{k} &= \frac{1}{0.0000286}\\
&\approx 34965 \text{ lbs}
\end{aligned}$$

43.
$$\begin{aligned}
\int_1^{\infty} x^r\, dx &= \lim_{b\to\infty}\int_1^b x^r\, dx\\
&= \lim_{b\to\infty}\left[\frac{1}{r+1}\; x^{r+1}\right]_1^b
\end{aligned}$$

In order for the limit to converge, the exponent $r+1$ must be negative, that is $r+1<0$.
Therefore, the integral is convergent for $r<-1$ and divergent otherwise.

45.
$$\begin{aligned}
\int_0^{\infty} te^{-kt}\, dt &= \lim_{b\to\infty}\int_0^b te^{-kt}\, dt\\
&= \lim_{b\to\infty}\left[\frac{-te^{-kt}}{k} - \frac{e^{-kt}}{k^2}\right]_0^b\\
&= \lim_{b\to\infty}\left[\frac{-be^{-bk}}{k} - e^{-bk} + 0 + \frac{1}{k^2}\right]\\
&= \frac{1}{k^2}
\end{aligned}$$

The total amount of the drug dosage that goes through the body is $\dfrac{1}{k^2}$.

47. If $y = \dfrac{1}{x^2}$ then $\displaystyle\int y\, dx = \frac{-1}{x}$

$$\begin{aligned}
\int_1^{\infty} y\, dx &= \lim_{b\to\infty}\left[\frac{-1}{b} + 1\right]\\
&= 1
\end{aligned}$$

If $y = \dfrac{1}{x}$ then $\displaystyle\int y\, dx = \ln x$

$$\begin{aligned}
\int_1^{\infty} y\, dx &= \lim_{b\to\infty}\left[\ln b - 0\right]\\
&= \infty
\end{aligned}$$

The region under $y = \dfrac{1}{x^2}$ could be painted. Since the integral of that y is convergent while the other integral is divergent.

49. $\displaystyle\int_1^{\infty} \frac{4}{1+x^2}\, dx = \pi$

Chapter 6

Matrices

Exercise Set 6.1

1.

$$A+B = \begin{bmatrix} 4+3 & -1+9 \\ 7+2 & -9+(-2) \end{bmatrix} = \begin{bmatrix} 7 & 8 \\ 9 & -11 \end{bmatrix}$$

$$B+A = \begin{bmatrix} 3+4 & 9+(-1) \\ 2+7 & -2+(-9) \end{bmatrix} = \begin{bmatrix} 7 & 8 \\ 9 & -11 \end{bmatrix}$$

3.

$$B+C = \begin{bmatrix} 3+8 & 9+3 \\ 2+0 & -2+(-3) \end{bmatrix} = \begin{bmatrix} 11 & 12 \\ 2 & -5 \end{bmatrix}$$

5. Not possible, dimensions not the same

7.

$$B2 = \begin{bmatrix} 3\cdot 2 & 9\cdot 2 \\ 2\cdot 2 & -2\cdot 2 \end{bmatrix} = \begin{bmatrix} 6 & 19 \\ 4 & -4 \end{bmatrix}$$

9.

$$A-B = \begin{bmatrix} 4-3 & -1-9 \\ 7-2 & -9-(-2) \end{bmatrix} = \begin{bmatrix} 1 & -10 \\ 5 & -7 \end{bmatrix}$$

$$\begin{aligned} A+(-1)B &= \begin{bmatrix} 4+(-1)(3) & -1+(-1)(9) \\ 7+(-1)(2) & -9+(-1)(-2) \end{bmatrix} \\ &= \begin{bmatrix} 1 & -10 \\ 5 & -7 \end{bmatrix} \end{aligned}$$

The answers are identical.

11.

$$\begin{aligned} AD &= \begin{bmatrix} 20-10 & -24-3 & 4+1 \\ 35+90 & -42-27 & 7+9 \end{bmatrix} \\ &= \begin{bmatrix} 10 & -27 & 5 \\ -55 & -69 & 16 \end{bmatrix} \end{aligned}$$

13.

$$\begin{aligned} A^3v &= \begin{bmatrix} 71 & -54 \\ 378 & -631 \end{bmatrix} \begin{bmatrix} 2 \\ -3 \end{bmatrix} \\ &= \begin{bmatrix} (71)(2)+(-54)(-3) \\ (378)(2)+(-631)(-3) \end{bmatrix} = \begin{bmatrix} 304 \\ 2649 \end{bmatrix} \end{aligned}$$

15.

$$\begin{aligned} B^3v &= \begin{bmatrix} 99 & 225 \\ 50 & -26 \end{bmatrix} \begin{bmatrix} 2 \\ -3 \end{bmatrix} \\ &= \begin{bmatrix} (99)(2)+(225)(-3) \\ (50)(2)+(-26)(-3) \end{bmatrix} = \begin{bmatrix} -477 \\ 178 \end{bmatrix} \end{aligned}$$

17.

$$DE = \begin{bmatrix} -35+18+2 & 20-12-1 \\ -70-9-2 & 40+6+1 \end{bmatrix} = \begin{bmatrix} -15 & 7 \\ -81 & 47 \end{bmatrix}$$

$$\begin{aligned} A+DE &= \begin{bmatrix} 4 & -1 \\ 7 & -9 \end{bmatrix} + \begin{bmatrix} -15 & 7 \\ -81 & 47 \end{bmatrix} \\ &= \begin{bmatrix} -11 & 6 \\ -74 & 38 \end{bmatrix} \end{aligned}$$

19.

$$BD = \begin{bmatrix} 15+90 & -18+27 & 3-9 \\ 10-20 & -12-6 & 2+2 \end{bmatrix} = \begin{bmatrix} 105 & 9 & -6 \\ -10 & -18 & 4 \end{bmatrix}$$

$$\begin{aligned} (BD)F &= \begin{bmatrix} 105 & 9 & -6 \\ -10 & -18 & 4 \end{bmatrix} \begin{bmatrix} -4 & 2 & 3 \\ 0 & -1 & 2 \\ -7 & -2 & 5 \end{bmatrix} \\ &= \begin{bmatrix} -420+0+42 & 210-9+12 & 315+18-30 \\ 40+0-28 & -20+18-8 & -30-36+20 \end{bmatrix} \\ &= \begin{bmatrix} -378 & 213 & 303 \\ 12 & -10 & -46 \end{bmatrix} \end{aligned}$$

$$\begin{aligned} DF &= \begin{bmatrix} -20+0-7 & 10+6-2 & 15-12+5 \\ -40+0+7 & 20-3+2 & 30+6-5 \end{bmatrix} \\ &= \begin{bmatrix} -27 & 14 & 8 \\ -33 & 19 & 31 \end{bmatrix} \end{aligned}$$

$$\begin{aligned} B(DF) &= \begin{bmatrix} 3 & 9 \\ 2 & -2 \end{bmatrix} \begin{bmatrix} -27 & 14 & 8 \\ -33 & 19 & 31 \end{bmatrix} \\ &= \begin{bmatrix} -81-297 & 42+171 & 24+279 \\ -54+66 & 28-38 & 16-62 \end{bmatrix} \\ &= \begin{bmatrix} -378 & 213 & 303 \\ 12 & -10 & -46 \end{bmatrix} \end{aligned}$$

21. $(B+C) = \begin{bmatrix} 11 & 12 \\ 2 & -5 \end{bmatrix}$

$$\begin{aligned} A(B+C) &= \begin{bmatrix} 4 & -1 \\ 7 & -9 \end{bmatrix} \begin{bmatrix} 11 & 12 \\ 2 & -5 \end{bmatrix} \\ &= \begin{bmatrix} 44-2 & 48+5 \\ 77-18 & 84+45 \end{bmatrix} = \begin{bmatrix} 42 & 53 \\ 59 & 129 \end{bmatrix} \end{aligned}$$

$$\begin{aligned} AB &= \begin{bmatrix} 12-2 & 36+2 \\ 21-18 & 63+18 \end{bmatrix} = \begin{bmatrix} 10 & 38 \\ 3 & 81 \end{bmatrix} \\ AC &= \begin{bmatrix} 32+0 & 12+3 \\ 56+0 & 21+27 \end{bmatrix} = \begin{bmatrix} 32 & 15 \\ 56 & 48 \end{bmatrix} \\ AB+AC &= \begin{bmatrix} 10+32 & 38+15 \\ 3+56 & 81+48 \end{bmatrix} = \begin{bmatrix} 42 & 53 \\ 59 & 129 \end{bmatrix} \end{aligned}$$

23.

$$AB = \begin{bmatrix} 0+30-12 & 4+0-4 & -2-18-2 \\ 0-10-6 & 16+0-2 & -8+6-1 \\ 0+15-30 & 0+0-10 & 0-9-5 \end{bmatrix} = \begin{bmatrix} 18 & 0 & -22 \\ -16 & 14 & -3 \\ -15 & -10 & -14 \end{bmatrix}$$

$$BA = \begin{bmatrix} 0+16+0 & 0-8-6 & 0-4+10 \\ 5+0+0 & 30+0-9 & -10+0+15 \\ 6+8+0 & 36-4+3 & -12-2-5 \end{bmatrix} = \begin{bmatrix} 16 & -14 & 6 \\ 5 & 21 & 5 \\ 14 & 35 & -19 \end{bmatrix}$$

Clearly, $AB \neq BA$

25.

$$A(B-C) = \begin{bmatrix} 1 & 6 & -2 \\ 4 & -2 & -1 \\ 0 & 3 & -5 \end{bmatrix}\begin{bmatrix} -4 & 0 & -2 \\ 7 & -3 & -11 \\ 5 & 5 & -5 \end{bmatrix} = \begin{bmatrix} -4+42-10 & 0-18-10 & -2-66+10 \\ 16-14-5 & 0+6-5 & -8+22+5 \\ 0+21-25 & 0-9-25 & 0-33+25 \end{bmatrix} = \begin{bmatrix} 28 & -28 & -58 \\ -35 & 1 & 19 \\ -4 & -34 & -8 \end{bmatrix}$$

$$AB = \begin{bmatrix} 18 & 0 & -22 \\ -16 & 14 & -3 \\ -15 & -10 & -14 \end{bmatrix}$$

$$AC = \begin{bmatrix} 4-12-2 & 4+18+6 & 0+48-12 \\ 16+4-1 & 16-6+3 & 0-16-6 \\ 0-6-5 & 0+9+15 & 0+24-30 \end{bmatrix} = \begin{bmatrix} -10 & 28 & 36 \\ 19 & 13 & -22 \\ -11 & 24 & -6 \end{bmatrix}$$

$$AB - BC = \begin{bmatrix} 18 & 0 & -22 \\ 16 & 14 & -3 \\ -15 & -10 & -14 \end{bmatrix} - \begin{bmatrix} -10 & 28 & 36 \\ 19 & 13 & -22 \\ -11 & 24 & -6 \end{bmatrix} = \begin{bmatrix} 28 & -28 & -58 \\ -35 & 1 & 19 \\ -4 & -34 & -8 \end{bmatrix}$$

27. a)

$$\begin{bmatrix} 0 & 4 & -2 \\ 5 & 0 & -3 \\ 6 & 2 & 1 \end{bmatrix}\begin{bmatrix} 1 \\ 4 \\ 0 \end{bmatrix} = \begin{bmatrix} 0+16+0 \\ 5+0+0 \\ 6+8+0 \end{bmatrix} = \begin{bmatrix} 16 \\ 5 \\ 14 \end{bmatrix}$$

b)

$$\begin{bmatrix} 0 & 4 & -2 \\ 5 & 0 & -3 \\ 6 & 2 & 1 \end{bmatrix}\begin{bmatrix} 6 \\ -2 \\ 3 \end{bmatrix} = \begin{bmatrix} 0-8-6 \\ 30+0-9 \\ 36-4+3 \end{bmatrix} = \begin{bmatrix} -14 \\ 21 \\ 35 \end{bmatrix}$$

c)

$$\begin{bmatrix} 0 & 4 & -2 \\ 5 & 0 & -3 \\ 6 & 2 & 1 \end{bmatrix}\begin{bmatrix} -2 \\ -1 \\ -5 \end{bmatrix} = \begin{bmatrix} 0-4+10 \\ -10+0+15 \\ -12-2-5 \end{bmatrix} = \begin{bmatrix} 6 \\ 5 \\ -19 \end{bmatrix}$$

d) The multiplication is only defined in the direction stated in parts a), b) and c)

29.

$$B - 4I = \begin{bmatrix} 0 & 4 & -2 \\ 5 & 0 & -3 \\ 6 & 2 & 1 \end{bmatrix} - \begin{bmatrix} 4 & 0 & 0 \\ 0 & 4 & 0 \\ 0 & 0 & 4 \end{bmatrix} = \begin{bmatrix} 0-4 & 4-0 & -2-0 \\ 5-0 & 0-4 & -3-0 \\ 6-0 & 2-0 & 1-4 \end{bmatrix} = \begin{bmatrix} -4 & 4 & -2 \\ 5 & -4 & -3 \\ 6 & 2 & -3 \end{bmatrix}$$

31. a) Left to the student

b) $\begin{bmatrix} 1.4 & 1.8 \\ 0.5 & 0.4 \end{bmatrix}\begin{bmatrix} 56 \\ 20 \end{bmatrix} = \begin{bmatrix} 114 \\ 36 \end{bmatrix}$

c) $\begin{bmatrix} 1.4 & 1.8 \\ 0.5 & 0.4 \end{bmatrix}\begin{bmatrix} 114 \\ 20 \end{bmatrix} = \begin{bmatrix} 225 \\ 72 \end{bmatrix}$

33. a) $\begin{bmatrix} 0.8 & 2 \\ 0.5 & 0.4 \end{bmatrix}$

b) $\begin{bmatrix} 0.8 & 2 \\ 0.5 & 0.4 \end{bmatrix}\begin{bmatrix} 1200 \\ 1520 \end{bmatrix} = \begin{bmatrix} 4000 \\ 1208 \end{bmatrix}$

c) $\begin{bmatrix} 0.8 & 2 \\ 0.5 & 0.4 \end{bmatrix}\begin{bmatrix} 4000 \\ 1208 \end{bmatrix} = \begin{bmatrix} 5616 \\ 2483 \end{bmatrix}$

35. a) Left to the student

b) $\begin{bmatrix} 15 & 7.5 \\ 0.8 & 0 \end{bmatrix}$

c) $\begin{bmatrix} 0.8 & 2 \\ 0.5 & 0.4 \end{bmatrix}\begin{bmatrix} 20 \\ 0 \end{bmatrix} = \begin{bmatrix} 300 \\ 16 \end{bmatrix}$

d) $\begin{bmatrix} 0.8 & 2 \\ 0.5 & 0.4 \end{bmatrix}\begin{bmatrix} 300 \\ 16 \end{bmatrix} = \begin{bmatrix} 4620 \\ 240 \end{bmatrix}$

37. a)

$$A^2 = \begin{bmatrix} 3 & -2 \\ 4 & 5 \end{bmatrix}\begin{bmatrix} 3 & -2 \\ 4 & 5 \end{bmatrix} = \begin{bmatrix} 9-8 & -6-10 \\ 12+20 & -8+25 \end{bmatrix} = \begin{bmatrix} 1 & -16 \\ 32 & 17 \end{bmatrix}$$

b)

$$\begin{aligned} B^2 &= \begin{bmatrix} -2 & 7 \\ 1 & -3 \end{bmatrix}\begin{bmatrix} -2 & 7 \\ 1 & -3 \end{bmatrix} \\ &= \begin{bmatrix} 4+7 & -14-21 \\ -2-3 & 7+9 \end{bmatrix} = \begin{bmatrix} 11 & -35 \\ -5 & 16 \end{bmatrix} \end{aligned}$$

c)

$$\begin{aligned} (A+B)^2 &= \begin{bmatrix} 1 & 5 \\ 5 & 2 \end{bmatrix}\begin{bmatrix} 1 & 5 \\ 5 & 2 \end{bmatrix} \\ &= \begin{bmatrix} 1+25 & 5+10 \\ 5+10 & 25+4 \end{bmatrix} = \begin{bmatrix} 26 & 15 \\ 15 & 29 \end{bmatrix} \end{aligned}$$

d)

$$\begin{aligned} A^2+2AB+B^2 &= \begin{bmatrix} 1 & -16 \\ 32 & 17 \end{bmatrix} + 2\begin{bmatrix} -8 & 27 \\ -3 & 13 \end{bmatrix} + \\ & \begin{bmatrix} 11 & -35 \\ -5 & 16 \end{bmatrix} \\ &= \begin{bmatrix} 1 & -16 \\ 32 & 17 \end{bmatrix} + \begin{bmatrix} -16 & 54 \\ -6 & 26 \end{bmatrix} + \\ & \begin{bmatrix} 11 & -35 \\ -5 & 16 \end{bmatrix} \\ &= \begin{bmatrix} 1-16+11 & -16+54-35 \\ 32-6-5 & 17+26+16 \end{bmatrix} \\ &= \begin{bmatrix} -4 & 3 \\ 21 & 59 \end{bmatrix} \end{aligned}$$

e - f) $(A+B)^2 = A^2 + AB + BA + B^2$ the middle two terms are not equal since matrix multiplication is not commutative in all cases. Therefore $(A+B)^2 \neq A^2 + 2AB + B^2$

39. **a)** A^T is $1 \times n$ and B^T is $n \times 1$

b) $A \cdot B$ will have dimensions of 1×1

$$\begin{aligned} A \cdot B &= \begin{bmatrix} a_{11} & a_{12} & a_{13} & \cdots & a_{1n} \end{bmatrix} \cdot \begin{bmatrix} b_{11} \\ b_{21} \\ b_{31} \\ \vdots \\ b_{n1} \end{bmatrix} \\ &= a_{11}b_{11} + a_{12}b_{21} + a_{13}b_{31} + \cdots + a_{1n}b_{n1} \end{aligned}$$

$B^T \cdot A^T$ will have dimensions of 1×1

$$\begin{aligned} B^T \cdot A^T &= \begin{bmatrix} b_{11} & b_{12} & b_{13} & \cdots & b_{1n} \end{bmatrix} \cdot \begin{bmatrix} a_{11} \\ a_{21} \\ a_{31} \\ \vdots \\ a_{n1} \end{bmatrix} \\ &= b_{11}a_{11} + b_{12}a_{21} + b_{13}a_{31} + \cdots + b_{1n}a_{n1} \end{aligned}$$

Thus, $A \cdot B = B^T \cdot A^T$

41. This is true due to the associative property of addition for real numbers

43. Since each entry in A is added to its additive inverse each entry in $A + (-A)$ will be zero

45. - 49.] Left to the student

51. $\begin{bmatrix} 200502 \\ 63757 \end{bmatrix}$

53. $\begin{bmatrix} 3.414 \times 10^7 \\ 1.399 \times 10^7 \end{bmatrix}$

55. $\begin{bmatrix} 7.644 \times 10^9 \\ 1.021 \times 10^9 \\ 1.568 \times 10^8 \end{bmatrix}$

Exercise Set 6.2

1. From the second equation we have $y = 2x$. Then

$$\begin{aligned} x + 2y &= 5 \\ x + 2(2x) &= 5 \\ x + 4x &= 5 \\ 5x &= 5 \\ x &= 1 \end{aligned}$$

Which means $y = 2(1) = 2$

3. From the first equation we have $z = 5w - 14$. Then

$$\begin{aligned} 2w + 3z &= 26 \\ 2w + 3(5w - 14) &= 26 \\ 2w + 15w - 42 &= 26 \\ 17w - 42 &= 26 \\ 17w &= 26 + 42 = 68 \\ w &= 4 \end{aligned}$$

Which means $z = 5(4) - 14 = 6$

5. From the first equation we have $s = t + 7$. Then

$$\begin{aligned} -2s + 2t &= -5 \\ -2(t+7) + 2t &= -5 \\ -2t - 14 + 2t &= -5 \\ -14 &= -5 \end{aligned}$$

This is a contradiction, which means there is no solution.

7. From the first equation we have $x = y + 7$. Then

$$\begin{aligned} -2x + 2y &= -14 \\ -2(y+7) + 2y &= -14 \\ -2y - 14 + 2y &= -14 \\ -14 &= -14 \end{aligned}$$

This is an identity, which means there are many solutions

9. - 15.] Left to the student

17. Write the augmented matrix

$$\left[\begin{array}{ccc|c} 0 & 1 & 3 & -1 \\ 1 & 0 & 6 & 37 \\ 0 & 2 & 1 & -2 \end{array}\right]$$

Interchange $R1$ and $R2$

$$\left[\begin{array}{ccc|c} 1 & 0 & 6 & 37 \\ 0 & 1 & 3 & -1 \\ 0 & 2 & 1 & -2 \end{array}\right]$$

$-2R2 + R3$

$$\left[\begin{array}{ccc|c} 1 & 0 & 6 & 37 \\ 0 & 1 & 3 & -1 \\ 0 & 0 & -5 & 0 \end{array}\right]$$

This means the $z = 0$
$y + 3(0) = -1 \rightarrow y = -1$
$x + 6(0) = 37 \rightarrow x = 37$

19. Write the augmented matrix

$$\left[\begin{array}{ccc|c} 7 & -1 & -9 & 1 \\ 2 & 0 & -4 & -4 \\ -4 & 0 & 6 & -3 \end{array}\right]$$

Interchange $R2$ and $R1$

$$\left[\begin{array}{ccc|c} 2 & 0 & -4 & -4 \\ 7 & -1 & -9 & 1 \\ -4 & 0 & 6 & -3 \end{array}\right]$$

$R1/2$

$$\left[\begin{array}{ccc|c} 1 & 0 & -2 & -2 \\ 7 & -1 & -9 & 1 \\ -4 & 0 & 6 & -3 \end{array}\right]$$

$-7R1 + R2$

$$\left[\begin{array}{ccc|c} 1 & 0 & -2 & -2 \\ 0 & -1 & 5 & 15 \\ -4 & 0 & 6 & -3 \end{array}\right]$$

$4R1 + R3$

$$\left[\begin{array}{ccc|c} 1 & 0 & -2 & -2 \\ 0 & -1 & 5 & 15 \\ 0 & 0 & -2 & -11 \end{array}\right]$$

$-R2$

$$\left[\begin{array}{ccc|c} 1 & 0 & -2 & -2 \\ 0 & 1 & -5 & -15 \\ 0 & 0 & -2 & -11 \end{array}\right]$$

This means $z = 11/2 = 5.5$
$y - 5(5.5) = -15 \rightarrow y = 12.5$
$x - 2(5.5) = -2 \rightarrow x = 9$

21. Write the augmented matrix

$$\left[\begin{array}{ccc|c} 2 & -2 & 3 & 3 \\ 4 & -3 & 3 & 2 \\ -1 & 1 & -1 & 4 \end{array}\right]$$

Interchange $R1$ and $-R3$

$$\left[\begin{array}{ccc|c} 1 & -1 & 1 & -4 \\ 4 & -3 & 3 & 2 \\ 2 & -2 & 3 & 3 \end{array}\right]$$

$-4R1 + R2$

$$\left[\begin{array}{ccc|c} 1 & -1 & 1 & -4 \\ 0 & 1 & -1 & 18 \\ 2 & -2 & 3 & 3 \end{array}\right]$$

$-2R1 + R3$

$$\left[\begin{array}{ccc|c} 1 & -1 & 1 & -4 \\ 0 & 1 & -1 & 18 \\ 0 & 0 & 1 & 11 \end{array}\right]$$

This means $z = 11$
$y - 11 = 18 \rightarrow y = 29$
$x - 29 + 11 = -4 \rightarrow x = 14$

23. - 27.] Left to the student

29. Write the augmented matrix

$$\left[\begin{array}{ccc|c} 1 & 1 & 2 & 5 \\ 1 & 1 & 1 & -10 \\ 2 & 3 & 4 & 2 \end{array}\right]$$

$-R1 + R2$

$$\left[\begin{array}{ccc|c} 1 & 1 & 2 & 5 \\ 0 & 0 & -1 & -15 \\ 2 & 3 & 4 & 2 \end{array}\right]$$

Interchange $-R2$ and $R3$

$$\left[\begin{array}{ccc|c} 1 & 1 & 2 & 5 \\ 2 & 3 & 4 & 2 \\ 0 & 0 & 1 & 15 \end{array}\right]$$

$-2R1 + R2$

$$\left[\begin{array}{ccc|c} 1 & 1 & 2 & 5 \\ 0 & 1 & 0 & -5 \\ 0 & 0 & 1 & 15 \end{array}\right]$$

This means $z = 15$
$y = -5$
$x + (-5) + 2(15) = 5 \rightarrow x = -17$

31. Write the augmented matrix

$$\left[\begin{array}{ccc|c} 1 & 1 & -2 & 4 \\ 4 & 7 & 3 & 3 \\ 14 & 23 & 5 & 17 \end{array}\right]$$

$-4R1 + R2$

$$\left[\begin{array}{ccc|c} 1 & 1 & -2 & 4 \\ 0 & 3 & 11 & -13 \\ 14 & 23 & 5 & 17 \end{array}\right]$$

$-14R1 + R3$

$$\left[\begin{array}{ccc|c} 1 & 1 & -2 & 4 \\ 0 & 3 & 11 & -13 \\ 0 & 9 & 33 & -39 \end{array}\right]$$

$-3R2 + R3$

$$\left[\begin{array}{ccc|c} 1 & 1 & -2 & 4 \\ 0 & 3 & 11 & -13 \\ 0 & 0 & 0 & 0 \end{array}\right]$$

This means the system has many solutions.
Let $z = z$, then $y = -\frac{11}{3}z - \frac{13}{3}$
$x = \frac{11}{3}z + \frac{13}{3} + 2z + 4 = \frac{17}{3}z + \frac{25}{3}$

33. Write the augmented matrix

$$\left[\begin{array}{ccc|c} 1 & -1 & 3 & 2 \\ 2 & 3 & -1 & 5 \\ -1 & -9 & 11 & 1 \end{array}\right]$$

$-2R1 + R2$

$$\left[\begin{array}{ccc|c} 1 & -1 & 3 & 2 \\ 0 & 5 & -7 & 1 \\ -1 & -9 & 11 & 1 \end{array}\right]$$

$R1 + R3$

$$\left[\begin{array}{ccc|c} 1 & -1 & 3 & 2 \\ 0 & 5 & -7 & 1 \\ 0 & -10 & 14 & 3 \end{array}\right]$$

$2R2 + R3$

$$\left[\begin{array}{ccc|c} 1 & -1 & 3 & 2 \\ 0 & 5 & -7 & 1 \\ 0 & 0 & 0 & 5 \end{array}\right]$$

This means the system has no solution

35. Write the augmented matrix

$$\left[\begin{array}{ccc|c} 1 & -2 & -5 & 0 \\ 2 & 3 & 15 & 0 \\ -2 & -1 & -8 & 1 \end{array}\right]$$

$-2R1 + R2$

$$\left[\begin{array}{ccc|c} 1 & -2 & -5 & 0 \\ 0 & 7 & 25 & 0 \\ -2 & -1 & -8 & 1 \end{array}\right]$$

$2R1 + R3$

$$\left[\begin{array}{ccc|c} 1 & -2 & -5 & 0 \\ 0 & 7 & 25 & 0 \\ 0 & -5 & -18 & 1 \end{array}\right]$$

$\frac{1}{7}R2$

$$\left[\begin{array}{ccc|c} 1 & -2 & -5 & 0 \\ 0 & 1 & \frac{25}{7} & 0 \\ 0 & -5 & -18 & 1 \end{array}\right]$$

$5R2 + R3$

$$\left[\begin{array}{ccc|c} 1 & -2 & -5 & 0 \\ 0 & 1 & \frac{25}{7} & 0 \\ 0 & 0 & -\frac{1}{7} & 1 \end{array}\right]$$

This means $z = -7$
$y = -\frac{25}{7} \cdot -7 = 25$
$x = 2(25) + 5(-7) = 15$

37. Write the augmented matrix

$$\left[\begin{array}{cccc|c} 1 & 1 & 1 & 1 & 5 \\ 1 & 0 & 1 & 1 & 6 \\ 0 & 1 & 1 & 1 & 4 \\ 1 & 0 & 1 & 0 & 3 \end{array}\right]$$

$-R1 + R2$

$$\left[\begin{array}{cccc|c} 1 & 1 & 1 & 1 & 5 \\ 0 & -1 & 0 & 0 & 1 \\ 0 & 1 & 1 & 1 & 4 \\ 1 & 0 & 1 & 0 & 3 \end{array}\right]$$

$-R1 + R4$ and $-R2$

$$\left[\begin{array}{cccc|c} 1 & 1 & 1 & 1 & 5 \\ 0 & 1 & 0 & & -1 \\ 0 & 1 & 1 & 1 & 4 \\ 0 & -1 & 0 & -1 & -2 \end{array}\right]$$

$-R2 + R3$

$$\left[\begin{array}{cccc|c} 1 & 1 & 1 & 1 & 5 \\ 0 & 1 & 0 & & -1 \\ 0 & 0 & 1 & 1 & 5 \\ 0 & -1 & 0 & -1 & -2 \end{array}\right]$$

$R2 + R4$

$$\left[\begin{array}{cccc|c} 1 & 1 & 1 & 1 & 5 \\ 0 & 1 & 0 & & -1 \\ 0 & 0 & 1 & 1 & 5 \\ 0 & 0 & 0 & -1 & -3 \end{array}\right]$$

This means $w = 3$
$z = -3 + 5 = 2$
$y = -1$
$x = 1 - 2 - 3 + 5 = 1$

39. Write the augmented matrix

$$\left[\begin{array}{cccc|c} -2 & -1 & 6 & -1 & 2 \\ -3 & -5 & 6 & 1 & -3 \\ 1 & 1 & -2 & 0 & 4 \\ 0 & 1 & -1 & 0 & 1 \end{array}\right]$$

Interchange $R1$ and $R3$

$$\left[\begin{array}{cccc|c} 1 & 1 & -2 & 0 & 4 \\ -3 & -5 & 6 & 1 & -3 \\ -2 & -1 & 6 & -1 & 2 \\ 0 & 1 & -1 & 0 & 1 \end{array}\right]$$

$3R1 + R2$

$$\left[\begin{array}{cccc|c} 1 & 1 & -2 & 0 & 4 \\ 0 & -2 & 0 & 1 & 9 \\ -2 & -1 & 6 & -1 & 2 \\ 0 & 1 & -1 & 0 & 1 \end{array}\right]$$

$2R1 + R3$

$$\left[\begin{array}{cccc|c} 1 & 1 & -2 & 0 & 4 \\ 0 & -2 & 0 & 1 & 9 \\ 0 & 1 & 2 & -1 & 10 \\ 0 & 1 & -1 & 0 & 1 \end{array}\right]$$

Interchange $R2$ and $R4$

$$\left[\begin{array}{cccc|c} 1 & 1 & -2 & 0 & 4 \\ 0 & 1 & -1 & 0 & 1 \\ 0 & 1 & 2 & -1 & 10 \\ 0 & -2 & 0 & 1 & 9 \end{array}\right]$$

$-R2 + R3$ and $2R2 + R4$

$$\left[\begin{array}{cccc|c} 1 & 1 & -2 & 0 & 4 \\ 0 & 1 & -1 & 0 & 1 \\ 0 & 0 & 3 & -1 & 9 \\ 0 & 0 & -2 & 1 & 11 \end{array}\right]$$

$R3 + R4$

$$\left[\begin{array}{cccc|c} 1 & 1 & -2 & 0 & 4 \\ 0 & 1 & -1 & 0 & 1 \\ 0 & 0 & 3 & -1 & 9 \\ 0 & 0 & 1 & 0 & 20 \end{array}\right]$$

This means that $z = 20$
$w = 3(20) - 9 = 51$
$y = 20 + 1 = 21$
$x = -21 + 2(20) + 4 = 23$

41. We are looking for the values of B and F for which $B' = 0$ and $F' = 0$. Thus,

$$0 = -0.01B + 0.1 \to B = 10$$

and

$$0 = 0.01(10) - 0.02F \to F = 5$$

43. Write the augmented matrix

$$\left[\begin{array}{ccc|c} 0 & 0.76 & 0.95 & 579 \\ 0.2 & 0 & 0 & 80 \\ 0 & 0.91 & 0.91 & 609 \end{array}\right]$$

Interchange the $R1$ and $R2$

$$\left[\begin{array}{ccc|c} 0.2 & 0 & 0 & 80 \\ 0 & 0.76 & 0.95 & 579 \\ 0 & 0.91 & 0.91 & 609 \end{array}\right]$$

$R1/0.2$

$$\left[\begin{array}{ccc|c} 1 & 0 & 0 & 400 \\ 0 & 0.76 & 0.95 & 579 \\ 0 & 0.91 & 0.91 & 609 \end{array}\right]$$

$R2/0.76$

$$\left[\begin{array}{ccc|c} 1 & 0 & 0 & 400 \\ 0 & 1 & 1.25 & 761.842 \\ 0 & 0.91 & 0.91 & 609 \end{array}\right]$$

$-0.91R2 + R3$

$$\left[\begin{array}{ccc|c} 1 & 0 & 0 & 400 \\ 0 & 1 & 1.25 & 761.842 \\ 0 & 0 & -0.2275 & -84.276 \end{array}\right]$$

$R3/-0.2275$

$$\left[\begin{array}{ccc|c} 1 & 0 & 0 & 400 \\ 0 & 1 & 1.25 & 761.842 \\ 0 & 0 & 1 & 370.45 \end{array}\right]$$

This means that $A = 370$, $S = 761.842 - 1.25(370) = 299$, and $H = 400$

45. Left to the student. [Answers may vary]

47. **a)** $AB = \begin{bmatrix} 1 & 2 & 3 \\ 16 & 20 & 24 \\ 7 & 8 & 9 \end{bmatrix}$

b) $AC = \begin{bmatrix} 4 & -2 & 7 \\ 4 & 4 & -12 \\ -4 & 3 & 6 \end{bmatrix}$

c) Matrix A is the identity matrix with second row multiplied by 4. Therefore, the effect of multiplying A on the left with another 3×3 matrix will give a matrix with the second row multiplied by 4.

49. **a)** Since (x_0, y_0, z_0) and (x_1, y_1, z_1) are solutions then

$$\begin{aligned} ax_0 + by_0 + cz_0 &= d \\ ax_1 + by_1 + cz_1 &= d \end{aligned}$$

Now consider $(tx_0 + (1-t)x_1, ty_0 + (1-t)y_1, tz_0 + (1-t)z_1)$

$a[tx_0+(1-t)x_1]+b[ty_0+(1-t)y_1]+c[tz_0+(1-t)z_1] = atx_0+ax_1-btx_1+aty_0+by_1-bty_1+ctz_0+cz_1-ctz_1$
Rearranging the terms we get

$$\begin{aligned} &= (atx_0 + bty_0 + ctz_0) + (ax_1 + by_1 + cz_1) \\ &\quad -(atx_1 + bty_1 + ctz_1) \\ &= t(ax_0 + by_0 + z_0) + (ax_1 + by_1 + cz_1) \\ &\quad -t(ax_1 + by_1 + cz_1) \\ &= t(d) + d - t(d) \\ &= d \end{aligned}$$

Therefore, $(tx_0+(1-t)x_1, ty_0+(1-t)y_1, tz_0+(1-t)z_1)$ is also a solution to $ax + by + cz = d$ for any value of t.

b) As seen in the previous part, since any value of t could be used to find a solution of the form $(tx_0 + (1-t)x_1, ty_0 + (1-t)y_1, tz_0 + (1-t)z_1)$, then if a system has two solutions then it has infinitely many solutions.

c) The answers in part b) may be generalized since the technique used in part a) can be easily extended to any number of variables.

51. - 55. Left to the student

Exercise Set 6.3

1.

$$\left[\begin{array}{cc|cc} 1 & 1 & 1 & 0 \\ -1 & 0 & 0 & 1 \end{array}\right] = \left[\begin{array}{cc|cc} 1 & 1 & 1 & 0 \\ 0 & 1 & 1 & 1 \end{array}\right] = \left[\begin{array}{cc|cc} 1 & 0 & 0 & -1 \\ 0 & 1 & 1 & 1 \end{array}\right]$$

Thus, the inverse is $\begin{bmatrix} 0 & -1 \\ 1 & 1 \end{bmatrix}$

3.

$$\left[\begin{array}{cc|cc} 0 & 1 & 1 & 0 \\ 1 & 0 & 0 & 1 \end{array}\right] = \left[\begin{array}{cc|cc} 1 & 0 & 0 & 1 \\ 0 & 1 & 1 & 0 \end{array}\right]$$

Thus, the inverse is $\begin{bmatrix} 0 & 1 \\ 1 & 0 \end{bmatrix}$

5. $det = (3 \cdot 7) - (5 \cdot 4) = 1$
Thus, the inverse is $\frac{1}{1} \cdot \begin{bmatrix} 7 & -4 \\ -5 & 3 \end{bmatrix} = \begin{bmatrix} 7 & -4 \\ -5 & 3 \end{bmatrix}$

7. $det = (3 \cdot -2) - (8 \cdot 7) = 62$
Thus, the inverse is $\frac{1}{-62} \cdot \begin{bmatrix} -2 & -7 \\ -8 & -3 \end{bmatrix} = \begin{bmatrix} 1/31 & 7/62 \\ 4/31 & -3/62 \end{bmatrix}$

9.

$$\left[\begin{array}{ccc|ccc} -2 & 2 & 1 & 1 & 0 & 0 \\ 1 & 2 & 0 & 0 & 1 & 0 \\ 0 & -1 & 0 & 0 & 0 & 1 \end{array}\right] = \left[\begin{array}{ccc|ccc} 1 & 2 & 0 & 0 & 1 & 0 \\ 0 & -1 & 0 & 0 & 0 & 1 \\ -2 & 2 & 1 & 1 & 0 & 0 \end{array}\right]$$

$$= \left[\begin{array}{ccc|ccc} 1 & 2 & 0 & 0 & 1 & 0 \\ 0 & -1 & 0 & 0 & 0 & 1 \\ 0 & 6 & 1 & 1 & 2 & 0 \end{array}\right]$$

$$= \left[\begin{array}{ccc|ccc} 1 & 2 & 0 & 0 & 1 & 0 \\ 0 & -1 & 0 & 0 & 0 & 1 \\ 0 & 0 & 1 & 1 & 2 & 6 \end{array}\right]$$

$$= \left[\begin{array}{ccc|ccc} 1 & 0 & 0 & 0 & 1 & 2 \\ 0 & -1 & 0 & 0 & 0 & 1 \\ 0 & 0 & 1 & 1 & 2 & 6 \end{array}\right]$$

$$= \left[\begin{array}{ccc|ccc} 1 & 0 & 0 & 0 & 1 & 2 \\ 0 & 1 & 0 & 0 & 0 & -1 \\ 0 & 0 & 1 & 1 & 2 & 6 \end{array}\right]$$

Thus, the inverse is $\begin{bmatrix} 0 & 1 & 2 \\ 0 & 0 & -1 \\ 1 & 2 & 6 \end{bmatrix}$

11. Technology was used. Inverse is $\begin{bmatrix} 2 & 1 & -2 \\ 2 & 0 & 1 \\ -1 & 0 & 0 \end{bmatrix}$

13. Technology was used. Inverse is $\begin{bmatrix} 8 & 2 & -1 \\ 8 & 3 & -3 \\ -3 & -1 & 1 \end{bmatrix}$

15. Technology was used. Inverse is $\begin{bmatrix} -20 & -27 & -14 \\ 7 & 10 & 5 \\ 10 & 13 & 7 \end{bmatrix}$

17. Technology was used. Inverse is $\begin{bmatrix} 0 & -3/2 & -1 \\ -1 & -3/2 & -5/2 \\ 0 & -1 & -1/2 \end{bmatrix}$

19. Technology was used. Inverse is $\begin{bmatrix} -7 & -5 & -6 \\ 1 & 1 & 2 \\ 13/2 & 9/2 & 9/2 \end{bmatrix}$

21. Technology was used. Inverse is $\begin{bmatrix} 8/3 & 4/3 & -3/2 \\ 7/3 & 5/3 & -3/2 \\ 19/3 & 8/3 & -7/2 \end{bmatrix}$

23. Technology was used. Inverse is $\begin{bmatrix} 7 & 7 & 37 & 8 \\ 10 & 10 & 53 & 10 \\ -3 & -3 & -16 & -3 \\ 2 & 3 & 13 & 4 \end{bmatrix}$

25. $det = (2 \cdot 6) - (3 \cdot 1) = 12 - 3 = 9$ The matrix is invertible

27. $det = (9 \cdot 3) - (-5 \cdot 2) = 27 + 10 = 37$ The matrix is invertible

29. $det = (2 \cdot 9) - (6 \cdot 3) = 18 - 18 = 0$ The matrix is not invertible

31.

$$\begin{aligned} det &= 1[(5)(2) - (-4)(-3)] - 8[(-2)(2) - (7)(-3)] + \\ &\quad 3[(-2)(-4) - (7)(5)] \\ &= -2 - 136 - 81 \\ &= -219 \end{aligned}$$

The matrix is invertible

33.

$$\begin{aligned} det &= 1[(2)(1) - (1)(3)] - 1[(2)(1) - (3)(1)] + \\ &\quad 1[(2)(3) - (2)(3)] \\ &= -1 + 1 - 0 \\ &= 0 \end{aligned}$$

The matrix is not invertible

35.

$$\begin{aligned} det &= 0[(6)(6) - (14)(2)] - 4[(-1)(6) - (-1)(2)] + \\ &\quad 2[(-1)(14) - (-1)(6)] \\ &= 0 + 16 - 16 \\ &= 0 \end{aligned}$$

The matrix is not invertible

37.

$$\begin{aligned} det &= 2[(-5)(5) - (-1)(7)] + 4[(-3)(5) - (-4)(7)] + \\ &\quad 8[(-3)(-1) - (-5)(-4)] \\ &= -36 + 52 - 136 \\ &= -120 \end{aligned}$$

The matrix is invertible

39. $det = -96$ (technology used) The matrix is invertible

41. $det = 12$ The matrix is invertible

43.

$$\begin{aligned} P_1 &= \begin{bmatrix} 0.5 & 1.25 \\ 0.75 & 0.25 \end{bmatrix}^{-1} \begin{bmatrix} 156 \\ 48 \end{bmatrix} \\ &= \begin{bmatrix} -4/13 & 20\ /13 \\ 12/13 & -8\ /13 \end{bmatrix} \begin{bmatrix} 156 \\ 48 \end{bmatrix} \\ &= \begin{bmatrix} 26 \\ 114 \end{bmatrix} \end{aligned}$$

45. **a)** $G^{-1} = \begin{bmatrix} 0 & 5/4 \\ 2/15 & -5/2 \end{bmatrix}$

b)

$$\begin{aligned} G_2 &= \begin{bmatrix} 0 & 5/4 \\ 2/15 & -5/ \end{bmatrix} \begin{bmatrix} 58815 \\ 3060 \end{bmatrix} \\ &= \begin{bmatrix} 3825 \\ 192 \end{bmatrix} \end{aligned}$$

c)

$$\begin{aligned} G_1 &= \begin{bmatrix} 0 & 5/4 \\ 2/15 & -5/2 \end{bmatrix} \begin{bmatrix} 3825 \\ 192 \end{bmatrix} \\ &= \begin{bmatrix} 240 \\ 30 \end{bmatrix} \end{aligned}$$

47. Left to the student

49. **a)** $AB = \begin{bmatrix} -8 & 25 \\ -10 & 41 \end{bmatrix}$ Thus

$$\begin{aligned} det(AB) &= -328 + 250 = -78 \\ &= -6 \cdot 13 \\ &= det(A)\ det(B) \end{aligned}$$

b) $AB = \begin{bmatrix} 8 & -26 & 20 \\ -2 & -7 & 0 \\ 10 & -34 & 16 \end{bmatrix}$ Thus

$$\begin{aligned} det(AB) &= 1032 \\ &= 12 \cdot 86 \\ &= det(A)\ det(B) \end{aligned}$$

c) Left to the student (answers vary)

51. - 53. Left to the student

55. - 57. Left to the student

Exercise Set 6.4

1.

$$\begin{aligned} \begin{bmatrix} 2 & 0 \\ 0 & 3 \end{bmatrix}\begin{bmatrix} 1 \\ 0 \end{bmatrix} &= \begin{bmatrix} 2 \\ 0 \end{bmatrix} \\ &= 2 \cdot \begin{bmatrix} 1 \\ 0 \end{bmatrix} \end{aligned}$$

Thus, the vector is an eigenvector with eigenvalue of 2

3.

$$\begin{bmatrix} 3 & 2 \\ 0 & 4 \end{bmatrix}\begin{bmatrix} 0 \\ 1 \end{bmatrix} = \begin{bmatrix} 2 \\ 4 \end{bmatrix}$$

The vector is not an eigenvector

5.

$$\begin{aligned} \begin{bmatrix} 5 & 0 \\ 3 & -2 \end{bmatrix}\begin{bmatrix} 0 \\ 1 \end{bmatrix} &= \begin{bmatrix} 0 \\ -2 \end{bmatrix} \\ &= -2 \cdot \begin{bmatrix} 0 \\ 1 \end{bmatrix} \end{aligned}$$

Thus, the vector is an eigenvector with eigenvalue -2

7.

$$\begin{bmatrix} -8.5 & -4.5 \\ 21 & 11 \end{bmatrix}\begin{bmatrix} 2 \\ -7 \end{bmatrix} = \begin{bmatrix} 15.5 \\ -35 \end{bmatrix}$$

The vector is not an eigenvector

9.

$$\begin{aligned} \begin{bmatrix} 5 & 0 & 0 \\ 0 & 3 & 0 \\ 0 & 0 & -2 \end{bmatrix}\begin{bmatrix} 0 \\ 1 \\ 0 \end{bmatrix} &= \begin{bmatrix} 0 \\ 3 \\ 0 \end{bmatrix} \\ &= 3 \cdot \begin{bmatrix} 1 \\ 0 \\ 0 \end{bmatrix} \end{aligned}$$

Thus, the vector is an eigenvector with eigenvalue 5

11.

$$\begin{bmatrix} -25 & 40 & 39 \\ -32 & 47 & 39 \\ 16 & -20 & -1 \end{bmatrix}\begin{bmatrix} -13 \\ -13 \\ 4 \end{bmatrix} = \begin{bmatrix} -39 \\ -39 \\ 48 \end{bmatrix}$$

The vector is not an eigenvector

13. $v = 3w + 4u$

15. $v = 2w + u$

17. $v = -w + 3u$

19.

$$\begin{aligned} \det(A - rI) &= 0 \\ \begin{vmatrix} 1\text{-r} & 0 \\ -1 & 2\text{-r} \end{vmatrix} &= 0 \\ (1-r)(2-r) - 0 &= 0 \\ r &= 1 \\ r &= 2 \end{aligned}$$

For $r = 1$

$$\begin{aligned} \begin{bmatrix} 0 & 0 \\ -1 & 1 \end{bmatrix}\begin{bmatrix} x \\ y \end{bmatrix} &= \begin{bmatrix} 0 \\ 0 \end{bmatrix} \\ \begin{bmatrix} 0 \\ -x+y \end{bmatrix} &= \begin{bmatrix} 0 \\ 0 \end{bmatrix} \\ x &= y \\ x &= t \\ y &= t \end{aligned}$$

Thus for the eigenvalue of $r = 1$ the eigenvector is $\begin{bmatrix} t \\ t \end{bmatrix}, (t \neq 0)$

For $r = 2$

$$\begin{aligned} \begin{bmatrix} -1 & 0 \\ -1 & 0 \end{bmatrix}\begin{bmatrix} x \\ y \end{bmatrix} &= \begin{bmatrix} 0 \\ 0 \end{bmatrix} \\ \begin{bmatrix} -x \\ -y \end{bmatrix} &= \begin{bmatrix} 0 \\ 0 \end{bmatrix} \\ x &= 0 \\ y &= t \end{aligned}$$

Thus, for the eigenvalue of $r = 2$ the eigenvector is $\begin{bmatrix} 0 \\ t \end{bmatrix}, (t \neq 0)$

21.

$$\begin{aligned} \det(A - rI) &= 0 \\ \begin{vmatrix} 5\text{-r} & 2 \\ -24 & -9\text{-r} \end{vmatrix} &= 0 \\ (5-r)(-9-r) - (-48) &= 0 \\ r^2 + 4r + 3 &= 0 \\ r &= -1 \\ r &= -3 \end{aligned}$$

For $r = -1$

$$\begin{bmatrix} 6 & 2 \\ -24 & -8 \end{bmatrix}\begin{bmatrix} x \\ y \end{bmatrix} = \begin{bmatrix} 0 \\ 0 \end{bmatrix}$$
$$\begin{bmatrix} 6x+2y \\ -24x-8y \end{bmatrix} = \begin{bmatrix} 0 \\ 0 \end{bmatrix}$$
$$3x = y$$

Thus for the eigenvalue of $r = -1$ the eigenvector is $\begin{bmatrix} t \\ 3t \end{bmatrix}, (t \neq 0)$

For $r = -3$

$$\begin{bmatrix} 8 & 2 \\ -24 & -6 \end{bmatrix}\begin{bmatrix} x \\ y \end{bmatrix} = \begin{bmatrix} 0 \\ 0 \end{bmatrix}$$
$$\begin{bmatrix} 8x+2y \\ -24x-6y \end{bmatrix} = \begin{bmatrix} 0 \\ 0 \end{bmatrix}$$
$$4x = y$$

Thus, for the eigenvalue of $r = -3$ the eigenvector is $\begin{bmatrix} t \\ -4t \end{bmatrix}, (t \neq 0)$

23.

$$\det(A - rI) = 0$$
$$\begin{vmatrix} -7.5-r & -15.75 \\ 6 & 12-r \end{vmatrix} = 0$$
$$(-7.5 - r)(12 - r) - (-94.5) = 0$$
$$r^2 - 4.5r + 4.5 = 0$$
$$r = 1.5$$
$$r = 3$$

For $r = 1.5$

$$\begin{bmatrix} -9 & -15.75 \\ 6 & 10.5 \end{bmatrix}\begin{bmatrix} x \\ y \end{bmatrix} = \begin{bmatrix} 0 \\ 0 \end{bmatrix}$$
$$\begin{bmatrix} -9x-15.75y \\ 6x+10.5y \end{bmatrix} = \begin{bmatrix} 0 \\ 0 \end{bmatrix}$$
$$4x = -7y$$

Thus for the eigenvalue of $r = 1.5$ the eigenvector is $\begin{bmatrix} 7t \\ -4t \end{bmatrix}, (t \neq 0)$

For $r = 3$

$$\begin{bmatrix} -10.5 & -15.75 \\ 6 & 9 \end{bmatrix}\begin{bmatrix} x \\ y \end{bmatrix} = \begin{bmatrix} 0 \\ 0 \end{bmatrix}$$
$$\begin{bmatrix} -10.5x-15.75y \\ 6x+9y \end{bmatrix} = \begin{bmatrix} 0 \\ 0 \end{bmatrix}$$
$$2x = -3y$$

Thus, for the eigenvalue of $r = 3$ the eigenvector is $\begin{bmatrix} 3t \\ -2t \end{bmatrix}, (t \neq 0)$

25.

$$\det(A - rI) = 0$$
$$\begin{vmatrix} 9.5-r & -4.5 \\ 15 & -7-r \end{vmatrix} = 0$$
$$(9.5 - r)(-7 - r) - (-67.5) = 0$$
$$r^2 - 2.5r + 1 = 0$$
$$r = 0.5$$
$$r = 2$$

For $r = 0.5$

$$\begin{bmatrix} 9 & -4.5 \\ 15 & -7.5 \end{bmatrix}\begin{bmatrix} x \\ y \end{bmatrix} = \begin{bmatrix} 0 \\ 0 \end{bmatrix}$$
$$\begin{bmatrix} 9x-4.5y \\ 15x-7.5y \end{bmatrix} = \begin{bmatrix} 0 \\ 0 \end{bmatrix}$$
$$2x = y$$

Thus for the eigenvalue of $r = 0.5$ the eigenvector is $\begin{bmatrix} t \\ 2t \end{bmatrix}, (t \neq 0)$

For $r = 2$

$$\begin{bmatrix} 7.5 & -4.5 \\ 15 & -9 \end{bmatrix}\begin{bmatrix} x \\ y \end{bmatrix} = \begin{bmatrix} 0 \\ 0 \end{bmatrix}$$
$$\begin{bmatrix} 7.5x-4.5y \\ 15x-9y \end{bmatrix} = \begin{bmatrix} 0 \\ 0 \end{bmatrix}$$
$$5x = 3y$$

Thus, for the eigenvalue of $r = 2$ the eigenvector is $\begin{bmatrix} 3t \\ 5t \end{bmatrix}, (t \neq 0)$

27. $\begin{bmatrix} 10-r & -4 & 15 \\ 8 & -2-r & 15 \\ -4 & 2 & -5-r \end{bmatrix}$

The characteristic equation is $r^3 - 3r^2 + 2r = 0$
The eigenvalues are $r = 0$, $r = 1$, and $r = 2$
For $r = 0$

$$\begin{bmatrix} 10 & -4 & 15 \\ 8 & -2 & 15 \\ -4 & 2 & -5 \end{bmatrix}\begin{bmatrix} x \\ y \\ z \end{bmatrix} = \begin{bmatrix} 0 \\ 0 \\ 0 \end{bmatrix}$$
$$10x - 4y + 15z = 0$$
$$8x - 2y - 15z = 0$$
$$-4x + 2y - 5z = 0$$

Solving the matrix above using any method discussed earlier this chapter yields that $x = 5t$, $y = 5t$, and $z = 2t$, which are the eigenvectors for $r = 0$.

For $r = 1$

$$\begin{bmatrix} 9 & -4 & 15 \\ 8 & -3 & 15 \\ -4 & 2 & -6 \end{bmatrix}\begin{bmatrix} x \\ y \\ z \end{bmatrix} = \begin{bmatrix} 0 \\ 0 \\ 0 \end{bmatrix}$$
$$9x - 4y + 15z = 0$$
$$8x - 3y - 15z = 0$$
$$-4x + 2y - 6z = 0$$

Solving the matrix above using any method discussed earlier this chapter yields that $x = 3t$, $y = 3t$, and $z = t$, which are the eigenvectors for $r = 1$.

For $r = 2$

$$\begin{bmatrix} 8 & -4 & 15 \\ 8 & -4 & 15 \\ -4 & 2 & -7 \end{bmatrix}\begin{bmatrix} x \\ y \\ z \end{bmatrix} = \begin{bmatrix} 0 \\ 0 \\ 0 \end{bmatrix}$$

$$\begin{aligned} 8x - 4y + 15z &= 0 \\ 8x - 4y - 15z &= 0 \\ -4x + 2y - 7z &= 0 \end{aligned}$$

Solving the matrix above using any method discussed earlier this chapter yields that $x = t$, $y = 2t$, and $z = 0$, which are the eigenvectors for $r = 2$.

29. $\begin{bmatrix} -5-r & 8 & 2 \\ -15 & 18-r & 4 \\ 45 & -48 & -10-r \end{bmatrix}$

The characteristic equation is $r^3 - 3r^2 + 2r = 0$
The eigenvalues are $r = 0$, $r = 1$, and $r = 2$
For $r = 0$

$$\begin{bmatrix} -5 & 8 & 2 \\ -15 & 18 & 4 \\ 45 & -48 & -10 \end{bmatrix}\begin{bmatrix} x \\ y \\ z \end{bmatrix} = \begin{bmatrix} 0 \\ 0 \\ 0 \end{bmatrix}$$

$$\begin{aligned} -5x + 8y + 2z &= 0 \\ -15x - 18y + 4z &= 0 \\ 45x - 48y - 10z &= 0 \end{aligned}$$

Solving the matrix above using any method discussed earlier this chapter yields that $x = 15t$, $y = -5t$, and $z = -2t$, which are the eigenvectors for $r = 0$.

For $r = 1$

$$\begin{bmatrix} -6 & 8 & 2 \\ -15 & 17 & 4 \\ 45 & -48 & -11 \end{bmatrix}\begin{bmatrix} x \\ y \\ z \end{bmatrix} = \begin{bmatrix} 0 \\ 0 \\ 0 \end{bmatrix}$$

$$\begin{aligned} -6x + 8y + 2z &= 0 \\ -15x + 17y + 4z &= 0 \\ 45x - 48y - 11z &= 0 \end{aligned}$$

Solving the matrix above using any method discussed earlier this chapter yields that $x = -t$, $y = -3t$, and $z = 9t$, which are the eigenvectors for $r = 1$.

For $r = 2$

$$\begin{bmatrix} -7 & 8 & 2 \\ -15 & 16 & 4 \\ 45 & -48 & -12 \end{bmatrix}\begin{bmatrix} x \\ y \\ z \end{bmatrix} = \begin{bmatrix} 0 \\ 0 \\ 0 \end{bmatrix}$$

$$\begin{aligned} -7x + 8y + 2z &= 0 \\ -15x + 16y + 4z &= 0 \\ 45x - 48y - 12z &= 0 \end{aligned}$$

Solving the matrix above using any method discussed earlier this chapter yields that $x = 0$, $y = -t$, and $z = 4t$, which are the eigenvectors for $r = 2$.

31. The characteristic equation is $r^3 + 2r^2 - r - 2 = 0$
The eigenvalues are $r = -2$, $r = -1$, and $r = 1$
The eigenvectors are $[-t, t, 2t]$, $[t, -2t, -4t]$, and $[0, t, t]$ respectively

33. The characteristic equation is $r^3 - 2r^2 - 16r + 32 = 0$
The eigenvalues are $r = -4$, $r = 2$, and $r = 4$
The eigenvectors are $[t, t, t]$, $[0, 0, t]$, and $[2t, t, 0]$ respectively

35. The characteristic equation is $r^3 - 2r^2 - r + 2 = 0$
The eigenvalues are $r = -1$, $r = 1$, and $r = 2$
The eigenvectors are $[t, t, -1]$, $[2t, t, 0]$, and $[-2t, -2t, t]$ respectively

37.

$$\begin{aligned} A^n w &= 2(2)^{10}\begin{bmatrix} 1 \\ 0 \end{bmatrix} + 3(1)^{10}\begin{bmatrix} 0 \\ 1 \end{bmatrix} \\ &= \begin{bmatrix} 2048 \\ 0 \end{bmatrix} + \begin{bmatrix} 0 \\ 3 \end{bmatrix} \\ &= \begin{bmatrix} 2048 \\ 3 \end{bmatrix} \end{aligned}$$

39.

$$\begin{aligned} A^n w &= 2(-2)^{10}\begin{bmatrix} 1 \\ 1 \end{bmatrix} + 3(0)^{10}\begin{bmatrix} 0 \\ 1 \end{bmatrix} \\ &= \begin{bmatrix} 2048 \\ 2048 \end{bmatrix} + \begin{bmatrix} 0 \\ 0 \end{bmatrix} \\ &= \begin{bmatrix} 2048 \\ 2048 \end{bmatrix} \end{aligned}$$

41. Long-term growth rate: 1.5
Long-term growth rate percentage: 50% (See page 483 for a reference; $Percentage = 100 * Long - termrate - 100\%$

43. Long-term growth rate: 2.1
Long-term growth rate percentage: 110% (See page 483 for a reference; $Percentage = 100 * Long - termrate - 100\%$

45. The characteristic equation for the Leslie matrix is

$$\begin{aligned} (0.5 - r)^2 - 1 &= 0 \\ r^2 - r + 0.25 - 1 &= 0 \\ r^2 - r - 0.75 &= 0 \\ r &= 1.5 \\ r &= -0.5 \end{aligned}$$

Thus, the long-term growth rate is 1.5

47. The characteristic equation for the Leslie matrix is

$$\begin{aligned} (15 - r)(0 - r) - 6 &= 0 \\ r^2 - 15r - 6 &= 0 \\ r &= 15.38986692 \\ r &= -0.3898669190 \end{aligned}$$

Thus, the long-term growth rate is ≈ 15.39

49. a)

$$\begin{aligned} p &= \begin{bmatrix} 3 \\ 4 \end{bmatrix} \\ G^n p &= 3(1.1)^n \begin{bmatrix} 1 \\ 1 \end{bmatrix} + 4(0.8)^n \begin{bmatrix} 2 \\ 3 \end{bmatrix} \\ \lim_{n\to\infty} \left(\frac{1}{1.1}\right)^n G^n p &= \lim_{n\to\infty} \left(\frac{1}{1.1}\right)^n 3(1.1)^n \begin{bmatrix} 1 \\ 1 \end{bmatrix} + \\ & \lim_{n\to\infty} \left(\frac{1}{1.1}\right)^n 4(0.8)^n \begin{bmatrix} 2 \\ 3 \end{bmatrix} \\ &= 3\begin{bmatrix} 1 \\ 1 \end{bmatrix} \end{aligned}$$

b) If the long term rate is not 1.1 then the $\lim_{n\to\infty} \left(\frac{1}{1.1}\right)^n G^n p$ may not converge

51. $B^{-1} = \begin{bmatrix} 3 & -5 \\ -1 & 2 \end{bmatrix}$

a) $BA = \begin{bmatrix} 2 & 5 \\ 0 & 1 \end{bmatrix}$

$$\begin{aligned} B^{-1}BA &= \begin{bmatrix} 3 & -5 \\ -1 & 2 \end{bmatrix}\begin{bmatrix} 2 & 5 \\ 0 & 1 \end{bmatrix} \\ &= \begin{bmatrix} 6 & 10 \\ -2 & -3 \end{bmatrix} \end{aligned}$$

b) For matrix $A: (1-r)(2-r) - 0 = r^2 - 3r + 2$
For matrix $B^{-1}AB: (6-r)(-3-r)+20 = r^2-3r+2$
The answers are identical

c) Since the two matrices have the same characteristic equation, they will have the same eigenvalues

53. Left to the student

55. a) $(2-r)(-5-r) - 3 = r^2 + 3r - 13$

b)

$$\begin{aligned} A^2 + 3A - 13I &= \begin{bmatrix} 7 & -9 \\ -3 & 28 \end{bmatrix} + \begin{bmatrix} 6 & 9 \\ 3 & -15 \end{bmatrix} - \\ & \begin{bmatrix} 13 & 0 \\ 0 & 13 \end{bmatrix} \\ &= \begin{bmatrix} 0 & 0 \\ 0 & 0 \end{bmatrix} \end{aligned}$$

c) Left to student

Exercise Set 6.5

1. Not linear

3. Linear and homogeneous

5. Linear and homogeneous

7. $a = 2$ and $b = 3$
$r^2 - 2r - 3 = 0$
$r = -1$ and $r = 3$
$x_n = c_1(-1)^n + c_2(3)^n$

9. $a = -1$ and $b = 6$
$r^2 + r - 6 = 0$
$r = -3$ and $r = 2$
$x_n = c_1(-3)^n + c_2(2)^n$

11. $a = 1$ and $b = -2$
$r^2 - r + 2 = 0$
$r = -1$ and $r = 2$
$x_n = c_1(-1)^n + c_2(2)^n$

13. $r^2 - r - 2 = 0$
$r = -1$ and $r = 2$
$x_n = c_1(-1)^n + c_2(2)^n$
For $n = 0$, $c_1 + c_2 = 2$
For $n = 1$, $-c_1 + 2c_2 = 1$
Solving the system gives $c_1 = 1$ and $c_2 = 1$
Therefore, $x_n = (-1)^n + (2)^n$

15. $r^2 - \frac{5}{2}r + 1 = 0$
$r = \frac{1}{2} = 2^{-1}$ and $r = 2$
$x_n = c_1(2)^{-n} + c_2(2)^n$
For $n = 0$, $c_1 + c_2 = 0$
For $n = 1$, $\frac{1}{2}c_1 + 2c_2 = -\frac{3}{2}$
Solving the system gives $c_1 = 1$ and $c_2 = -1$
Therefore, $x_n = (2)^{-n} - (2)^n$

17. $r^2 - 2r - 8 = 0$
$r = -2$ and $r = 4$
$x_n = c_1(-2)^n + c_2(4)^n$
For $n = 0$, $c_1 + c_2 = 1$
For $n = 1$, $-2c_1 + 4c_2 = -2$
Solving the system gives $c_1 = 1$ and $c_2 = 0$
Therefore, $x_n = (-2)^n$

19. $r^2 + \frac{3}{2}r - 1 = 0$
$r = -2$ and $r = \frac{1}{2} = 2^{-1}$
$x_n = c_1(-2)^n + c_2(2)^{-n}$
For $n = 0$, $c_1 + c_2 = 5$
For $n = 1$, $\frac{1}{2}c_1 - 2c_2 = -\frac{5}{2}$
Solving the system gives $c_1 = 2$ and $c_2 = 3$
Therefore, $x_n = 2(-2)^n + 3(2)^{-n}$

21. $c = 5c - 6c + 8 \to c = 4$
Homogeneous case:
$r^2 - 5r + 6 = 0$
$r = 2$ and $r = 3$
Therefore, $x_n = c_1(2)^n + c_2(3)^n + 4$

23. $c = 9c - 20c + 12 \to c = 1$
Homogeneous case:
$r^2 - 9r + 20 = 0$
$r = 4$ and $r = 5$
Therefore, $x_n = c_1(4)^n + c_2(5)^n + 1$

25. $c = 7c - 12c + 12 \to c = 2$
Homogeneous case:
$r^2 - 7r + 12 = 0$
$r = 3$ and $r = 4$
Therefore, $x_n = c_1(3)^n + c_2(4)^n + 2$

27. $c = -2c + 8c - 45 \to c = 9$
Homogeneous case:
$r^2 + 2r - 8 = 0$
$r = -4$ and $r = 2$
$x_n = c_1(-4)^n + c_2(2)^n + 9$
For $n = 0$, $c_1 + c_2 + 9 = 11$
For $n = 1$, $-2c_1 + 3c_2 + 9 = -5$
Solving the system gives $c_1 = 3$ and $c_2 = -1$
Therefore, $x_n = 3(-4)^n - (2)^n + 9$

29. $c = 9c - 18c + 20 \to c = 2$
Homogeneous case:
$r^2 - 9r + 18 = 0$
$r = 3$ and $r = 6$
$x_n = c_1(3)^n + c_2(6)^n + 2$
For $n = 0$, $c_1 + c_2 + 2 = 7$
For $n = 1$, $3c_1 + 6c_2 + 2 = 26$
Solving the system gives $c_1 = 2$ and $c_2 = 3$
Therefore, $x_n = 2(3)^n + 3(6)^n + 2$

31. $a = 0.35$ and $b = 0.45$
Since $a + b < 1$
The population decreases exponentially to 0

33. $a = 2.3$ and $b = 1.5$
Since $a + b > 1$
The population grows exponentially to ∞

35. $a = 0.01$ and $b = 1.5$
Since $a + b > 1$
The population grows exponentially to ∞

37. **a)** $r^2 - 0.92r - 0.15 = 0$
$r = -0.14133$ and $r = 1.06133$
$x_n = c_1(-0.14133)^n + c_2(1.06133)^n$
For $n = 0$, $c_1 + c_2 = 0$
For $n = 1$, $-0.14133c_1 + 1.06133c_2 = 50$
Solving the system gives
$c_1 = -41.57451$ and $c_2 = 41.57451$
Therefore,
$x_n = -41.57451(-0.14133)^n + 41.57451(1.06133)^n$

b) Since $0.92 + 0.15 > 1$, the population will grow

39.

$$\begin{aligned} x_{n+1} &= ax_n + az_n \\ &= ax_n + ab(M - x_{n-1}) \\ &= ax_n - abx_{n-1} + abM \end{aligned}$$

41. $x_{n+1} = 0.95x_n - 0.095x_{n-1} + 95$

a) $c = 0.95c - 0.095c + 95 \to c = 655.17241$
Homogeneous case:
$r^2 - 0.95r + 0.095 = 0$
$r = 0.11358$ and $r = 0.83642$
$x_n = c_1(0.11358)^n + c_2(0.83642)^n + 655.17241$

b) For $n = 0$, $c_1 + c_2 + 655.17241 = 50$
For $n = 1$, $0.11358c_1 + 0.83642c_2 + 655.17241 = 130$
Solving the system gives
$c_1 = 26.27632$ and $c_2 = -631.44873$
Therefore,
$x_n = 26.27632(0.11385)^n - 631.44873(0.83642)^n + 655.17241$

c) Since $0.95 + 0.095 > 1$ the limit grows to ∞

43. **a)** $x_{n+1} = x_{n-1}$

b) $a = 0$ and $b = 1$
$r^2 - 1 = 0$
$x_n = c_1(-1)^n + c_2(1)^n$
$x_0 = 2 \to c_1 + c_2 = 2$
$x_1 = 4 \to -c_1 + c_2 = 4$
Solving the system gives
$c_1 = -1$ and $c_2 = 3$
Thus,
$x_n = 3(1)^n - (-1)^n$

45. $a = 4$ and $b = -4$
$r^2 - 4r + 4 = 0$
$r = 2$ Repeated
$x_n = c_1(2)^n + c_2n(2)^n$

47. $c = -2c - c + 12 \to c = 3$ $a = -2$ and $b = -1$
$r^2 + 2r + 1 = 0$
$r = -1$ Repeated
$x_n = c_1(-1)^n + c_2n(-1)^n + 3$

49. - .53 Left to the student

55. **a)** $n = 1$, $M_1 = 1.2987$
$n = 2$, $M_2 = 1.6861$
$n = 3$, $M_3 = 2.1883$
$n = 4$, $M_4 = 2.8385$
$n = 5$, $M_5 = 3.6797$

b) The results from part (a) are very close since the ratio $\dfrac{(1+\rho)M}{1+\theta M} \approx 1.3$

57. **a)**

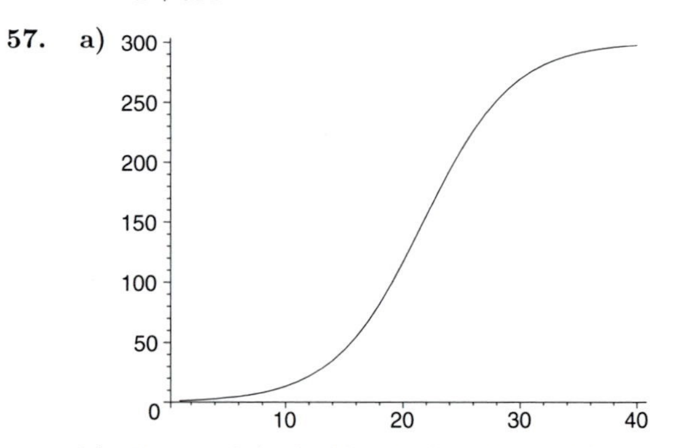

b) The graph looks like the logistic model discussed in section 4.3

Chapter 7

Functions of Several Variables

Exercise Set 7.1

1. $f(x,y) = x^2 - 2xy$

$f(0,-2) = 0^2 - 2(0)(-2) = 0 - 0 = 0$

$f(2,3) = 2^2 - 2(2)(3) = 4 - 12 = -8$

$f(10,-5) = 10^2 - 2(10)(-5) = 100 + 100 = 200$

3. $f(x,y) = 3^x + 7xy$

$f(0,-2) = 3^0 + 7(0)(-2) = 1 + 0 = 1$

$f(-2,1) = 3^{-2} + 7(-2)(1) = \frac{1}{9} - 14 = -\frac{125}{9}$

$f(2,1) = 3^2 + 7(2)(1) = 9 + 14 = 23$

5. $f(x,y) = sin\ x\ tan\ y$

$f(\pi/2, 0) = sin\ \pi/2\ tan\ 0 = (1)(0) = 0$

$f(3\pi/4, 2\pi/3) = sin\ 3\pi/4\ tan\ 2\pi/3 = (-1)(\frac{\sqrt{3}}{2}) = -\frac{\sqrt{3}}{2}$

$f(\pi/6, \pi/4) = sin\ \pi/6\ tan\ \pi/4 = (\frac{1}{2})(1) = \frac{1}{2}$

7. $f(x,y,z) = x^2 - y^2 + z^2$

$f(-1,2,3) = (-1)^2 - (2)^2 + (3)^2 = 1 - 4 + 9 = 6$

$f(2,-1,3) = (2)^2 - (-1)^2 + (3)^2 = 4 - 1 + 9 = 12$

9. $S(h,w) = \frac{\sqrt{hw}}{60}$

$$\begin{aligned} S(165,80) &= \frac{\sqrt{165 \cdot 80}}{60} \\ &= \frac{\sqrt{13200}}{60} \\ &\approx 1.915\ m^2 \end{aligned}$$

11. $w(x,s,h,m,r,p) = x(9.38 + 0.264s + 0.000233hm + 4.62r[p+1])$

a) $w(275,1,160,71,0.068,0)$

$$\begin{aligned} w(275,1,160,71,0.068,0) &= 275(9.38 + 0.264(1) \\ &\quad +0.000233(160)(71) + \\ &\quad 4.62(0.068)[0+1]) \\ &= 3466.386 \end{aligned}$$

b) $w(282,-1,171,76,0.085,3)$

$$\begin{aligned} w(282,-1,171,76,0.085,3) &= 282(9.38 + 0.264(-1) \\ &\quad +0.000233(171)(76) + \\ &\quad 4.62(0.085)[3+1]) \\ &= 3771.589 \end{aligned}$$

13. $S(d,V,a) = \frac{aV}{0.51d^2}$

$$\begin{aligned} S(100,1600000,0.78) &= \frac{0.78 \cdot 1600000}{0.51(100)^2} \\ &= \frac{1248000}{5100} \\ &= 244.706 \end{aligned}$$

15. $V(L,p,R,r,v) = \frac{p}{4Lv}(R^2 - r^2)$

$$\begin{aligned} V(1,100,0.0075,0.0025,0.05) &= \frac{100}{4(1)(0.05)} \times \\ &\quad [(0.0075)^2 - (0.0025)^2] \\ &= 500(0.00005) \\ &= 0.025 \end{aligned}$$

17. Left to the student (answer may vary)

19. $W, 40, 20) = -22$

21.

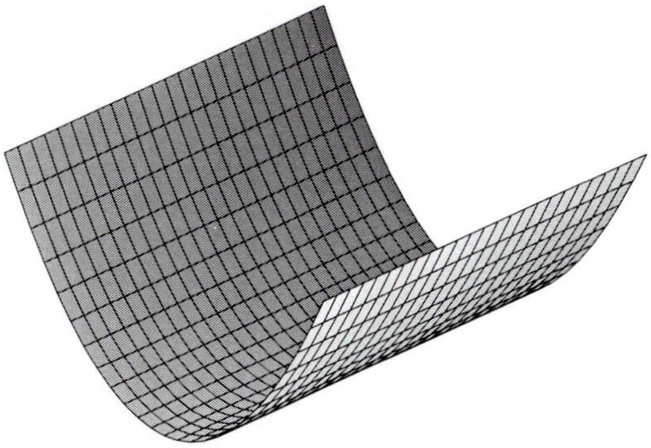

23.

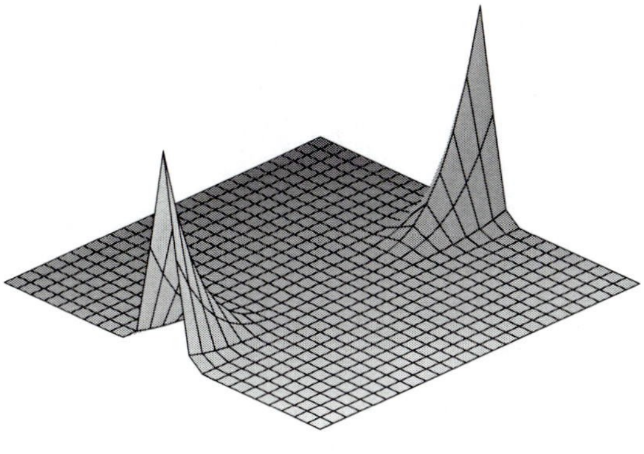

25.

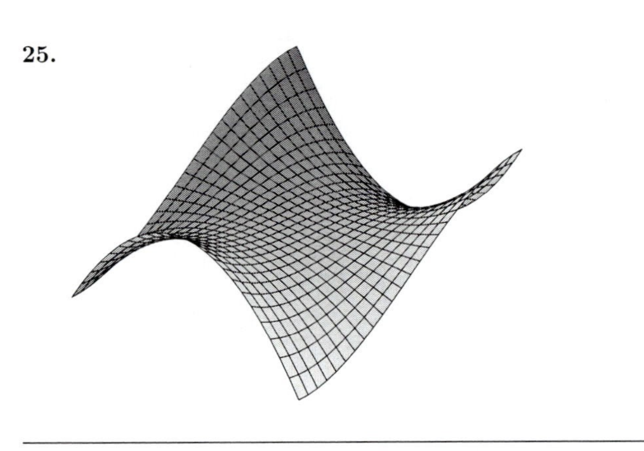

Exercise Set 7.2

1. $z = 2x - 3xy$

$\frac{dz}{dx} = 2 - 3y$

$\frac{dz}{dy} = -3x$

$\frac{dz}{dx}|_{(-2,-3)} = 2 - 3(-3) = -11$

$\frac{dz}{dy}|_{(0,-5)} = -3(0) = 0$

3. $z = 3x^2 - 2xy + y$

$\frac{dz}{dx} = 6x - 2y$

$\frac{dz}{dx} = -2x + 1$

$\frac{dz}{dx}|_{(-2,-3)} = 6(-2) - 2(-3) = -12 + 6 = -6$

$\frac{dz}{dy}|_{(0,-5)} = -2(0) + 1 = 0 + 1 = 1$

5. $f(x, y) = 2x - 3y$

$f_x = 2$ (a constant independent of x or y)

$f_y = -3$ (a constant independent of x or y)

$f_x(-2, 1) = 2$

$f_y(-3, -2) = -3$

7. $f(x, y) = (x^2 + y^2)^{1/2}$

$f_x = \frac{1}{2}(x^2 + y^2)^{-1/2}(2x) = \frac{x}{\sqrt{x^2+y^2}}$

$f_x = \frac{1}{2}(x^2 + y^2)^{-1/2}(2y) = \frac{y}{\sqrt{x^2+y^2}}$

$f_x(-2, 1) = \frac{-2}{\sqrt{4+1}} = \frac{-2}{\sqrt{5}}$

$f_y(-3, -2) = \frac{-2}{\sqrt{9+4}} = \frac{-2}{\sqrt{13}}$

9. $f(x, y) = 2x - 3y$

$f_x = 2$

$f_y = -3$

11. $f(x, y) = \sqrt{x} + sin(xy)$

$f_x = \frac{1}{2\sqrt{x}} + cos(xy)(y) = \frac{1}{2\sqrt{x}} + y\ cos(xy)$

$f_y = cos(xy)(x) = x\ cos(xy)$

13. $f(x, y) = x\ ln\ y$

$f_x = ln\ y$

$f_y = x \cdot \frac{1}{y} = \frac{x}{y}$

15. $f(x, y) = x^3 - 4xy + y^2$

$f_x = 3x^2 - 4y$

$f_y = -4x + 2y$

17. $f(x, y) = (x^2 + 2y + 2)^4$

$f_x = 4(x^2 + 2y + 2)^3(2x) = 8x(x^2 + 2y + 2)^3$

$f_y = 4(x^2 + 2y + 2)^3(2) = 8(x^2 + 2y + 2)$

19. $f(x, y) = sin(e^{x+y})$

$f_x = cos(e^{x+y})(e^{x+y}(1)) = e^{x+y}\ cos(e^{x+y})$

$f_x = cos(e^{x+y})(e^{x+y}(1)) = e^{x+y}\ cos(e^{x+y})$

21. $f(x, y) = \frac{e^x}{y^2+1}$

$f_x = \frac{1}{y^2+1} \cdot e^x = \frac{e^x}{y^2+1}$

$f_y = e^x[-1(y^2 + 1)^{-2}(2y)] = \frac{-2ye^x}{(y^2+1)^2}$

23. $f(x, y) = [x^5 + tan(y^2)]^4$

$f_x = 4[x^5 + tan(y^2)]^3(5x^4) = 20x^4[x^5 + tan(y^2)]^3$

$f_y = 4[x^5 + tan(y^2)]^3(sec^2(y^2)(2y))$
$= 8y\ sec^2(y^2)\ [x^5 + tan(y^2)]^3$

25. $f(x, y) = \frac{y\ ln\ x}{y^3-1}$

$f_x = \frac{y}{y^3-1} \cdot \frac{1}{x} = \frac{y}{x(y^3-1)}$

$f_y = \frac{(y^3-1)(ln\ x) - y\ ln\ x(3y^2)}{(y^3-1)^2}$
$= \frac{-2y^3\ -1)\ ln\ x}{(y^3-1)^2}$

27. $f(b,m) = (m+b-4)^2 + (2m+b-5)^2 + 3m+b-6)^2$

$$\begin{aligned} \frac{\partial f}{\partial b} &= 2(m+b-4)(1) + 2(2m+b-5)(1) + \\ &\quad 2(3m+b-6)(1) \\ &= 2m+2b-8+4m+2b-10+6m+2b-12 \\ &= 12m+6b-30 \end{aligned}$$

$$\begin{aligned} \frac{\partial f}{\partial m} &= 2(m+b-4)(1) + 2(2m+b-5)(2) + \\ &\quad 2(3m+b-6)(3) \\ &= 2m+2b-8+8m+4b-20+18m+6b-36 \\ &= 28m+12b-64 \end{aligned}$$

29. $z = \frac{x^2+t^2}{x^2-t^2}$

$$\begin{aligned} z_x &= \frac{(x^2-t^2)(2x)-(x^2+t^2)(2x)}{(x^2-t^2)^2} \\ &= \frac{2x^3-2t^2x-2x^3-2t^2x}{(x^2-t^2)^2} \\ &= \frac{-4t^2x}{(x^2-t^2)^2} \end{aligned}$$

$$\begin{aligned} z_t &= \frac{(x^2-t^2)(2t)-(x^2+t^2)(-2t)}{(x^2-t^2)^2} \\ &= \frac{2tx^2-2t^3+2tx^2+2t^3}{(x^2-t^2)^2} \\ &= \frac{4tx^2}{(x^2-t^2)^2} \end{aligned}$$

31. $z = \frac{2\sqrt{x}-2\sqrt{t}}{1+2\sqrt{t}}$

$$\begin{aligned} z_x &= \frac{2}{1+\sqrt{t}} \cdot \frac{1}{2\sqrt{x}} \\ &= \frac{1}{(1+2\sqrt{t})\sqrt{x}} \end{aligned}$$

$$\begin{aligned} z_t &= \frac{(1+2\sqrt{t})(-2\cdot\frac{1}{2\sqrt{t}}) - (2\sqrt{x}-2\sqrt{t})(-2\cdot\frac{1}{2\sqrt{t}})}{(1+2\sqrt{t})^2} \\ &= \frac{2\sqrt{x}+2\sqrt{t}-1-2\sqrt{t}}{\sqrt{t}\,(1+2\sqrt{t})^2} \\ &= \frac{2\sqrt{x}-1}{\sqrt{t}\,(1+2\sqrt{t})^2} \end{aligned}$$

33. $z = (x^3t^5)^{1/4}$

$$\begin{aligned} z_x &= \frac{1}{4}(x^3t^5)^{-3/4}(3x^2t^5) \\ &= \frac{3x^2t^5}{4\sqrt[4]{(x^3t^5)^3}} \end{aligned}$$

$$\begin{aligned} z_t &= \frac{1}{4}(x^3t^5)^{-3/4}(5x^3t^4) \\ &= \frac{5x^3t^4}{4\sqrt[4]{(x^3t^5)^3}} \end{aligned}$$

35. $f(x,y) = x+3y$, $g(x,y) = x-2y$

$$\begin{aligned} J &= \begin{bmatrix} \partial f/\partial x & \partial f/\partial y \\ \partial g/\partial x & \partial g/\partial y \end{bmatrix} \\ &= \begin{bmatrix} 1 & 3 \\ 1 & \text{-}2 \end{bmatrix} \end{aligned}$$

37. $f(x,y) = \sqrt{x+3y}$, $g(x,y) = e^{-x-y}$

$$\begin{aligned} J &= \begin{bmatrix} \partial f/\partial x & \partial f/\partial y \\ \partial g/\partial x & \partial g/\partial y \end{bmatrix} \\ &= \begin{bmatrix} \frac{1}{2}(x+3y)^{-1/2}(1) & \frac{1}{2}(x+3y)^{-1/2}(3) \\ e^{-x-y}(-1) & e^{-x-y}(-1) \end{bmatrix} \\ &= \begin{bmatrix} \frac{1}{2\sqrt{x+3y}} & \frac{3}{2\sqrt{x+3y}} \\ -e^{-x-y} & -e^{-x-y} \end{bmatrix} \end{aligned}$$

39. We will use the results from Exercise 9.

$f_x = 2 \rightarrow f_{xx} = 0$ and $f_{xy} = 0$

$f_y = -3 \rightarrow f_{yx} = 0$ and $f_{yy} = 0$

41. We will use the results from Exercise 11.

$f_x = \frac{1}{2}x^{-1/2} + y\ cos(xy) \rightarrow$

$f_{xx} = \frac{1}{4}x^{-3/2} - y(sin(xy)(y)) = \frac{1}{4\sqrt{x^3}} - y^2 sin(xy)$

$f_{xy} = y[-sin(xy)(x)] + cos(xy)(1) = -xy\ sin(xy) + cos(xy)$

$f_y = x\ cos(xy) \rightarrow$

$f_{yx} = x[-sin(xy)(y)] + cos(xy)(1) = -xy\ sin(xy) + cos(xy)$

$f_{yy} = x[-sin(xy)(x)] = -x^2\ sin(xy)$

43. We will use the results from Exercise 13.

$f_x = ln\ y \rightarrow f_{xx} = 0$

$f_{xy} = \frac{1}{y}$

$f_y = \frac{x}{y} \rightarrow f_{yx} = \frac{1}{y}$

$f_{yy} = x \cdot -1y^{-2} = \frac{-x}{y^2}$

45. We will use the results from Exercise 15.

$f_x = 3x^2 - 4y \rightarrow f_{xx} = 6x$ and $f_{xy} = -4$

$f_y = -4x + 2y \rightarrow f_{yx} = -4$ and $f_{yy} = 2$

47. $f(x,y,z) = x^2y^3z^4$

$f_x = 2xy^3z^4$

$f_y = 3x^2y^2z^4$

$f_z = 4x^2y^3z^3$

49. $f(x,y,z) = e^{x+y^2+z^3}$

$f_x = e^{x+y^2+z^3}(1) = e^{x+y^2+z^3}$

$f_y = e^{x+y^2+z^3}(2y) = 2y\ e^{x+y^2+z^3}$

$f_z = e^{x+y^2+z^3}(3z^2) = 3z^2\ e^{x+y^2+z^3}$

51. $z = f(x,y) = xy^2$

$f(2,3) = (2)(3)^2 = 4 \cdot 9 = 36$
$f_x = y^2$, so, $f_x(2,3) = 3^2 = 9$
$f_y = 2xy$, so, $f_y(2,3) = 2(2)(3) = 12$

$$\begin{aligned} z(x,y) &= f(a,b) + f_x(a,b)(x-a) + f_y(a,b)(y-b) \\ z(2.01,3.02) &= f(2,3) + f_x(2,3)(2.01-2) + \\ &\quad f_y(2,3)(3.02-3) \\ &= 36 + 9(0.01) + 12(0.02) \\ &= 18.33 \end{aligned}$$

53. $z = f(x,y) = x\ sin(xy)$

$f(1,0) = (1)\ sin(0) = 0$
$f_x = x[ycos(xy)] + sin(xy)$, so $f_x(1,0) = 1[(0)cos(0)] + sin(0) = 0$
$f_y = x[cos(xy)(x)] = x^2\ cos(xy)$, so $f_(1,0) = 1^2\ cos(0) = 1$

$$\begin{aligned} z(x,y) &= f(a,b) + f_x(a,b)(x-a) + f_y(a,b)(y-b) \\ z(0.99,0.02) &= f(1,0) + f_x(1,0)(0.99-1) + \\ &\quad f_y(1,0)(0.02-0) \\ &= 0 + 0(-0.01) + 1(0.02) \\ &= 0.02 \end{aligned}$$

55. $S(h,w) = \sqrt{hw}/60$

a) $S(100,28)$

$$\begin{aligned} S(100,28) &= \sqrt{(100)(28)}/60 \\ &= \sqrt{2800}/60 \\ &= 0.881917 \end{aligned}$$

b) Use linearization to estimate $S(102,30)$

$S(100,28) = 0.881917$
$S_h = \frac{w}{120\sqrt{hw}}$, so
$S_h(100,28) = \frac{28}{120\ \sqrt{2800}} = 0.00441$
$S_w = \frac{h}{120\sqrt{hw}}$, so
$S_w(100,28) = \frac{100}{120\ \sqrt{2800}} = 0.01575$

$$\begin{aligned} S(102,30) &= S(100,28) + S_h(100,28)(102-100) + \\ &\quad S_w(100,28)(30-28) \\ &= 0.88917 + (0.00441)(2) + (0.01575)(2) \\ &= 0.9294 \end{aligned}$$

57. $w(x,s,h,m,r,p) = x(9.38 + 0.264s + 0.000233hm + 4.62r[p+1])$

a) $w(280,1,150,65,0.08,0)$

$$\begin{aligned} w &= 280(9.38 + 0.264(1) + 0.000233(150)(65) + \\ &\quad 4.62(0.08)[0+1]) \\ &= 280(9.38 + 0.264 + 2.27175 + 0.3696 \\ &= 3439.9 \end{aligned}$$

b) $w_x = (9.38 + 0.264s + 0.000233hm + 4.62r[p+1])$
$w_r = 4.62x[p+1]$

$$\begin{aligned} w_x(280,1,150,65,0.08,0) &= 9.38 + 0.264(1) + \\ &\quad 0.0000233(150)(65) + 4.62(0.08)[0+ \\ &= 9.38 + 0.264 + 0.227175 + 0.3696 \\ &= 12.28535 \end{aligned}$$

$$\begin{aligned} w_r(280,1,150,65,0.08,0) &= 4.62(280)[0+1] \\ &= 1293.6 \end{aligned}$$

$$\begin{aligned} w(276,1,150,65,0.081,0) &= 3439.9 + 12.28535(276-280) \\ &\quad +1293.6(0.081-0.08) \\ &= 3439.9 - 49.1414 + 1.2936 \\ &= 3392.05 \end{aligned}$$

59. $P(m,T) = \frac{1}{1+e^{3.222-31.669m+0.083T}}$

a) $P(0.15,20)$

$$\begin{aligned} P(0.15,20) &= \frac{1}{1+e^{3.222-31.669(0.15)+0.083(20)}} \\ &= \frac{1}{1+e^{0.13165}} \\ &= 0.467135 \end{aligned}$$

b) $P_m = -(1+e^{3.222-31.669m+0.083T})^{-2}(31.669) = \frac{-31.669}{(1+e^{3.222-31.669m+0.083T})^2}$
$P_T = -(1+e^{3.222-31.669m+0.083T})^{-2}(0.083) = \frac{-0.083}{(1+e^{3.222-31.669m+0.083T})^2}$

$$P_m(0.15,20) = \frac{-31.669}{(1+e^{3.222-31.669(0.15)+0.083(20)})^2} = \frac{31.669}{(2.140709)^2} = 6.91065$$

$$P_T(0.15,20) = \frac{-0.083}{(1+e^{3.222-31.669(0.15)+0.083(20)})^2} = \frac{-0.083}{(2.140709)^2} = -0.018112$$

$$\begin{aligned} P(0.155,19) &= 0.467135 + 6.91065(0.155-0.15) - 0.018112(19.5-20) \\ &= 0.467135 + 0.03455325 + 0.009056 \\ &= 0.5107 \end{aligned}$$

61.

$$\begin{aligned} T_h(90,1) &= 1.98(90) - 1.09(1-1)(90-58) - 56.9 \\ &= 121.3 \end{aligned}$$

63. $\frac{\partial T_h}{\partial H} = 1.09(T-58)$
This means that for every 1 point change in humidity at a specific temperature T, the Temperature-Humidity Heat index changes by $1.09(T-58)$

65.

$$\begin{aligned} E(146,5) &= 206.835 - 0.846(146) - 1.015(5) \\ &= 206.835 - 123.516 - 5.075 \\ &= 78.244 \end{aligned}$$

67. $\frac{\partial E}{\partial w} = -0.846$. This means that for every increase of one syllable there is a decrease of 0.846 in the reading ease of a 100-word section.

69. $f(x,y) = ln\ (x^2+y^2)$

$f_x = \frac{2x}{(x^2+y^2)}$
$f_x x = \frac{(x^2+y^2)(2)-2x(2x)}{(x^2+y^2)^2} = \frac{2(y^2-x^2)}{(x^2+y^2)^2}$

$f_y = \frac{2y}{(x^2+y^2)}$
$f_x x = \frac{(x^2+y^2)(2)-2y(2y)}{(x^2+y^2)^2} = \frac{2(x^2-y^2)}{(x^2+y^2)^2}$

$$\frac{\partial^2 f}{\partial x^2} + \frac{\partial^2 f}{\partial x^2} = \frac{2(y^2-x^2)}{(x^2+y^2)^2} + \frac{2(x^2-y^2)}{(x^2+y^2)^2} = \frac{2y^2-2x^2+2x^2-2y^2}{(x^2+y^2)^2} = 0$$

71. a)

$$\lim_{h\to 0}\frac{f(h,y)-f(0,y)}{h} = \lim_{h\to 0}\frac{\frac{hy(h^2-y^2)}{(h^2+y^2)}-0}{h} = \lim_{h\to 0}\frac{y(h^2-y^2)}{h^2+y^2} = \frac{-y^3}{y^2} = -y$$

a)

$$\lim_{h\to 0}\frac{f(h,y)-f(0,y)}{h} = \lim_{h\to 0}\frac{\frac{hy(h^2-y^2)}{(h^2+y^2)}-0}{h} = \lim_{h\to 0}\frac{y(h^2-y^2)}{h^2+y^2} = \frac{-y^3}{y^2} = -y$$

b)

$$\lim_{h\to 0}\frac{f(x,h)-f(x,0)}{h} = \lim_{h\to 0}\frac{\frac{xh(x^2-h^2)}{(x^2+h^2}-0}{h} = \lim_{h\to 0}\frac{x(x^2-h^2)}{x^2+h^2} = \frac{x^3}{x^2} = x$$

c) Using the results from the previous two parts we have $f_y(x,0) = x$, which means $f_{yx}(x,0) = 1$ and thus $f_{yx}(0,0) = 1$. Also, $f_y(0,y) = -y$, which means $f_{yx}(0,y) = -1$ and thus $f_{xy}(0,0) = -1$. We see that $f_{xy}(0,0) = -f_{yx}(0,0)$.

Exercise Set 7.3

1. $f(x,y) = x^2 + xy + y^2 - y$

- Find the partial derivatives:
  $f_x = 2x + y,\ f_{xx} = 2$

  $f_y = x + 2y - 1,\ f_{yy} = 2$

  $f_{xy} = 1$
- We solve $f_x = 0$ and $f_y = 0$. We use the substitution method, from $f_x = 0$ we can write the $y = -2x$. Thus

$$\begin{aligned} x + 2y - 1 &= 0 \\ x + 2(-2x) - 1 &= 0 \\ x - 4x &= 1 \\ x &= -\frac{1}{3} \end{aligned}$$

and therefore

$$y = 2\cdot -\frac{1}{3} = -\frac{2}{3}$$

- Find D for $(-1/3, -2/3)$

$$\begin{aligned} D &= f_{xx}(-1/3,-2/3)\cdot f_{yy}(-1/3,-2/3) - \\ &\quad [f_{xy}(-1/3,-2/3)]^2 \\ &= 2\cdot 2 - [1]^2 \\ &= 4-1 \\ &= 3 \end{aligned}$$

- Since $D > 0$ and $f_{xx}(-1/3,-2/3) > 0$, $f(x,y)$ has a relative minimum at $(-1/3,-2/3)$

3. $f(x,y) = 2xy - x^3 - y^2$

- Find the partial derivatives:
  $f_x = 2y - 3x^2$, $f_{xx} = -6x$

  $f_y = 2x - 2y$, $f_{yy} = -2$

  $f_{xy} = 2$
- We solve $f_x = 0$ and $f_y = 0$. We use the substitution method, from $f_y = 0$ we can write the $y = x$. Thus

$$\begin{aligned} 2y - 3x^2 &= 0 \\ 2(x) - 3x^2 &= 0 \\ x(2-3x) &= 1 \\ x &= 0 \\ \text{and therefore} & \\ y &= 0 \\ \text{and} & \\ x &= 2/3 \\ \text{and therefore} & \\ y &= 2/3 \end{aligned}$$

- Find D for $(0,0)$ and $(2/3, 2/3)$

$$\begin{aligned} D &= f_{xx}(0,0)\cdot f_{yy}(0,0) - [f_{xy}(0,0)]^2 \\ &= 0\cdot -2 - [2]^2 \\ &= -4 \\ D &= f_{xx}(2/3,2/3)\cdot f_{yy}(2/3,2/3) - \\ &\quad [f_{xy}(2/3,2/3)]^2 \\ &= -4\cdot -2 - [2]^2 \\ &= 4 \end{aligned}$$

- Since $D < 0$ at $(0,0)$, $f(x,y)$ has a saddle at $(0,0)$
  Since $D > 0$ and $f_{xx}(2/3,2/3) < 0$, $f(x,y)$ has a relative maximum at $(2/3, 2/3)$

5. $f(x,y) = x^3 + y^3 - 3xy$

- Find the partial derivatives:
  $f_x = 3x^2 - 3y$, $f_{xx} = 6x$

  $f_y = 3y^2 - 3x$, $f_{yy} = 6y$

  $f_{xy} = -3$
- We solve $f_x = 0$ and $f_y = 0$. We use the substitution method, from $f_x = 0$ we can write the $y = x^2$. Thus

$$\begin{aligned} -3x + 3y^2 &= 0 \\ -3x + 3(x^2) &= 0 \\ 3x(x-1) &= 0 \\ x &= 0 \\ \text{and therefore} & \\ y &= 0 \\ \text{and} & \\ x &= 1 \\ \text{and therefore} & \\ y &= 1 \end{aligned}$$

- Find D for $(0,0)$ and $(1,1)$

$$\begin{aligned} D &= f_{xx}(0,0)\cdot f_{yy}(0,0) - [f_{xy}(0,0)]^2 \\ &= 0\cdot 0 - [-3]^2 \\ &= -9 \\ D &= f_{xx}(1,1)\cdot f_{yy}(1,1) - \\ &\quad [f_{xy}(1,1)]^2 \\ &= 6\cdot 6 - [-3]^2 \\ &= 27 \end{aligned}$$

- Since $D < 0$ at $(0,0)$, $f(x,y)$ has a saddle at $(0,0)$
  Since $D > 0$ and $f_{xx}(1,1) > 0$, $f(x,y)$ has a relative minimum at $(1,1)$

7. $f(x,y) = x^2 + y^2 - 2x + 4y - 2$

- Find the partial derivatives:
  $f_x = 2x - 2$, $f_{xx} = 2$

  $f_y = 2y + 4$, $f_{yy} = 2$

  $f_{xy} = 0$
- We solve $f_x = 0$ and $f_y = 0$.

$$\begin{aligned} 2x - 2 &= 0 \\ 2x &= 2 \\ x &= 1 \\ \text{and} & \\ 2y + 4 &= 0 \\ 2y &= -4 \\ y &= -2 \end{aligned}$$

- Find D for $(1,-2)$

$$\begin{aligned} D &= f_{xx}(1,-2)\cdot f_{yy}(1,-2) - [f_{xy}(1,-2)]^2 \\ &= 2\cdot 2 - [0]^2 \\ &= 4 \end{aligned}$$

- Since $D > 0$ and $f_{xx}(1,-2) > 0$, $f(x,y)$ has a relative minimum at $(1,-2)$

9. $f(x,y) = x^2 + y^2 + 2x - 4y$

- Find the partial derivatives:
  $f_x = 2x + 2,\ f_{xx} = 2$

  $f_y = 2y - 4,\ f_{yy} = 2$

  $f_{xy} = 0$
- We solve $f_x = 0$ and $f_y = 0$.

$$\begin{aligned} 2x+2 &= 0 \\ 2x &= -2 \\ x &= -1 \\ 2y-4 &= 0 \\ 2y &= 4 \\ y &= 2 \end{aligned}$$

- Find D for $(-1,2)$

$$\begin{aligned} D &= f_{xx}(-1,2)\cdot f_{yy}(-1,2) - [f_{xy}(-1,2)]^2 \\ &= 2\cdot 2 - [0]^2 \\ &= 4 \end{aligned}$$

- Since $D > 0$ and $f_{xx}(-1,2) > 0$, $f(x,y)$ has a relative minimum at $(-1,2)$

11. $f(x,y) = 4x^2 - y^2$

- Find the partial derivatives:
  $f_x = 8x,\ f_{xx} = 8$

  $f_y = -2y,\ f_{yy} = -2$

  $f_{xy} = 0$
- We solve $f_x = 0$ and $f_y = 0$.

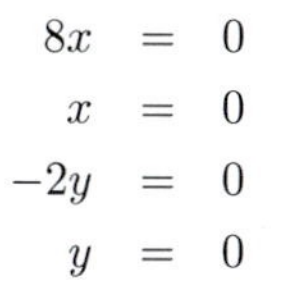

$$\begin{aligned} 8x &= 0 \\ x &= 0 \\ -2y &= 0 \\ y &= 0 \end{aligned}$$

- Find D for $(0,0)$

$$\begin{aligned} D &= f_{xx}(0,0)\cdot f_{yy}(0,0) - [f_{xy}(0,0)]^2 \\ &= 8\cdot -2 - [0]^2 \\ &= -16 \end{aligned}$$

- Since $D < 0$ at $(0,0)$, $f(x,y)$ has a saddle at $(0,0)$

13. $f(x,y) = e^{x^2+y^2+1}$

- Find the partial derivatives:
  $f_x = 2x\ e^{x^2+y^2+1},\ f_{xx} = 4x^2\ e^{x^2+y^2+1} + e^{x^2+y^2+1}$

  $f_y = 2y\ e^{x^2+y^2+1},\ f_{yy} = 4y^2\ e^{x^2+y^2+1} + e^{x^2+y^2+1}$

  $f_{xy} = 4xy\ e^{x^2+y^2+1}$
- We solve $f_x = 0$ and $f_y = 0$.

$$\begin{aligned} 2x\ e^{x^2+y^2+1} &= 0 \\ 2x &= 0 \\ x &= 0 \\ 2y\ e^{x^2+y^2+1} &= 0 \\ 2y &= 0 \\ y &= 0 \end{aligned}$$

- Find D for $(0,0)$

$$\begin{aligned} D &= f_{xx}(0,0)\cdot f_{yy}(0,0) - [f_{xy}(0,0)]^2 \\ &= 1\cdot 1 - [0]^2 \\ &= 1 \end{aligned}$$

- Since $D > 0$ and $f_{xx}(0,0) > 0$, $f(x,y)$ has a relative minimum at $(0,0)$

15. We need to find the point at which $P_x = 0$ and $P_y = 0$

$P_x = 0.0345 - 0.000230x + 0.109y$

$P_y = 25.6 + 0.109x - 126.4y$

Solve $P_x = 0$ and $P_y = 0$. We can use the substitution method. From $P_x:\ y = 0.00211x - 0.31651$

$$\begin{aligned} 25.6 + 0.109x - 126.4y &= 0 \\ 25.6 + 0.109x - 126.2(0.00211x - 0.31651) &= 0 \\ -0.1577x + 65.60686 &= 0 \\ x &= 416.0127 \\ \text{and therefore} & \\ 0.00211(416.0127) - 0.31651 &= y \\ &= 0.56218 \end{aligned}$$

17. $P = -5a^2 - 3n^2 + 48a - 4n + 2an + 300$

$P_a = -10a + 48 + 2n$
$P_{aa} = -10$

$P_{an} = 2$

$P_n = -6n - 4 + 2a$
$P_{nn} = -6$

Solve $P_a = 0$ and $P_n = 0$. From $P_a = 0$, $n = 5a - 24$

$$\begin{aligned} -6n - 4 + 2a &= 0 \\ -6(5a-24) - 4 + 2a &= 0 \\ -30a + 144 - 4 + 2a &= 0 \\ -28a &= -140 \\ a &= 5 \\ \text{therefore} & \\ n &= 5(5) - 24 \\ &= 1 \end{aligned}$$

Find D for $(5,1)$

$$\begin{aligned} D &= -10 \cdot -6 - [2]^2 \\ &= 60 - 4 \\ &= 56 \end{aligned}$$

Since $D > 0$ and $P_{aa}(5,1) < 0$ then $P(a,n)$ has a maximum at $(5,1)$. To find the maximum value we have to find $P(5,1)$

$$\begin{aligned} P(5,1) &= -5(5)^2 - 3(1)^2 + 48(5) + \\ &\quad -4(1) + 2(5)(1) + 300 \\ &= -125 - 3 + 240 - 4 + 10 + 300 \\ &= 418 \end{aligned}$$

19. $T(x,y) = x^2 + 2y^2 - 8x + 4y$

$T_x = 2x - 8$
$T_{xx} = 2$

$T_{xy} = 0$

$T_y = 4y + 4$
$T_{yy} = 4$

Solve $T_x = 0$ and $T_y = 0$

$$\begin{aligned} 2x - 8 &= 0 \\ 2x &= 8 \\ x &= 4 \\ 4y + 4 &= 0 \\ 4y &= -4 \\ y &= -1 \end{aligned}$$

Find D for $(4,-1)$

$$\begin{aligned} D &= 2 \cdot 4 - [0]^2 \\ &= 8 \end{aligned}$$

Since $D > 0$ and $T_{xx}(4,-1) > 0$ then $T(x,y)$ has a minimum at $(4,-1)$. To find the value of the minimum we need to find $T(4,-1)$

$$\begin{aligned} T(4,-1) &= (4)^2 + 2(-1)^2 - 8(4) + 4(-1) \\ &= 16 + 2 - 32 - 4 \\ &= -18 \end{aligned}$$

There is no maximum value.

21. $f(x,y) = e^x + e^y - e^{x+y}$

$f_x = e^x - e^{x+y}$
$f_{xx} = e^x - e^{x+y}$

$f_{xy} = -e^{x+y}$

$f_y = e^y - e^{x+y}$
$f_{yy} = e^y - e^{x+y}$

Solve $f_x = 0$ and $f_y = 0$

$$\begin{aligned} e^x - e^{x+y} &= 0 \\ e^x &= e^{x+y} \\ x &= x + y \\ 0 &= y \\ e^y - e^{x+y} &= 0 \\ e^y &= e^{x+y} \\ y &= x + y \\ 0 &= x \end{aligned}$$

Find D for $(0,0)$

$$\begin{aligned} D &= f_{xx}(0,0) \cdot f_{yy}(0,0) - [f_{xy}(0,0)]^2 \\ &= 0 \cdot 0 - [-1]^2 \\ &= -1 \end{aligned}$$

Since $D < 0$ at $(0,0)$ then $f(x,y)$ has a saddle point at $(0,0)$

23. $f(x,y) = 2y^2 + x^2 - x^2 y$

$f_x = 2x - 2xy$
$f_{xx} = 2 - 2y$

$f_{xy} = -2x$

$f_y = 4y - x^2$
$f_{yy} = 4$

Solve $f_x = 0$ and $f_y = 0$

$$\begin{aligned} 2x - 2xy &= 0 \\ 2x(1 - y) &= 0 \\ x &= 0 \\ &\text{and} \\ y &= 1 \end{aligned}$$

From $f_y = 0$ we get $y = x^2/4$ which means when $x = 0$, $y = 0$ and when $y = 1$, $x = \pm 2$. Find D for $(0,0)$

$$\begin{aligned} D &= 2 \cdot 4 - [0]^2 \\ &= 8 \end{aligned}$$

Since $D > 0$ and $f_{xx}(0,0) > 0$ then $f(x,y)$ has a relative minimum at $(0,0)$ Find D for $(2,1)$

$$\begin{aligned} D &= 0 \cdot 4 - [-4]^2 \\ &= -16 \end{aligned}$$

Since $D < 0$ at $(2,1)$ then $f(x,y)$ has a saddle point at $(2,1)$ Find D for $(-2,1)$

$$\begin{aligned} D &= 0 \cdot 4 - [4]^2 \\ &= -16 \end{aligned}$$

Since $D < 0$ at $(-2,1)$ then $f(x,y)$ has a saddle point at $(-2,1)$

25. The D-Test is a method similar to the second derivative test for functions of one variables. It computes the points were the first partial derivatives are zero and then computes the value of D to determine the nature of the zeros f the first partial derivatives

27. $R = e^{-1.236+1.35\ cos\ l\ cos\ s\ -1.707\ sin\ l\ sin\ s}$

a) $\partial R/\partial l$

$$\begin{aligned}\partial R/\partial l &= e^{-1.236+1.35\ cos\ l\ cos\ s\ -1.707\ sin\ l\ sin\ s} \times \\ &\quad (-1.35\ sin\ l\ cos\ s\ - 1.707\ cos\ l\ sin\ s)\end{aligned}$$

b) The sign of $\partial R/\partial l$ is determined by the sign of the term in parenthesis in part a) since e^{anything} is positive for all values permissible. Since the trigonometric functions are positive in the first quadrant (the permissible values for l and s fall in the first quadrant) then the term in parenthesis will always be negative. Thus, $\partial R/\partial l$ is always negative for the permissible values of l and s

c - d) Since $\partial R/\partial l$ is negative for $0 \leq l \leq 90$ then the function is decreasing in the l -direction which means that the maximum occurs when $l = 0$ (The beginning of the interval)

29. Relative minimum of -5 at $(0, 0)$

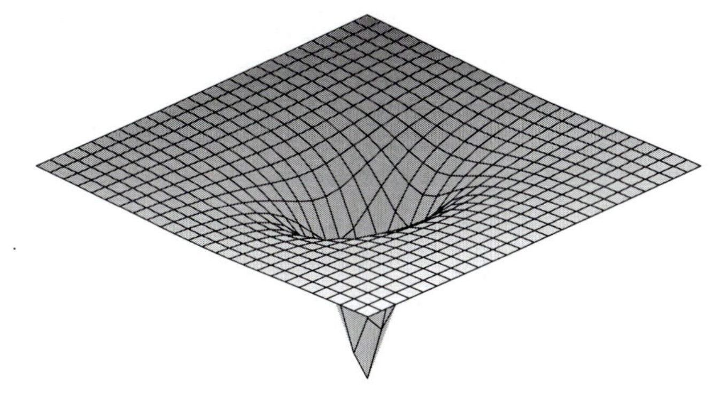

Exercise Set 7.4

1. $\overline{x} = \dfrac{0+1+\cdots+6}{7} = 3$

$\overline{y} = \dfrac{33.49+34.72+\cdots+47.70}{7} = 40.75$

$m = \dfrac{(0-3)(33.49-40.75)+\cdots+(6-3)(47.70-40.75)}{(0-3)^2+(1-3)^2+\cdots+(6-3)^2}$

$m = 2.618$

a) The regression line is

$$\begin{aligned} y-\overline{y} &= m(x-\overline{x}) \\ y-40.75 &= 2.618(x-3) \\ y &= 2.618x-7.854+40.75 \\ y &= 2.618x+32.896 \end{aligned}$$

b) Find y when $x = 16$

$$\begin{aligned} y &= 2.618(16)+32.896 \\ &= 74.78 \end{aligned}$$

Find y when $x = 21$

$$\begin{aligned} y &= 2.618(21)+32.896 \\ &= 87.87 \end{aligned}$$

3. $\overline{x} = \dfrac{0+10+\cdots+50}{6} = 25$

$\overline{y} = \dfrac{71.1+73.1+\cdots+79.5}{6} = 75.8$

$m = \dfrac{(0-25)(71.1-75.8)+\cdots+(50-25)(79.5-75.8)}{(10-25)^2+(20-25)^2+\cdots+(50-25)^2}$

$m = 0.177$

a) The regression line is

$$\begin{aligned} y-\overline{y} &= m(x-\overline{x}) \\ y-75.8 &= 0.177(x-25) \\ y &= 0.177x-4.425+75.8 \\ y &= 0.177x+71.375 \end{aligned}$$

b) Find y when $x = 60$

$$\begin{aligned} y &= 0.177(60)+71.375 \\ &= 81.995 \end{aligned}$$

Find y when $x = 65$

$$\begin{aligned} y &= 0.177(65)+71.375 \\ &= 82.88 \end{aligned}$$

5. $\overline{x} = \dfrac{0+1+\cdots+4}{5} = 2$

$\overline{y} = \dfrac{86+83.6+\cdots+79.5}{5} = 82.28$

$m = \dfrac{(0-2)(86-82.28)+\cdots+(4-2)(79.5-82.28)}{(0-2)^2+(1-2)^2+\cdots+(4-2)^2}$

$m = -1.63$

a) The regression line is

$$\begin{aligned} y-\overline{y} &= m(x-\overline{x}) \\ y-82.28 &= -1.63(x-2) \\ y &= -1.63x+3.26+82.28 \\ y &= -1.63x+85.54 \end{aligned}$$

b) Find y when $x = 14$

$$\begin{aligned} y &= -1.63(14)+85.54 \\ &= 62.72 \end{aligned}$$

7. $\overline{x} = \dfrac{70+60+85}{3} = 71.67$

$\overline{y} = \dfrac{75+62+89}{3} = 75.34$

$$\begin{aligned} m &= \frac{(70-71.67)(75-75.34)+(60-71.67)(62-75.34)}{(70-75.67)^2+(60-71.67)^2+(85-71.67)^2} \\ &+\frac{(85-75.67)(89-71.34)}{(70-75.67)^2+(60-71.67)^2+(85-71.67)^2} \\ &= 1.07 \end{aligned}$$

a) The regression line is

$$\begin{aligned} y-\overline{y} &= m(x-\overline{x}) \\ y-75.34 &= 1.07(x-71.67) \\ y &= 1.07x-76.69+75.34 \\ y &= 1.07x-1.35 \end{aligned}$$

b) Find y when $x=81$

$$\begin{aligned} y &= 1.07(81)-1.35 \\ &= 85.32 \end{aligned}$$

9. Linear regression is a method for finding an equation to model a data set obtained from an experiment.

11. a)

$X=\log\ x$	$Y=\log\ y$
1.4771	1.3979
2.301	1.4771
4.301	1.9031
7.3979	2.2304
9	2.3979

b) $\overline{x}=4.89542$ and $\overline{y}=1.88131$

$$\begin{aligned} m &= \frac{(1.4771-4.89542)(1.3979-1.88131)}{(1.4771-4.89542)^2+\cdots(9-4.89542)^2} \\ &+\cdots+\frac{(9-4.89542)(2.3979-1.88131)}{(1.4771-4.89542)^2+\cdots(9-4.89542)^2} \\ &= 0.13568 \end{aligned}$$

The regression line is

$$\begin{aligned} Y-\overline{Y} &= m(X-\overline{X}) \\ Y-1.88131 &= 0.13568(X-4.89542) \\ Y &= 0.13568X-0.66421+1.88131 \\ Y &= 0.13568X+1.21710 \end{aligned}$$

c)

$$\begin{aligned} Y &= 0.13568X+1.21710 \\ \log(y) &= 0.13568\log(x)+1.21710 \\ \log(y) &= \log(x^0.13568)+1.21710 \\ \log(y)-\log(x^{0.13568}) &= 1.21710 \\ \log\frac{y}{x^{0.13568}} &= 1.2170 \\ \frac{y}{x^{0.13568}} &= 10^{1.21710} \\ \frac{y}{x^{0.13568}} &= 16.48542 \\ y &= 16.48542\ x^{0.13568} \end{aligned}$$

d) Find y when $x=1000000$

$$\begin{aligned} y &= 16.48542(1000000)^{0.13568} \\ &= 107 \end{aligned}$$

13. a) $y=-0.005938x+15.571914$

b) In 2010
$y=-0.005938(2010)+15.571914=3.636534=3:38:19$
In 2015
$y=-0.005938(2015)+15.571914=3.606844=3:36:41$

c) In 1999, the predicted value for the record is
$y=-0.005938(1999)+15.571914=3.701852=3:42:11$

Exercise Set 7.5

1.

$$\begin{aligned} \int_0^1\int_0^1 2y\ dxdy &= \int_0^1 2yx\ |_0^1\ dy \\ &= \int_0^1 2y(1-0)\ dy \\ &= \int_0^1 2y\ dy \\ &= y^2\ |_0^1 \\ &= 1-0 \\ &= 1 \end{aligned}$$

3.

$$\begin{aligned} \int_{-1}^1\int_x^1 xy\ dydx &= \int_{-1}^1 x\frac{y^2}{2}\ |_x^1\ dx \\ &= \int_{-1}^1 (\frac{x}{2}-\frac{x^3}{2})\ dx \\ &= \left(\frac{x^2}{4}-\frac{x^4}{8}\right|_{-1}^1 \\ &= (\frac{1}{4}-\frac{1}{8})-(\frac{1}{4}-\frac{1}{8}) \\ &= 0 \end{aligned}$$

5.

$$\begin{aligned} \int_0^1\int_{-1}^3 (x+y)\ dydx &= \int_0^1 \left(xy+\frac{y^2}{2}\right|_{-1}^3\ dx \\ &= \int_0^1 \left(3x+\frac{9}{2}-x-\frac{1}{2}\right)\ dx \\ &= \int_0^1 (2x+4)\ dx \\ &= \left(x^2+4x\right|_0^1 \\ &= 1+4-0 \\ &= 5 \end{aligned}$$

7.

$$\int_0^1\int_{x^2}^{x}(x+y)\,dydx = \int_0^1\left(xy+\frac{y^2}{2}\Big|_{x^2}^{x}\right)dx = \int_0^1\left(x^2+\frac{x^2}{2}-x^3-\frac{x^4}{2}\right)dx = \int_0^1\left(\frac{3x^2}{2}-x^3-\frac{x^4}{2}\right)dx = \left(\frac{x^3}{2}-\frac{x^4}{4}-\frac{x^5}{10}\Big|_0^1\right) = \frac{1}{2}-\frac{1}{4}-\frac{1}{10} = \frac{3}{20}$$

9.

$$\int_0^1\int_1^{e^x}\frac{1}{y}\,dydx = \int_0^1 \ln\, y|_1^{e^x}\,dx = \int_0^1(x-0)\,dx = \int_0^1 x\,dx = \left(\frac{x^2}{2}\Big|_0^1\right) = \frac{1}{2}-0 = \frac{1}{2}$$

11.

$$\int_0^2\int_0^x(x+y^2)\,dydx = \int_0^2\left(xy+\frac{y^3}{3}\Big|_0^x\right)dx = \int_0^2\left(x^2+\frac{x^3}{3}\right)dx = \left(\frac{x^3}{3}+\frac{x^4}{12}\Big|_0^2\right) = \frac{8}{3}+\frac{4}{3} = 4$$

13.

$$\int_0^1\int_0^{1-x^2}(1-y-x^2)\,dydx = \int_0^1\left(y-\frac{y^2}{2}-x^2y\Big|_0^{1-x^2}\right)dx = \int_0^1(1-x^2)dx - \int_0^1\frac{(1-x^2)^2}{2}dx - \int_0^1(1-x^2)x^2dx = \int_0^1\left(1-x^2-\frac{1}{2}+x^2\right)dx + \int_0^1(-x^4-x^2+x^4)\,dx = \int_0^1\left(\frac{1}{2}-x^2\right)dx = \left(\frac{x}{2}-\frac{x^3}{3}\Big|_0^1\right) = \frac{1}{2}-\frac{1}{3} = \frac{1}{6}$$

15.

$$\int_0^1\int_x^{3x} y\,e^{x^3}dydx = \int_0^1\left(\frac{y^2}{2}\,e^{x^3}\Big|_x^{3x}\right)dx = \int_0^1\left(\frac{9x^2\,e^{x^3}}{2}-\frac{x^2\,e^{x^3}}{2}\right)dx = 4x^2\,e^{x^3}dx = \left(\frac{4}{3}\,e^{x^3}\Big|_0^1\right) = \frac{4}{3}e-\frac{4}{3}$$

17.

$$\int_0^3\int_0^{3x-x^2}(7-2x)y\,dydx = \int_0^3\left(\frac{(7-2x)y^2}{2}\Big|_0^{3x-x^2}\right)dx = \int_0^3\left(\frac{(7-2x)}{2}(3x-x^2)^2\right)dx = \frac{1}{2}\int_0^3 4(3x-x^2)2\,dx + \frac{1}{2}\int_0^3(3-2x)(3x-x^2)^2\,dx = \frac{1}{2}\int_0^3(36x^2-24x^3+4x^4)\,dx + \frac{1}{2}\int_0^3(3-2x)(3x-x^2)^2\,dx = \frac{1}{2}\left[12x^3-6x^4+\frac{4}{5}x^5\Big|_0^3\right.+\left[\frac{(3x-x^2)^3}{3}\Big|_0^3\right. = 6(3)^3-3(3)^3+\frac{2(3)^5}{5}+\frac{\frac{1}{2}(9-9)^3}{3}-0 = 16.2$$

19.

$$\int_0^1\int_0^{1-x}x^2y\,dydx = \int_0^1\left(\frac{x^2y^2}{2}\Big|_0^{1-x}\right)dx = \frac{1}{2}\int_0^1 x^2(1-x)^2\,dx = \frac{1}{2}\int_0^1\left[x^2-2x^3+x^4\right] = \left(\frac{x^3}{6}-\frac{x^4}{6}+\frac{x^5}{10}\Big|_0^1\right.$$

$$
\begin{aligned}
&= \frac{1}{6} - \frac{1}{4} + \frac{1}{10} \\
&= \frac{1}{60} \\
&= 0.01\overline{6}
\end{aligned}
$$

21.

$$
\begin{aligned}
\int_0^{\pi/2}\int_{-sin\ x}^{sin\ x} y^2\ cos\ x\ dydx &= \int_0^{\pi/2}\left(\frac{y^3}{3}cos\ x\Big|_{-sin\ x}^{sin\ x}\right. dx \\
&= \int_0^{\pi/2} \frac{2}{3}sin^3(x)\ cos\ x\ dx \\
&= \left(\frac{sin^4(x)}{6}\right|_0^{\pi/2} \\
&= \frac{1}{6} - 0 \\
&= \frac{1}{6}
\end{aligned}
$$

23. a)

$$
\begin{aligned}
A &= \int_2^4 (x^3 - x)\ dx \\
&= \left(\frac{x^4}{4} - \frac{x^2}{2}\right|_2^4 \\
&= (64 - 8) - (4 - 2) \\
&= 54
\end{aligned}
$$

b)

$$
\begin{aligned}
V &= \int_2^4\int_x^{x^3} 1\ dydx \\
&= \int_2^4 y|_x^{x^3}\ dx \\
&= \int_2^4 (x^3 - x)\ dx \\
&= \left(\frac{x^4}{4} - \frac{x^2}{2}\right|_2^4 \\
&= (64 - 8) - (4 - 2) \\
&= 54
\end{aligned}
$$

c) The area found in part (a) equals the volume in part (b) because the "thickness" is 1. The volume equals the product of the area and the thickness.

25. $\int_0^1\int_1^3\int_{-1}^2 (2x + 3y - z)\ dxdydz$

$$
\begin{aligned}
&= \int_0^1\int_1^3 x^2 + 3x - xz\Big|_{-1}^2 dydz \\
&= \int_0^1\int_1^3 (3 + 3y - z)\ dydz \\
&= \int_0^1 \left(3y + \frac{3}{2}y - yz\right|_1^3 dz \\
&= \int_0^1 15 - 2z\ dz \\
&= 15z - z^2\Big|_0^1 \\
&= (15 - 1) - (0 - 0) \\
&= 14
\end{aligned}
$$

27.

$$
\begin{aligned}
\int_0^1\int_0^{1-x}\int_0^{2-x} xyz\ dzdydx &= \int_0^1 \left(\frac{xyz^2}{2}\right|_0^{2-x} \\
&= \int_0^1\int_0^{1-x} \frac{x(2-x)^2y}{2}\ dydx \\
&= \int_0^1 \left(\frac{x(2-x)^2y^2}{4}\right|_0^{1-x} dx \\
&= \int_0^1 \frac{x(2-x)^2(1-x)^2}{4}\ dx \\
&= \int_0^1 \left(\frac{2x - 8x^2 + 11x^3}{4}\right) + \\
&\quad \int_0^1 \left(\frac{-6x^4 + x^5}{4}\right)\ dx \\
&= \left(\frac{x^2 - \frac{8}{3}x^3 + \frac{11}{4}x^4}{4}\right|_0^1 \\
&\quad + \left(\frac{-\frac{6}{5}x^5 + \frac{1}{6}x^6}{4}\right|_0^1 \\
&= \frac{1}{80}
\end{aligned}
$$

29. The geometric meaning of the multiple integral of a function of two variables is the volume of the solid generated from the function bounded by the limits of the multiple integrals.

31. 2.957335369

33. 0.3353157821

Chapter 8

First Order Differential Equations

Exercise Set 8.1

1. $y' = 4x^3$

$$\begin{aligned} y &= \int y'\,dx \\ &= \int 4x^3\,dx \\ &= x^4 + C \end{aligned}$$

3. $y' = \frac{3}{x} - x^2 + x^5$

$$\begin{aligned} y &= \int y'\,dx \\ &= \int \frac{3}{x} - x^2 + x^5\,dx \\ &= 3\ \ln x - \frac{1}{3}x^3 + \frac{1}{6}x^6 + C \end{aligned}$$

5. $y' = 4e^{3x} + \sqrt{x}$

$$\begin{aligned} y &= \int y'\,dx \\ &= \int 4e^{3x} + x^{1/2}\,dx \\ &= \frac{4}{3}e^{3x} + \frac{2}{3}x^{3/2} + C \end{aligned}$$

7. $y' = x^2\sqrt{3x^3 - 5}$

$$\begin{aligned} y &= \int y'\,dx \\ &= \int x^2\sqrt{3x^3 - 5}\,dx \\ &= \frac{2}{27}(3x^3 - 5)^{3/2} + C \end{aligned}$$

9. $y' = \frac{sin\ 2x}{(4 + cos\ 2x)^3}$

$$\begin{aligned} y &= \int y'\,dx \\ &= \int sin\ 2x(4 + cos\ 2x)^{-3}\,dx \\ &= \frac{1}{4}(4 + cos\ 2x)^{-2} + C \end{aligned}$$

11. $y' = \frac{1}{1 - x^2}$

$$\begin{aligned} y &= \int y'\,dx \\ &= \int \frac{1}{1 - x^2}\,dx \\ &= \int \frac{1}{2(1+x)} + \frac{1}{2(1-x)}\,dx \\ &= \frac{\ln|1+x|}{2} - \frac{\ln|1-x|}{2} + C \\ &= \frac{1}{2}\ln\left|\frac{1+x}{1-x}\right| + C \end{aligned}$$

13. $y' = x^2 + 2x - 3$

$$\begin{aligned} y &= \int y'\,dx \\ &= \int x^2 + 2x - 3\,dx \\ &= \frac{1}{3}x^3 + x^2 - 3x + C \\ 4 &= 0 + 0 - 0 + C \\ 4 &= C \\ y &= \frac{1}{3}x^3 + x^2 - 3x + 4 \end{aligned}$$

15. $y' = e^{3x} + 1$

$$\begin{aligned} y &= \int y'\,dx \\ &= \int e^{3x} + 1\,dx \\ &= \frac{1}{3}\ e^{3x} + x + C \\ 2 &= \frac{1}{3} + 0 + C \\ \frac{5}{3} &= C \\ y &= \frac{1}{3}\ e^{3x} + x + \frac{5}{3} \end{aligned}$$

17. $f'(x)' = x^{2/3} - x$

$$\begin{aligned} f(x) &= \int f'(x)\,dx \\ &= \int x^{2/3} - x\,dx \\ &= \frac{3}{5}x^{5/3} - \frac{1}{2}\ x^2 + C \\ -6 &= \frac{3}{5} - \frac{1}{2} + C \\ -\frac{61}{10} &= C \\ f(x) &= \frac{3}{5}\ x^{5/3} - \frac{1}{2}\ x^2 - \frac{61}{10} \end{aligned}$$

19. $y' = x\sqrt{x^2+1}$

$$\begin{aligned} y &= \int y'\,dx \\ &= \int x\sqrt{x^2+1}\,dx \\ &= \frac{1}{3}\,(x^2+1)^{3/2}+C \\ 3 &= \frac{1}{3}+C \\ \frac{8}{3} &= C \\ y &= \frac{1}{3}\,(x^2+1)^{3/2}+\frac{8}{3} \end{aligned}$$

21. $y' = x^3/(x^4+1)^2$

$$\begin{aligned} y &= \int y'\,dx \\ &= \int x^3(x^4+1)^{-2}\,dx \\ &= -\frac{1}{4}(x^4+1)^{-1} \\ -2 &= \frac{-1}{8}+C \\ -\frac{16}{8} &= C \\ y &= -\frac{1}{4(x^4+1)}-\frac{15}{8} \end{aligned}$$

23. $y' = xe^x$

$$\begin{aligned} y &= \int y'\,dx \\ &= \int xe^x\,dx \\ &= xe^x-e^x+C \\ 2 &= -1+C \\ 3 &= C \\ y &= xe^x-e^x+3 \end{aligned}$$

25. $y' = \ln x$

$$\begin{aligned} y &= \int y'\,dx \\ &= \int \ln x\,dx \\ &= x\,\ln x-x+C \\ 2 &= 0-1+C \\ 3 &= C \\ y &= x\,\ln x-x+3 \end{aligned}$$

27. $y' = x\,sin(x^2)$

$$\begin{aligned} y &= \int y'\,dx \\ &= \int x\,sin(x^2)\,dx \\ &= -\frac{1}{2}\,cos(x^2)+C \\ 3 &= -\frac{1}{2}+C \\ -\frac{5}{2} &= C \\ y &= -\frac{1}{2}\,cos(x^2)-\frac{5}{2} \end{aligned}$$

29. $f''(x) = 2$

$$\begin{aligned} f'(x) &= \int f''(x)\,dx \\ &= \int 2\,dx \\ &= 2x+C \\ 4 &= 0+C \\ 4 &= C \\ f'(x) &= 2x+4 \\ f(x) &= \int f'(x)\,dx \\ &= \int 2x+4\,dx \\ &= x^2+4x+K \\ 3 &= 0+0+K \\ 3 &= K \\ f(x) &= x^2+4x+3 \end{aligned}$$

31. $f''(x) = x+1/x^3$

$$\begin{aligned} f'(x) &= \int f''(x)\,dx \\ &= \int x+x^{-3}\,dx \\ &= \frac{1}{2}\,x^2-\frac{1}{2}\,x^{-2}+C \\ 0 &= 2-\frac{1}{8}+C \\ -\frac{15}{8} &= C \\ f'(x) &= \frac{1}{2}\,x^2-\frac{1}{2}\,x^{-2}-\frac{15}{8} \\ f(x) &= \int f'(x)\,dx \\ &= \int \frac{1}{2}\,x^2-\frac{1}{2}\,x^{-2}-\frac{15}{8}\,dx \\ &= \frac{1}{6}\,x^3+\frac{1}{2x}-\frac{15}{8}\,x+K \\ 1 &= \frac{8}{6}+\frac{1}{4}-\frac{15}{4}+K \\ \frac{19}{6} &= K \\ f(x) &= \frac{1}{6}\,x^3+\frac{1}{2x}-\frac{15}{8}\,x+\frac{19}{6} \end{aligned}$$

33. $f''(x) = sin\,3x$

$$\begin{aligned} f'(x) &= \int f''(x)\,dx \\ &= \int sin\,3x\,dx \end{aligned}$$

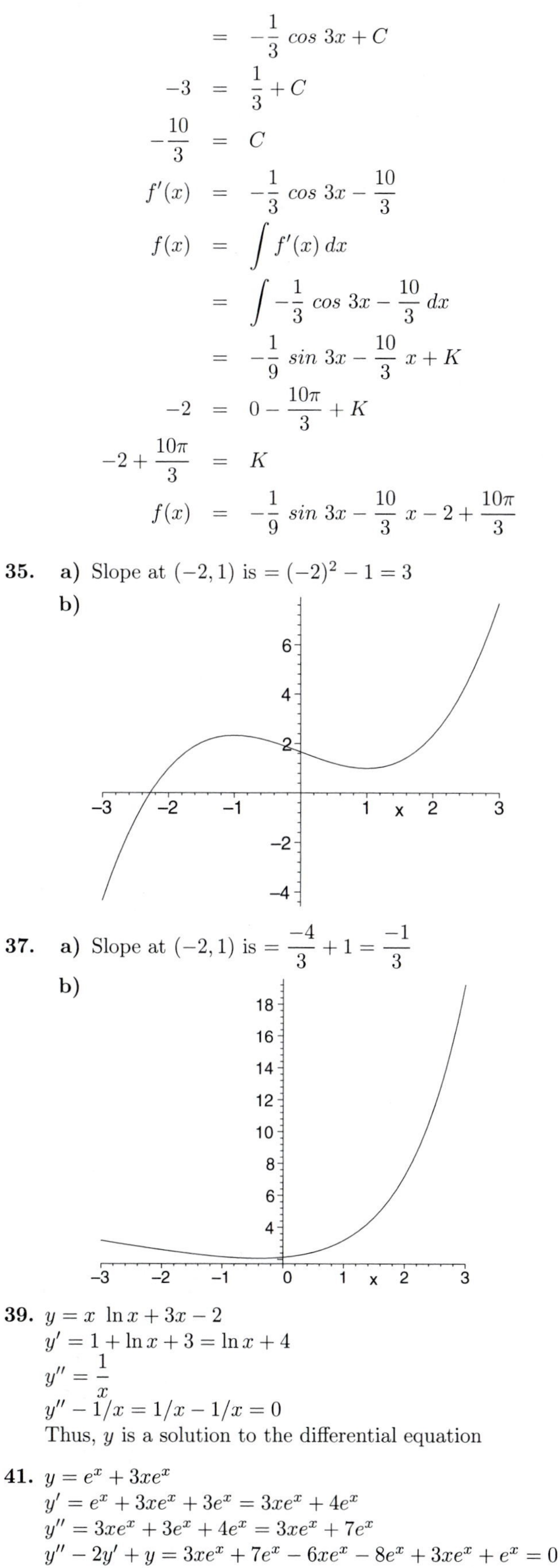

$$\begin{aligned}
&= -\frac{1}{3}\cos 3x + C \\
-3 &= \frac{1}{3} + C \\
-\frac{10}{3} &= C \\
f'(x) &= -\frac{1}{3}\cos 3x - \frac{10}{3} \\
f(x) &= \int f'(x)\,dx \\
&= \int -\frac{1}{3}\cos 3x - \frac{10}{3}\,dx \\
&= -\frac{1}{9}\sin 3x - \frac{10}{3}x + K \\
-2 &= 0 - \frac{10\pi}{3} + K \\
-2 + \frac{10\pi}{3} &= K \\
f(x) &= -\frac{1}{9}\sin 3x - \frac{10}{3}x - 2 + \frac{10\pi}{3}
\end{aligned}$$

35. **a)** Slope at $(-2,1)$ is $= (-2)^2 - 1 = 3$

b)

37. **a)** Slope at $(-2,1)$ is $= \frac{-4}{3} + 1 = \frac{-1}{3}$

b)

39. $y = x\ \ln x + 3x - 2$
$y' = 1 + \ln x + 3 = \ln x + 4$
$y'' = \frac{1}{x}$
$y'' - 1/x = 1/x - 1/x = 0$
Thus, y is a solution to the differential equation

41. $y = e^x + 3xe^x$
$y' = e^x + 3xe^x + 3e^x = 3xe^x + 4e^x$
$y'' = 3xe^x + 3e^x + 4e^x = 3xe^x + 7e^x$
$y'' - 2y' + y = 3xe^x + 7e^x - 6xe^x - 8e^x + 3xe^x + e^x = 0$
Thus, y is a solution to the differential equation

43. $y = 2e^x - 7e^{3x}$
$y' = 2e^x - 21e^{3x}$
$y'' = 2e^x - 61e^{3x}$
$y'' - 4y' + 3y = 2e^x - 61e^{3x} - 8e^x + 84e^{3x} + 6e^x - 21e^{3x} = 0$
Thus, y is a solution to the differential equation

45. $y = e^x \sin 2x$
$y' = 2e^x \cos 2x + e^x \sin 2x$
$y'' = -4e^x \sin 2x + 2e^x \cos 2x + 2e^x \cos 2x + e^x \sin 2x$
$y'' = 4e^x \cos 2x - 3e^x \sin 2x$
$y'' - 2y' + 5y$

$$\begin{aligned}
&= 4e^x\cos 2x - 3e^x\sin 2x - 4e^x\cos 2x - \\
&\quad 2e^x\sin 2x + 5e^x\sin 2x \\
&= 0
\end{aligned}$$

Thus, y is a solution to the differential equation

47.

$$\begin{aligned}
\frac{dR}{dS} &= \frac{k}{S} \\
dR &= \frac{k}{S}\,dS \\
R &= \int \frac{k}{S}\,dS \\
&= k\ln|S| + C \\
0 &= k\ln|S_0| + C \\
-k\ln|S_0| &= C \\
R(S) &= k\ln|S| - k\ln|S_0|
\end{aligned}$$

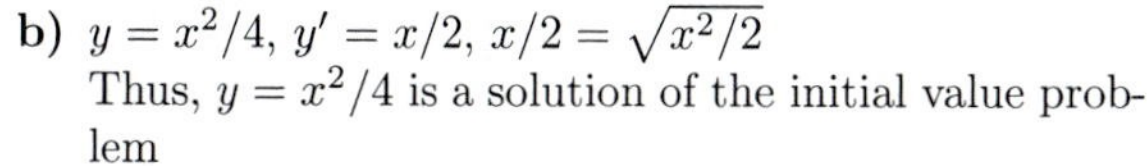

Thus, $R(S) = k\ln\left|\frac{S}{S_0}\right|$

49. **a)** $y = 0$, $y' = 0$, $0 = \sqrt{0}$
Thus, $y = 0$ is a solution of the initial value problem

b) $y = x^2/4$, $y' = x/2$, $x/2 = \sqrt{x^2/2}$
Thus, $y = x^2/4$ is a solution of the initial value problem

c)

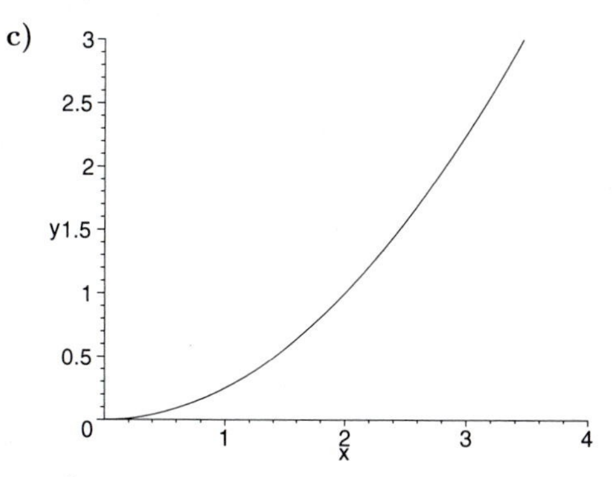

d) $x^2/4 = 0 \to x = 0 \to y = 0$
The initial value problem does not have a unique solution

e) $\partial f/\partial y = \frac{1}{2\sqrt{y}}$

f) The function does not satisfy the continuity criteria near the origin since the function is does not have "real" values to the left of the origin

51. a) $x^2 - 1 = 0 \to x = -1$ and $x = 1$
b) The slope of the tangent line at the point where $y' = 0$ is 0
c) Left to the student

53. a) $2x/3 + y = 0 \to y = -2x/3$
b) The slope of the tangent line at the point where $y' = 0$ is 0
c) Left to the student

Exercise Set 8.2

1. Both $-x^2$ and x^3 are continuous for all real numbers. Therefore the unique solution will exist on all real numbers

3. Discontinuities at $x = \pm\frac{\pi}{2}$. Therefore the unique solution will exist on $(-\pi/2, \pi/2)$

5. Discontinuities at $x = 2$ and $x = -1$. Therefore the unique solution will exist on $(2, \infty)$

7.

$$\begin{aligned} \int -3\,dx &= -3x + C \\ G(x) &= e^{-3x} \\ e^{-3x}y &= \int 0\,dx \\ e^{-3x}y &= C \\ y &= C\,e^{3x} \end{aligned}$$

9.

$$\begin{aligned} \int \cos 2x\,dx &= \frac{1}{2}\sin 2x + C \\ G(x) &= e^{1/2\,\sin 2x} \\ e^{1/2\,\sin 2x}y &= \int \cos 2x\,e^{1/2\,\sin 2x}\,dx \\ e^{1/2\,\sin 2x}y &= e^{1/2\,\sin 2x} + C \\ y &= C\,e^{-1/2\,\sin 2x} + 1 \end{aligned}$$

11.

$$\begin{aligned} \int -2t\,dt &= -t^2 + C \\ G(t) &= e^{-t^2} \\ e^{-t^2}y &= \int e^{-t^2}(2t)\,dt \\ e^{-t^2}y &= -e^{-t^2} + C \\ y &= Ce^{t^2} - 1 \end{aligned}$$

13.

$$\begin{aligned} \int -\,dt &= -t + C \\ G(t) &= e^{-t} \\ e^{-t}y &= \int e^{-t}e^{t}\,dt \\ e^{-t}y &= t + C \\ y &= t\,e^{t} + Ce^{t} \end{aligned}$$

15. $y' - \frac{4}{x}\,y = x^5e^{x^2} + 3x^3 - 6x^{-2}$

$$\begin{aligned} \int \frac{4}{x}\,dx &= -4\,\ln|\,x\,| + C = \ln x^{-4} + C \\ G(x) &= x^{-4} \\ x^{-4}y &= \int xe^{x^2} + \frac{3}{x} - 6x^{-6}\,dx \\ x^{-4}y &= \frac{1}{2}\,e^{x^2} + 3\,\ln|\,x\,| + \frac{6}{5}\,x^{-5} + C \\ y &= \frac{1}{2}\,x^4e^{x^2}3x^4\,\ln|\,x\,| + \frac{6}{5x} + cx^4 \end{aligned}$$

17. $y' + \frac{3}{x}\,y = \ln x$

$$\begin{aligned} \int \frac{3}{x}\,dx &= 3\,\ln|\,x\,| + C = \ln|\,x^3\,| + C \\ G(x) &= x^3 \\ x^3y &= \int x^3\,\ln x\,dx \\ x^3y &= \frac{1}{4}\,x^4\,\ln x - \frac{1}{16}\,x^4 + C \\ y &= \frac{1}{4}\,x\ln x - \frac{1}{16}\,x + Cx^{-3} \end{aligned}$$

19.

$$\begin{aligned} \int dt &= t + C \\ G(t) &= e^{t} \\ e^{t}y &= \int \ln t\,dt \\ e^{t}y &= t\,\ln t - t + C \\ y &= (t\,\ln t - t + C)e^{-t} \end{aligned}$$

21.

$$\begin{aligned} \int 4\,dx &= 4x + C \\ G(x) &= e^{4x} \\ e^{4x}y &= \int 6e^{4x}\,dx \\ e^{4x}y &= \frac{3}{2}\,e^{4x} + C \\ y &= \frac{3}{2} + Ce^{-4x} \\ 2 &= \frac{3}{2} + C \\ \frac{1}{2} &= C \\ y &= \frac{3}{2} + \frac{1}{2}\,e^{-4x} \end{aligned}$$

The solution exists for all real numbers

23. $y' + \frac{\cos x}{1 + \sin x}\,y = \cos x$

$$\begin{aligned} \int \frac{\cos x}{1 + \sin x}\,dx &= \ln|\,1 + \sin x\,| + C \\ G(x) &= 1 + \sin x \end{aligned}$$

$$\begin{aligned}
(1+\sin x)y &= \int (1+\sin x)\cos x\,dx \\
(1+\sin x)y &= \frac{1}{2}(1+\sin x)^2 + C \\
y &= \frac{1}{2}(1+\sin x) + \frac{C}{1+\sin x} \\
-1 &= 1+\frac{C}{2} \\
-4 &= C \\
y &= \frac{1}{2}(1+\sin x) - \frac{4}{1+\sin x} \\
&= \frac{2\sin^2 x + 2\sin x - 7}{2(1+\sin x)}
\end{aligned}$$

The solution exists on $(-\pi/2, 3\pi/2)$

25. $y' + \frac{1}{t}\,y = t^2$

$$\begin{aligned}
\int \frac{1}{t}\,dt &= \ln|\,t\,| + C \\
G(t) &= t \\
ty &= \int t^3\,dt \\
ty &= \frac{1}{4}\,t^4 + C \\
y &= \frac{1}{4}\,t^3 + \frac{C}{t} \\
5 &= 2+\frac{C}{2} \\
6 &= C \\
y &= \frac{1}{4}\,t^3 - \frac{6}{t}
\end{aligned}$$

The solution exists on $(0,\infty)$

27.

$$\begin{aligned}
\int -5\,dx &= -5x + C \\
G(x) &= e^{-5x} \\
e^{-5x}y &= \int xe^{-5x} + e^{-5x}\,dt \\
e^{-5x}y &= \frac{-1}{5}\,xe^{-5x} - \frac{6}{25}\,e^{-5x} + C \\
y &= \frac{-1}{5}\,x - \frac{6}{25} + C\,e^{5x} \\
1 &= \frac{6}{25} + C \\
\frac{31}{25} &= C \\
y &= \frac{-1}{5}\,x - \frac{6}{25} + \frac{31}{25}\,e^{5x}
\end{aligned}$$

The solution exists for all real numbers

29. $y' + \frac{e^x}{(e^x-2)}\,y = \frac{2e^{-2x} - e^{-3x}}{e^x-2}$

$$\begin{aligned}
\int \frac{e^x}{e^x-2}\,dx &= \ln|\,e^x-2\,| + C \\
G(x) &= e^x - 2 \\
(e^x-2)y &= \int 2e^{-2x} - e^{-3x}\,dx \\
(e^x-2)y &= -e^{-2x} + \frac{1}{3}\,e^{-3x} + C \\
y &= \frac{e^{-3x}}{3(e^x-2)} - \frac{e^{-2x}}{e^x-2} + \frac{C}{e^x-2} \\
3 &= \frac{-1}{3} + 1 - C \\
-\frac{7}{3} &= C \\
y &= \frac{e^{-3x}}{3(e^x-2)} - \frac{e^{-2x}}{e^x-2} - \frac{7}{3(e^x-2)} \\
&= \frac{e^{-3x} - 3e^{-2x} - 7}{3e^x - 6}
\end{aligned}$$

The solution exists on $(-\infty, \ln 2)$

31.

$$\begin{aligned}
\int 3t^2\,dt &= t^3 + C \\
G(t) &= e^{t^3} \\
e^{t^3}y &= \int t^2\,e^{2t^3}\,dt \\
e^{t^3}y &= \frac{1}{6}\,e^{2t^3} + C \\
y &= \frac{1}{6}\,e^{2t^3} + C\,e^{-t^3} \\
3 &= \frac{1}{6} + C \\
\frac{17}{6} &= C \\
y &= \frac{1}{6}\,e^{2t^3} + \frac{17}{6}\,e^{-t^3}
\end{aligned}$$

The solution exists for all real numbers

33.

$$\begin{aligned}
\int 4t\,dt &= 2t^2 + C \\
G(t) &= e^{2t^2} \\
e^{2t^2}y &= \int t\,e^{2t^2}\,dx \\
e^{2t^2}y &= \frac{1}{4}\,e^{2t^2} + C \\
y &= \frac{1}{4} + C\,e^{-2t^2} \\
3 &= \frac{1}{4} + C \\
\frac{11}{4} &= C \\
y &= \frac{1}{4} + \frac{11}{4}\,e^{-2t^2}
\end{aligned}$$

The solution exists for all real numbers

35. $P' = -0.2P + 3 \to P' + 0.2P = 3$

$$\begin{aligned} \int 0.2\,dt &= 0.2t + C \\ G(t) &= e^{0.2t} \\ e^{0.2t}P &= \int 3e^{0.2t}\,dt \\ e^{0.2t}P &= 15e^{0.2t} + C \\ P &= 15 + C\,e^{-0.2t} \end{aligned}$$

37. $Q' = -0.1Q - 5 \to Q' + 0.1Q = -5$

$$\begin{aligned} \int 0.1\,dt &= 0.1t + C \\ G(t) &= e^{0.1t} \\ e^{0.1t}Q &= \int -5e^{0.1t}\,dt \\ e^{0.1t}Q &= -50e^{0.1t} + C \\ Q &= -50 + C\,e^{-0.1t} \end{aligned}$$

39. $P' = kP \to P' - kP = 0$

$$\begin{aligned} \int -k\,dt &= -kt + C \\ G(t) &= e^{-kt} \\ e^{-kt}P &= \int 0\,dt \\ e^{-kt}P &= C \\ P &= C\,e^{kt} \\ P_0 &= C \\ P &= P_0e^{kt} \end{aligned}$$

41. $S' = -0.005S + 4 \to S' + 0.005S = 4$

$$\begin{aligned} \int 0.005\,dt &= 0.005t + C \\ G(t) &= e^{0.005t} \\ e^{0.005t}S &= \int 4e^{0.005t}\,dt \\ e^{0.005t}S &= 800e^{0.005t} + C \\ S &= 800 + C\,e^{-0.005t} \\ 100 &= 800 + C \\ -700 &= C \\ S &= 800 - 700\,e^{-0.005t} \end{aligned}$$

After two hours, $t = 120$

$$\begin{aligned} S(120) &= 800 - 700e^{-0.005(120)} \\ &= 415.83 \text{ lbs} \end{aligned}$$

43. $T' + kT = kC$

a)

$$\begin{aligned} \int k\,dt &= kt + C \\ G(t) &= e^{kt} \\ e^{kt}T &= \int kCe^{kt}\,dt \\ e^{kt}T &= C\,e^{kt} + K \\ T &= C + Ke^{-kt} \\ T_0 &= C + K \\ K &= T_0 - C \\ T &= C + (T_0 - C)e^{-kt} \end{aligned}$$

b)

$$\begin{aligned} 117 &= 70 + (143 - 70)e^{-30k} \\ \frac{47}{73} &= e^{-30k} \\ \ln\frac{47}{73} &= -30k \\ k &= \frac{1}{-30}\ln\frac{47}{73} \\ &= 0.0146771 \\ T &= C + (T_0 - C)e^{-0.0146771t} \\ 90 &= 70 + (143 - 70)e^{-0.0146771t} \\ \ln\frac{20}{73} &= -0.0146771t \\ t &= \frac{1}{-0.0146771}\ln\frac{20}{73} \\ &= 88.2 \text{ min} \end{aligned}$$

45. $y' + \frac{2}{x}\,y = 5x^2$

a) The solution will exist on $(0, \infty)$

b)

$$\begin{aligned} \int \frac{2}{x}\,dx &= 2\ln x + C = \ln x^2 + C \\ G(x) &= x^2 \\ x^2y &= \int 5x^4\,dx \\ x^2y &= x^5 + C \\ y &= x^3 + Cx^{-2} \\ 1 &= 1 + C \\ 0 &= C \\ y &= x^3 \end{aligned}$$

c) The domain of x^3 is all real numbers where the domain of $x^3 + \frac{1}{4x^2}$ is $(0, \infty)$

d) Left to the student

47. There is a net flow into the tank of 3 gallons per minute and no salt is added. Therefore

$$S' = 0 - \frac{2S}{500+3t} = -\frac{2}{500+3t}\,S$$

is the differential equation associated with this problem

$$\begin{aligned}
S' &= -\frac{2}{500+3t}\,S \\
\int \frac{dS}{S} &= \int \frac{2}{500+3t}\,dt \\
\ln S &= -\frac{2}{3}\,\ln(500+3t)+C \\
\ln S &= \ln(500+3t)^{-2/3}+C \\
S &= \frac{C}{(500+3t)^{2/3}} \\
200 &= \frac{C}{(500)^{2/3}} \\
S &= \frac{200(500)^{2/3}}{(500+3t)^{2/3}}
\end{aligned}$$

It will take $2000 = 500+3t \to t = 500$ minutes to fill the tank

$$\begin{aligned}
S(500) &= \frac{200(500)^{2/3}}{(2000)^{2/3}} \\
&= 79.37 \text{ pounds}
\end{aligned}$$

49.

$$\begin{aligned}
P' &= 1.2(1-e^{-1.2t}-P) \\
P'+1.2P &= 1.2-1.2e^{-1.2t} \\
\int 1.2\,dt &= 1.2t+C \\
G(t) &= e^{1.2t} \\
e^{1.2t}P &= \int 1.2e^{1.2t}-1.2\,dt \\
e^{1.2t}P &= e^{1.2t}-1.2t+C \\
P &= 1-1.2te^{-1.2t}+Ce^{-1.2t} \\
0 &= 1-0+C \\
C &= -1 \\
P &= 1-1.2te^{-1.2t}-e^{-1.2t}
\end{aligned}$$

51. a) Left to the student

b) $Q'+bQ=a$

$$\begin{aligned}
\int b\,dt &= bt+C \\
G(t) &= e^{bt} \\
e^{bt}Q &= \int ae^{bt}\,dt \\
e^{bt}Q &= \frac{a}{b}\,e^{bt}+C \\
Q &= \frac{a}{b}+C\,e^{-bt} \\
0 &= \frac{a}{b}+C \\
C &= -\frac{a}{b} \\
Q &= \frac{a}{b}-\frac{a}{b}\,e^{-bt}
\end{aligned}$$

c)

$$\begin{aligned}
Q'+bQ &= a \\
0+bQ &= a \\
Q &= \frac{a}{b}
\end{aligned}$$

Exercise Set 8.3

1. a) $2-y=0 \to y=2$

b) $y'' = -1 < 0$ Therefore the equilibrium point is asymptotically stabe

c) No inflection points since y'' does not change signs

d)

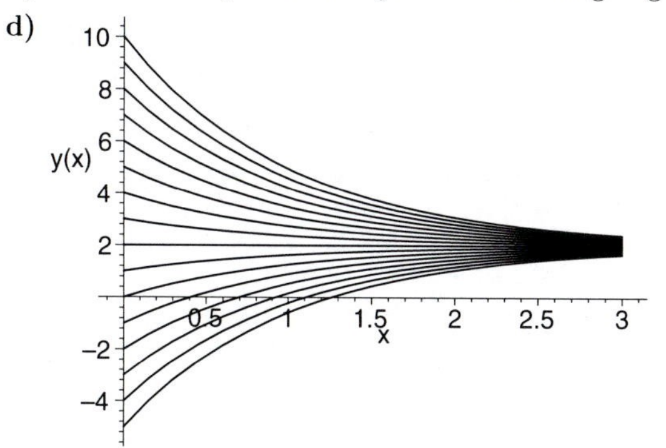

3. a)

$$\begin{aligned}
y^2-5y+4 &= 0 \\
(y-1)(y-4) &= 0 \\
y &= 1 \\
y &= 4
\end{aligned}$$

b) $y'' = 2y-5$
$y''(1) = -3$ Therefore $y=1$ is an asymptotically stable equilibrium point
$y''(4) = 3$ Therefore $y=4$ is an unstable equilibrium point

c) $2y-5=0 \to y=\frac{5}{2}$ Inflection point at $y=\frac{5}{2}$

d)

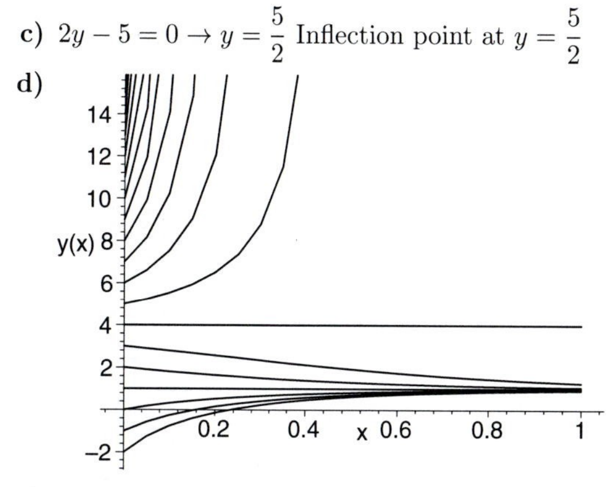

5. a)

$$\begin{aligned}
y^3-2y^2 &= 0 \\
y^2(y-2) &= 0 \\
y &= 0 \\
y &= 2
\end{aligned}$$

b) $y'' = 3y^2 - 4y$
$y''(0) = 0$ Therefore $y = 0$ is a semistable equilibrium point
$y''(2) = 4$ Therefore $y = 2$ is an unstable equilibrium point

c) $y(3y-4) = 0 \to y = 0,\ y = \dfrac{4}{3}$ Inflection point at $y = \dfrac{4}{3}$

d)

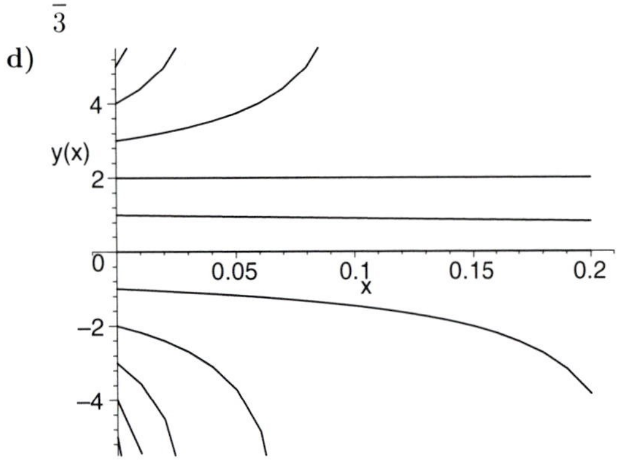

7. **a)**

$$\begin{aligned} y^3 + 8y^2 + 15y &= 0 \\ y(y+3)(y+5) &= 0 \\ y &= -5 \\ y &= -3 \\ y &= 0 \end{aligned}$$

b) $y'' = 3y^2 + 16y + 15$
$y''(-5) = 10$ Therefore $y = -5$ is an unstable equilibrium point
$y''(-3) = -6$ Therefore $y = -3$ is an asymptotically stable equilibrium point
$y''(0) = 15$ Therefore $y = 0$ is an unstable equilibrium point

c) $3y^2 + 16y + 15 = 0 \to y = \dfrac{-8 \pm \sqrt{19}}{3}$
Inflection points at $y = \dfrac{-8 \pm \sqrt{19}}{3}$

d)

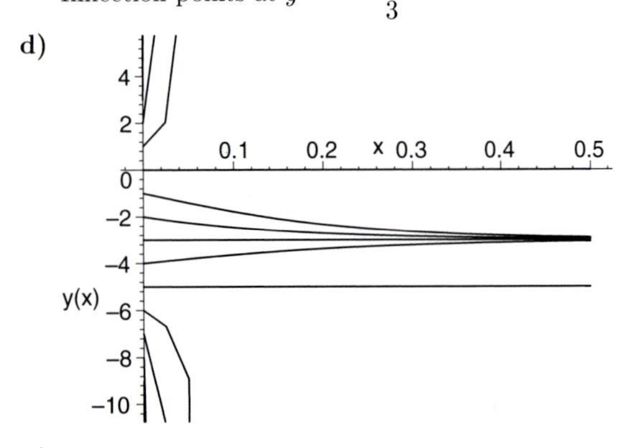

9. **a)**

$$\begin{aligned} e^{2y} - e^y &= 0 \\ e^y(e^y - 1) &= 0 \\ y &= 0 \end{aligned}$$

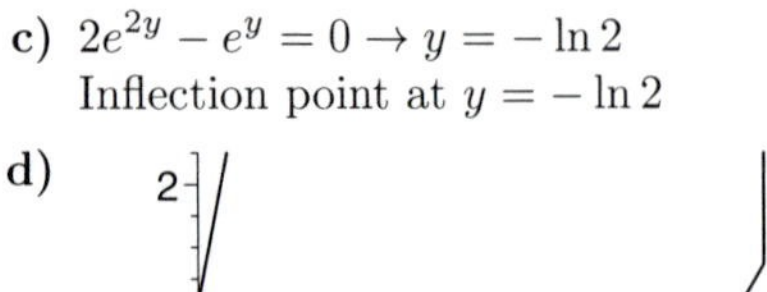

b) $y'' = 2e^{2y} - e^y$
$y''(0) = -1$ Therefore $y = -3$ is an unstable equilibrium point

c) $2e^{2y} - e^y = 0 \to y = -\ln 2$
Inflection point at $y = -\ln 2$

d)

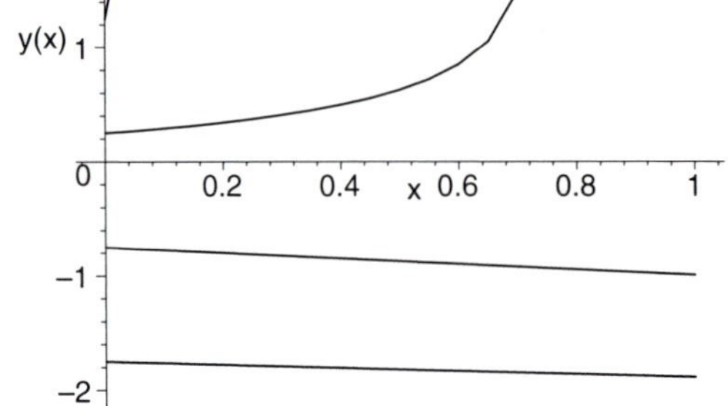

11. The population satisfies the logistic growth model with $k = 0.4$ and $L = 12500$.

a)

$$\begin{aligned} P(t) &= \frac{12500(9)}{9 + (12500 - 9)e^{-0.4t}} \\ &= \frac{112500}{9 + 12491e^{-0.4t}} \\ P(7) &= \frac{112500}{9 + 12491e^{-0.4(7)}} \\ &= 146.37 \text{ million} \end{aligned}$$

b)

$$\begin{aligned} 900 &= \frac{112500}{9 + 12491e^{-0.4t}} \\ \frac{112500}{900} &= 9 + 12491e^{-0.4t} \\ \frac{112500}{900} - 9 &= 12491e^{-0.4t} \\ 116 &= 12491e^{-0.4t} \\ -0.4t &= \ln\frac{116}{12491} \\ t &= \frac{-1}{0.4}\ln\frac{116}{12491} \\ &= 11.70 \text{ hours} \end{aligned}$$

13. $y' = ky\left(1 - \left[\frac{y}{L}\right]\right)^{\theta}$

a) Left to the student

b) The per capita graowth rate is $k\left(1 - \left[\frac{y}{L}\right]^{\theta}\right)$

c)

$$\begin{aligned} ky\left(1 - \left[\frac{y}{L}\right]^{\theta}\right) &= 0 \\ y &= 0 \\ \left(1 - \left[\frac{y}{L}\right]^{\theta}\right) &= 0 \\ y &= L \end{aligned}$$

$y'' = k - k\left(\frac{y}{L}\right)^\theta - k\left(\frac{y}{L}\right)^\theta \theta$
$y''(0) = k > 0$
Therefore $y = 0$ is an unstable equilibrium point
$y''(L) = -k\theta < 0$
Therefore $y = L$ is an asyptotically stable equilibrium point

d)

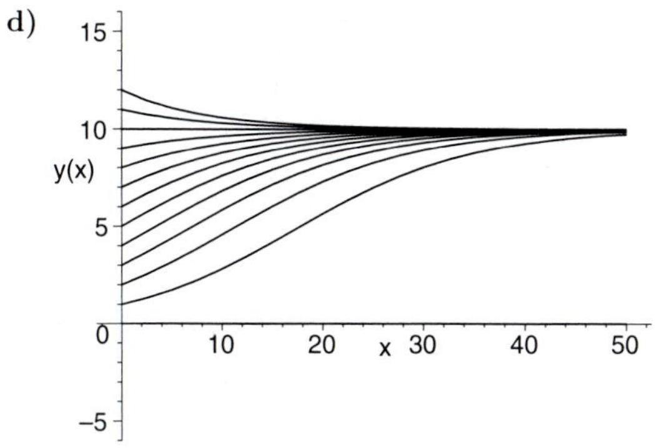

15. $y' = ky(P - y)$

a)

$$\begin{aligned} ky(P-y) &= 0 \\ y &= 0 \\ y &= P \end{aligned}$$

$y'' = kP - 2ky$
$y''(0) = kP > 0$
Therefore $y = 0$ is an unstable equilibrium point
$y''(P) = -kP < 0$
Therefore $y = P$ is an asymptotically stabe equilibrium point

b)

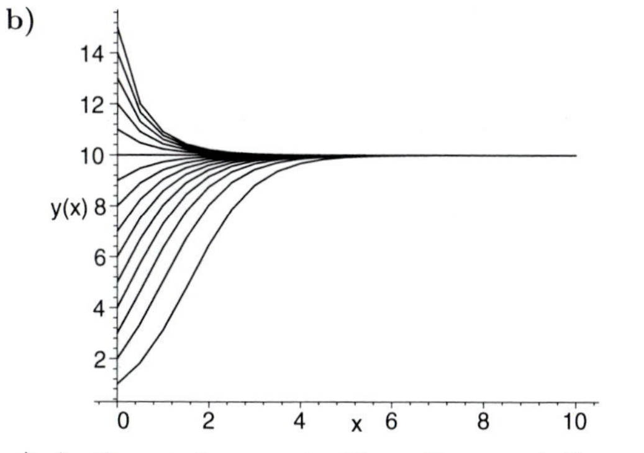

c) As the graph suggests, the entire population gets infected

17. **a)** $y'' = k - s - 2\frac{k}{L}y$
$y''(0) = k - s > 0$, thus $y = 0$ is an unstable equilibrium point
$y''(c) = -2(k-s) < 0$, thus $y = c$ is an asymptotically stable equilibrium point

b)

$$\begin{aligned} H &= sc \\ &= s \cdot \frac{(k-s)L}{k} \\ &= Ls - \frac{Ls^2}{k} \end{aligned}$$

c) $H' = L - \frac{2Ls}{k} = 0 \to s = \frac{k}{2}$

19. $y' = y^2 - 6y$

$$\begin{aligned} y^2 - 6y &= 0 \\ y(y-6) &= 0 \\ y &= 0 \\ y &= 6 \\ y'' &= 2y - 6 \\ y''(0) &= -6 < 0 \text{ stable} \\ y''(6) &= 6 > 0 \text{ unstable} \end{aligned}$$

$\lim_{t\to\infty} y(t) = 0$

21. **a)** By Theorem 4 it follows that c is semistable

b) The result follows from Part (a) and the definition on page 564

23. $y' = \dfrac{ky^2}{T} - ky - \dfrac{ky^3}{LT} + ky^2/L$

$$y'' = \frac{2ky}{T} - k - \frac{3ky^2}{LT} + \frac{2ky}{L}$$

$y''(0) = -k < 0$ asymptotically stabe

$$y''(L) = \frac{2kL}{T} - k - \frac{3kL}{T} + 2k =$$

$k - \dfrac{kL}{T} < 0$ asymptotically stable

25.

$$\begin{aligned} \frac{2ky}{T} - k - \frac{3ky^2}{LT} + \frac{2ky}{L} &= 0 \\ \frac{3k}{LT}y^2 - \left(\frac{2k}{T} + \frac{2k}{L}\right)y + k &= 0 \\ \frac{\frac{2k}{T} + \frac{2k}{L} \pm \sqrt{\left(\frac{2k}{T} + \frac{2k}{L}\right)^2 - \frac{12k^2}{LT}}}{\frac{6k}{LT}} &= y \\ \frac{L + T \pm \sqrt{L^2 - LT + T^2}}{3} &= \end{aligned}$$

27. **a)**

$$\begin{aligned} \frac{k}{L}y^2 - ky + s &= 0 \\ \frac{k \pm \sqrt{k^2 - \frac{4ks}{L}}}{\frac{2k}{L}} &= y \end{aligned}$$

The radicand is negative and therefore there are no real equilibrium points

b) $s > kL/4 \to s - kL/4 > 0$ but from Part (a) we know that there are no equilibrium points therefore the most the right hand side can get is $-(s - kl/4)$

c) Because 0 is the only physical value the population can approach.

29.

$$\begin{aligned} y'' &= nky^{n-1} - k \\ y''(0) &= -k < 0 \end{aligned}$$

$y = 0$ is an asymptotically stable equilibrium point

31.

$$\begin{aligned} nky^{n-1} - k &= 0 \\ y^{n-1} &= \frac{1}{n} \\ y &= \left(\frac{1}{n}\right)^{1/(n-1)} \end{aligned}$$

33. a)

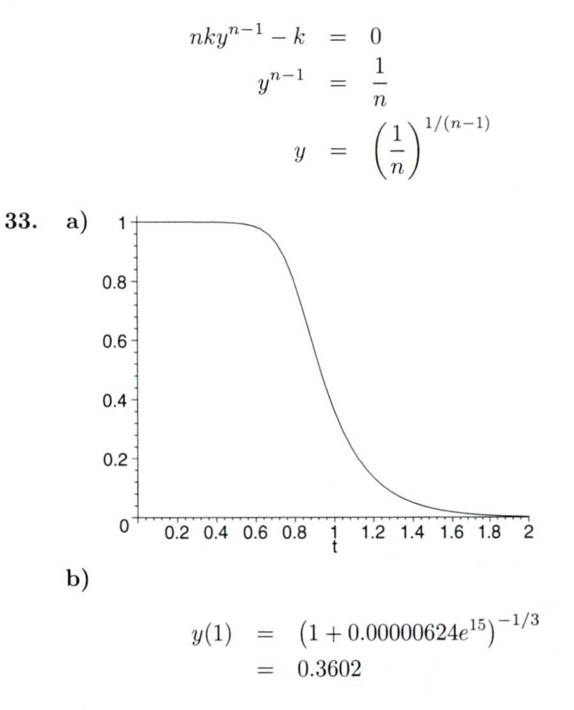

b)

$$\begin{aligned} y(1) &= \left(1 + 0.00000624e^{15}\right)^{-1/3} \\ &= 0.3602 \end{aligned}$$

c)

$$\begin{aligned} 0.5 &= \left(1 + 0.00000624e^{15t}\right)^{-1/3} \\ 0.5^{-3} - 1 &= 0.00000624e^{15t} \\ \frac{0.5^{-3} - 1}{0.00000624} &= e^{15t} \\ \ln\left(\frac{0.5^{-3} - 1}{0.00000624}\right) &= 15t \\ \frac{1}{15}\ln\left(\frac{0.5^{-3} - 1}{0.00000624}\right) &= t \\ 0.93 &= t \end{aligned}$$

Exercise Set 8.4

1.

$$\begin{aligned} \frac{dy}{dx} &= 4x^3 y \\ \frac{dy}{y} &= 4x^3\,dx \\ \int \frac{dy}{y} &= \int 4x^3\,dx \\ \ln y &= x^4 + C \\ y &= Ce^{x^4} \end{aligned}$$

3.

$$\begin{aligned} \frac{dy}{dx} &= \frac{x}{2y} \\ 2y\,dy &= x\,dx \\ \int 2y\,dy &= \int x\,dx \\ y^2 &= \frac{1}{2}\,x^2 + C \\ y &= \pm\sqrt{\frac{1}{2}\,x^2 + C} \end{aligned}$$

5.

$$\begin{aligned} \frac{dy}{dx} &= \frac{x\sqrt{y^2+1}}{y} \\ \frac{y}{\sqrt{y^2+1}}\,dy &= x\,dx \\ \sqrt{y^2+1} &= \frac{1}{2}\,x^2 + C \\ y^2 &= \left(\frac{1}{2}\,x^2 + C\right)^2 - 1 \\ y &= \pm\sqrt{\left(\frac{1}{2}\,x^2 + C\right)^2 - 1} \end{aligned}$$

7.

$$\begin{aligned} \frac{dy}{dx} &= \frac{x^2 \sec y}{(x^3+1)^3} \\ \cos y\,dy &= \frac{x^2}{(x^3+1)^3}\,dx \\ \int \cos y\,dy &= \int \frac{x^2}{(x^3+1)^3}\,dx \\ \sin y &= -\frac{3}{2}\,(x^3+1)^{-2} + C \\ \sin y + \frac{3}{2(x^3+1)^2} &= C \end{aligned}$$

9.

$$\begin{aligned} \frac{dy}{dt} &= \frac{y^3+1}{y^2} \\ \frac{y^2}{y^3+1}\,dy &= dt \\ \int \frac{y^2}{y^3+1}\,dy &= \int dt \\ \frac{1}{3}\,\ln|\,y^3+1\,| &= t + C \\ \ln|\,(y^3+1)^{1/3}\,| &= t + C \\ (y^3+1)^{1/3} &= Ce^t \\ y^3 + 1 &= Ce^{3t} \\ y &= \sqrt[3]{Ce^{3t} - 1} \end{aligned}$$

11.

$$\begin{aligned}
\frac{dy}{dx} &= x\ cos^2 y \\
sec^2 y\ dy &= x\ dx \\
\int sec^2 y\ dy &= \int x\ dx \\
tan\ y &= \frac{1}{2}\ x^2 + C \\
tan\ y - \frac{1}{2}\ x^2 &= C
\end{aligned}$$

NOTE: $y = \dfrac{(2n+1)\pi}{2}$ are constant solutions as well

13.

$$\begin{aligned}
\frac{dy}{dx} &= \frac{\sqrt{x}}{sin\ y + cos\ y} \\
sin\ y + cos\ y\ dy &= \sqrt{x}\ dx \\
\int sin\ y + cos\ y\ dy &= \int \sqrt{x}\ dx \\
-cos\ y + sin\ y &= \frac{2}{3}\ x^{3/2} + C \\
sin\ y - cos\ y - \frac{2}{3}\ x^{3/2} &= C
\end{aligned}$$

15.

$$\begin{aligned}
\frac{dy}{dt} &= \frac{t}{(t^2+1)(y^4+1)} \\
y^4 + 1\ dy &= \frac{t}{t^2+1}\ dt \\
\frac{1}{5}\ y^5 + y &= \frac{1}{2}\ \ln(t^2+1) + C \\
\frac{1}{5}\ y^5 + y - \frac{1}{2}\ \ln(t^2+1) &= C
\end{aligned}$$

17.

$$\begin{aligned}
\frac{dy}{dx} &= 3x^2(y-2)^2 \\
(y-2)^{-2}\ dy &= 3x^2\ dx \\
\int (y-2)^{-2}\ dy &= \int 3x^2\ dx \\
\frac{-1}{(y-2)} &= x^3 + C \\
y - 2 &= \frac{-1}{x^3+C} \\
y &= 2 - \frac{1}{x^3+C}
\end{aligned}$$

NOTE $y = 2$ is a constant solution as well

19.

$$\begin{aligned}
\int 3y^2\ dy &= \int 2x\ dx \\
y^3 &= x^2 + C \\
125 &= 4 + C \\
C &= 121 \\
y^3 &= x^2 + 121 \\
y &= \sqrt[3]{x^2+121}
\end{aligned}$$

21.

$$\begin{aligned}
\int csc^2 y\ dy &= \int e^{2t}\ dt \\
-cot\ y &= \frac{1}{2}\ e^{2t} + C \\
-1 &= \frac{1}{2} + C \\
-\frac{3}{2} &= C \\
cot\ y + \frac{1}{2}\ e^{2t} &= \frac{3}{2}
\end{aligned}$$

23.

$$\begin{aligned}
\int y^{-2}\ dy &= \int \frac{\ln t}{t}\ dt \\
-\frac{1}{y} &= \frac{1}{2}\ \ln^2 t + C \\
\frac{1}{4} &= C \\
\frac{1}{y} &= -\frac{1}{2}\ \ln^2 t - \frac{1}{4} \\
y &= \frac{-4}{2\ln^2 t + 1}
\end{aligned}$$

25.

$$\begin{aligned}
\int 3y^2(y^3+2)^{-2}\ dy &= \int x\ dx \\
-(y^3+2)^{-1} &= \frac{1}{2}\ x^2 + C \\
-\frac{1}{10} &= \frac{1}{2} + C \\
-\frac{3}{5} &= C \\
\frac{1}{y^3+2} &= -\frac{1}{2}\ x^2 + \frac{3}{5} \\
\frac{1}{y^3+2} &= \frac{-5x^2+6}{10} \\
y^3 + 2 &= \frac{-10}{6-5x^2} \\
y^3 &= \frac{-10+10x^2-12}{6-5x^2} \\
y &= \sqrt[3]{\frac{10x^2-2}{6-5x^2}}
\end{aligned}$$

27.

$$\begin{aligned}
\int \frac{\cos y}{(2+\sin y)^2}\,dy &= \int t^2\,dt \\
-\frac{1}{2+\sin y} &= \frac{1}{3}\,t^3+C \\
-\frac{1}{5/2} &= -\frac{1}{3}+C \\
-\frac{1}{15} &= C \\
\frac{1}{2+\sin y} &= -\frac{1}{3}\,t^3+\frac{1}{15} \\
\frac{1}{2+\sin y} &= \frac{-5t^3+1}{15} \\
\sin y+2 &= \frac{15}{1-5t^3}
\end{aligned}$$

29.

$$\begin{aligned}
\int e^{4y}+e^{5y}\,dy &= \int \sqrt{t} \\
\frac{1}{4}\,e^{4y}+\frac{1}{5}\,e^{5t} &= \frac{2}{3}\,t^{3/2}+C \\
\frac{1}{4}+\frac{1}{5} &= \frac{2}{3}+C \\
\frac{-13}{60} &= C \\
\frac{1}{4}\,e^{4y}+\frac{1}{5}\,e^{5y} &= \frac{2}{3}\,t^{3/2}-\frac{13}{60}
\end{aligned}$$

31.

$$\begin{aligned}
\int y\,dy &= \int x\,dx \\
\frac{1}{2}\,y^2 &= \frac{1}{2}\,x^2+C \\
\frac{21}{2}-\frac{25}{2} &= C \\
-2 &= C \\
\frac{1}{2}\,y^2 &= \frac{1}{2}\,x^2-2 \\
y &= \sqrt{x^2-4}
\end{aligned}$$

Which has a domain of $(2,\infty)$

33. a)

$$\begin{aligned}
\int \frac{dy}{y} &= \int k\,dt \\
\ln|\,y\,| &= kt+C \\
y &= Ce^{kt} \\
y_0 &= C \\
y &= y_0e^{kt}
\end{aligned}$$

b) Left to the student

35.

$$\begin{aligned}
y &= y_0e^{kt} \\
4.404 &= 5e^{11k} \\
\frac{4.404}{5} &= e^{11k} \\
\frac{\ln 4.404}{5} &= 11k \\
\frac{1}{11}\,\ln\frac{4.404}{5} &= k \\
-0.01154 &= k
\end{aligned}$$

The decay rate is 1.154% per day

37. - 39. Left to the student

41. Left to the student

43. a)

$$\begin{aligned}
\int \frac{L}{R(L-R)}\,dR &= \int k\,dt \\
\frac{1}{L}\,\ln\left(\frac{R}{L-R}\right) &= kt+C \\
\frac{1}{L}\,\ln\left(\frac{R_0}{L-R_0}\right) &= C \\
kt+\frac{1}{L}\,\ln\left(\frac{R_0}{L-R_0}\right) &= \frac{1}{L}\,\ln\left(\frac{R}{L-R}\right) \\
\frac{1}{L}\left[\ln\left(\frac{R}{L-R}\right)-\ln\left(\frac{R_0}{L-R_0}\right)\right] &= kt \\
\frac{1}{L}\left[\ln\frac{\left(\frac{R}{L-R}\right)}{\left(\frac{R_0}{L-R_0}\right)}\right] &= kt \\
\left[\ln\frac{\left(\frac{R}{L-R}\right)}{\left(\frac{R_0}{L-R_0}\right)}\right] &= Lkt \\
\frac{\left(\frac{R}{L-R}\right)}{\left(\frac{R_0}{L-R_0}\right)} &= e^{Lkt} \\
R(L-R_0) &= LR_0e^{Lkt} \\
R(L-R_0)+RR_0e^{Lkt} &= LR_0e^{Lkt} \\
R\left[(L-R_0)+R_0e^{Lkt}\right] &= LR_0e^{Lkt} \\
\frac{LR_0e^{Lkt}}{(L-R_0)+R_0e^{Lkt}} &= R \\
\frac{LR_0}{(L-R_0)e^{-Lkt}+R_0} &=
\end{aligned}$$

b) Left to the student

44. **a)**

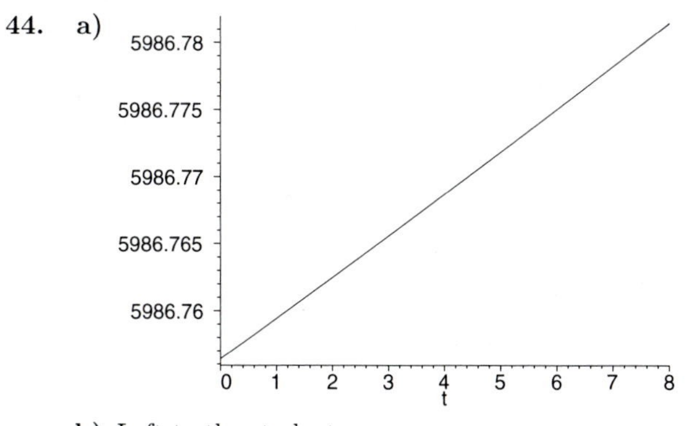

b) Left to the student

Exercise Set 8.5

1. **a)** $y(1) \approx -1.2$

b) $y(1) \approx -1$

c)

$$\begin{aligned} \int dy &= \int 2x\, dx \\ y &= x^2 + C \\ -2 &= C \\ y &= x^2 - 2 \end{aligned}$$

d) The exact value is $y(1) = -1$

3. **a)** $y(2) \approx 18.235$

b) $y(2) \approx 40.067$

c)

$$\begin{aligned} \int \frac{dy}{y} &= \int 2x\, dx \\ \ln y &= x^2 + C \\ y &= Ce^{x^2} \\ 2 &= Ce \\ C &= 2/e \\ y &= 2e^{x^2-1} \end{aligned}$$

d) The exact value is $y(2) = 40.171$

5. **a)** $y(2) \approx 28.8988$

b) $y(2) \approx 34.1710$

c) $y' - y = 2x$

$$\begin{aligned} \int -1\, dx &= -x + C \\ G(x) &= e^{-x} \\ e^{-x}y &= \int 2xe^{-x}\, dx \\ e^{-x}y &= -2xe^{-x} - 2e^{-x} + C \\ y &= -2x - 2 + Ce^{x} \\ 2 &= 2 - 2 + Ce{-1} \\ 2e &= C \\ y &= 2e^{x+1} - 2xe^{-x} - 2e^{-x} \end{aligned}$$

d) The exact value is $y(2) = 34.1711$

7. **a)** $y(2) \approx 9.304$

b) $y(2) \approx 14.390$

c)

$$\begin{aligned} \int \frac{dy}{y} &= \int x^2\, dx \\ \ln y &= \frac{x^3}{3} + C \\ 0 &= C \\ \ln y &= \frac{x^3}{3} \\ y &= e^{x^3/3} \end{aligned}$$

d) The exact value is $y(2) = e^{8/3} \approx 14.392$

9. **a)** $P(5) \approx 101.782$
$P(8) \approx 404.5580472$

b) $P(5) \approx 108.307$
$P(8) \approx 436.449$

c) $P(5) = 108.307$
$P(8) = 436.449$

11. **a)** $P(2) \approx 0.6975$
$P(4) \approx 0.9589$

b) $P(2) \approx 0.6916$
$P(4) \approx 0.9523$

c) $P(2) = 0.6916$
$P(4) = 0.9523$

13. **a)** $Q(6) \approx 3.0746$

b) $Q(6) \approx 3.1553$

15. **a)** No

b) No

c) Yes

d) $\Delta x = \frac{1}{n}$ where n is an integer

e) $\Delta x = \frac{a}{n}$ where n is an integer

17. **a)** $y(2) = 2.47523873$

b) $y(2) = 2.47522970$

c) $y(2) = 2.47522913$

d) $\mid 2.47522913 - 2.47523873 \mid = 0.00000917$
$\mid 2.47522913 - 2.47522970 \mid = 0.000000577$

e) 16

Chapter 9

Higher Order and Systems of Differential Equations

Exercise Set 9.1

1.

$$\begin{aligned} r^2 - 6r + 5 &= 0 \\ (r-1)(r-5) &= 0 \\ r &= 1 \\ r &= 5 \\ y &= C_1 e^x + C_2 e^{5x} \end{aligned}$$

3.

$$\begin{aligned} r^2 - r - 2 &= 0 \\ (r-2)(r+1) &= 0 \\ r &= -1 \\ r &= 2 \\ y &= C_1 e^{-x} + C_2 e^{2x} \end{aligned}$$

5.

$$\begin{aligned} r^2 + 3r + 2 &= 0 \\ (r+1)(r+2) &= 0 \\ r &= -1 \\ r &= -2 \\ y &= C_1 e^{-2x} + C_2 e^{-1x} \end{aligned}$$

7.

$$\begin{aligned} 2r^2 - 5r + 2 &= 0 \\ (2r-1)(r-2) &= 0 \\ r &= \frac{1}{2} \\ r &= 2 \\ y &= C_1 e^{x/2} + C_2 e^{2x} \end{aligned}$$

9.

$$\begin{aligned} r^2 - 9 &= 0 \\ (r-3)(r+3) &= 0 \\ r &= -3 \\ r &= 3 \\ y &= C_1 e^{-3x} + C_2 e^{3x} \end{aligned}$$

11.

$$\begin{aligned} r^2 + 10r + 25 &= 0 \\ (r+5)^2 &= 0 \\ r &= -5 \text{ repeated} \\ y &= C_1 e^{-5x} + C_2 x e^{-5x} \end{aligned}$$

13.

$$\begin{aligned} 4r^2 + 12r + 9 &= 0 \\ (2r+3)^2 &= 0 \\ r &= -\frac{3}{2} \text{ repeated} \\ y &= C_1 e^{-3x/2} + C_2 x e^{-3x/2} \end{aligned}$$

15.

$$\begin{aligned} r^3 + r^2 + 4r + 4 &= 0 \\ r^2(r+1) + 4(r+1) &= 0 \\ (r+1)(r^2+4) &= 0 \\ r &= -1 \\ r &= \pm 2i \\ y &= C_1 e^{-x} + C_2 \ sin \ 2x + C_3 \ cos \ 2x \end{aligned}$$

17.

$$\begin{aligned} r^3 + 6r^2 + 12r + 8 &= 0 \\ (r+2)^3 &= 0 \\ r &= -2 \text{ repeated twice} \\ y &= C_1 e^{-2x} + C_2 x e^{-2x} + C_3 x^2 e^{-2x} \end{aligned}$$

19.

$$\begin{aligned} r^3 - 6r^2 + 3r - 18 &= 0 \\ (r-6)(r^2+3) &= 0 \\ r &= 6 \\ r &= \pm i\sqrt{3} \\ y &= C_1 e^{6x} + C_2 \ sin \ \sqrt{3}x + C_3 \ cos \ \sqrt{3}x \end{aligned}$$

21.

$$\begin{aligned} r^4 - 5r^3 + 4r^2 &= 0 \\ r^2(r-4)(r-1) &= 0 \\ r &= 0 \text{ repeated} \\ r &= 4 \\ r &= 1 \\ y &= C_1 + C_2 x + C_3 e^x + C_4 e^{4x} \end{aligned}$$

23.

$$\begin{aligned} r^2 + 36 &= 0 \\ r &= \pm 6i \\ y &= C_1 \ sin \ 6x + C_2 \ cos \ 6x \end{aligned}$$

25.

$$\begin{aligned} r^2+8r+41 &= 0 \\ r &= 4\pm 5i \\ y &= e^{4x}\left(C_1\ sin\ 5x + C_2\ cos\ 5x\right) \end{aligned}$$

27.

$$\begin{aligned} r^3+2r^2+5r &= 0 \\ r(r^2+2r+5) &= 0 \\ r &= 0 \\ r &= -1\pm 2i \\ y &= C_1 + e^{-x}\left(C_1\ sin\ 2x + C_2\ cos\ 2x\right) \end{aligned}$$

29.

$$\begin{aligned} r^3-1 &= 0 \\ (r-1)(r^2+r+1) &= 0 \\ r &= 1 \\ r &= -\frac{1}{2}\pm\frac{i\sqrt{3}}{2} \\ y &= C_1e^x + C_2e^{-x/2}\ sin\ \frac{\sqrt{3}x}{2} + \\ & C_3e^{-x/2}\ cos\ \frac{\sqrt{3}x}{2} \end{aligned}$$

31.

$$\begin{aligned} 2r^2+2r-5 &= 0 \\ r &= -\frac{1\pm\sqrt{11}}{2} \\ y &= C_1e^{(-1-\sqrt{11})x/2} + C_2e^{(-1+\sqrt{11})x/2} \end{aligned}$$

33.

$$\begin{aligned} 3r^2-2r+10 &= 0 \\ r &= \frac{2\pm i\sqrt{116}}{6} \\ y &= e^{x/3}\left(C_1\ sin\ \frac{\sqrt{116}x}{6} + C_2\ cos\ \frac{\sqrt{116}x}{6}\right) \\ &= e^{x/3}\left(C_1\ sin\ \frac{\sqrt{29}x}{3} + C_2\ cos\ \frac{\sqrt{29}x}{3}\right) \end{aligned}$$

35.

$$\begin{aligned} r(r-1) &= 0 \\ r &= 0 \\ r &= 1 \\ y &= C_1 + C_2e^x \\ y' &= C_2e^x \\ y(0) &= 0 \to C_1 = -C_2 \\ y'(0) &= -1 \to C_2 = -1 \\ \to & \\ C_1 &= 1 \\ y &= 1-e^x \end{aligned}$$

37.

$$\begin{aligned} r^2-1 &= 0 \\ r &= \pm 1 \\ y &= C_1e^{-x} + C_2e^x \\ y' &= -C_1e^{-x} + C_2e^x \\ y(0) &= 1 \to C_1 + C_2 = 1 \\ y'(0) &= 2 \to -C_1 + C_2 = 3 \end{aligned}$$

Solving the system

$$\begin{aligned} C_1 &= -1 \\ C_2 &= 2 \\ y &= 2e^x - e^{-x} \end{aligned}$$

39.

$$\begin{aligned} (r+2)^2 &= 0 \\ r &= -2 \text{ repeated} \\ y &= C_1e^{-2x} + C_2xe^{-2x} \\ y' &= -2C_1e^{-x} - 2C_2xe^{-x} + C_2e^{-x} \\ y(0) &= 2 \to C_1 = 2 \\ y'(0) &= 3 \to C_2 = 7 \\ y &= e^x + xe^x \end{aligned}$$

41.

$$\begin{aligned} r(r^2+1) &= 0 \\ r &= 0 \\ r &= \pm i \\ y &= C_1 + C_2\ sin\ x + C_3\ cos\ x \\ y' &= C_2\ cos\ x - C_3\ sin\ x \\ y'' &= -C_2\ sin\ x - C_3\ cos\ x \\ y(\pi) &= 1 \to C_1 - C_2 = 1 \\ y'(\pi) &= 8 \to C_2 = 4 \\ y''(\pi) &= 4 \to C_3 = -8 \\ \to & \\ C_1 = 5 & \\ y &= 5 - 8\ sin\ x + 4\ cos\ x \end{aligned}$$

43.

$$\begin{aligned} r^2-3r+2 &= 0 \\ r &= 1 \\ r &= 2 \\ N &= C_1e^t + C_2e^{2t} \\ N' &= C_1e^t + 2C_2e^{2t} \\ N(0) &= 5 \to C_1 + C_2 = 5 \\ N'(0) &= 15 \to C_1 + 2C_2 = 15 \end{aligned}$$

Solving the system

$$\begin{aligned} C_1 &= -5 \\ C_2 &= 10 \\ N(t) &= 10e^{2t} - 5e^t \\ N(4) &= 10e^8 - 5e^4 \\ &= 29537 \end{aligned}$$

45. $2.7 + 3.94 + 2(0.0576) = 6.7552$
$0.0576(2.7 + 3.94 + 0.0576) = 0.38578176$

$$\begin{aligned} r^2 + 6.7552r + 0.38578176 &= 0 \\ r &= -0.0576 \\ r &= -6.6976 \\ N(t) &= C_1\ e^{-0.0576t} + C_2\ e^{-6.6976t} \end{aligned}$$

47. $0.31 + 1.64 + 2(0.51) = 2.97$
$0.51(0.31 + 1.64 + 0.51) = 1.2546$

$$\begin{aligned} r^2 + 2.97r + 1.2546 &= 0 \\ r &= -0.51 \\ r &= -2.46 \\ N(t) &= C_1\ e^{-0.51t} + C_2\ e^{-2.46t} \end{aligned}$$

49.

$$\begin{aligned} y(x) &= C_1e^{7x} + C_2e^{9x} + C_3e^{2x}\ sin\ 4x + \\ &\quad C_4e^{2x}\ cos\ 4x + C_5xe^{2x}\ sin\ 4x + C_6xe^{2x}\ cos\ 4x \end{aligned}$$

51. $y = C_1y_1 + C_2y_2$

$y' = C_1y_1' + C_2y_2'$

$y'' = C_1y_1'' + C_2y_2''$

$$\begin{aligned} ay'' + by' + cy &= a(C_1y_1'' + C_2y_2'') + \\ &\quad b(C_1y_1' + C_2y_2') + \\ &\quad c(C_1y_1'' + C_2y_2'') \\ &= aC_1y_1'' + bC_1y_1' + cC_1y + \\ &\quad aC_2y_2'' + bC_2y_2' + cC_2y_2 \\ &= 0 + 0 \\ &= 0 \end{aligned}$$

53. **a)** If the quadratic equation has only one root then the discriminant must equal 0. Thus

$$\begin{aligned} x &= \frac{-b \pm \sqrt{0}}{2a} \\ &= \frac{-b}{2a} \end{aligned}$$

b) Since $x = r$ is a solution to the equation then substitution r for x in the equation $ax^2 + bx + c$ yields the desired result.

c)

$$\begin{aligned} r &= \frac{-b}{2a} \\ 2ar &= -b \\ 2ar + b &= 0 \end{aligned}$$

d)

$$\begin{aligned} y &= xe^{rx} \\ y' &= rxe^{rx} + e^{rx} \\ y'' &= r^2xe^{rx} + re^{rx} + re^{rx} \\ ay'' + by' + cy &= \\ &= a(r^2xe^{rx} + re^{rx} + re^{rx}) + \\ &\quad b(rxe^{rx} + e^{rx}) + cxe^{rx} \\ &= (ar^2 + br + c)xe^{rx} + (2ar + b)e^{rx} \\ &= 0 + 0 \\ &= 0 \end{aligned}$$

Exercise Set 9.2

1.

$$\begin{aligned} r^2 + 1 &= 0 \\ r &= \pm\ i \\ y_h &= C_1\ sin\ x + C_2\ cos\ x \\ y_p &= A \\ y_p' &= 0 \\ y_p'' &= 0 \\ y_p'' + y_p &= 7 \\ 0 + A &= 7 \\ y_p &= 7 \\ y &= C_1\ sin\ x + C_2\ cos\ x + 7 \end{aligned}$$

3.

$$\begin{aligned} r^2 - 2r + 1 &= 0 \\ r &= 1 \text{ repeated} \\ y_h &= C_1\ e^x + C_2x\ e^x \\ y_p &= A \\ y_p' &= 0 \\ y_p'' &= 0 \\ y_p'' - 2y_p' + y_p &= 3 \\ 0 - 0 + A &= 3 \\ y_p &= 3 \\ y &= C_1\ e^x + C_2x\ e^x + 3 \end{aligned}$$

5.

$$\begin{aligned} r^2 + 4r + 4 &= 0 \\ r &= -2 \text{ repeated} \\ y_h &= C_1\ e^{-2x} + C_2x\ e^{-2x} \\ y_p &= Ax + B \\ y_p' &= A \\ y_p'' &= 0 \\ 0 + 4(A) + 4(Ax + B) &= 8 - 12x \\ 4Ax + (4A + 4B) &= 8 - 12x \\ 4A &= -12 \rightarrow A = -3 \\ -12 + 4B &= 8 \rightarrow B = 5 \\ y_p &= 5 - 3x \\ y &= C_1\ e^{-2x} + C_2x\ e^{-2x} - 3x + 5 \end{aligned}$$

7.

$$\begin{aligned}
r^2 - 4r + 3 &= 0 \\
r &= -3 \\
r &= -1 \\
C_1\, e^{-3x} + C_2\, e^{-x} &= y_h \\
y_p &= Ax^2 + Bx + C \\
y_p' &= 2Ax + B \\
y_p'' &= 2A \\
2A - 4(2Ax + B) + 3(Ax^2 + Bx + C) &= 6x^2 - 4 \\
3Ax^2 + (-8A + 3B)x + (2A - 4B + 3C) &= 6x^2 - 4 \\
3A &= 6 \to A = 2 \\
-8A + 3B &= 0 \to B = -\frac{16}{3} \\
2A - 4B + 3C &= -4 \to C = \frac{40}{9} \\
2x^2 - \frac{16}{3}\, x + \frac{40}{9} &= y_p \\
C_1\, e^{-3x} + C_2\, e^{-x} + 2x^2 - \frac{16}{3}\, x + \frac{40}{9} &= y
\end{aligned}$$

9.

$$\begin{aligned}
r^2 - r - 2 &= 0 \\
r &= 2 \\
r &= -1 \\
C_1\, e^{-x} + C_2\, e^{2x} &= y_h \\
Ax^3 + Bx^2 + Cx + D &= y_p \\
3Ax^2 + 2Bx + C &= y_p' \\
6Ax + 2B &= y_p'' \\
x^3 - 1 &= 6Ax + 2B - 3Ax^2 - 2Bx - \\
&\quad C + 2(Ax^3 + Bx^2 + Cx + D) \\
x^3 - 1 &= 2Ax^3 + (2B - 3A)x^2 + \\
&\quad (2C - 2B + 6A)x + \\
&\quad (2D + 2B - C) \\
A &= \frac{1}{2} \\
2B - 3A &= 0 \to B = \frac{3}{4} \\
2C - 2B + 6A &= 0 \to C = \frac{-9}{4} \\
2D - C + 2B &= -1 \to D = \frac{19}{8} \\
y_p &= \frac{x^3}{2} + \frac{3x^2}{4} - \frac{9x}{4} + \frac{19}{8} \\
y &= C_1\, e^{-x} + C_2\, e^{2x} + \\
&\quad \frac{x^3}{2} + \frac{3x^2}{4} - \frac{9x}{4} + \frac{19}{8}
\end{aligned}$$

11.

$$\begin{aligned}
r^2 - 3r &= 0 \\
r &= 0 \\
r &= 3 \\
y_h &= C_1 + C_2\, e^{3x} \\
y_p &= Ax + B \\
y_p' &= A \\
y_p'' &= 0 \\
0 - 3A &= 4 \\
y_p &= -\frac{4}{3} \\
y &= C_1 + C_2\, e^{3x} - \frac{4x}{3}
\end{aligned}$$

Note: The particular solution had to be $Ax + B$ since the homogeneous solution already contained the constant solution.

13.

$$\begin{aligned}
r^3 + r^2 &= 0 \\
r &= 0 \text{ repeated} \\
r &= -1 \\
y_h &= C_1 + C_2 x + C_3\, e^{-x} \\
y_p &= Ax^2 + Bx + C \\
y_p' &= 2Ax + B \\
y_p'' &= 2A \\
y_p''' &= 0 \\
0 + 2A &= -2 \\
A &= -1 \\
y_p &= -x^2 \\
y &= C_1 + C_2 x + C_3\, e^{-x} - x^2
\end{aligned}$$

Note: The particular solution had to be $Ax^2 + Bx + C$ since the homogeneous solution already contained the linear solution.

15.

$$\begin{aligned}
r^3 + 4r^2 + 20r &= 0 \\
r &= 0 \\
r &= -2 \pm\, 4i \\
y_h &= C_1 +\, e^{-2x}\, (C_1\; sin\; 4x + C_2\; cos\; 4x) \\
y_p &= Ax^2 + Bx + C \\
y_p' &= 2Ax + B \\
y_p'' &= 2A \\
y_p''' &= 0 \\
0 + 8A + 40Ax + 20B &= 40x - 12 \\
40Ax + (8A + 20B) &= 40x - 12 \\
40A &= 40 \to A = 1 \\
8A + 20B &= -12 \to B = -1 \\
y_p &= x^2 - x \\
y &= C_1 +\, e^{-2x}\, (C_2\; sin\; 4x + C_3\; cos\; 4x) \\
&\quad + x^2 - x
\end{aligned}$$

Note: The particular solution had to be $Ax^2 + Bx + C$ since the associated auxiliary equation has 0 as a root.

17.

$$\begin{aligned}
r^2 - r - 2 &= 0\\
r &= 2\\
r &= -1\\
y_h &= C_1\,e^{-x} + C_2\,e^{2x}\\
y_p &= Ax + B\\
y_p' &= A\\
y_p'' &= 0\\
0 - A - 2Ax - 2B &= 2x - 1\\
-2Ax + (-A - 2B) &= 2x - 1\\
A &= -1\\
B &= 1\\
y_p &= -x + 1\\
y &= C_1\,e^{-x} + C_2\,e^{2x} - x\\
y' &= -C_1\,e^{-x} + 2C_2\,e^{2x} - 1\\
y(0) &= 6 \to C_1 + C_2 = 6\\
y'(0) &= 0 \to -C_1 + 2C_2 = 0
\end{aligned}$$

Solving the system

$$\begin{aligned}
C_1 &= 4\\
C_2 &= 2\\
y &= 4\,e^{-x} + 2\,e^{2x} - x
\end{aligned}$$

19.

$$\begin{aligned}
r^2 + 2r + 1 &= 0\\
r &= -1 \text{ repeated}\\
y_h &= C_1\,e^{-x} + C_2 x\,e^{-x}\\
y_p &= Ax^2 + Bx + C\\
y_p' &= 2Ax + B\\
y_p'' &= 2A\\
2A + 4Ax + 2B + Ax^2 + Bx + C &= x^2\\
Ax^2 + (4A + B)x + (2A + 2B + C) &= x^2\\
A &= 1\\
4A + B &= 0 \to B = -4\\
2A + 2B + C &= 0 \to C = 6\\
y_p &= x^2 - 4x + 6\\
C_1\,e^{-x} + C_2 x\,e^{-x} + x^2 - 4x + 6 &= y\\
-C_1\,e^{x} - C_2 x\,e^{x} + C_2\,e^{x} + 2x - 4 &= y'\\
C_1 = -5 \leftarrow 1 &= y(0)\\
4 = -C_1 + C_2 \leftarrow 0 &= y'(0)\\
C_2 &= -1\\
-5\,e^{x} - x\,e^{x} + x^2 - 4x + 6 &= y
\end{aligned}$$

21.

$$\begin{aligned}
r^2 + 4r &= 0\\
r &= 0\\
r &= -4\\
y_h &= C_1 + C_2\,e^{-4x}\\
y_p &= Ax^2 + Bx + C\\
y_p' &= 2Ax + B\\
y_p'' &= 2A\\
2A + 8Ax + 4B &= 16x\\
8Ax + (2A + 4B) &= 16x\\
8A &= 16 \to A = 2\\
2A + 4B &= 0 \to B = -1\\
y_p &= 2x^2 - x\\
y &= C_1 + C_2\,e^{-4x} + 2x^2 - x\\
y' &= -4C_2\,e^{-4x} + 4x - 1\\
y(0) &= 2 \to C_1 + C_2 = 2\\
y'(0) &= -3 \to -4C_2 = -2\\
C_2 &= \frac{1}{2}\\
C_1 &= \frac{3}{2}\\
y &= \frac{3}{2} + \frac{1}{2}\,e^{-4x} + 2x - 1
\end{aligned}$$

Note: $r = 0$ is a root to the auxiliary equation and the effected our selection of y_p

23.

$$\begin{aligned}
r^2 + 2r + 2 &= 0\\
r &= -1 \pm\, i\\
y_h &= e^{-x}\,(C_1\,sin\,x + C_2\,cos\,x)\\
y_p &= Ax + B\\
y_p' &= A\\
y_p'' &= 0\\
0 + 2A + 2Ax + 2B &= 2\\
2Ax + (2A + 2B) &= 2\\
A &= 0\\
2A + 2B &= 2 \to B = 1\\
y_p &= 1\\
y &= e^{-x}\,(C_1\,sin\,x + C_2\,cos\,x) + 1\\
y' &= e^{-x}\,(C_1\,cos\,x - C_2\,sin\,x)\\
&\quad -e^{-x}\,(C_1\,sin\,x + C_2\,cos\,x)\\
y(0) &= 2 \to C_2 + 1 = 2 \to C_2 = 1\\
y'(0) &= 1 \to C_1 - C_2 = 1 \to C_1 = 2\\
y &= e^{-x}\,(2\,sin\,x + cos\,x) + 1
\end{aligned}$$

25.

$$\begin{aligned}
r^3 + 4r^2 + 5r &= 0\\
r &= 0\\
r &= -2 \pm\, i\\
y_h &= C_1 + e^{-2x}\,(C_2\,sin\,x + C_3\,cos\,x)\\
y_p &= Ax^2 + Bx + C\\
y_p' &= 2Ax + B\\
y_p'' &= 2A\\
y_p''' &= 0\\
0 + 8A + 10Ax + 5B &= 25x - 5\\
10Ax + (8A + 5B) &= 25x - 5
\end{aligned}$$

$$\begin{aligned}
10A &= 25 \to A = \frac{5}{2} \\
8A + 5B &= -5 \to B = 5 \\
y_p &= \frac{5}{2}\,x^2 + 5x \\
y &= C_1 + e^{-2x}\,(C_2 \; sin \; x + C_3 \; cos \; x) \\
&\quad + \frac{5}{2}x^2 + 5x \\
y' &= e^{-2x}\,(C_2 \; cos \; x - C_3 \; sin \; x) \\
&\quad -2\,e^{-2x}\,(C_2 \; sin \; x + C_3 \; cos \; x) \\
&\quad +5x \\
y'' &= 4\,e^{-2x}\,(C_2 \; sin \; x + C_3 \; cos \; x) - \\
&\quad 4\,e^{-2x}\,(C_2 \; cos \; x - C_3 \; sin \; x) \\
&\quad +\,e^{-2x}\,(-C_2 \; sin \; x - C_3 \; cos \; x) \\
&\quad +5 \\
y(0) &= 0 \to C_1 + C_3 = 0 \\
y'(0) &= 0 \to C_2 - 2C_3 = 0 \\
y''(0) &= 1 \to 4C_3 - 4C_2 - C_3 = 1
\end{aligned}$$

Solving the system

$$\begin{aligned}
C_1 &= \frac{21}{5} \\
C_2 &= -\frac{7}{5} \\
C_3 &= -\frac{16}{5} \\
y &= \frac{21}{5} + e^{-2x}\left(-\frac{7}{5} \; sin \; x - \frac{16}{5} \; cos \; x\right) \\
&\quad + \frac{5}{2}\,x^2 + 5x
\end{aligned}$$

27. $x'' + 16x = 1$

$$\begin{aligned}
r^2 + 16 &= 0 \\
r &= \pm\,4i \\
x_h &= C_1 \; sin \; 4t + C_2 \; cos \; 4t \\
x_p &= A \\
x_p' &= 0 \\
x_p'' &= 0 \\
0 + 16A &= 1 \to A = \frac{1}{16} \\
x_p &= \frac{1}{16} \\
x &= C_1 \; sin \; 4t + C_2 \; cos \; 4t + \frac{1}{16} \\
x' &= 4C_1 \; cos \; 4t - 4C_2 \; sin \; 4t \\
x(0) &= 0 \to C_2 + \frac{1}{16} \to C_2 = -\frac{1}{16} \\
x'(0) &= 0 \to 4C_1 = 0 \to C_1 = 0 \\
x &= \frac{1}{16} - \frac{1}{16} \; cos \; 4t \\
&= \frac{1 - cos \; 4t}{16}
\end{aligned}$$

29. $x'' + 2x' + 5x = 10$

$$\begin{aligned}
r^2 + 2r + 5 &= 0 \\
r &= -1 \pm 2i \\
e^{-t}\,(C_1 \; sin \; 2t + C_2 \; cos \; 2t) &= x_h \\
x_p &= A \\
x_p' &= 0 \\
x_p'' &= 0 \\
0 + 0 + 5A = 10 \to A &= 2 \\
x_p &= 2 \\
e^{-t}\,(C_1 \; sin \; 2t + C_2 \; cos \; 2t) + 2 &= x \\
e^{-t}\,(2C_1 \; cos \; 2t - 2C_2 \; sin \; 2t) - & \\
e^{-t}\,(C_1 \; sin \; 2t + C_2 \; cos \; 2t) &= x' \\
0 = C_2 + 2 \leftarrow 0 &= x(0) \\
C_2 &= -2 \\
0 = 2C_1 - C_2 \leftarrow 0 &= x'(0) \\
C_1 &= -1 \\
e^{-t}\,(-sin \; 2t - 2 \; cos \; 2t) + 2 &= x
\end{aligned}$$

31. $cm \; x'' + x = cpd + x_0$

a)

$$\begin{aligned}
cm \; r^2 + 1 &= 0 \\
r &= \pm\frac{1}{\sqrt{cm}}\;i \\
x_h &= C_1 \; sin \; \frac{t}{\sqrt{cm}} + C_2 \; cos \; \frac{t}{\sqrt{cm}} \\
x_p &= A \\
x_p' &= 0 \\
x_p'' &= 0 \\
A &= cpd + x_0 \\
x_p &= cpd + x_0 \\
x &= C_1 \; sin \; \frac{t}{\sqrt{cm}} + C_2 \; cos \; \frac{t}{\sqrt{cm}} \\
&\quad + cpd + x_0
\end{aligned}$$

b)

$$\begin{aligned}
\frac{C_1}{\sqrt{cm}} \; cos \; \frac{t}{\sqrt{cm}} - \frac{C_2}{\sqrt{cm}} \; sin \; \frac{t}{\sqrt{cm}} &= x' \\
C_2 + cpd + x_0 = 0 \leftarrow 0 &= x(0) \\
-cpd - x_0 &= C_2 \\
0 = \frac{C_1}{\sqrt{cm}} \leftarrow 0 &= x'(0) \\
C_1 &= 0 \\
cpd + x_0 - (cpd + x_0) \; cos \; \frac{t}{\sqrt{cm}} &= x \\
(cpd + x_0)\left(1 - cos \; \frac{t}{\sqrt{cm}}\right) &=
\end{aligned}$$

33. $F'' + 1.05F' + 0.05F = 0.05$

$$\begin{aligned}
r^2 + 1.05r + 0.05 &= 0 \\
r &= -\frac{1}{20} \\
r &= -1 \\
F_h &= C_1\, e^{-t} + C_2\, e^{-t/20} \\
F_p &= A \\
F_p' &= 0 \\
F_p'' &= 0 \\
0.05A &= 0.05 \to A = 1 \\
F &= C_1\, e^{-t} + C_2\, e^{-t/20} + 1 \\
F' &= -C_1\, e^{-t} - \frac{C_2}{20}\, e^{-t/20} \\
F(0) &= 0 \to C_1 + C_2 = -1 \\
F'(0) &= 0.05 \to -C_1 - \frac{C_2}{20} = 0.05
\end{aligned}$$

Solving the system

$$\begin{aligned}
C_1 &= 0 \\
C_2 &= -1 \\
F &= 1 - e^{-t/20}
\end{aligned}$$

35. $y' + 2y = x^2$

$$\begin{aligned}
r + 2 &= 0 \\
r &= -2 \\
y_h &= C\, e^{-2x} \\
y_p &= Ax^2 + Bx + C \\
y_p' &= 2Ax + B \\
y_p'' &= 2A \\
x^2 &= 2Ax + B + 2Ax^2 + 2Bx + 2C \\
A &= \frac{1}{2} \\
2A + 2B &= 0 \to B = -\frac{1}{2} \\
B + 2C &= 0 \to C = \frac{1}{4} \\
y &= C\, e^{-2x} + \frac{1}{2}\, x^2 - \frac{1}{2}\, x + \frac{1}{4}
\end{aligned}$$

Exercise Set 9.3

1.

$$\begin{aligned}
x'' &= -2x + 3x' \\
x'' - 3x' + 2x &= 0 \\
r^2 - 3r + 2 &= 0 \\
r &= 2 \\
r &= 1 \\
x &= C_1\, e^t + C_2\, e^{2t} \\
y &= x' = C_1\, e^t + 2C_2\, e^{2t}
\end{aligned}$$

3. $y = x' - 2x$ and $x'' = 2x + y'$

$$\begin{aligned}
x'' &= 2x' + 3x + 4(x' - 2x) \\
x'' - 6x' + 5x &= 0 \\
r^2 - 6r + 5 &= 0 \\
r &= 5 \\
r &= 1 \\
x &= C_1\, e^t + C_2\, e^{5t} \\
x' &= C_1\, e^t + 5\, C_2\, e^{5t} \\
y &= x' - 2x \\
&= 3\, C_2\, e^{5t} - C_1\, e^t
\end{aligned}$$

5. $x' = -0.5x + y \to y = x' + 0.5x$ and $y' = 0.5x$

$$\begin{aligned}
x'' &= -0.5x' + y' \\
x'' &= -0.5x' + 0.5x \\
x'' + 0.5x' - 0.5x &= 0 \\
r^2 + 0.5r - 0.5 &= 0 \\
r &= 1/2 \\
r &= -1 \\
x &= C_1\, e^{t/2} + C_2\, e^{-t} \\
x &= \frac{C_1}{2}\, e^{t/2} - C_2\, e^{-t} \\
y &= x' + 0.5x \\
&= C_1\, e^{t/2} - C_2\, e^{-t}
\end{aligned}$$

7. $y = x' - 4x$

$$\begin{aligned}
x'' &= 4x' + y' \\
&= 4x' - x + 2y \\
x'' &= 4x' - x + 2x' - 8x \\
x'' - 6x' + 9x &= 0 \\
r^2 - 6r + 9 &= 0 \\
r &= 3 \text{ repeated} \\
x &= C_1\, e^{-3t} + C_2 t\, e^{-3t} \\
x' &= -3C_1\, e^{-3t} - 3C_2 t\, e^{-3t} + C_2\, e^{-3t} \\
y &= x' - 4x \\
&= (C_2 - C_1)\, e^{-3t} - 3C_2 t\, e^{-3t}
\end{aligned}$$

9. $y = \dfrac{x'}{2}$

$$\begin{aligned}
x'' &= 2y' = 2(-18x) \\
x'' + 36x &= 0 \\
r^2 + 36 &= 0 \\
r &= \pm\, 6i \\
x &= C_1\, sin\ 6x + C_2\, cos\ 6x \\
x' &= 6C_1\, cos\ 6x - C_2\, sin\ 6x \\
y &= \frac{x}{2} \\
&= 3C_1\, cos\ 6x - 3C_2\, sin\ 6x
\end{aligned}$$

11. $5y = 3x - x'$

$$\begin{aligned} x'' &= 3x' - 5(x - y) \\ &= 3x' - 5x + 3x - x' \\ x'' - 2x' + 2 &= 0 \\ r^2 - 2r + 2 &= 0 \\ r &= 1 \pm i \\ x &= e^t\left(C_1 \sin t + C_2 \cos t\right) \\ x' &= e^t\left(C_1 \cos t - C_2 \sin t\right) \\ &\quad + e^t\left(C_1 \sin t + C_2 \cos t\right) \\ y &= \frac{e^t\left[(2C_1 - C_2)\sin t + (2C_2 - C_1)\cos t\right]}{5} \end{aligned}$$

13. $y = 5x - x'$

$$\begin{aligned} x'' &= 5x' - (2x + (10x - 2x') + 4) \\ x'' - 7x' + 12 &= -4 \\ r^2 - 7r + 12 &= 0 \\ r &= 3 \\ r &= 4 \\ x_h &= C_1 e^{3t} + C_2 e^{4t} \\ x_p &= A \\ x_p' &= 0 \\ x_p'' &= 0 \\ 0 - 0 + 12A &= -4 \\ A = -\frac{1}{3} & \\ x_p &= -\frac{1}{3} \\ x &= C_1 e^{3t} + C_2 e^{4t} - \frac{1}{3} \\ x' &= 3C_1 e^{3t} + 4C_2 e^{4t} \\ y &= 5x - x' \\ &= 2C_1 e^{3t} + C_2 e^{4t} - \frac{5}{3} \end{aligned}$$

15.

$$\begin{aligned} x'' &= y' + 2 \\ &= -x + 4t - 2 + 2 \\ x'' + x &= 4t \\ r^2 + 1 &= 0 \\ r &= \pm i \\ x_h &= C_1 \sin t + C_2 \cos t \\ x_p &= At + B \\ x_p' &= A \\ x_p'' &= 0 \\ 0 + At + B &= 4t \\ A &= 4 \\ B &= 0 \\ x_p &= 4t \\ x &= C_1 \sin t + C_2 \cos t + 4t \\ x' &= C_1 \cos t - C_2 \sin t + 4 \\ y &= x' - 2t - 3 \\ &= C_1 \cos t - C_2 \sin t - 2t + 1 \end{aligned}$$

17.

$$\begin{aligned} x'' &= 2x' - (2x + 10x - 5x') \\ x'' - 7x' + 12x &= 0 \\ r^2 - 7r + 12 &= 0 \\ r &= 3 \\ r &= 4 \\ x &= C_1 e^{3t} + C_2 e^{4t} \\ x' &= 3C_1 e^{3t} + 4C_2 e^{4t} \\ y &= 2x - x' \\ &= -C_1 e^{3t} - 2C_2 e^{4t} \\ x(0) &= 3 \to C_1 + C_2 = 3 \\ y(0) &= -5 \to -C_1 - 2C_2 = -5 \end{aligned}$$

Solving the system

$$\begin{aligned} C_1 &= 1 \\ C_2 &= 2 \\ x &= e^{3t} + 2e^{4t} \\ y &= -e^{3t} - 4e^{4t} \end{aligned}$$

19.

$$\begin{aligned} x'' &= 2x' + 3(-3x + 8y) \\ x'' &= 2x' - 9x + 8x' - 16x \\ x'' - 10x' + 25x &= 0 \\ r^2 - 10r + 25 &= 0 \\ r &= 5 \text{ repeated} \\ x &= C_1 e^{5t} + C_2 t e^{5t} \\ x' &= 5C_1 e^{5t} + 5C_2 t e^{5t} + C_2 e^{5t} \\ y &= \frac{x' - 2x}{3} \\ &= \frac{(3C_1 + C_2)e^{5t}}{3} + C_2 t e^{5t} \\ x(0) &= 1 \to C_1 + C_2 = 1 \\ y(0) &= 1 \to C_1 + \frac{1}{3}C_2 = 1 \end{aligned}$$

Solving the system

$$\begin{aligned} C_1 &= 1 \\ C_2 &= 0 \\ x &= e^{5t} \\ y &= e^{5t} \end{aligned}$$

21.

$$\begin{aligned} x'' &= -4x \\ x'' + 4x &= 0 \\ r^2 + 4 &= 0 \\ r &= \pm 2i \\ x &= C_1 \sin 2x + C_2 \cos 2x \\ y &= x' = 2C_1 \cos 2x - 2C_2 \sin 2x \\ x(0) &= 1 \to C_2 = 1 \\ y(0) &= 2 \to C_1 = 1 \\ x &= \sin 2x + \cos 2x \\ y &= 2\cos 2x - 2\sin 2x \end{aligned}$$

23.

$$\begin{aligned}
x'' &= 2x' + (5x - 2y + 12)\\
&= 2x' + 5x - 2x' + 4x + 6 + 12\\
x'' - 9x &= 18\\
r^2 - 9 &= 0\\
r &= -3\\
r &= 3\\
x_h &= C_1\, e^{-3t} + C_2\, e^{3t}\\
x_p &= A\\
x_p' &= 0\\
x_p'' &= 0\\
0 - 9A &= 18\\
A &= -2\\
x &= C_1\, e^{-3t} + C_2\, e^{3t} - 2\\
x' &= -3C_1\, e^{-3t} + 3\, e^{3t}\\
y &= x' - 2x - 3\\
&= -5C_1\, e^{-3t} + C_2\, e^{3t} + 1\\
x(0) &= 6 \to C_1 + C_2 = 8\\
y(0) &= -3 \to -5C_1 + C_2 = -4
\end{aligned}$$

Solving the system

$$\begin{aligned}
C_1 &= 2\\
C_2 &= 6\\
x &= 2\, e^{-3t} + 6\, e^{3t} - 2\\
y &= -10\, e^{3-t} + 6\, e^{3t} + 1
\end{aligned}$$

25. $P' = -3P + 2Q$ and $Q' = 3P - 2Q$

$$\begin{aligned}
P'' &= -3P' + 2(3P - 2Q)\\
P'' &= -3P' + 6P - 2(P' + 3P)\\
P'' + 5P' &= 0\\
r^2 + 5r &= 0\\
r &= 0\\
r &= -5\\
P &= C_1 + C_2\, e^{-5t}\\
P' &= -5C_2\, e^{-5t}\\
Q &= \frac{P' + 3P}{2}\\
&= \frac{3}{2}\, C_1 + C_2\, e^{-5t}
\end{aligned}$$

27. $x' = -2x + 2x - 10$ and $y' = 2x - 2y$

$$\begin{aligned}
x'' &= -2x' + 2y'\\
&= -2x' + 4x - 2(x' + 2x + 10)\\
x'' + 4x' &= -20\\
r^2 + 4r &= 0\\
r &= 0\\
r &= -4\\
x_h &= C_1 + C_2\, e^{-4t}\\
x_p &= At + B\\
x_p' &= A\\
x_p'' &= = 0\\
0 + 4A &= -20\\
A &= -5\\
x_p &= -5t\\
x &= C_1 + C_2\, e^{-4t} - 5t\\
x' &= -4C_2\, e^{-4t} - 5\\
y &= \frac{x' + 2x + 10}{2}\\
&= C_1 - C_2\, e^{-4t} - 5t + \frac{5}{2}
\end{aligned}$$

29. **a)** $A(t)/2$ per hour and $B(t)/2$ per hour

b) Left to the student

c) It follows from the statement of the problem

d)

$$\begin{aligned}
A'' &= -0.5A' + 0.5B'\\
&= -0.5A' + 0.5(0.5A - 0.5B)\\
&= -0.5A' + 0.25A -\\
&\quad 0.25\left(\frac{A' + 0.5A'}{0.5}\right)\\
&= -0.5A' + 0.25A - 0.5A' - 0.25A\\
A'' + A' &= 0\\
r^2 + r &= 0\\
r &= 0\\
r &= -1\\
A &= C_1 + C_2\, e^{-t}\\
A' &= -C_2\, e^{-t}\\
B &= 2A' + A\\
&= C_1 - C_2\, e^{-t}\\
A(0) &= 2000 \to C_1 + C_2 = 2000\\
B(0) &= 1000 \to C_1 - C_2 = 1000
\end{aligned}$$

Solving the system

$$\begin{aligned}
C_1 &= 1500\\
C_2 &= 500\\
A(t) &= 1500 + 500\, e^{-t}\\
B(t) &= 1500 - 500\, e^{-t}
\end{aligned}$$

e)

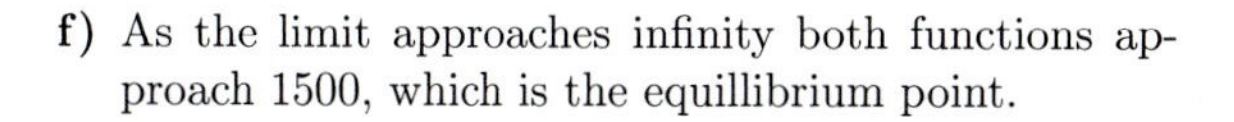

f) As the limit approaches infinity both functions approach 1500, which is the equillibrium point.

31. **a)** Left to the student

b) $P' = -3P + 0.6L - 2.4P$
$\rightarrow P' = -5.4P + 0.6L$ and $L' = 3P - 0.6L$

c)

$$\begin{aligned}
-5.4P' + 0.6(3P - 0.6L) &= P'' \\
-5.4P' + 1.8P - 0.6(-5.4P - P') &= \\
-4.8P' + 5.04P &= \\
P'' + 4.8P' - 5.04P &= 0 \\
r^2 + 4.8r - 5.04 &= 0 \\
0.886 &= r \\
-5.686 &= r \\
C_1\, e^{-5.686t} + C_2\, e^{-0.886t} &= P \\
-5.686C_1\, e^{-5.686t} - 0.886C_2\, e^{-0.886t} &= P' \\
\frac{P' + 5.4P}{0.6} &= L \\
-0.47667C_1\, e^{-5.686t} + 7.52333C_2\, e^{-0.886t} &= \\
C_1 + C_2 = 0 \leftarrow 0 &= P(0) \\
-0.47667C_1 + 7.52333C_2 = 241/15 \leftarrow 241/15 &= L(0) \\
\text{Solving the system} & \\
2.28004 &= C_1 \\
-2.28004 &= C_2 \\
2.28004\, e^{-5.686t} - 2.28004\, e^{-0.886t} &= P \\
1.086827\, e^{-5.686t} + 17.15349\, e^{-0.886t} &= L
\end{aligned}$$

33. **a)** Left to the student

b) Left to the student

c)

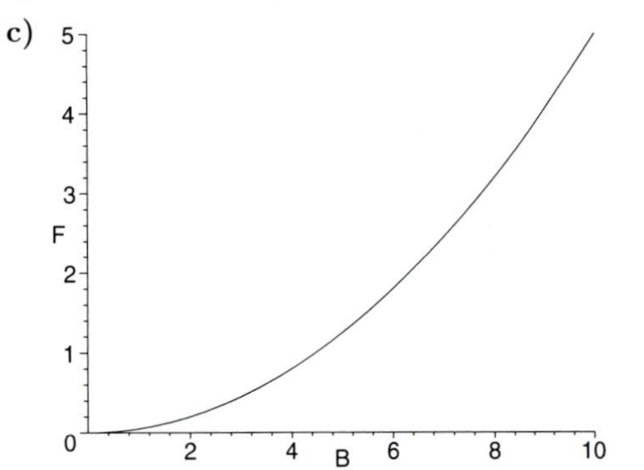

35. $L' = -aL - mL + cM + p$ and $M' = mL - aM - cM$
$L' = -(a+m)L + cM + p$ and $M' = mL - (a+c)M$

37.

$$\begin{aligned}
L_p &= A \\
L'_p &= 0 \\
L''_p &= 0 \\
0 + 0 + a(a+c+m)A &= (a+c)p \\
A &= \frac{(a+c)p}{a(a+c+m)}
\end{aligned}$$

39. $N'' + (c+m+2a)N' + a(c+m+a)N = 0$

$$\begin{aligned}
L(t) - \frac{(c+a)p}{a(c+m+a)} &= N(t) \\
N'(t) &= L'(t) \\
N'' &= L''(t)
\end{aligned}$$

Thus, $L'' + (c+m+2a)L' + a(c+m+a)\left(L - \frac{(c+a)p}{a(c+m+a)}\right) \rightarrow$

$$L'' + (c+m+2a)L' + a(c+m+a)L = (c+a)p$$

Which has L_p as a solution

41. Left to the student

43.

$$\begin{aligned}
b(t) &= \frac{C_1 + C_2\, e^{(q-p)t}}{V_B} \\
&= \frac{C_1}{V_B} + \frac{C_2}{V_B}\, e^{(q-p)t} \\
&= C_1 + C_2\, e^{-rt}
\end{aligned}$$

45.

$$\begin{aligned}
C_1 + C_2 &= -\frac{qD}{rV_B} + \frac{pD}{rV_B} \\
C_1 + C_2 &= \frac{D(p-q)}{rV_B} \\
C_1 + C_2 &= \frac{D}{V_B} \\
V_B &= \frac{D}{C_1 + C_2}
\end{aligned}$$

$$\begin{aligned}
q &= -\frac{C_1 r V_B}{D} \\
&= -\frac{C_1 r}{C_1 + C_2} \\
p &= \frac{C_2 r V_B}{D} \\
&= \frac{C_2 r}{C_1 + C_2}
\end{aligned}$$

Exercise Set 9.4

1. $\begin{bmatrix} x \\ y \end{bmatrix}' = \begin{bmatrix} 1 & -1 \\ 3 & -2 \end{bmatrix} \begin{bmatrix} x \\ y \end{bmatrix}$

3. $\begin{bmatrix} x \\ y \end{bmatrix}' = \begin{bmatrix} 4 & 2 \\ 0 & 1 \end{bmatrix} \begin{bmatrix} x \\ y \end{bmatrix}$

5. $x' = x + 3y$ and $y' = 5x + 7y$

7. $x' = 3y$ and $y' = x - 2y$

9. $\begin{bmatrix} x \\ y \end{bmatrix}' = \begin{bmatrix} (2e^{3t} - e^{2t})' \\ (-2e^{3t} + 2e^{2t})' \end{bmatrix} = \begin{bmatrix} 6e^{3t} - 2e^{2t} \\ -6e^{3t} + 4e^{2t} \end{bmatrix}$

$\begin{bmatrix} 4 & 1 \\ -2 & 1 \end{bmatrix} \begin{bmatrix} 2e^{3t} - e^{2t} \\ -2e^{3t} + 2e^{2t} \end{bmatrix} = \begin{bmatrix} 6e^{3t} - 2e^{2t} \\ -6e^{3t} + 4e^{2t} \end{bmatrix}$

11. $\begin{bmatrix} x \\ y \end{bmatrix}' = \begin{bmatrix} (-2e^t \sin 2t)' \\ (3e^t \sin 2t + e^t \cos 2t)' \end{bmatrix} =$

$\begin{bmatrix} -4e^t \cos 2t - 2e^t \sin 2t \\ 6e^t \cos 2t + 3e^t \sin 2t - e^t \sin t + e^t \cos t \end{bmatrix}$

$\begin{bmatrix} -5 & -4 \\ 10 & 7 \end{bmatrix} \begin{bmatrix} -2e^t \sin 2t \\ 3e^t \sin 2t + e^t \cos 2t \end{bmatrix} =$

$\begin{bmatrix} -4e^t \cos 2t - 2e^t \sin 2t \\ 6e^t \cos 2t + 3e^t \sin 2t - e^t \sin t + e^t \cos t \end{bmatrix}$

13. $A = \begin{bmatrix} 0 & -2 \\ 1 & 3 \end{bmatrix}$ $\det(A) = 2$ and $\text{trace}(A) = 3$

$$\begin{aligned} r^2 - 3r + 2 &= 0 \\ (r-1)(r-2) &= 0 \\ r &= 1 \\ r &= 2 \end{aligned}$$

Then (by Theorem 9 of Chapter 6) the eigenvectors

For $r = 1$ are $\begin{bmatrix} -2 \\ 1 \end{bmatrix}$

For $r = 2$ are $\begin{bmatrix} -1 \\ 1 \end{bmatrix}$

Therefore, the general solution is given by

$$\begin{aligned} \begin{bmatrix} x \\ y \end{bmatrix} &= C_1 e^t \begin{bmatrix} -2 \\ 1 \end{bmatrix} + C_2 e^{2t} \begin{bmatrix} -2 \\ 1 \end{bmatrix} \\ &= \begin{bmatrix} -2C_1 e^t - C_2 e^{2t} \\ C_1 e^t + C_2 e^{2t} \end{bmatrix} \end{aligned}$$

15. $A = \begin{bmatrix} 2 & 4 \\ 3 & -2 \end{bmatrix}$ $\det(A) = -16$ and $\text{trace}(A) = 0$

$$\begin{aligned} r^2 - 16 &= 0 \\ (r-4)(r+4) &= 0 \\ r &= -4 \\ r &= 4 \end{aligned}$$

Then (by Theorem 9 of Chapter 6) the eigenvectors

For $r = -4$ are $\begin{bmatrix} -4 \\ 6 \end{bmatrix} = \begin{bmatrix} -2 \\ 3 \end{bmatrix}$

For $r = 4$ are $\begin{bmatrix} 6 \\ 3 \end{bmatrix} = \begin{bmatrix} 2 \\ 1 \end{bmatrix}$

Therefore, the general solution is given by

$$\begin{aligned} \begin{bmatrix} x \\ y \end{bmatrix} &= C_1 e^{-4t} \begin{bmatrix} -2 \\ 3 \end{bmatrix} + C_2 e^{4t} \begin{bmatrix} 2 \\ 1 \end{bmatrix} \\ &= \begin{bmatrix} -2C_1 e^{-4t} + 2C_2 e^{4t} \\ 3C_1 e^{-4t} + C_2 e^{4t} \end{bmatrix} \end{aligned}$$

17. $A = \begin{bmatrix} -2 & 4 \\ -1 & -7 \end{bmatrix}$ $\det(A) = 18$ and $\text{trace}(A) = -9$

$$\begin{aligned} r^2 - 9r + 18 &= 0 \\ r &= 3 \\ r &= 6 \end{aligned}$$

Then (by Theorem 9 of Chapter 6) the eigenvectors

For $r = 3$ are $\begin{bmatrix} -4 \\ -1 \end{bmatrix}$

For $r = 6$ are $\begin{bmatrix} -1 \\ 1 \end{bmatrix}$

Therefore, the general solution is given by

$$\begin{aligned} \begin{bmatrix} x \\ y \end{bmatrix} &= C_1 e^{3t} \begin{bmatrix} -4 \\ -1 \end{bmatrix} + C_2 e^{6t} \begin{bmatrix} -1 \\ 1 \end{bmatrix} \\ &= \begin{bmatrix} -4C_1 e^{3t} - C_2 e^{6t} \\ -C_1 e^{3t} + C_2 e^{6t} \end{bmatrix} \end{aligned}$$

19. $A = \begin{bmatrix} -5 & 10 \\ -4 & 7 \end{bmatrix}$ $\det(A) = 5$ and $\text{trace}(A) = 2$

$$\begin{aligned} r^2 - 2r + 5 &= 0 \\ r &= 1 \pm 2i \\ x &= e^t (C_1 \sin 2t + C_2 \cos 2t) \\ x' &= e^t (2C_1 \cos 2t - 2C_2 \sin 2t) \\ &\quad + e^t (C_1 \sin 2t + C_2 \cos 2t) \\ y &= \frac{x' + 5x}{10} \\ &= \frac{(3C_1 - C_2)}{5} e^t \sin 2t + \frac{(C_1 + 3C_2)}{5} e^t \cos 2t \end{aligned}$$

21. $A = \begin{bmatrix} 0 & -1 \\ 1 & 2 \end{bmatrix}$ $\det(A) = 1$ and $\text{trace}(A) = 2$

$$\begin{aligned} r^2 - 2r + 1 &= 0 \\ r &= 1 \text{ repeated} \\ x &= C_1 e^t + C_2 t e^t \\ x' &= C_1 e^t + C_2 t e^t + C_2 e^t \\ y &= -x' = -(C_1 + C_2) e^t - C_2 t e^t \end{aligned}$$

23. Since the eigenvalues have different signs, then the origin is an unstable saddle point

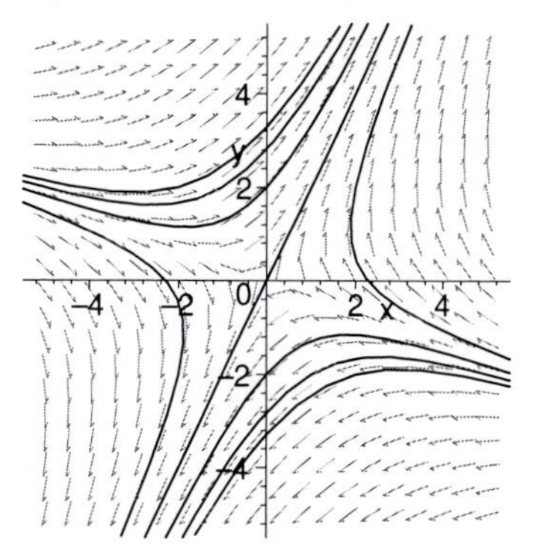

25. Since the eigenvalues are both positive, then the origin is an unstable node

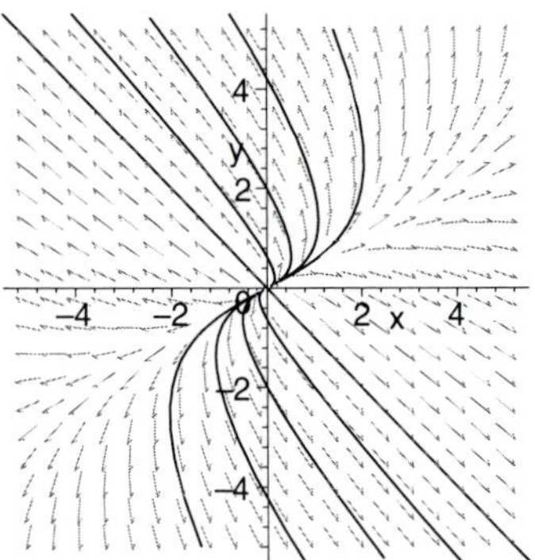

27. Since the eigenvalues are both negative, then the origin is an asymptotically stable node

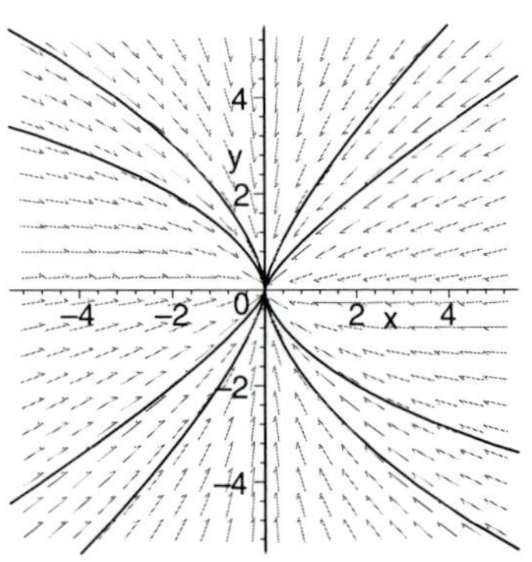

29. Since the eigenvalues are both positive, then the origin is an unstable node

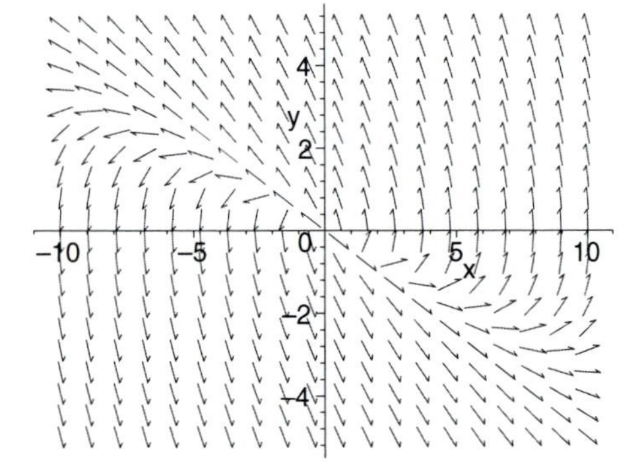

31. Since the eigenvalues have different signs, then the origin is an unstable saddle point

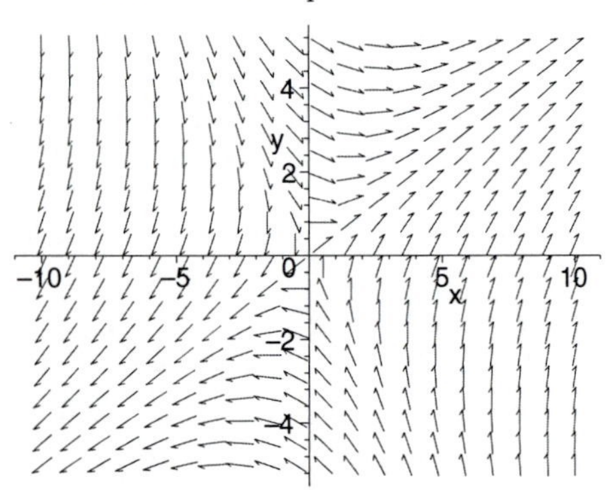

33. Since the eigenvalues are both positive, then the origin is an asymptotically stable node

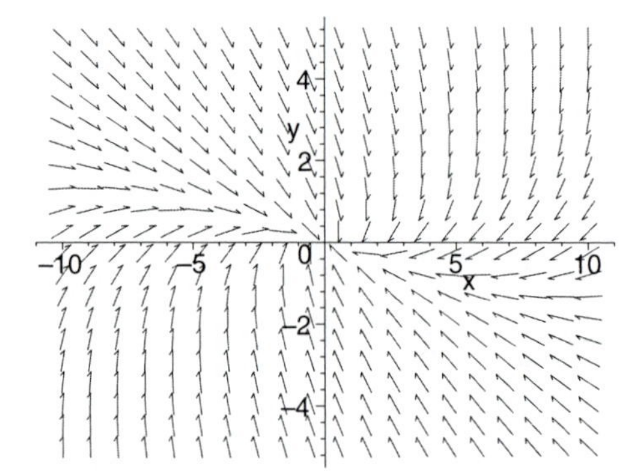

35. Since the eigenvalues are complex with a positive real part, then the prigin is an unstable spiral point

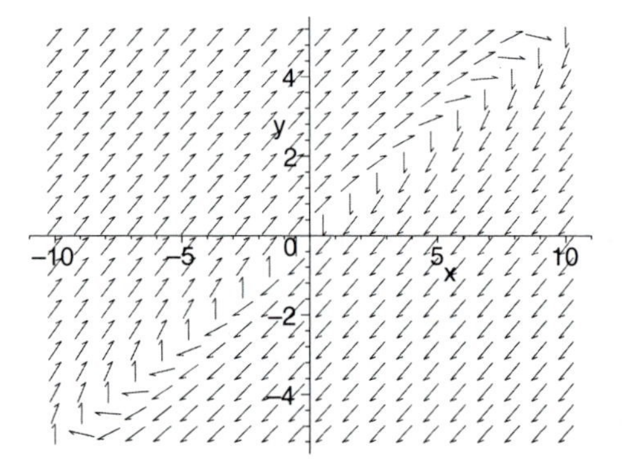

37. Since the eigenvalues are positive and equal, then the origin is an unstable improper node

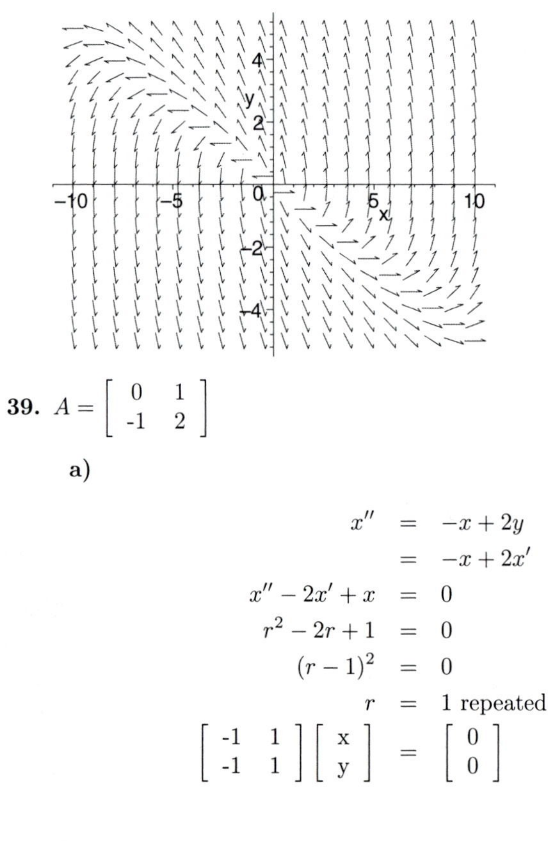

39. $A = \begin{bmatrix} 0 & 1 \\ -1 & 2 \end{bmatrix}$

a)

$$\begin{aligned} x'' &= -x + 2y \\ &= -x + 2x' \\ x'' - 2x' + x &= 0 \\ r^2 - 2r + 1 &= 0 \\ (r-1)^2 &= 0 \\ r &= 1 \text{ repeated} \end{aligned}$$

$$\begin{bmatrix} -1 & 1 \\ -1 & 1 \end{bmatrix} \begin{bmatrix} x \\ y \end{bmatrix} = \begin{bmatrix} 0 \\ 0 \end{bmatrix}$$

$$\begin{bmatrix} x \\ y \end{bmatrix} = \begin{bmatrix} 1 \\ 1 \end{bmatrix}$$

b) Left to the student

41.

$$\begin{aligned} x'' &- ax' + by' \\ &= ax' + b(cx + dy) \\ &= ax' + bcx + d(x' - ax) \\ &= ax' + bcx + dx' - adx \\ &= (a + d)x' - (ad - bc)x \\ x'' - (a + d)x' + (ad - bc)x &= 0 \\ r^2 - (a + d)r + (ad - bc) &= 0 \end{aligned}$$

43. a)

$$\begin{aligned} x'' &= -2x' + 6(2x - 6y) \\ &= -2x' + 12x + 6(-x' - 2x) \\ &= -2x' + 12x - 6x' - 12x \\ x'' + 8x' &= 0 \\ r^2 + 8r &= 0 \\ r &= -8 \\ r &= 0 \end{aligned}$$

b) The eigenvectors associated with $r = -8$ are $\begin{bmatrix} -1 \\ 1 \end{bmatrix}$ and the eigenvectors associated with $r = 0$ are $\begin{bmatrix} 3 \\ 1 \end{bmatrix}$. Thus

$$\begin{aligned} \begin{bmatrix} x \\ y \end{bmatrix} &= \begin{bmatrix} 3 \\ 1 \end{bmatrix} C_1 + \begin{bmatrix} -1 \\ 1 \end{bmatrix} C_2\, e^{-8} \\ &= \begin{bmatrix} 3C_1 - C_2\, e^{-8t} \\ C_1 + C_2\, e^{-8t} \end{bmatrix} \end{aligned}$$

c)

$$\begin{aligned} y + x &= 3C_1 - C_2\, e^{-8t} + C_1 + C_2\, e^{-8t} \\ y + x &= 4C_1 \\ y &= 4C_1 - x \\ y &= b - x \end{aligned}$$

d)

$$\begin{aligned} x' &= 0 \\ -2x + 6y &= 0 \\ y &= \frac{x}{3} \end{aligned}$$

Similarly,

$$\begin{aligned} y' &= 0 \\ 2x - 6y &= 0 \\ y &= \frac{x}{3} \end{aligned}$$

e)

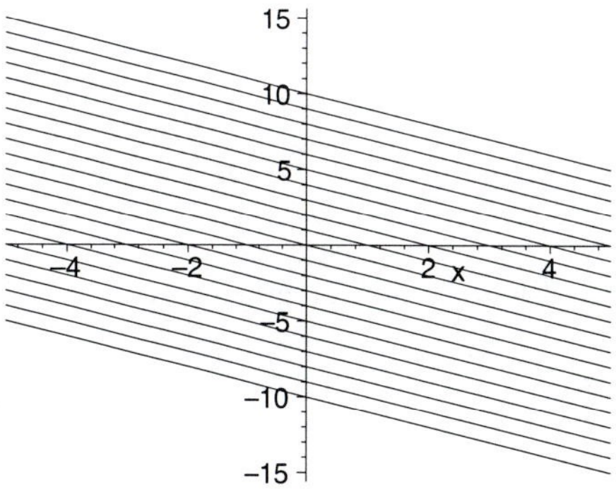

45. $A = \begin{bmatrix} 1 & -3 & 0 \\ -3 & -3 & -4 \\ 3 & 5 & 6 \end{bmatrix}$

The eigenvalues of matrix A are $r = -2$, $r = 2$ and $r = 4$ which have the following eigenvectors $\begin{bmatrix} -1 \\ -1 \\ 1 \end{bmatrix}$, $\begin{bmatrix} -3 \\ 1 \\ 1 \end{bmatrix}$, and $\begin{bmatrix} 1 \\ -1 \\ 1 \end{bmatrix}$ respectively. Therefore, the general solution is given by

$$\begin{bmatrix} -C_1\, e^{-2t} - 3C_2\, e^{2t} + C_3\, e^{4t} \\ -C_1\, e^{-2t} + C_2\, e^{2t} - C_3\, e^{4t} \\ C_1\, e^{-2t} + C_2\, e^{2t} + C_3\, e^{4t} \end{bmatrix}$$

47. $A = \begin{bmatrix} 3 & 0 & 2 \\ -1 & 2 & -2 \\ -1 & 0 & 0 \end{bmatrix}$

The eigenvalues of matrix A are $r = 1$, $r = 2$ and $r = 2$ which have the following eigenvectors $\begin{bmatrix} -1 \\ 1 \\ 1 \end{bmatrix}$, $\begin{bmatrix} 0 \\ 1 \\ 0 \end{bmatrix}$, and $\begin{bmatrix} -2 \\ 0 \\ 1 \end{bmatrix}$ respectively. Therefore, the general solution is given by

$$\begin{bmatrix} C_1\, e^{t} - 2C_3\, e^{2t} \\ C_1\, e^{t} + C_2\, e^{2t} \\ C_1\, e^{t} + C_3\, e^{2t} \end{bmatrix}$$

49. $A = \begin{bmatrix} 2 & 0 & 6 \\ 1 & 1 & 0 \\ -1 & 1 & -4 \end{bmatrix}$

The eigenvalues of matrix A are $r = -2$, $r = -1$ and $r = 2$ which have the following eigenvectors $\begin{bmatrix} -3 \\ 1 \\ 2 \end{bmatrix}$, $\begin{bmatrix} -2 \\ 1 \\ 1 \end{bmatrix}$, and $\begin{bmatrix} 1 \\ 1 \\ 0 \end{bmatrix}$ respectively. Therefore, the general solution is given by

$$\begin{bmatrix} -3C_1\, e^{-2t} - 2C_2\, e^{-t} + C_3\, e^{2t} \\ C_1\, e^{-2t} + C_2\, e^{-t} + C_3\, e^{2t} \\ 2C_1\, e^{-2t} + C_2\, e^{-t} \end{bmatrix}$$

51. $S' = -aS$
$B' = aS - (b+c)B + dH$
$H' = cB - dH$

53. Since all the eigenvalues are negative then the origin will be an asymptotically stable node

55. The eigenvalues -8, -5, and $\frac{-1}{2}$ have the eigenvectors

$$\begin{bmatrix} 0 \\ -2 \\ 1 \end{bmatrix}, \begin{bmatrix} 1 \\ 40/27 \\ -35/27 \end{bmatrix}, \text{ and } \begin{bmatrix} 0 \\ 1 \\ 7 \end{bmatrix}$$ respectively. Therefore, the general solution is given by

$$\begin{bmatrix} C_2\, e^{-5t} \\ -2C_1\, e^{-8t} + 40/27C_2\, e^{-5t} + C_3\, e^{-t/2} \\ C_1\, e^{-8t} - 35/27C_2\, e^{-5t} + 7C_3\, e^{t/2} \end{bmatrix}$$

Applying the initial conditions

$$\begin{aligned} 1 &= C_2 \\ -2C_1 + \frac{40}{27}C_2 + C_3 &= 0 \\ C_1 - \frac{35}{27}C_2 + 7C_3 &= 0 \end{aligned}$$

Solving the system

$$\begin{aligned} C_1 &= \frac{7}{9} \\ C_2 &= 1 \\ C_3 &= \frac{2}{27} \end{aligned}$$

Thus the solution is

$$S(t) = e^{-5t}$$

$$B(t) = --\frac{14}{9}\, e^{-8t} + \frac{40}{27}\, e^{-5t} + \frac{2}{7}\, e^{-t/2}$$

$$H(t) = \frac{7}{9}\, e^{-8t} - \frac{35}{27}\, e^{-5t} + \frac{14}{27}\, e^{-t/2}$$

Exercise Set 9.5

1. Jacobian $J = \begin{bmatrix} 2x+y+4 & \text{x} \\ \text{y} & x-4y+1 \end{bmatrix}$

$$\begin{aligned} x(x+y+4) &= 0 \\ x &= 0 \\ y(-2y+1) &= 0 \\ y &= 0 \\ y &= \frac{1}{2} \\ x+y+4 &= 0 \\ y &= -x-4 \\ (-x-4)(x+2x+8+1) &= 0 \\ x &= -4 \\ y &= 0 \\ x &= -3 \\ y &= -1 \end{aligned}$$

The equilibrium points are:
$(0,0)$ $\to$ Jacobian has two positive eigenvalues which means the equilibrium point is an unstable node
$(-4,0)$ $\to$ Jacobian has two negative eigenvalues which means the equilibrium point is an asymptotically stable node
$(-3,-1)$ $\to$ Jacobian has eigenvalues with opposite signs which means the equilibrium point is an unstable saddle point
$\left(0, \frac{1}{2}\right)$ $\to$ Jacobian has eigenvalues with opposite signs which means the equilibrium point is an unstable saddle point

3. Jacobian $= \begin{bmatrix} \text{2x} & \text{2y} \\ 1 & 1 \end{bmatrix}$

$$\begin{aligned} y &= 7-x \\ x^2 + (7-x)^2 - 25 &= 0 \\ x^2 + 49 - 14x + x^2 - 25 &= 0 \\ 2x^2 - 14x + 24 &= 0 \\ x &= 3 \to y = 4 \\ x &= 4 \to y = 3 \end{aligned}$$

The equilibrium points are:
$(3,4)$ $\to$ Jacobian has eigenvalues with opposite signs which means the equilibrium point is an unstable saddle point
$(4,3)$ $\to$ Jacobian has eigenvalues with opposite signs which means the equilibrium point is an unstable saddle point

5. Jacobian $= \begin{bmatrix} -\sqrt{y} & (2-x)/\sqrt{y} \\ \text{y} & \text{x} \end{bmatrix}$

$$\begin{aligned} (2-x)\sqrt{y}7 &= 0 \\ x &= 0 \to y = 4 \\ y &= 0 \to \text{ no solution} \end{aligned}$$

The equilibrium point is $(2,4)$ $\to$ Jacobian has eigenvalues with opposite signs which means the equilibrium point is an unstable saddle point

7. Jacobian $= \begin{bmatrix} \text{-1} & 2y \\ 1 & 4y^3 \end{bmatrix}$

$$\begin{aligned} -x + y^2 &= 0 \\ x + y^4 - 2 &= 0 \\ y^4 + y^2 - 2 &= 0 \\ (y^2-1)(y^2+2) &= 0 \\ y &= -1 \to x = 1 \\ y &= 1 \to x = 1 \end{aligned}$$

The equilibrium point are:
$(1,1)$ $\to$ Jacobian has eigenvalues with opposite signs which means the equilibrium point is an unstable saddle point
$(1,-1)$ $\to$ Jacobian has two negative eigenvalues which means the equilibrium point is an asymptotically stable node

9. Jacobian $= \begin{bmatrix} 1 & -e^y \\ 1 & 2e^{2y} \end{bmatrix}$

$$\begin{aligned} x - e^y &= 0 \\ x + e^{2y} - 2 &= 0 \\ e^{2y} + e^y - 2 &= 0 \\ (e^y - 1)(e^y + 2) &= 0 \\ e^y &= -2 \text{ no solution} \\ e^y - 1 &= 0 \\ y &= 0 \to x = 1 \end{aligned}$$

The equilibrium point is $(1, 0) \to$ Jacobian has complex eigenvalues with positive real part which means the equilibrium point is an unstable spiral point

11. $J = \begin{bmatrix} \text{0.1-0.02x -0.005y} & \text{-0.005x} \\ \text{-0.001y} & \text{0.05-0.001x-0.004y} \end{bmatrix}$

a)

$$\begin{aligned} x(0.1 - 0.01x - 0.005y) &= 0 \\ y(0.05 - 0.001x - 0.002y) &= 0 \\ x &= 0 \\ y(0.05 - 0.002y) &= 0 \\ y &= 0 \\ y &= 25 \\ y &= 0 \\ x(0.1 - 0.01x) &= 0 \\ x &= 0 \\ x &= 10 \\ 0.1 - 0.01x - 0.005y &= 0 \\ 20 - 2x &= y \\ (20 - 2x)(0.05 - 0.001x - 0.002(20 - 2x)) &= 0 \\ y = 0 \leftarrow x &= 10 \\ \text{Only non-negative values accepted} \leftarrow x &= \frac{-4}{3} \end{aligned}$$

The non-negative equilibrium points are:
$(0, 0) \to$ Jacobian has two positive eigenvalues which means the equilibrium point is an unstable node
$(0, 25) \to$ Jacobian has two negative eigenvalues which means the equilibrium point is an asym[totically stable node
$(10, 0) \to$ Jacobian has eigenvalues with opposite signs which means the equilibrium point is an unstable saddle point

b) Only the second species survives

13. $J = \begin{bmatrix} \text{0.1-0.01x -0.002y} & \text{-0.002x} \\ \text{-0.001y} & \text{0.05-0.001x-0.004y} \end{bmatrix}$

a)

$$\begin{aligned} x(0.1 - 0.005x - 0.002y) &= 0 \\ y(0.05y - 0.001x - 0.002y) &= 0 \\ x &= 0 \\ y(0.05 - 0.002y) &= 0 \\ y &= 0 \\ y &= 25 \\ y &= 0 \\ x(0.1 - 0.01x) &= 0 \\ x &= 0 \\ x &= 10 \\ 0.1 - 0.005x - 0.002y &= 0 \\ 50 - 2.5x &= y \\ (50 - 2.5x)(0.05 - 0.001x - 0.002(50 - 2.5x)) &= 0 \\ (50 - 2.5x)(-0.5 + 0.04x) &= 0 \\ y = 0 \leftarrow x &= 25 \\ y = 18.75 \leftarrow x &= 12.25 \end{aligned}$$

The non-negative equilibrium points are:
$(0, 0) \to$ Jacobian has two positive eigenvalues which means the equilibrium point is an unstable node
$(0, 25) \to$ Jacobian has two negative eigenvalues which means the equilibrium point is an asymptotically stable node
$(10, 0) \to$ Jacobian has eigenvalues with opposite signs which means the equilibrium point is an unstable saddle point
$(12.5, 18.75) \to$ Jacobian has two negative eigenvalues which means the equilibrium point is an asymptotically stable node

b) The stable equilibrium point indicates there is coexistence

15. $J = \begin{bmatrix} \text{0.1-0.000x-0.0008y} & \text{-0.0008x} \\ \text{-0.002y} & \text{0.1-0.002x-0.01y} \end{bmatrix}$

a) The non-negative equilibrium points are:
$(0, 0) \to$ Jacobian has two positive eigenvalues which means the equilibrium point is an unstable node
$(0, 20) \to$ Jacobian has eigenvalues with opposite signs which means the equilibrium point is an unstable saddle point
$(100, 0) \to$ Jacobian has two negative eigenvalues which means the equilibrium point is an asymptotically stable node
$(300, -100) \to$ Jacobian has eigenvalues with opposite signs which means the equilibrium point is an unstable saddle point

b) The stable equilibrium point indicates only the first species survives

17. $J = \begin{bmatrix} \text{0.1 -0.02x -0.005y} & \text{-0.005x} \\ \text{-0.015x} & \text{0.2-0.015x-0.04y} \end{bmatrix}$

$$\begin{aligned} x(0.1 - 0.01x - 0.005y) &= 0 \\ y(0.2 - 0.015x - 0.02y) &= 0 \\ x &= 0 \to y(0.2 - 0.02y) = 0 \\ y &= 0 \\ y &= 10 \\ y &= 0 \to x(0.1 - 0.01x) = 0 \\ x &= 0 \\ x &= 10 \end{aligned}$$

$$\begin{aligned} 0 &= (0.1 - 0.01x - 0.005y) \\ y &= 20 - 2x \\ 0 &= (20 - 2x)(0.2 - 0.015x - \\ &\quad 0.02(20 - 2x)) \\ x &= 10 \\ x &= 8 \to y = 4 \end{aligned}$$

$(0,0)$ gives an unstable node
$(0,10)$ gives an unstable node
$(10,0)$ gives an unstable saddle point $(8,4)$ gives an asymptotically stable node, and it indicates coexistence

19. $J = \begin{bmatrix} \text{0.1-0.2x -0.005y} & \text{-0.005x} \\ \text{-0.015y} & \text{0.2-0.015x -0.4y} \end{bmatrix}$

a) $x(0.1 - 0.01x - 0.005y) - 0.08x = 0$
$y(0.2 - 0.015x - 0.02y) = 0$

$$\begin{aligned} 0.2 - 0.015x - 0.02y &= 0 \\ 10 - 0.75x &= y \\ x(0.1 - 0.01x - 0.02(10 - 0.75x) - 0.08x &= 0 \\ x\,[0.1 - 0.01x - 0.2 + 0.015x - 0.08] &= 0 \\ x &= -4.8 \\ \text{not acceptable} & \\ y &= 0 \\ x &= 0 \\ x &= 2 \end{aligned}$$

There are no coexistence equilibrium points

b) $(0,0)$ gives an unstable node
$(0,10)$ gives an asymptotically stable node
$(2,0)$ gives an unstable saddle point

c) The excessive fishing (of the first species) helps eliminate the first species

21. $J = \begin{bmatrix} \text{0.6-0.3y} & \text{-0.3x} \\ \text{0.2y} & \text{-1+0.2x} \end{bmatrix}$

$$\begin{aligned} x(0.6 - 0.3y) &= 0 \\ -y(1 - 0.2x) &= 0 \\ x &= 0 \to y = 0 \\ y &= 2 \to x = 5 \end{aligned}$$

$(0,0)$ gives an unstable saddle point
$(5,2)$ gives a center

23. $J = \begin{bmatrix} \text{0.5-0.2y} & \text{-0.2x} \\ \text{0.1y} & \text{-0.4+0.1x} \end{bmatrix}$

$$\begin{aligned} x(0.5 - 0.2y) &= 0 \\ -y(0.4 - 0.1x) &= 0 \\ x &= 0 \to y = 0 \\ y &= 2.5 \to x = 4 \end{aligned}$$

$(0,0)$ gives an unstable saddle point
$(4,2.5)$ gives a center

25. $J = \begin{bmatrix} \text{0.6-0.06x -0.3y} & \text{-0.3x} \\ \text{0.2y} & \text{-1+0.2x} \end{bmatrix}$

$$\begin{aligned} x\,[0.6 - 0.06x - 0.3y] &= 0 \\ -y(1 - 0.2x) &= 0 \\ x &= 0 \to y = 0 \\ x &= 5 \to y = 1.5 \\ x &= 20 \to y = 0 \end{aligned}$$

$(0,0)$ gives an unstable saddle point
$(5,1.5)$ gives a center $(20,0)$ gives an unstable node

27. $J = \begin{bmatrix} \text{0.5-0.1x -0.2y} & \text{-0.2x} \\ \text{0.1y} & \text{-0.4+0.1x} \end{bmatrix}$

$$\begin{aligned} x\,[0.5 - 0.05x - 0.2y] &= 0 \\ -y(0.4 - 0.1x) &= 0 \\ x &= 0 \to y = 0 \\ x &= 4 \to y = 1.5 \\ x &= 10 \to y = 0 \end{aligned}$$

$(0,0)$ gives an unstable saddle point
$(4,1.5)$ gives an asymptotically stable spiral point $(10,0)$ gives an unstable saddle point

29. **a)** When $x = 0$ and $y = 0$, then $ax - by = 0$ and $-dy + \dfrac{cx}{\sqrt{c^2x^2+1}} = 0$

b) $x = \dfrac{\sqrt{65}}{12}$, $y = \dfrac{\sqrt{65}}{9}$

c) Left to the student

31. $J = \begin{bmatrix} 0.2 - 0.1x - 0.02y & -0.02x \\ -0.01y & 0.1 - 0.01x - 0.04y \end{bmatrix}$

a) The eigenvalues for $(0,5)$ are -0.1 and 0.1 which have the corresponding eigenvectors $[0,1]$ and $[4,-1]$ respectively. The trajectories seems to approach parallel to $[0,1]$ and parallel to $[4,-1]$ as they leave

b) The eigenvalues for $(4,0)$ are -0.2 and 0.06 which have the corresponding eigenvectors $[1,0]$ and $[-4,13]$ respectively. The trajectories seems to approach parallel to $[1,0]$ and parallel to $[-4,13]$ as they leave

c) The eigenvalues for $(2.5, 3.75)$ are -0.05 and -0.15 which have the corresponding eigenvectors $[-2,3]$ and $[2,1]$ respectively. The trajectories seems to approach parallel to $[-2,3]$

33. **a)** In Exercises 11, 13, 15 and 16, a_1a_2 is larger and in Exercises 12 and 14 b_1b_2 is larger

b) Left to the student

35. If $a = 0$ then

$$\begin{aligned} x' &= px\left(1 - \frac{x}{L}\right) - \frac{qxy}{1+0} \\ &= px\left(1 - \frac{x}{L}\right) - qxy \\ y' &= -ry + \frac{sxy}{1+0} \\ &= -ry + sxy \end{aligned}$$

37. $x' = 0.2x\left(1 - \dfrac{x}{10}\right) - \dfrac{0.04xy}{1+0.1x}$

$y' = -0.05y + \dfrac{0.03xy}{1+0.1x}$

The coexistence equilibrium point is $(2, 5.7)$ which is an unstable spiral point

39. **a)** Left to the student

b) When $x = \dfrac{r}{s}$ is plugged into the formula the numerator becomes zero

c) When $y = \dfrac{p}{q}$ is plugged into the formula the denominator becomes zero

d) One population is maximized when the other population agrees with the coexistence equilibrium point value

41. $x' = 2x - xy$ and $y' = -y + 0.4xy$
This means $p = 2$, $q = 1$, $r = 1$, and $s = 0.4 \rightarrow$

$$\begin{aligned} 2\,\ln y - y + \ln x - 0.4x &= C \\ 2\,\ln 1 - 1 + \ln 5 - 0.4(5) &= C \\ -1 + \ln 5 - 2 &= C \\ \ln 5 - 3 &= C \\ 2\,\ln y - y + \ln x - 0.4x &= \ln 5 - 3 \end{aligned}$$

43. $x = 1.0159393$ and $x = 5$

45. $x = 1.6312622$ and $x = 3.6330767$

Exercise Set 9.6

1. $x(1) \approx 2.42368$
$y(1) \approx 1.048$

3. $x(2) \approx 38.21927$
$y(2) \approx 83.47852$

5. $x(3) \approx 75.15387$
$y(3) \approx 212.32202$

7. **a)**

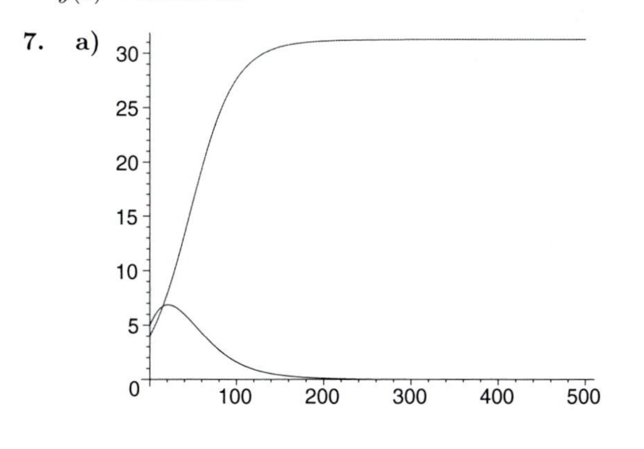

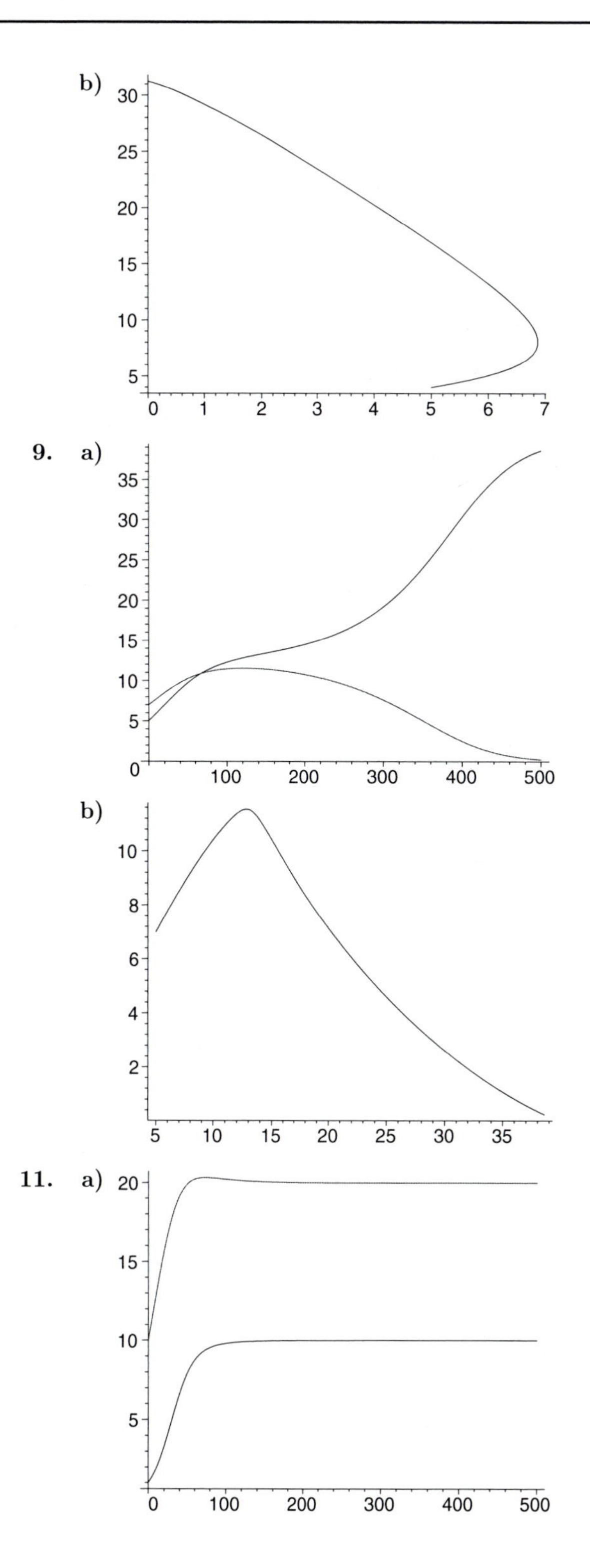

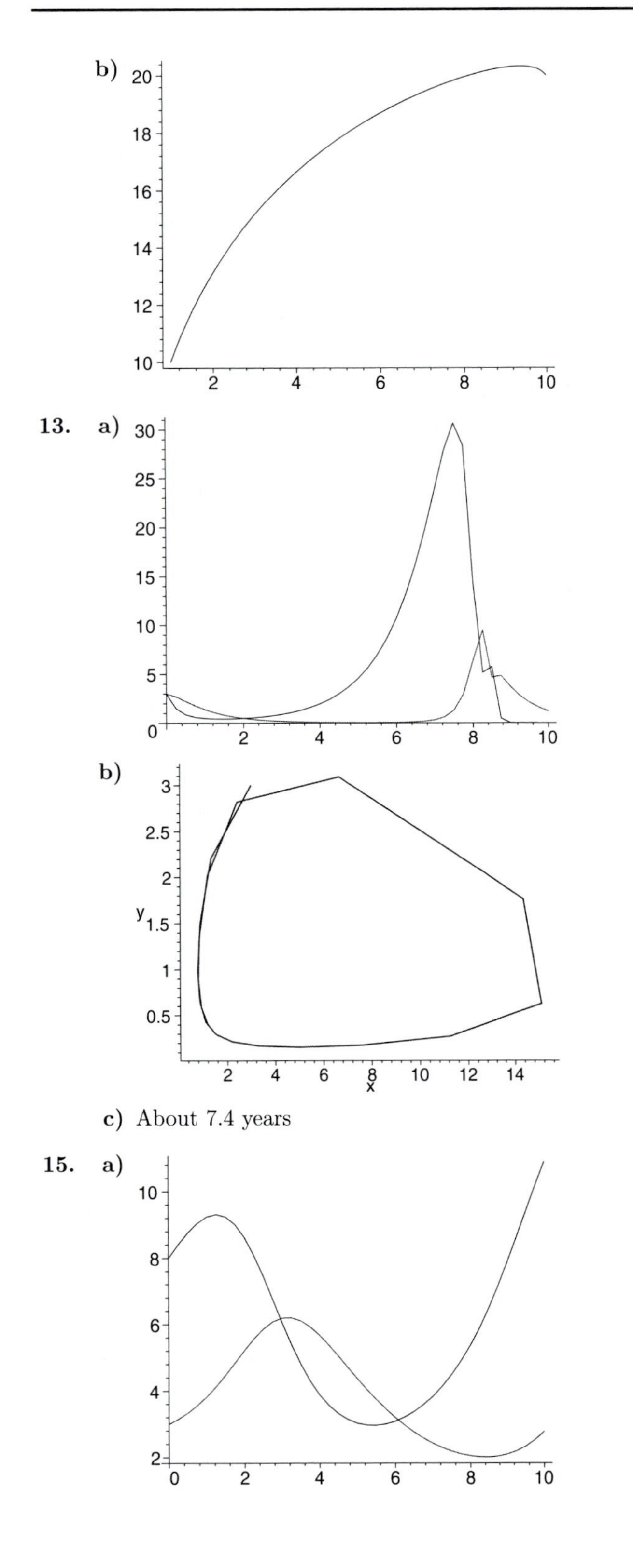

b)

13. a)

b)

c) About 7.4 years

15. a)

b)

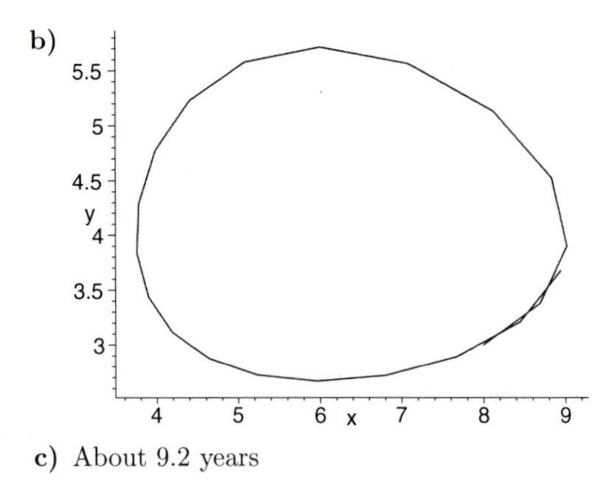

c) About 9.2 years

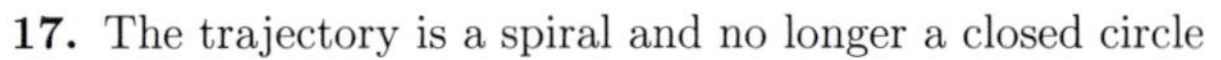

17. The trajectory is a spiral and no longer a closed circle

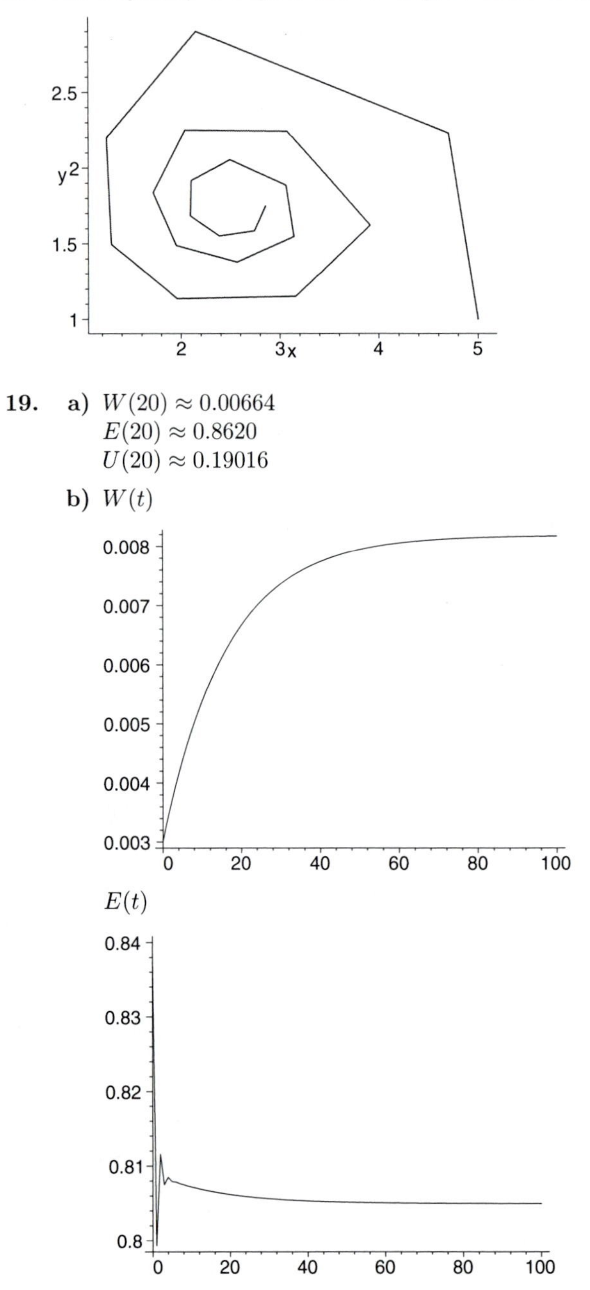

19. a) $W(20) \approx 0.00664$
$E(20) \approx 0.8620$
$U(20) \approx 0.19016$

b) $W(t)$

$E(t)$

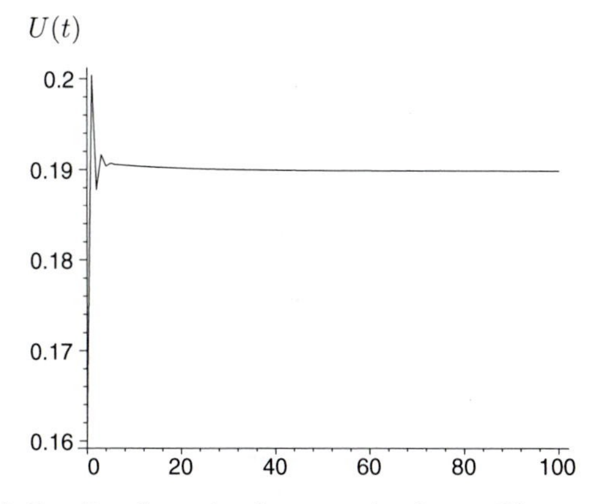

c) Reading form the three graphs the equilibrium points are approximately 0.0082, 0.8050, 0.1899 respectively

21. $J = \begin{bmatrix} 0.1 - 0.02x - 0.004y & -0.004x \\ -0.001y & 0.5 - 0.001x - 0.0032y \end{bmatrix}$

$$\begin{aligned} x(0.1 - 0.01x - 0.004y) &= 0 \\ x &= 0 \\ y &= 0 \\ (0.05 - 0.0016y) &= 0 \\ 31.25 &= y \\ 25 - 2.5x &= y \\ (25 - 2.5x)(0.05 - 0.001x - 0.0016(25 - 2.5x)) &= 0 \\ x &= 10 \\ y &= 0 \\ x &= -\frac{10}{3} \end{aligned}$$

Equilibrium points are:
$(0, 0)$ $\to$ Jacobian has two positive eigenvalues which means the equilibrium point is an unstable node
$(0, 31.25)$ $\to$ Jacobian has two negative eigenvalues which means the equilibrium point is an asymptotically stable node
$(10, 0)$ $\to$ Jacobian has eigenvalues with opposite signs which means the equilibrium point is an unstable saddle point

23. $J = \begin{bmatrix} 0.1 - 0.012x - 0.002y & -0.002x \\ -0.001y & 0.05 - 0.001x - 0.004y \end{bmatrix}$

$$\begin{aligned} x(0.1 - 0.006x - 0.0002y) &= 0 \\ x &= 0 \\ y &= 0 \\ (0.05 - 0.002y) &= 0 \\ y &= 25 \\ y &= 50 - 3x \\ (50 - 3x)(0.05 - 0.001x - 0.002(50 - 0.5x)) &= 0 \\ x &= \frac{50}{3} \\ y &= 0 \\ x &= 10 \\ y &= 20 \end{aligned}$$

Equilibrium points are:
$(0, 0)$ $\to$ Jacobian has two positive eigenvalues which means the equilibrium point is an unstable node
$(0, 25)$ $\to$ Jacobian has two positive eigenvalues which means the equilibrium point is an unstable node
$(10, 20)$ $\to$ Jacobian has two negative eigenvalues which means the equilibrium point is an asymptotically stable node
$(\frac{50}{3}, 0)$ $\to$ Jacobian has eigenvalues with opposite signs which means the equilibrium point is an unstable saddle point

25. $J = \begin{bmatrix} 0.8 - 0.2y & -0.2x \\ 0.1y & -0.6 + 0.1x \end{bmatrix}$

The equilibrium points are:
$(0, 0)$ $\to$ Jacobian has eigenvalues with opposite signs which means the equilibrium point is an unstable saddle point
$(6, 4)$ $\to$ Jacobian has two imaginary eigenvalues which means the equilibrium point is a center

27. Left to the student.

Chapter 10

Probability

Exercise Set 10.1

1. $\frac{3}{50} = 0.06$

3. $\frac{14}{50} = 0.28$

5. $\frac{23}{50} = 0.46$

7. $\frac{10+14}{50} = \frac{24}{50} = 0.48$

9. $\frac{50-14}{50} = \frac{36}{50} = 0.72$

11. The possible outcomes are **HH,HT,TH,TT**

a.

$$\begin{aligned} P(A)\cdot P(B) &= \frac{1}{2}\cdot\frac{1}{2} \\ &= \frac{1}{4} \\ P(A \,\&\, B) &= \frac{1}{4} \end{aligned}$$

Therefore, events A and B are independent

b. Events A and B are not disjoint. The outcome HT is common between them

13. The possible outcomes are $\{1, 2, 3, 4, 5, 6\}$

a.

$$\begin{aligned} P(A)\cdot P(B) &= \frac{1}{6}\cdot\frac{1}{6} \\ &= \frac{1}{36} \\ P(A \,\&\, B) &= 0 \end{aligned}$$

Therefore, events A and B are independent

b. Events A and B are disjoint

15. **a.**

$$\begin{aligned} P(A)\cdot P(B) &= \frac{4}{52}\cdot\frac{3}{51} \\ &= \frac{1}{221} \\ P(A \,\&\, B) &= \frac{12}{1326} = \frac{2}{221} \end{aligned}$$

Therefore, events A and B are not independent

b. Events A and B are not disjoint

17. **a.**

$$\begin{aligned} P(A)\cdot P(B) &= \frac{4}{52}\cdot\frac{4}{52} \\ &= \frac{1}{169} \\ P(A \,\&\, B) &= \frac{1}{52} \end{aligned}$$

Therefore, events A and B are not independent

b. Events A and B are disjoint

19. $\frac{1}{5} = 0.2$

21. $\frac{4}{5} = 0.8$

23. $\frac{2}{5} = 0.4$

25. $\frac{1}{5}\cdot\frac{1}{4} = 0.05$

27. $\frac{1}{5}\cdot\frac{3}{4} = 0.15$

29. $\frac{1}{5}\cdot\frac{1}{5} = 0.04$

31. $\frac{1}{5}\cdot\frac{4}{5} = 0.16$

33. $\frac{4}{10} = 0.4$

35. $\frac{10-1}{10} = \frac{9}{10} = 0.9$

37. $\frac{3+4}{10} = \frac{7}{10} = 0.7$

39. $\frac{2}{10}\cdot\frac{1}{9} = \frac{1}{45}$

41. $\frac{1}{10}\cdot\frac{6}{9} = \frac{1}{18}$

43. $\frac{2}{10}\cdot\frac{2}{10} = 0.4$

45. $\frac{1}{10}\cdot\frac{6}{10} = 0.06$

47. $\frac{1}{6}\cdot\frac{1}{6}\cdot\frac{1}{6} = \frac{1}{216}$

49. $1 - \frac{1}{216} = \frac{215}{216}$

51. $\left(\frac{1}{2}\right)^5 = \frac{1}{32}$

53. $1 - \frac{1}{32} = \frac{31}{32}$

55. $\frac{1}{6}\cdot\frac{1}{6}\cdot\frac{1}{6}\cdot\frac{1}{6} = \frac{1}{1296}$

57. $(6\cdot 5\cdot 4\cdot 3)\cdot\frac{1}{1296} = \frac{5}{18}$

59.

$$\begin{aligned} P(6 \text{ or } 4) &= P(6) + P(4) - P((6,4)) \\ &= \frac{6}{36} + \frac{6}{36} - \frac{1}{36} \\ &= \frac{11}{36} \end{aligned}$$

61.

$$\begin{aligned} P(Ace\ or\ King) &= P(Ace) + P(King) - P((Ace, King)) \\ &= \frac{4}{52} + \frac{4}{52} - \frac{4}{52} \cdot \frac{4}{51} \\ &= 0.14781 \end{aligned}$$

63. Answers Vary.

Exercise Set 10.2

1. The possible outcomes are $(1,2)$ or $(3,2)$ Thus the probability is
$\frac{1}{3} \cdot \frac{1}{2} + \frac{1}{3} \cdot \frac{1}{2} = \frac{1}{3}$

3. The possible outcomes are $(1,3)$ or $(3,1)$ Thus the probability is
$\frac{1}{3} \cdot \frac{1}{2} + \frac{1}{3} \cdot \frac{1}{2} = \frac{1}{3}$

5. The possible outcomes are $(1,2)$, $(2,2)$, or $(3,2)$ Thus the probability is
$\frac{1}{3} \cdot \frac{1}{3} + \frac{1}{3} \cdot \frac{1}{3} + \frac{1}{3} \cdot \frac{1}{3} = \frac{1}{3}$

7. The possible outcomes are $(1,3)$, $(2,2)$, or $(3,1)$ Thus the probability is
$\frac{1}{3} \cdot \frac{1}{3} + \frac{1}{3} \cdot \frac{1}{3} + \frac{1}{3} \cdot \frac{1}{3} = \frac{1}{3}$

9. $\frac{1}{2} \cdot \frac{1}{2} \cdot \frac{1}{2} \cdot \frac{1}{2} = \frac{1}{16}$

11. $2 \cdot \frac{1}{16} = \frac{1}{8}$

13. $4 \cdot \frac{1}{2} \cdot \frac{1}{2} \cdot \frac{1}{2} \cdot \frac{1}{2} = \frac{1}{4}$

15. $\frac{13}{52} \cdot \frac{39}{51} + \frac{39}{52} \cdot \frac{13}{51} = \frac{13}{34} = 0.38235$

17. **a.)** Possible genotype outcomes are $\{FF,\ Ff,\ ff\}$. Since FF cannot occur when we cross Ff and ff its probability is 0. and the other two outcomes have a probability of 1/2

b.) The genotypes in Example 2 did not allow for the ff type while in this Exercise the ff genotype is allowed. If a cross test is performed with the sperm allele transmitted from the ff plant and the ovum allele transmitted from the FF plant then we will be able to distinguish between the FF and the Ff genotypes.

19. **a.)** $\frac{3}{4} \cdot \frac{3}{4} = \frac{9}{16}$

b.) $\frac{1}{4} \cdot \frac{3}{4} = \frac{3}{16}$

c.) $\frac{3}{4} \cdot \frac{1}{4} = \frac{3}{16}$

d.) $\frac{1}{4} \cdot \frac{1}{4} = \frac{1}{16}$

21. **a.)** $P(T+|D+) = \frac{37}{42} = 0.881$

b.) $P(T-|D-) = \frac{106}{110} = 0.964$

c.) $P(D+\ and\ T-) = (0.2)(0.119) = 0.0238$
$P(T-) = (0.2)(0.119) + (0.8)(0.964) = 0.795$
$P(D+|T-) = \frac{P(D+andT+)}{P(T+)} = \frac{0.0238}{0.795} = 0.0299$

23. **a.)** $P(D+\ and\ T+) = (0.1)(0.67) = 0.067$
$P(T+) = (0.1)(0.67) + (0.9)(0.24) = 0.283$
$P(D+|T+) = \frac{0.067}{0.283} = 0.237$

b.) $P(D+\ and\ T+) = (0.5)(0.67) = 0.335$
$P(T+) = (0.5)(0.67) + (0.5)(0.24) = 0.455$
$P(D+|T+) = \frac{0.335}{0.455} = 0.736$

25. - 27. Left to the student

29.

$$\begin{aligned} P(DD|S) &= \frac{P(DD\ and\ S)}{P(S)} \\ &= \frac{(0.25)(0.96)}{(0.25)(0.96) + (0.5)(0.96) + (0.25)(0.2)} \\ &= 0.312 \end{aligned}$$

$$\begin{aligned} P(Dd|S) &= \frac{P(Dd\ and\ S)}{P(S)} \\ &= \frac{(0.5)(0.96)}{(0.25)(0.96) + (0.5)(0.96) + (0.25)(0.2)} \\ &= 0.623 \end{aligned}$$

$$\begin{aligned} P(dd|S) &= \frac{P(dd\ and\ S)}{P(S)} \\ &= \frac{(0.25)(0.2)}{(0.25)(0.96) + (0.5)(0.96) + (0.25)(0.2)} \\ &= 0.065 \end{aligned}$$

Exercise Set 10.3

1. $\binom{4}{2} = 6$

3. $\binom{3}{3} = 1$

5. $\binom{8}{0} = 1$

7. $\binom{5}{6}$ is not possible

9. $\binom{9}{-4}$ is not possible

11. $\binom{16}{10} = 8008$

13.

$$\begin{aligned} P(X=2) &= \binom{3}{2} \cdot \frac{1}{3}^2 \cdot \frac{2}{3}^1 \\ &= 3 \cdot \frac{1}{9} \cdot \frac{2}{3} \\ &= \frac{2}{9} \end{aligned}$$

15.

$$\begin{aligned} P(X+3) &= \binom{4}{3}(0.8)^3(0.2)^1 \\ &= 4 \cdot 0.512 \cdot 0.2 \\ &= 0.4096 \end{aligned}$$

17.

$$\begin{aligned} P(X=5) &= \binom{6}{5}(0.9)^5(0.1)^1 \\ &= 6 \cdot 0.59049 \cdot 0.1 \\ &= 0.35429 \end{aligned}$$

19.

$$\begin{aligned} P(X=10) &= \binom{20}{10}(0.6)^{10}(0.4)^{10} \\ &= 184756 \cdot 0.0060466176 \cdot 0.0001048576 \\ &= 0.11714 \end{aligned}$$

21. $P(X=5)=0$ since we cannot choose 5 out of 4 items

23.

25.

27.

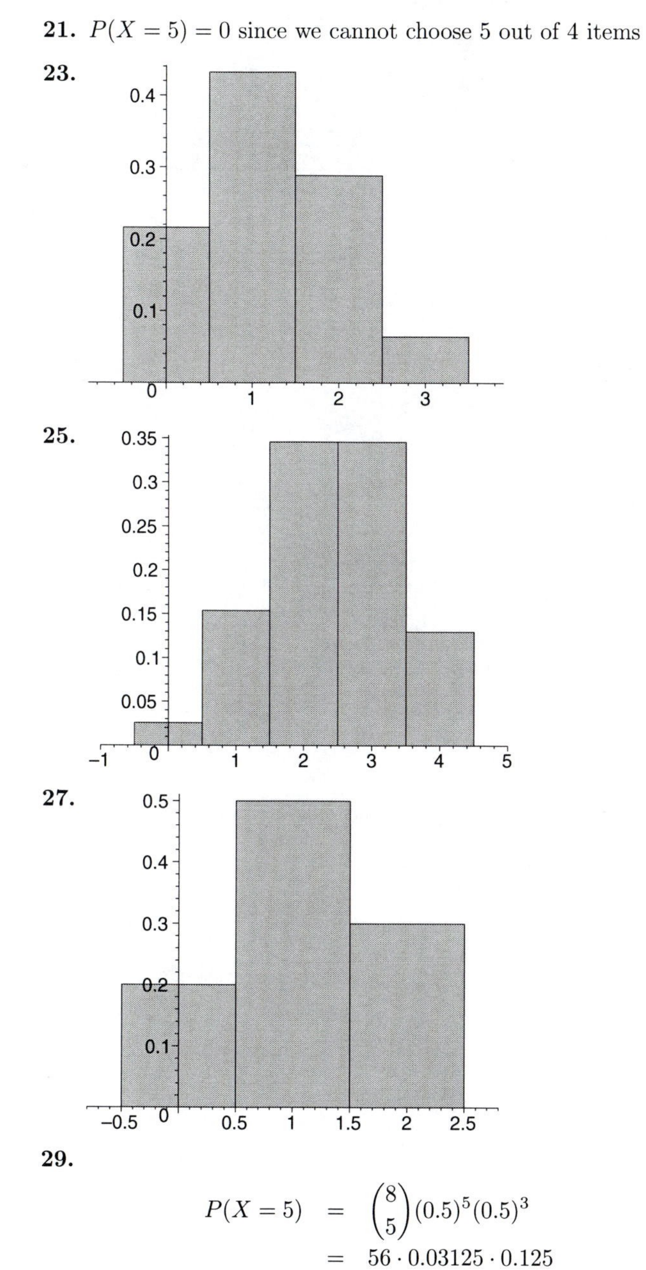

29.

$$\begin{aligned} P(X=5) &= \binom{8}{5}(0.5)^5(0.5)^3 \\ &= 56 \cdot 0.03125 \cdot 0.125 \\ &= 0.21875 \end{aligned}$$

31.

$$\begin{aligned} P(X<2) &= P(X=1)+P(X=0) \\ &= \binom{6}{1}(\frac{1}{2})^1(\frac{1}{2})^5 + \binom{6}{0}(\frac{1}{2})^0(\frac{1}{2})^6 \\ &= \frac{6}{64}+\frac{1}{64} \\ &= \frac{7}{64} \end{aligned}$$

33.

$$\begin{aligned} P(X=2) &= \binom{5}{2}\cdot(\frac{1}{6})^2\cdot(\frac{5}{6})^3 \\ &= 10\cdot\frac{1}{36}\cdot\frac{125}{216} \\ &= 0.16075 \end{aligned}$$

35.

$$\begin{aligned} P(X\geq 1) &= 1-P(X=0) \\ &= 1-\binom{10}{0}\cdot(\frac{1}{6})^0\cdot(\frac{5}{6})^{10} \\ &= 1-0.16151 \\ &= 0.0.838494 \end{aligned}$$

37.

$$\begin{aligned} P(X>6) &= P(X=7)+P(X=8)+P(X=9) \\ &= \binom{9}{7}\cdot(\frac{1}{6})^7\cdot(\frac{5}{6})^2 + \\ &\quad \binom{9}{8}\cdot(\frac{1}{6})^8\cdot(\frac{5}{6})^1 + \binom{9}{9}\cdot(\frac{1}{6})^9\cdot(\frac{5}{6})^0 \\ &= 0.000089306+0.0000044653+ \\ &\quad 0.000000099229 \\ &= 0.000093871 \end{aligned}$$

39.

$$\begin{aligned} P(X=30) &= \binom{40}{30}\cdot(\frac{319159}{534694})^{30}\cdot(\frac{215535}{534694})^{10} \\ &= 0.018172 \end{aligned}$$

41.

$$\begin{aligned} P(X=2) &= \binom{3}{2}\cdot(0.80)^2\cdot(0.20)^1 \\ &= 0.384 \end{aligned}$$

43. a.)

$$\begin{aligned} P(X\leq 2) &= P(X=2)+P(X=1)+P(X=0) \\ &= \binom{12}{2}\cdot(0.346)^2\cdot(0.654)^{10} + \\ &\quad \binom{12}{1}\cdot(0.346)^1\cdot(0.654)^{11} + \\ &\quad \binom{12}{0}\cdot(0.346)^0\cdot(0.654)^{12} \\ &= 0.11310+0.03887+0.0061226 \\ &= 0.15809 \end{aligned}$$

b.)

$$\begin{aligned} P(X\geq 9) &= P(X=9)+P(X=10)+ \\ &\quad P(X=11)+P(X=12) \\ &= \binom{12}{9}\cdot(0.346)^9\cdot(0.654)^3 + \end{aligned}$$

$$
\begin{aligned}
& \binom{12}{10} \cdot (0.346)^{10} \cdot (0.654)^2 + \\
& \binom{12}{11} \cdot (0.346)^{11} \cdot (0.654)^1 + \\
& \binom{12}{12} \cdot (0.346)^{12} \cdot (0.654)^0 \\
= \ & 0.004374 + 0.00069416 + \\
& 0.000066772 + 0.00000294383 \\
= \ & 0.005138
\end{aligned}
$$

45.

$$
\begin{aligned}
P(X = CC) &= \binom{2}{2} \cdot (0.08)^2 \cdot (0.92)^0 \\
&= 0.0064 \\
P(X = CG) &= \binom{2}{1} \cdot (0.08)^1 \cdot (0.92)^1 \\
&= 0.1472 \\
P(X = GG) &= \binom{2}{2} \cdot (0.08)^0 \cdot (0.92)^2 \\
&= 0.8464
\end{aligned}
$$

47. We seek the value of $2pq$

$p^2 = \dfrac{1}{2500} \to p = \dfrac{1}{50}$

$q = 1 - \dfrac{1}{50} = \dfrac{49}{50}$

Therefore, $2pq = 2(\dfrac{1}{50})(\dfrac{49}{50}) = 0.0392$

49. We seek the value of $2pq$

$p^2 = \dfrac{1}{360000} \to p = \dfrac{1}{600}$

$q = 1 - \dfrac{1}{600} = \dfrac{599}{600}$

Therefore, $2pq = 2(\dfrac{1}{600})(\dfrac{599}{600}) = 0.033278$

51.

$$
\begin{aligned}
P(X \geq 3) &= P(X = 3) + P(X = 4) + P(X = 5) \\
&= \binom{5}{3} \cdot (0.40)^3 (0.6)^2 + \binom{5}{4} \cdot (0.40)^4 (0.6)^1 \\
&\quad + \binom{5}{5} \cdot (0.40)^5 (0.6)^0 \\
&= 0.2304 + 0.0768 + 0.01024 \\
&= 0.31744
\end{aligned}
$$

53. a.) $X = \{0, 1, 2\}$

b.) $P(X = 0) = \dfrac{39}{52} \cdot \dfrac{38}{51} = 0.558824$

$P(X = 1) = \dfrac{13}{52} \cdot \dfrac{39}{51} + \dfrac{39}{52} \cdot \dfrac{13}{51} = 0.382353$

$P(X = 2) = \dfrac{13}{52} \cdot \dfrac{12}{51} = 0.0588235$

c.)

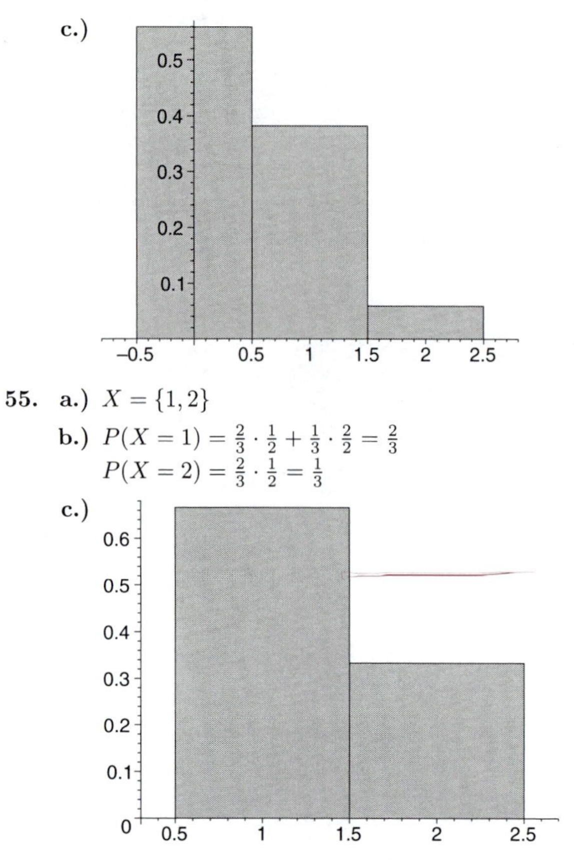

55. a.) $X = \{1, 2\}$

b.) $P(X = 1) = \frac{2}{3} \cdot \frac{1}{2} + \frac{1}{3} \cdot \frac{2}{2} = \frac{2}{3}$

$P(X = 2) = \frac{2}{3} \cdot \frac{1}{2} = \frac{1}{3}$

c.)

57. Left to the student

59. Left to the student

61. a.) $\binom{8}{5} = 56$ Eighth row fifth column

b.) $\binom{8}{5} = \dfrac{8!}{3! \cdot 5!} = 56$

63.

$$
\begin{aligned}
\binom{n}{k} &= \frac{n!}{(n-k)! \cdot k!} \\
\binom{n}{n} &= \frac{n!}{(n-n)! \cdot n!} \\
&= \frac{n!}{0! \cdot n!} \\
&= \frac{n!}{1 \cdot n!} \\
&= 1
\end{aligned}
$$

65. Left to the student

67. **a.)** $\{SSSS, SSSF, SSFS, SSFF,$
$SFSS, SFSF, SFFS, SFFF,$
$FSSS, FSSF, FSFS, FSFF,$
$FFSS, FFSF, FFFS, FFFF\}$

b.) 6

c.) 4

d.) $\binom{4}{2} = 6$

e.) $\binom{4}{3} = 4$

69.

$$\begin{aligned} P(i,j,k) &= \frac{n!}{i!j!k!}(P^2)^i(2pq)^j(q^2)^k \\ P(2,5,3) &= \frac{10!}{2!5!3!}(0.41^2)^2(2\cdot 0.41\cdot 0.59)^5(0.59^2)^3 \\ &= 2520\cdot 0.02825761\cdot 0.02650509\cdot 0.04218053 \\ &= 0.0796118 \end{aligned}$$

71. **a.)** $P(Aa) = 2pq$, $P(aa) = q^2$

$P(A|Aa) = 1/2$, $P(a|Aa) = 1/2$

$P(A|aa) = 0$, $P(a|aa) = 1$

b.)

$$\begin{aligned} P(A) &= p^2\cdot 1 + 2pq\cdot\frac{1}{2} + q^2\cdot 0 \\ &= P^2 + p(1-p) + 0 \\ &= p^2 + p - p^2 \\ &= p \end{aligned}$$

c.) $q = 1 - p$

$$\begin{aligned} P(a) &= p^2\cdot 0 + 2pq\cdot\frac{1}{2} + q^2\cdot 1 \\ &= 0 + (1-q)q + q^2 \\ &= q - q^2 + q^2 \\ &= q \end{aligned}$$

d.) Left to the student

Exercise Set 10.4

1.

$$\begin{aligned} E(X) &= 0\cdot P(X=0) + 1\cdot P(X=1) + 2\cdot P(X=2) \\ &= 0(0.2) + 1(0.3) + 2(0.5) \\ &= 0 + 0.3 + 1 \\ &= 1.3 \end{aligned}$$

3. From Exercise 1, $\mu = E(X) = 1.3$

$$\begin{aligned} Var(X) &= \sum (k-\mu)^2 P(X=k) \\ &= (0-1.3)^2(0.2) + (1-1.3)^2(0.3) + \\ &\quad (2-1.3)^2(0.5) \\ &= 0.338 + 0.027 + 0.245 \\ &= 0.61 \\ SD(X) &= \sqrt{Var(X)} \\ &= \sqrt{0.61} \\ &= 0.781025 \end{aligned}$$

5. **a)** $E = 1\cdot 0.25 + 1\cdot 0.25 = 0.5$
b) $E = np = 2\cdot 0.25 = 0.5$

7. **a)** $E = 1\cdot 0.5 + 1\cdot 0.5 + 1\cdot 0.5 = 1.5$
b) $E = np = 3\cdot 0.5 = 1.5$

9. $SD = \sqrt{npq} = \sqrt{2(0.25)(0.75)} = 0.61237$

11. $SD = \sqrt{npq} = \sqrt{2(0.25)(0.75)} = 0.$

13. $E(X) = np = 6\cdot 0.2 = 1.2$
$SD(X) = \sqrt{npq} = \sqrt{6(0.2)(0.8)} = 0.979796$

15. $E(X) = np = 20\cdot 0.1 = 2$
$SD(X) = \sqrt{npq} = \sqrt{20(0.1)(0.9)} = 1.34164$

17. $E(X) = np = 50\cdot 0.4 = 20$
$SD(X) = \sqrt{npq} = \sqrt{50(0.4)(0.6)} = 3.46410$

19.

$$\begin{aligned} z &= \frac{x - E(X)}{SD(X)} \\ &= \frac{20-16}{2} \\ &= 2 \end{aligned}$$

21.

$$\begin{aligned} z &= \frac{x - E(X)}{SD(X)} \\ &= \frac{13.1-13.5}{0.24} \\ &= \frac{-0.4}{0.24} \\ &= \frac{-5}{3} \end{aligned}$$

23.

$$\begin{aligned} z &= \frac{x - E(X)}{SD(X)} \\ &= \frac{29.3-20.3}{4.5} \\ &= 2 \end{aligned}$$

25. a) $E(X) = np = 12(0.73) = 8.76$
$SD(X) = \sqrt{npq} = \sqrt{12(0.73)(0.27)} = 1.53792$

b)

$$\begin{aligned} P(X = k) &= \binom{n}{k} p^k q^{n-k} \\ P(X = 8) &= \binom{12}{8} (0.73)^8 (0.27)^4 \\ &= 495 \cdot 0.0806460092 \cdot 0.00531441 \\ &= 0.21215 \end{aligned}$$

27. $E(X) = 12 \cdot 0.638 = 7.656$
$SD(X) = \sqrt{12(0.638)(0.362)} = 1.66477$

29.

$$\begin{aligned} Var(aX) &= E(a^2X^2) - [E(ax)^2] \\ &= a^2E(X^2) - E(aX)E(aX) \\ &= a^2E(X^2) - aE(X)aE(X) \\ &= a^2E(X^2) - a^2E^2(X) \\ &= a^2\left(E(X^2) - E^2(X)\right) \\ &= a^2Var(X) \end{aligned}$$

Exercise Set 10.5

1. $f(x) = 2x \geq 0$ on $[0, 1]$

$$\begin{aligned} \int_0^1 2x dx &= \left(x^2\right|_0^1 \\ &= 1^2 - 0^2 = 1 \end{aligned}$$

3. $f(x) = \frac{3}{26}x^2 \geq 0$ on $[1, 3]$

$$\begin{aligned} \int_1^3 \frac{3}{26}x^2 dx &= \left(\frac{1}{26}x^3\right|_1^3 \\ &= \frac{27}{26} - \frac{1}{26} = 1 \end{aligned}$$

5.

$$\begin{aligned} P(1/4 \leq X \leq 3/4) &= \int_{1/4}^{3/4} 4x^3 dx \\ &= \left(x^4\right|_{1/4}^{3/4} \\ &= \left(\frac{3}{4}\right)^4 - \left(\frac{1}{4}\right)^4 \\ &= \frac{81}{256} - \frac{1}{256} \\ &= \frac{80}{256} = \frac{5}{16} \end{aligned}$$

7.

$$\begin{aligned} P(1/4 \leq X) &= \int_{1/4}^1 20x(1-x)^3 dx \\ &= \int_{1/4}^1 20x(1 - 3x + 3x^2 - x^3) dx \\ &= \int_{1/4}^1 20x - 60x^2 + 60x^3 - 20x^4 dx \\ &= \left(10x^2 - 20x^3 + 15x^4 - 4x^5\right|_{1/4}^1 \\ &= [10 - 20 + 15 - 4] - [10(0.25)^2 - 20(0.25)^3 \\ &\quad +15(0.25)^4 - 4(0.25)^5] \\ &= 0.63281 \end{aligned}$$

9.

$$\begin{aligned} P(X \geq \pi/6) &= \int_{\pi/6}^{\pi/4} sec^2 x dx \\ &= \left(tan(x)\right|_{\pi/6}^{\pi/4} \\ &= tan(\pi/4) - tan(\pi/6) \\ &= 0.42265 \end{aligned}$$

11.

$$\begin{aligned} P(0 \leq X \leq 1/2) &= \int_0^{1/2} \frac{e^x}{e-1} dx \\ &= \left(\frac{e^x}{e-1}\right|_0^{1/2} \\ &= -\frac{1}{2}cos(\pi/4) + \frac{1}{2}cos(0) \\ &= 0.14645 \end{aligned}$$

13.

$$\begin{aligned} P(X \geq 16) &= \int_{16}^{\infty} \frac{dx}{x^2} \\ &= \lim_{b \to \infty} \int_{16}^{b} \frac{dx}{x^2} \\ &= \lim_{b \to \infty} \left[\frac{-1}{x}\right]_{16}^{b} \\ &= \lim_{b \to \infty} \left[-\frac{1}{b} - \frac{-1}{16}\right] \\ &= \frac{1}{16} \end{aligned}$$

15.

$$\begin{aligned} P(X \geq 3.4) &= \int_{3.4}^{\infty} e^{-x} dx \\ &= \lim_{b \to \infty} \left[-e^{-x}\right]_{3.4}^{b} \\ &= \lim_{b \to \infty} \left[-e^{-b} - (-e^{-3.4})\right] \\ &= 0.03337 \end{aligned}$$

17.

$$\begin{aligned} P(X \geq 4) &= \int_4^\infty \frac{2x}{(x^2+1)^2} dx \\ &= \lim_{b\to\infty}\left[\frac{-1}{(x^2+1)}\right]_4^b \\ &= \lim_{b\to\infty}\left[\frac{-1}{(b^2+1)} + \frac{1}{17}\right] \\ &= 0.058824 \end{aligned}$$

19.

$$\begin{aligned} P(X \geq c) &= \int_c^{\infty \frac{dx}{x^2}} \\ &= \lim_{b\to\infty}\int_c^b \frac{dx}{x^2} \\ &= \lim_{b\to\infty}\left[\frac{-1}{x}\right]_c^b \\ &= \lim_{b\to\infty}\left[\frac{-1}{b} + \frac{1}{c}\right] \\ &= \frac{1}{c} \end{aligned}$$

Thus,

$$\begin{aligned} \frac{1}{c} &= 0.05 \\ c &= \frac{1}{0.05} \\ &= 20 \end{aligned}$$

21.

$$\begin{aligned} P(X \geq c) &= \int_c^\infty e^{-x}\, dx \\ &= \lim_{b\to\infty}\int_c^b e^{-x}\, dx \\ &= \lim_{b\to\infty}\left[-e^{-x}\right]_c^b \\ &= \lim_{b\to\infty}\left[-e^{-b} + e^{-c}\right] \\ &= e^{-c} \end{aligned}$$

Thus,

$$\begin{aligned} e^{-c} &= 0.05 \\ c &= -\ln(0.05) \\ &= 2.99573 \end{aligned}$$

23.

$$\begin{aligned} P(X \geq c) &= \int_c^\infty (3.8/x^3 + 33.6/x^5)\, dx \\ &= \lim_{b\to\infty}\int_c^b (3.8/x^3 + 33.6/x^5)\, dx \\ &= \lim_{b\to\infty}\left[-1.9/x^2 - 8.4/x^4\right]_c^b \\ &= \lim_{b\to\infty}\left[\frac{-1.9}{b^2} - \frac{8.4}{b^4} + \frac{1.9}{c^2} + \frac{8.4}{c^4}\right] \\ &= \frac{1.9}{c^2} + \frac{8.4}{c^4} \end{aligned}$$

Thus,

$$\begin{aligned} \frac{1.9}{c^2} + \frac{8.4}{c^4} &= 0.05 \\ 0.05c^4 - 1.9c^2 - 8.4 &= 0 \\ &\to \\ c^2 &= -6.4807 \text{ not acceptable} \\ c^2 &= 6.4807 \to c = 2.5457 \end{aligned}$$

25. $f(x)$ is positive for all x in $[1,3]$

$$\begin{aligned} 1 &= \int_1^3 kx\, dx \\ &= \left[\frac{k}{2}x^2\right]_1^3 \\ &= \frac{9k}{2} - \frac{k}{2} \\ 1 &= 4k \end{aligned}$$

Thus,

$$k = \frac{1}{4}$$

27. $f(x)$ is positive for all x in $[-1,1]$

$$\begin{aligned} 1 &= \int_{-1}^1 kx^2\, dx \\ &= \left[\frac{k}{3}x^3\right]_{-1}^1 \\ &= \frac{k}{3} - \frac{-k}{3} \\ 1 &= \frac{2k}{3} \end{aligned}$$

Thus,

$$k = \frac{3}{2}$$

29. Since $f(x)$ is negative on $[0,1]$, we cannot make a probability density function of $f(x)$ on $[0,2]$

31. **a)** $f(x)$ is positive for all x in $[0,2]$

$$\begin{aligned} 1 &= \int_0^2 k(2-x)^3\, dx \\ &= k\int_0^2 (8 - 12x + 6x^2 - x^3)\, dx \\ &= k\left[8x - 6x^2 + 2x^3 - \frac{1}{4}x^4\right]_0^2 \\ 1 &= k\,[16 - 24 + 16 - 4] \\ 1 &= 4k \end{aligned}$$

Thus,

$$\begin{aligned} k &= \frac{1}{4} \\ P(X = x) &= \frac{(2-x)^3}{4} \end{aligned}$$

b)

$$\begin{aligned}
P(X \leq 1) &= \int_0^1 \frac{1}{4}(2-x)^3\ dx \\
&= \frac{1}{4}\int_0^1 (8 - 12x + 6x^2 - x^3)\ dx \\
&= \frac{1}{4}\left[8x - 6x^2 + 2x^3 - \frac{1}{4}x^4\right]_0^1 \\
&= \frac{1}{4}\left[8 - 6 + 2 - \frac{1}{4}\right] \\
&= \frac{15}{16}
\end{aligned}$$

33. **a)** $f(x)$ is positive for all x in $[\pi/6, \pi/2]$

$$\begin{aligned}
1 &= \int_{\pi/6}^{\pi/2} k\ sin(x)\ dx \\
&= -k\left[cos(x)\right]_{\pi/6}^{\pi/2} \\
1 &= -k\left[0 - \frac{\sqrt{3}}{2}\right] \\
1 &= \frac{\sqrt{3}}{2}k
\end{aligned}$$

Thus,

$$\begin{aligned}
k &= \frac{2\sqrt{3}}{3} \\
P(X = x) &= \frac{2\sqrt{3}}{3}\ sin(x)
\end{aligned}$$

b)

$$\begin{aligned}
P(X \leq 1) &= \int_{\pi/3}^{\pi/2} \frac{2\sqrt{3}}{3}\ sin(x)\ dx \\
&= \frac{2\sqrt{3}}{3}\int_{\pi/3}^{pi/2} sin(x)\ dx \\
&= \frac{2\sqrt{3}}{3}\left[-cos(x)\right]_{\pi/3}^{\pi/2} \\
&= \frac{2\sqrt{3}}{3}\left[0 - (-0.5)\right] \\
&= 0.57735
\end{aligned}$$

35. **a)** $f(x)$ is positive for all x in $[0, 5]$

$$\begin{aligned}
1 &= \int_0^5 k\frac{2x+3}{x2 + 3x + 4}\ dx \\
&= k\left[\ln(x^2 + 3x + 4)\right]_0^5 \\
&= k\left[\ln(44) - \ln(4)\right] = \ln(11)
\end{aligned}$$

Thus,

$$\begin{aligned}
k &= \frac{1}{\ln(11)} \\
P(X = x) &= \frac{2x+3}{\ln(11)(x^2 + 3x + 4)}
\end{aligned}$$

b)

$$\begin{aligned}
P(X \leq 3) &= \int_2^4 \frac{2x+3}{\ln(11)(x^2+3x+4)}\ dx \\
&= \frac{1}{\ln(11)}\left[\ln(x^2 + 3x + 4)\right]_2^4 \\
&= \frac{1}{\ln(11)}\left[\ln(32) - \ln(14)\right] \\
&= 0.34475
\end{aligned}$$

37. **a)** $f(x)$ is positive for all x in $[-\pi, \pi)$

$$\begin{aligned}
1 &= \int_{-\pi}^{\pi} kx\ sin(x) \\
&= k\left[sin(x) - x\ cos(x)\right]_{-\pi}^{\pi} \\
&= k\left[0 - \pi(-1) - (0 - (-\pi)(-1))\right] \\
1 &= 2\pi k
\end{aligned}$$

Thus,

$$\begin{aligned}
k &= \frac{1}{2\pi} \\
P(X = x) &= \frac{x\ sin(x)}{2\pi}
\end{aligned}$$

b)

$$\begin{aligned}
P(X \leq 3) &= \frac{1}{2pi}\int_{-pi}^{\pi/2} (x\ sin(x))\ dx \\
&= \frac{1}{2\pi}\left[sin(x) - x\ cos(x)\right]_{-\pi}^{\pi/2} \\
&= 0.65915
\end{aligned}$$

39. $f(v) = \frac{1}{4-2} = \frac{1}{2}$

$$\begin{aligned}
P(3 \leq X \leq 4) &= \int_3^4 \frac{1}{2}\ dx \\
&= \left[\frac{1}{2}x\right]_3^4 \\
&= 2 - \frac{3}{2} \\
&= \frac{1}{2}
\end{aligned}$$

41. $f(v) = \frac{1}{500-100} = \frac{1}{400}$

$$\begin{aligned}
P(200 \leq X \leq 350) &= \int_{200}^{350} \frac{1}{400}\ dx \\
&= \left[\frac{1}{400}x\right]_{200}^{350} \\
&= \frac{350}{400} - \frac{200}{400} \\
&= \frac{3}{8}
\end{aligned}$$

43.

$$\begin{aligned}
E(X) &= \int_0^3 \frac{2x^2}{9}\ dx \\
&= \left[\frac{2x^3}{27}\right]_0^3 \\
&= \frac{2}{27}(27 - 0) \\
&= 2
\end{aligned}$$

$$
\begin{aligned}
Var(X) &= \int_0^3 (x-2)^2 \frac{2x}{9}\, dx \\
&= \frac{2}{9}\int_0^3 (x^3 - 4x^2 + 4x)\, dx \\
&= \frac{2}{9}\left[\frac{x^4}{4} - \frac{4}{3}x^3 + 2x^2\right]_0^3 \\
&= \frac{2}{9}(2.25 - 0) \\
&= 0.5 \\
SD(X) &= \sqrt{0.5} \\
&= 0.70711
\end{aligned}
$$

45.

$$
\begin{aligned}
E(X) &= \int_1^2 \frac{dx}{\ln(2)} \\
&= \left[\frac{1}{\ln(2)}x\right]_1^2 \\
&= \frac{1}{\ln(2)}(2-1) \\
&= \frac{1}{\ln(2)} \\
&= 1.4427
\end{aligned}
$$

$$
\begin{aligned}
Var(X) &= \int_1^2 (x - \frac{1}{\ln(2)})^2 \frac{1}{x\ \ln(2)}\, dx \\
&= \frac{1}{\ln(2)}\int_1^2 (x - \frac{2}{\ln(2)} + \frac{1}{x\ \ln^2(2)})\, dx \\
&= \frac{1}{\ln(2)}\left[\frac{x^2}{2} - \frac{2x}{\ln(2)} + \frac{1}{\ln^2(2)}\ \ln(x)\right]_1^2 \\
&= \frac{1}{\ln(2)}\left[2 - \frac{4}{\ln(2)} + \frac{1}{\ln^2(2)}\ln(2) - \frac{1}{2}\right] \\
&\quad + \frac{1}{ln(2)}\left[\frac{2}{\ln(2)} - \frac{1}{\ln^2(2)}\ln(1)\right] \\
&= 0.082674 \\
SD(X) &= \sqrt{0.082674} \\
&= 0.28753
\end{aligned}
$$

47.

$$
\begin{aligned}
E(X) &= \frac{4}{21}\int_1^2 (x^2 + x^4)\, dx \\
&= \frac{4}{21}\left[\frac{x^3}{3} + \frac{x^5}{5}\right]_1^2 \\
&= \frac{4}{21}\left(\frac{8}{3} + \frac{32}{5} - \frac{1}{3} - \frac{1}{5}\right) \\
&= 1.625396825
\end{aligned}
$$

$$
\begin{aligned}
Var(X) &= \frac{4}{21}\int_1^2 (x - 1.625396825)^2(x + x^3)\, dx \\
&= \frac{4}{21}\int_1^2 (3.641914839x^3 + x^5 - 3.25079365x^2 \\
&\quad -3.25079365x^4 + 2.641914839x)\, dx \\
&= \frac{4}{21}\left[0.9104787098x^4 + \frac{1}{6}x^6 - 1.083597883x^3\right]_1^2 \\
&\quad -\frac{4}{21}\left[0.65015873x^5 + 1.32095742x\right]_1^2 \\
&= \frac{4}{21}(0.3799470914) \\
&= 0.07237087455 \\
SD(X) &= \sqrt{0.07237087455} \\
&= 0.26990183536
\end{aligned}
$$

49.

$$
\begin{aligned}
E(X) &= \frac{1}{2}\int_0^\pi xsin\ x\ dx \\
&= \frac{1}{2}\left[sin\ x - xcos\ x\right]_0^\pi \\
&= \frac{1}{2}(0 - (-\pi) - 0 + 0) \\
&= \frac{\pi}{2}
\end{aligned}
$$

$$
\begin{aligned}
Var(X) &= \frac{1}{2}\int_0^\pi (x - \frac{\pi}{2})^2 sin\ x\ dx \\
&= \frac{1}{2}\int_0^\pi (x^2 sin\ x - \pi xsin\ x + \frac{\pi^2}{4}sin\ x)\ dx \\
&= \frac{1}{2}\left[-x^2cos\ x + 2cos\ x + 2xsin\ x\right]_0^\pi - \\
&\quad \frac{1}{2}\left[\pi sin\ x + \pi xcos\ x - \frac{\pi^2}{4}cos\ x\right]_0^\pi \\
&= \frac{1}{2}\left[(\pi^2 - 2 + 0 - 0 - \pi^2 - \frac{\pi^2}{4})\right] \\
&\quad -\frac{1}{2}\left[(0 + 2 + 0 - 0 + 0 - \frac{\pi^2}{4})\right] \\
&= \frac{1}{2}\left[\frac{\pi^2}{2} - 4\right] \\
&= 0.467401 \\
SD(X) &= \sqrt{0.467401} \\
&= 0.683667
\end{aligned}
$$

51.

$$
\begin{aligned}
E(X) &= \frac{3}{14}\int_0^3 x(x+1)^{1/2})\ dx \\
&= \frac{3}{14}\left[\frac{2}{5}(x+1)^{5/2} - \frac{2}{3}(x+1)^{3/2}\right]_0^3 \\
&= \frac{3}{14}\left(\frac{2}{5}(32) + \frac{2}{3}(8)\right) \\
&= 1.657142857
\end{aligned}
$$

$$
\begin{aligned}
Var(X) &= \frac{3}{14}\int_0^3 (x - 1.657142857)^2(x+1)^{1/2})\ dx \\
&= \frac{3}{14}\left[\frac{2}{7}(x+1)^{7/2} - \frac{524}{75}(x+1)^{5/2}\right]_0^3 \\
&\quad + \frac{3}{14}\left[\frac{34322}{675}(x+1)^{3/2}\right]_0^3
\end{aligned}
$$

$$
\begin{aligned}
&= \frac{3}{14}(3.337142857) \\
&= 0.7151020408 \\
SD(X) &= \sqrt{0.7151020408} \\
&= 0.84564
\end{aligned}
$$

53.

$$
\begin{aligned}
E(X) &= \int_0^\infty 3x\, e^{-3x}\, dx \\
&= \lim_{b\to\infty} \int_0^b 3x\, e^{-3x}\, dx \\
&= \lim_{b\to\infty} \left[-x\, e^{-3x} - \frac{1}{3}\, e^{-3x} \right]_0^b \\
&= \frac{1}{3}
\end{aligned}
$$

$$
\begin{aligned}
Var(X) &= \int_0^\infty 3\, e^{-3x} \left(x - \frac{1}{3} \right)^2 dx \\
&= \lim_{b\to\infty} \int_0^b 3\, e^{-3x} \left(x - \frac{1}{3} \right)^2 dx \\
&= \lim_{b\to\infty} \left[-x^3\, e^{-3x} - \frac{x^2\, e^{-3x}}{3} - \frac{x\, e^{-3x}}{3} - \frac{e^{-3x}}{9} \right]_0^b \\
&= \frac{1}{9} \\
SD(X) &= \sqrt{\frac{1}{9}} \\
&= \frac{1}{3}
\end{aligned}
$$

55.

$$
\begin{aligned}
E(X) &= \frac{3+9}{2} \\
&= 6 \\
Var(X) &= \frac{(9-3)^2}{12} \\
&= 3 \\
SD(X) &= \sqrt{3} \\
&= 1.73205
\end{aligned}
$$

57.

$$
\begin{aligned}
E(X) &= \frac{10+20}{2} \\
&= 15 \\
Var(X) &= \frac{(20-10)^2}{12} \\
&= \frac{100}{12} = \frac{25}{3} \\
&\approx 8.3333333333 \\
SD(X) &= \sqrt{8.3333333333} \\
&= 2.88675
\end{aligned}
$$

59.

$$
\begin{aligned}
P &= \int_0^3 \frac{1}{5}\, dx \\
&= \left[\frac{x}{5} \right]_0^3 \\
&= \frac{3}{5}
\end{aligned}
$$

61. $70 - 60 = 10$

a)

$$
\begin{aligned}
P &= \int_{60}^{65} \frac{1}{10}\, dx \\
&= \left[\frac{x}{10} \right]_{60}^{65} \\
&= \frac{1}{2}
\end{aligned}
$$

b)

$$
\begin{aligned}
P &= \int_{68}^{70} \frac{1}{10}\, dx \\
&= \left[\frac{x}{10} \right]_{68}^{70} \\
&= \frac{1}{5}
\end{aligned}
$$

63.

$$
\begin{aligned}
\int_0^b x^3\, dx &= 1 \\
\left.\frac{x^4}{4}\right|_0^b &= 1 \\
\frac{b^4}{4} &= 1 \\
b &= \sqrt[4]{4} = \sqrt{2}
\end{aligned}
$$

65. **a)**

$$
\begin{aligned}
E(X) &= \int_a^b x\, f(x)\, dx \\
&= \int_a^b \frac{1}{b-a}\, x\, dx \\
&= \frac{1}{2(b-a)} \cdot \left. x^2 \right|_a^b \\
&= \frac{1}{2(b-a)} \cdot (b^2 - a^2) \\
&= \frac{1}{2(b-a)} \cdot (b-a)(b+a) \\
&= \frac{b+a}{2}
\end{aligned}
$$

b)

$$
\begin{aligned}
E(X^2) &= \int_a^b x^2\, f(x)\, dx \\
&= \int_a^b \frac{x^2}{b-a}\, dx \\
&= \left. \frac{x^3}{3(b-a)} \right|_a^b \\
&= \frac{b^3 - a^3}{3(b-a)} \\
&= \frac{(b-a)(b^2 + ab + b^2)}{3(b-a)} \\
&= \frac{1}{3}\,(b^2 + ab + a^2)
\end{aligned}
$$

c)

$$
\begin{aligned}
Var(X) &= \int_a^b (x-\mu)^2\, f(x)\, dx \\
&= \int_a^b \left(x - \frac{(b+a)}{2}\right)^2 \frac{1}{b-a}\, dx \\
&= \frac{1}{b-a} \int_a^b x^2 - (b+a)x + \frac{(b+a)^2}{4}\, dx \\
&= \frac{1}{b-a} \left[\frac{x^3}{3} - \frac{(b+a)x^2}{2} + \frac{(b+a)^2}{4}\right]_a^b \\
&= \frac{1}{b-a} \left[\frac{(b^3-a^3}{3} - \frac{(b+a)(b^2-a^2)}{2}\right] \\
&\quad + \frac{1}{b-a}\left[\frac{(b+a)(b-a)}{4}\right] \\
&= \frac{b^2+ab+a^2}{3} - \frac{(b+a)^2}{2} + \frac{(b+a)^2}{4} \\
&= \frac{b^2+ab+a^2}{3} - \frac{(b+a)^2}{4} \\
&= \frac{4b^2+4ab+4a^2-3b^2-6ab-3a^2}{12} \\
&= \frac{b^2-2ab+a^2}{12} \\
&= \frac{(b-a)^2}{12}
\end{aligned}
$$

67. a)

$$
\begin{aligned}
\int_1^{85} 1152.9k\, e^{0.051476x}\, dx &= 1 \\
1152.9k \int_1^{85} e^{0.051476x}\, dx &= 1 \\
1152.9k \cdot \left.\frac{e^{0.051476x}}{0.051476}\right|_1^{85} &= 1 \\
1152.9k \cdot 1523.497676 &= 1 \\
k &= \frac{1}{1152.9 \cdot 1523.497676} \\
&= 5.69333 \times 10^{-7} \\
&= 0.000000569333
\end{aligned}
$$

b) $k\ f(x) = 0.00065638\ e^{0.051476x}$

$$
\begin{aligned}
E(X) &= \int_1^{85} 0.00065638x\, e^{0.051476x}\, dx \\
&= (0.012751185x\ e0.051476x - \\
&\quad \left. 0.2477112639\, e^{0.051476x} \right|_1^{85} \\
&= 66.701
\end{aligned}
$$

c)

$$
\begin{aligned}
Var(X) &= \int_1^{85} 0.00065638\, (x-66.701)^2\, e^{0.051476x}\, dx \\
&= 0.012751x^2\, e^{0.051476x} - 2.19645xe^{0.051476x} \\
&\quad \left. + 99.39922\, e^{0.051476x} \right|_1^{85} \\
&= 281.3908120 \\
SD(X) &= \sqrt{281.3908120} \\
&= 16.7747
\end{aligned}
$$

69. a) Left to the student

b) $P(X \le 2)$

$$
\begin{aligned}
&= \int_0^2 \frac{0.68}{0.89} \left(\frac{x}{0.89}\right)^{-0.32} e^{-(x/0.89)^{0.68}}\, dx \\
&= \frac{0.68}{0.89} \left[-1.308823528\, e^{-1.082467325x^{0.68}}\right|_0^2 \\
&= 0.82347
\end{aligned}
$$

71. 0.028596

73. 0.018997

75. 0.004701

77. 0.265026

79. 9.488

81. 12.592

Exercise Set 10.6

1.

$$
\begin{aligned}
P(X=0) &= \frac{2.2^0}{0!}\, e^{-2.2} \\
&= 0.1108
\end{aligned}
$$

3.

$$
\begin{aligned}
P(X=1) &= \frac{1.3^1}{1!}\, e^{-1.3} \\
&= 0.0.3543
\end{aligned}
$$

5.

$$
\begin{aligned}
P(X=7) &= \frac{4^7}{7!}\, e^{-4} \\
&= 0.05954
\end{aligned}
$$

7.

$$
\begin{aligned}
P(X \le 2) &= P(X=0) + P(X=1) + P(X=2) \\
&= 0.2240 + 0.1494 + 0.04979 \\
&= 0.4232
\end{aligned}
$$

9.

$$
\begin{aligned}
P(X \ge 3) &= 1 - P(X \le 3) \\
&= 1 - \Big[P(X=0) + P(X=1) + P(X=2) \\
&\quad + P(X=3)\Big] \\
&= 1 - (0.22313 + 0.33470 + 0.25102 + 0.12551) \\
&= 0.0656
\end{aligned}
$$

11.

$$\begin{aligned} P(3 \le X \le 5) &= P(X=3)+P(X=4)+P(X=5) \\ &= 0.08674+0.02602+0.00625 \\ &= 0.1388 \end{aligned}$$

13. $E(X) = 2.3$

15. $SD(X) = \sqrt{4} = 2$

17.

$$\begin{aligned} P(0 \le X \le 2) &= \int_0^2 e^{-x}\,dx \\ &= -e^{-x}\Big|_0^2 \\ &= 0.8647 \end{aligned}$$

19.

$$\begin{aligned} P(0 \le X \le 2) &= \int_0^2 3\,e^{-3x}\,dx \\ &= -e^{-3x}\Big|_0^2 \\ &= 0.9975 \end{aligned}$$

21.

$$\begin{aligned} P(2 \le X) &= 1-P(2>X) \\ &= 1-\int_0^2 2e^{-2x}\,dx \\ &= 1-e^{-2x}\Big|_0^2 \\ &= 1-0.9817 \\ &= 0.0183 \end{aligned}$$

23.

$$\begin{aligned} P(X \le 0.5) &= \int_0^{0.5} 2.5\,e^{-2.5x}\,dx \\ &= -e^{-2.5x}\Big|_0^{0.5} \\ &= 0.7135 \end{aligned}$$

25. $E(X) = \frac{1}{4}$

27. $SD(X) = \frac{1}{4}$

29. - 39. Left to the student

41. - 47. Left to the student

49. $E(X) = 4$
$Var(X) = 4$
$SD(X) = \sqrt{4} = 2$

51. $E(X) = \frac{1}{3}$
$Var(X) = \frac{1}{9}$
$SD(X) = \sqrt{\frac{1}{9}} = \frac{1}{3}$

53. $E(X) = 22 \to \lambda = \frac{1}{22}$

$$\begin{aligned} P(X \le 20) &= \int_0^{20} \frac{1}{22}\,e^{-\frac{x}{22}}\,dx \\ &= -e^{-\frac{x}{22}}\Big|_{)}^{20} \\ &= 0.5971 \end{aligned}$$

55. **a)** $\lambda = 3$, $\mu = 3(2) = 6$

$$\begin{aligned} P(Y=1) &= \frac{6^1}{1!}\,e^{-6} \\ &= 0.01487 \end{aligned}$$

b) $\mu = 3(3) = 9$

$$\begin{aligned} P(Y=5) &= \frac{9^5}{5!}\,e^{-9} \\ &= 0.0607 \end{aligned}$$

c) $\mu = 3$

$$\begin{aligned} P(Y=0) &= \frac{3^0}{0!}\,e^{-3} \\ &= 0.0498 \end{aligned}$$

57. **a)** $\lambda = 3$, $\mu = \frac{4}{3}$

$$\begin{aligned} P(Y \ge 1) &= 1-P(Y=0) \\ &= 1-0.2636 \\ &= 0.7364 \end{aligned}$$

b) $\mu = \frac{2}{3}$

$$\begin{aligned} P(Y=0) &= \frac{\left(\frac{2}{3}\right)^0}{0!}\,e^{-2/3} \\ &= 0.5134 \end{aligned}$$

c) $\mu = \frac{1}{3}$

$$\begin{aligned} P(Y=0) &= \frac{\left(\frac{1}{3}\right)^0}{0!}\,e^{-1/3} \\ &= 0.7165 \end{aligned}$$

d) $\mu = \frac{1}{3}$

$$\begin{aligned} P(Y=1) &= \frac{\left(\frac{1}{3}\right)^1}{1!}\,e^{-1/3} \\ &= 0.2388 \end{aligned}$$

e) $\mu = \frac{1}{3}$

$$\begin{aligned} P(Y \ge 2) &= 1-P(Y<2) \\ &= 1-(P(Y=0)+P(Y=1)) \\ &= 1-(0.7165-0.2388) \\ &= 0.0447 \end{aligned}$$

59. a) $E(X) = 6$
$SD(X) = \sqrt{6} = 2.4495$

b)

$$P(Y = 0) = \frac{6^0}{0!} e^{-6} = 0.00248$$

61. First, note that $f(x) > 0$ for all x values and positive $\lambda > 0$
Second,

$$\begin{aligned}\int_0^\infty f(x)\,dx &= \int_0^\infty \lambda\, e^{-\lambda\, x}\,dx \\ &= -\,e^{-\lambda\, x}\Big|_0^\infty \\ &= 0-(-1) \\ &= 1\end{aligned}$$

Therefore the function $f(x) = \lambda\, e^{-\lambda\, x}$ is a probability density function

63.

$$\begin{aligned}\int_0^m \frac{1}{2}\,x\,dx &= \frac{1}{2} \\ \frac{x^2}{4}\Big|_0^m &= \frac{1}{2} \\ \frac{m^2}{4} &= \frac{1}{2} \\ m^2 &= 2 \\ m &= \sqrt{2}\end{aligned}$$

65. $E(X) = \frac{3}{2}$

$$\begin{aligned}\int_0^m \frac{1}{3}\,dx &= \frac{1}{2} \\ \frac{x}{3}\Big|_0^m &= \frac{1}{2} \\ \frac{m}{3} &= \frac{1}{2} \\ m &= \frac{3}{2}\end{aligned}$$

The mean and median are equal

67. a)

$$\begin{aligned}\int_0^m e^{-x}\,dx &= \frac{1}{2} \\ -e^{-x}\Big|_0^m &= \frac{1}{2} \\ -e^{-m}+1 &= \frac{1}{2} \\ e^{-m} &= \frac{1}{2} \\ -m &= \ln\frac{1}{2} = -\ln 2 \\ m &= \ln 2\end{aligned}$$

b) The median is larger than the mean $E(X) = 1$. This makes sense since at $m = \ln 2$, for exponential given, the value of $f(x) = 0.5$

69. Left to the student

71. Binomial: $P(X = 4) = 0.168284$
Poisson: $P(X = 4) = 0.168031$

73. Binomial: $P(X = 2) = 0.256561$
Poisson: $P(X = 2) = 0.256516$

75. Binomial: $P(X = 2) = 0.183949$
Poisson: $P(X = 2) = 0.183940$

Exercise Set 10.7

1.

$$\begin{aligned}P(-2 \le Z \le 2) &= P(Z = 2) + P(Z = -2) \\ &= 0.47725 + 0.47725 \\ &= 0.9545\end{aligned}$$

3.

$$\begin{aligned}P(-0.26 \le Z \le 0.7) &= P(Z = 0.7) + P(Z = -0.26) \\ &= 0.25804 + 0.10257 \\ &= 0.3606\end{aligned}$$

5.

$$\begin{aligned}P(1.26 \le Z \le 1.43) &= P(Z = 1.43) - P(Z = 1.26) \\ &= 0.42364 - 0.396165 \\ &= 0.0275\end{aligned}$$

7.

$$\begin{aligned}P(-2.47 \le Z \le -0.38) &= P(Z = -0.38) - P(Z = -2.47) \\ &= 0.493244 - 0.148027 \\ &= 0.3452\end{aligned}$$

9.

$$\begin{aligned}P(0 \le Z < \infty) &= P(Z = \infty) - P(Z = 0) \\ &= 0.5 - 0 \\ &= 0.5\end{aligned}$$

11.

$$\begin{aligned}P(1.17 \le Z < \infty) &= P(Z = \infty) - P(Z = 1.17) \\ &= 0.5 - 0.378999 \\ &= .1210\end{aligned}$$

13. **a)** $\mu = 2$

b) $\sigma = 5$

c) $Z = \dfrac{4-2}{5} = 0.4$

$$\begin{aligned} P(Z \geq 0.4) &= 1 - P(Z < 0.4) \\ &= 1 - 0.6554 \\ &= 0.3446 \end{aligned}$$

d) $Z = \dfrac{3-2}{5} = 0.2$

$$P(Z \leq 0.3) = 0.5793$$

e) $Z = \dfrac{-8-2}{5} = -2$, $Z = \dfrac{7-2}{5} = 1$

$$\begin{aligned} P(-2 \leq Z \leq 1) &= P(Z = -2) + P(Z = 1) \\ &= 0.4772 + 0.3413 \\ &= 0.8185 \end{aligned}$$

15. **a)** $Z = \dfrac{-0.2-0}{0.1} = -2$

$$\begin{aligned} P(Z \geq -2) &= 1 - P(Z < -2) \\ &= 1 - 0.0228 \\ &= 0.9972 \end{aligned}$$

b) $Z = \dfrac{-0.05-0}{0.1} = -0.5$

$$P(Z \leq -0.5) = 0.3085$$

c) $Z = \dfrac{-0.08-0}{0.1} = -0.8$

$Z = \dfrac{0.09-0}{0.1} = 0.9$

$$\begin{aligned} &P(-0.8 \leq Z \leq 0.9) \\ &= P(Z = -0.8) + P(Z = 0.9) \\ &= 0.2881 + 0.3159 \\ &= 0.6041 \end{aligned}$$

17. - 31. Left to the student (see answers to 1. - 15.)

33. $\mu = 100(0.2) = 20$

$\sigma = \sqrt{100(0.2)(0.8)} = 4$

$Z = \dfrac{18-20}{4} = -0.5$

$$\begin{aligned} P(Z \geq -0.5) &= 1 - P(Z < -0.5) \\ &= 1 - 0.3085 \\ &= 0.6915 \end{aligned}$$

35. $\mu = 64(0.5) = 32$

$\sigma = \sqrt{64(0.5)(0.5)} = 4$

$Z = \dfrac{30-32}{4} = -0.5$

$$P(Z \leq -0.5) = 0.3085$$

37. $\mu = 1000(0.3) = 300$

$\sigma = \sqrt{1000(0.3)(0.7)} = 14.4914$

$Z = \dfrac{280-300}{14.4914} = -1.38$

$Z = \dfrac{300-300}{14.4914} = 0$

$$\begin{aligned} P(-1.38 \leq Z \leq 0) &= P(Z = -1.38) + P(Z = 0) \\ &= 0.4162 + 0 \\ &= 0.4162 \end{aligned}$$

39. **a)** $Z = \dfrac{5-4}{1} = 1$

$$\begin{aligned} P(Z \geq 1) &= 1 - P(Z < 1) \\ &= 1 - 0.8413 \\ &= 0.1587 \end{aligned}$$

b) $Z = \dfrac{2-4}{1} = -2$

$$\begin{aligned} P(Z \geq -2) &= 1 - P(Z < -2) \\ &= 1 - 0.0228 \\ &= 0.0.9772 \end{aligned}$$

c) $Z = \dfrac{1-4}{1} = -3$

$$P(Z \leq -3) = 0.0013$$

41. $\mu = (300)(0.25) = 75$

$\sigma = \sqrt{300(0.75)(0.25)} = 7.5$

$Z = \dfrac{80-75}{7.5} = 0.67$

$$\begin{aligned} P(Z \geq 0.67) &= 1 - P(Z < 0.67) \\ &= 1 - 0.7486 \\ &= 0.2514 \end{aligned}$$

43. $\mu = (70)(0.25) = 17.5$

$\sigma = \sqrt{70(0.25)(0.75)} = 3.6228$

$Z = \dfrac{20-17.5}{3.6228} = 0.69$

$$P(Z < 0.69) = 0.7549$$

45. $p = \dfrac{238351}{362470} = 0.6576$

$\mu = 120(0.6576) = 78.91$

$\sigma = \sqrt{120(0.6576)(0.3424)} = 5.20$

$Z = \dfrac{80-78.91}{5.20} = 0.21$

$$\frac{90-78.91}{5.20}=2.13$$

$$\begin{aligned} P(0.21 \le Z \le 2.13) &= P(Z < 2.13) - P(Z = 0.21) \\ &= 0.4834 - 0.0832 \\ &= 0.4002 \end{aligned}$$

47. **a)** $\mu = 1000(9/19) = 473.7$

$$\sigma = \sqrt{1000(9/19)(10/19)} = 15.79$$

$$Z = \frac{501-473.7}{15.79} = 1.73$$

$$\begin{aligned} P(Z \ge 1.73) &= 1 - P(Z < 1.73) \\ &= 1 - 0.9582 \\ &= 0.0418 \end{aligned}$$

b) The probability in part a)was larger. This makes sense since the probability of winning is less than the probability of winning.

49. **a)** $35\% \approx X = -0.38$

$$\begin{aligned} -0.38 &= \frac{X'-507}{111} \\ -0.38(111)+507 &= X' \\ 464 &= \end{aligned}$$

b) $60\% \approx X = 0.25$

$$\begin{aligned} 0.25 &= \frac{X'-507}{111} \\ 0.25(111)+507 &= X' \\ 535 &= \end{aligned}$$

c) $92\% \approx X = 1.41$

$$\begin{aligned} 1.41 &= \frac{X'-507}{111} \\ 1.41(111)+507 &= X' \\ &= 664 \end{aligned}$$

51. Significance Level

$$\begin{aligned} &= 1 - P(X \ge 0.93) \\ &= 1 - 0.8238 \\ &= 0.1762 \end{aligned}$$

53. Significance Level

$$\begin{aligned} &= 1 - P(X \ge 2.33) \\ &= 1 - 0.9901 \\ &= 0.0099 \end{aligned}$$

55. Significance Level

$$\begin{aligned} &= P(X \le -1.24) \\ &= 0.1075 \end{aligned}$$

57. Critical Value is 2.33

59. Critical Value is -1.64

61. Critical Value is -2.58

63. $Z_1 = \frac{9-10}{SD} = \frac{-1}{SD}$

$$Z_2 = \frac{10-10}{SD} = 0$$

$$\begin{aligned} P(Z_1 \le Z \le 0) &= 0.4922 \\ Z_2 &= -2.42 \\ -2.42 &= \frac{-1}{SD} \\ SD &= \frac{-1}{-2.42} \\ &= .4132231405 \end{aligned}$$

65. $Z_1 = \frac{-6-(-3)}{SD} = \frac{-3}{SD}$, $Z_2 = \frac{0-(-3)}{SD} = \frac{3}{SD}$

$$\begin{aligned} P(-\frac{3}{SD} \le Z \le \frac{3}{SD}) &= 0.3108 \\ P(0 \le Z \le \frac{3}{SD}) &= \frac{0.3108}{2} = 0.1554 \\ \frac{3}{SD} &= 0.40 \\ SD &= \frac{3}{0.40} \\ &= 7.5 \\ Var &= (7.5)^2 = 56.25 \end{aligned}$$

67.

$$\begin{aligned} P(Z \ge -1) &= 0.409 \\ P(Z < -1) &= 1 - 0.409 \\ &= 0.591 \\ Z &= 0.23 \\ 0.23 &= \frac{-1-E}{2} \\ E &= -1 - 2(0.23) \\ &= -1.46 \end{aligned}$$

69. $Z_1 = \frac{2.8-E}{SD}$, $Z_2 = \frac{10.3-E}{SD}$

$$\begin{aligned} P(Z \ge Z_1) &= 1 - P(Z < Z_1) \\ &= 0.5 \\ Z_1 &= 0 \\ P(Z \le Z_2) &= 0.8944 \\ Z_2 &= 1.25 \end{aligned}$$

Thus,

$$\frac{2.8-E}{SD} = 0$$

$$\begin{aligned} E &= 2.8 \\ \frac{10.3-E}{SD} &= 1.25 \\ 1.25SD + 2.8 &= 10.3 \\ SD &= 6 \end{aligned}$$

71. $Z_1 = \dfrac{-3.1 - E}{SD}$, $Z_2 = \dfrac{1.4 - E}{SD}$

$$\begin{aligned} P(Z \geq Z_1) &= 1 - P(Z < Z_1) \\ &= 0.1539 \\ Z_1 &= 0.4 \\ P(Z \leq Z_2) &= 0.8461 \\ Z_2 &= 1 \end{aligned}$$

Thus,

$$\begin{aligned} \frac{-3.1 - E}{SD} = 0.4 & \\ 0.4SD + E &= -3.1 \\ \frac{1.4 - E}{SD} &= 1 \\ SD + E &= 1.4 \end{aligned}$$

Solving the system

$$\begin{aligned} E &= -0.85 \\ SD &= 2.25 \end{aligned}$$

73. The reason is the symmetric nature of the Normal(μ, σ) distribution about μ

75. **a)** $\mu = 2000$, $\sigma = 40$
$P(1949.5 \leq X \leq 2000.5) = 0.40160$

b) $Z_1 = \dfrac{1950 - 2000}{40} = -1.25$
$Z_2 = \dfrac{2000 - 2000}{4} = 0$
$P(-1.25 \leq Z \leq 0) = 0.3944$

c) 0.00720

77. As n increases the difference between probability with the continuity correction and the probability without the continuity correction becomes smaller.

79. $n = 2 \rightarrow 0.341529$
$n = 4 \rightarrow 0.341355$
$n = 6 \rightarrow 0.341347$
$n = 8 \rightarrow 0.341345$